D1171715

Ecological Studies

Analysis and Synthesis

Volume 27

Production Ecology of British Moors and Montane Grasslands

Edited by

O. W. Heal and D. F. Perkins

Assisted by Wendy M. Brown

With 132 Figures

Springer-Verlag Berlin Heidelberg New York 1978

Dr. O. W. Heal
The Institute of Terrestrial Ecology
Merlewood Research Station
Grange-over-Sands, Cumbria LA11 6JU
Great Britain

Dr. D. F. Perkins
The Institute of Terrestrial Ecology
Penrhos Road
Bangor, Gwynedd, LL57 2LQ/Great Britain

For explanation of the cover motive see legends to Fig. 6 (p. 42), Fig. 6 (p. 68), Fig. 9 (p. 262), and Fig. 2 (p. 358).

ISBN 3-540-08457-6 Springer-Verlag Berlin Heidelberg New York
ISBN 0-387-08457-6 Springer-Verlag New York Heidelberg Berlin

Library of Congress Cataloging in Publication Data. Main entry under title: Production ecology of British moors and montane grasslands. (Ecological studies; v. 27). Bibliography: p. Includes index. 1. Biological productivity—Great Britain. 2. Grassland ecology—Great Britain. 3. Moor ecology—Great Britain. 4. Mountain ecology—Great Britain. I. Heal, O.W. II. Perkins, Donald Francis, 1935–. III. Brown, Wendy M. IV. Series. QH137.P75. 574.5'264. 77-17853.

Typesetting, printing, and binding: Brühlsche Universitätsdruckerei, Lahn-Gießen.
2131/3130—543210

Preface

The International Biological Programme (IBP) was a cooperative effort on the part of scientists throughout the world, whose goal was an integrated study of the basic processes of biological productivity. The challenge of meeting the increasing food needs of a growing population demands optimum productivity from natural and managed ecosystems, which has not hitherto appeared to be compatible with the maintenance of environmental quality. The basic problem in natural resource development is how to transfer the high productivity and stability characteristic of natural ecosystems to managed ecosystems whose yield is in more useable form. The IBP studies aimed to investigate the basic production parameters of natural ecosystems, for use as base lines to assess the factors which control agricultural production (Worthington, 1975).

It was realised that much was to be gained by close cooperation between the countries within IBP, to describe global patterns of production and to utilise fully the limited financial resources and scientifically qualified personel available in the various disciplines in individual countries.

Within the Terrestrial Productivity section (PT) four major habitat types (Biomes) have been recognised—woodland, grassland, aridland and tundra. In each Biome the ecosystem structure and production, the interrelationships of the various components, and the factors influencing the operation of the systems, have been analysed.

Tundra is variously defined as comprising areas of permafrost or as approximating to areas with a mean annual temperature $<0°$ C at 1.5–2.0 m altitude. Although not strictly falling within the category of tundra by either of these definitions, it has become increasingly clear that the severe climatic conditions encountered on the exposed high moorland sites at Moor House and the biota have more in common with those of tundra than with other biomes, and close links were established between the Moor House and the International Tundra group. The data obtained by the various national groups are being compared and synthesised to produce a Biome-wide ecosystem analysis (Wielgolaski and Rosswall, 1972; Bliss and Wielgolaski, 1973; Holding et al., 1974; Rosswall and Heal, 1975; Moore, in prep).

At Moor House the bog ecosystem has been analysed in considerable detail, providing a description of carbon and mineral cycling with emphasis on the development of mathematical models to simulate the ecosystem and its components. Results from supporting research studies on other moorland and lowland heath systems (Chap. 12–14) have contributed valuable additional information to provide a broader appraisal of the structure and function of dwarf shrub ecosystems.

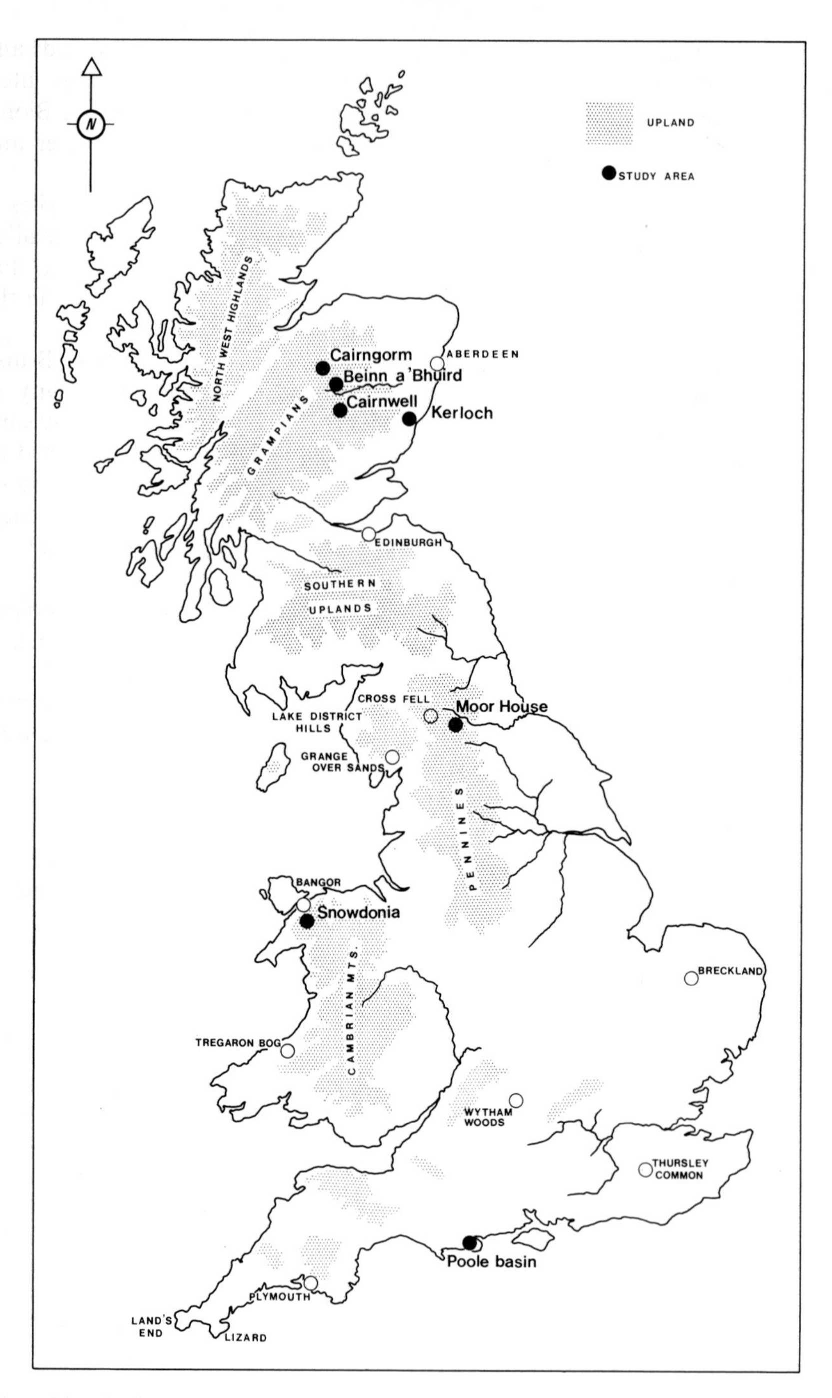

Fig. 1. The distribution of the main study areas and other places mentioned in the volume

The Snowdonia sites are sheep-grazed, semi-natural montane grasslands and research has concentrated on plant and herbivore production and their interaction. The Snowdon sites are linked with the International Grassland Biome whose synthesis will be published in Coupland (in prep) and Breymeyer and van Dyne (in prep).

A general comparison of the UK IBP grassland sites and moorland sites is given in Heal and Perkins (1976). The emphasis in this volume is on detailed description of the ecosystem structure at the sites, an approach similar to that adopted in the volumes on the West German and Scandinavian projects in the Ecological Studies series (Ellenberg, 1971; Wielgolaski, 1975).

The coordination of the UK contribution to IBP was through the British National Committee with a Sub-Committee for Studies on Productivity of Terrestrial Communities. The Royal Society provided considerable advisory, administrative and financial assistance and the research was also supported by funds and facilities from the Natural Environment Research Council, Universities, Nature Conservancy, and other Institutes. The cooperative effort of these organisations, and the many administrative, scientific and support staff, is gratefully acknowledged.

We are grateful to the publishing house of Springer for bringing the projects into the Ecological Studies series, thus combining the results in one volume, so that each aspect can be viewed as part of the whole and not in isolation.

The editors would also like to apologise to those authors who submitted their chapters at an early date for the long delay in publishing, which resulted from problems in the completion of the final manuscripts.

January, 1978

O. W. Heal
D. F. Perkins

Contents

Contributors

Ball, D. F.	Institute of Terrestrial Ecology, Penrhos Road, Bangor, Gwynedd, LL57 2LQ/Great Britain
Brasher, S.	c/o Institute of Terrestrial Ecology, Penrhos Road, Bangor, Gwynedd, LL57 2LQ/Great Britain
Chapman, S. B.	Institute of Terrestrial Ecology, Furzebrook Research Station, Wareham, Dorset, BH20 5AS/Great Britain
Clymo, R. S.	Department of Botany, Westfield College, Kidderpore Avenue, Hampstead, London, N.W. 3/Great Britain
Collins, Vera G.	Cragside, Atkinson Court, Fell Foot, Newby Bridge, Ulverston, Cumbria, LA12 8NW/Great Britain
Coulson, J. C.	Department of Zoology, University of Durham, South Road, Durham, DH1 3LE/Great Britain
Dale, J.	Institute of Terrestrial Ecology, Penrhos Road, Bangor, Gwynedd, LL57 2LQ/Great Britain
Forrest, G. I.	Forestry Commission, Northern Research Station, Roslin, Midlothian/Great Britain
Gore, A. J. P.	Institute of Terrestrial Ecology, Monks Wood Experimental Station, Abbots Ripton, Huntingdon, PE17 2LS/Great Britain
Grace, J.	Department of Forestry and Natural Resources, University of Edinburgh, King's Buildings, Mayfield Road, Edinburgh, EH9 3JU/Great Britain
Heal, O. W.	Institute of Terrestrial Ecology, Merlewood Research Station, Grange-over-Sands, Cumbria, LA11 6JU/Great Britain
Holding, A. J.	Department of General Microbiology, University of Edinburgh, West Mains Road, Edinburgh, EH9 3JG/Great Britain
Howson, Gillian	Institute of Terrestrial Ecology, Merlewood Research Station, Grange-over-Sands, Cumbria, LA11 6JU/Great Britain
Hughes, R. E.	Institute of Terrestrial Ecology, Penrhos Road, Bangor, Gwynedd, LL57 2LQ/Great Britain

JONES, HELEN E.	Institute of Terrestrial Ecology, Merlewood Research Station, Grange-over-Sands, Cumbria, LA11 6JU/Great Britain
JONES, VERNA	Institute of Terrestrial Ecology, Penrhos Road, Bangor, Gwynedd, LL57 2LQ/Great Britain
LATTER, PAMELA M.	Institute of Terrestrial Ecology, Merlewood Research Station, Grange-over-Sands, Cumbria, LA11 6JU/Great Britain
LUTMAN, JANE	c/o Institute of Terrestrial Ecology, Penrhos Road, Bangor, Gwynedd, LL57 2LQ/Great Britain
MARKS, T. C.	Department of Biology, Liverpool Polytechnic, Byrom Street, Liverpool, L3 3AF/Great Britain
MARTIN, N. J.	Department of Microbiology, West of Scotland Agricultural College, Auchincruive, Ayr, Scotland/Great Britain
MILLAR, R. O.	Institute of Terrestrial Ecology, Penrhos Road, Bangor, Gwynedd, LL57 2LQ/Great Britain
MILLER, G. R.	Institute of Terrestrial Ecology, Hill of Brathens, Glassel, Banchory, Kincardineshire, AB3 4BY/Great Britain
NEEP, P.	Institute of Terrestrial Ecology, Penrhos Road, Bangor, Gwynedd, LL57 2LQ/Great Britain
PERKINS, D. F.	Institute of Terrestrial Ecology, Penrhos Road, Bangor, Gwynedd, LL57 2LQ/Great Britain
RAWES, M.	Nature Conservancy Council, Moor House Field Station, Garrigill, Alston, Cumbria/Great Britain
SMITH, ROSALIND, A. H.	Nature Conservancy Council, 12 Hope Terrace, Edinburgh, EH9 2AS/Great Britain
SUMMERS, C. F.	Institute for Marine Environmental Research, Seals Research Division, Madingley Road, Cambridge, CB3 OET/Great Britain
D'SYLVA, BARBARA T. (now Mrs. B. T. COLIGAN)	c/o State Electricity Department, P.O. Box 2629, Doha, Qatar, Arabian Gulf
WATSON, A.	Institute of Terrestrial Ecology, Blackhall, Banchory, Kincardineshire, AB3 3PS/Great Britain
WEBB, N. R.	Institute of Terrestrial Ecology, Furzebrook Research Station, Wareham, Dorset, BH20 5AS/Great Britain
WHITTAKER, J. B.	Department of Biological Sciences, University of Lancaster, Bailrigg, Lancaster/Great Britain

I. The Moor House Programme

Moor House National Nature Reserve. The dark areas are *Calluneto-Eriophoretum*. The stream in the foreground is bordered by light-coloured *Sphagneto-Juncetum effusi*. Eroding peat is in the middle distance with Bog Hill bordered by peat haggs (photo K. J. F. Park).

1. Introduction and Site Description

O. W. HEAL and R. A. H. SMITH

The main site for IBP research on moorland ecosystems in the UK was the Moor House National Nature Reserve in northern England. The research aims and approach are outlined and the general characteristics of the area described.

1.1 Introduction

The Moor House National Nature Reserve was selected in 1965 by the UK IBP PT subcommittee as the main site for studies on moorland communities. The comprehensive IBP study on the production and dynamics of the moorland, initiated under the direction of J.B.Cragg, was built on, and has made extensive use of, results of non-IBP sponsored University and Nature Conservancy research studies (Cragg, 1961). The IBP work was funded by a grant from the Royal Society and administered by NERC and the Nature Conservancy.

The Reserve, acquired in 1952, is part of one of the largest areas of blanket bog-covered moorland in Great Britain. Although there are several vegetation and soil types, the programme has concentrated on the blanket bog ecosystem, because this occupies the greater part (57%) of the Reserve, and because grassland studies were being carried out in Snowdonia. However, the contrast between the ecology of the blanket bog and grassland sites at Moor House helps to explain some of the factors influencing biological processes under the same general climatic regime.

The aims of the studies have been to define (1) the levels of primary and secondary production and the main organisms involved, (2) the major pathways and rates of circulation of carbon and dry matter within the ecosystem, and (3) the influence of the major factors which affect production and circulation.

The area, close to or above the treeline, is one of the most isolated parts of England, and has always been a marginal habitat for man. The blanket bog, in particular, is little used, being of low agricultural and silvicultural value, and it was decided at an early stage that applied studies should not be developed. The approach adopted in the study of productivity was largely one of measurement of field populations of flora and fauna supported by limited laboratory and field experimental studies. The research was concentrated on one main site, Sike Hill, but studies on other blanket bog sites at Moor House provide estimates of variability in production and populations over a range of blanket bog conditions (Appendix).

The papers in this volume, each a synthesis of results from several projects, represent the cooperative efforts of many research workers, both in making data available and in their analysis and interpretation; they examine first the components of a bog ecosystem (Chaps. 1–7), then the total system (Chaps. 8–10).

1.2 Site Description

1.2.1 Topography

The Reserve (54° 65′ N, 2° 45′ W) is an area of nearly 4000 ha of moorland which straddles the summit ridge of the northern Pennines, a chain of hills running north-south in central northern England. To the west it includes a steep scarp slope which rises from about 300 m on the edge of the Vale of Eden, and the summit ridge itself at 600–845 m. The eastern part of the Reserve is a gently sloping, high-level plateau-like area at 500–600 m which has been dissected into east-west trending ridges by small, shallow, rapidly flowing streams of the Tees catchment (Fig. 1).

1.2.2 Geology

The Moor House area of the northern Pennines forms part of a fault-bounded structural unit, the Alston Block, which has been active since Devonian times. The floor or basement of the block is formed of Silurian or Ordovician strata, which outcrop at the foot of the escarpment. These rocks are overlain by a series of alternating, almost horizontal beds of limestone, sandstone and shale of Carboniferous age, which now dominate the surface geology. In the late Carboniferous a sill of quartz dolerite, the Whin Sill, intruded into the sedimentary rocks and at a later date mineralising fluids gave rise to lead and zinc veins. In the Tertiary, uplift of the Alston Block produced a high, steep, westward-facing escarpment; subsequent differential weathering on its variable strata has resulted in a stepped topography. To the east the gently dipping surface of the Block forms the high-level plateau-like area with relatively subdued relief. Glaciation of the area during the Pleistocene has resulted in a mantle of drift which obscures the solid geology over much of the Reserve. On the eastern slopes thick till is almost continuous, while on the summit ridge and the scarp slope drift is thin and discontinuous.

1.2.3 Soils

The dominant soil over more than 50% of the Reserve is the deep (usually *c* 1 m) layer of peat which overlies the boulder clay. Widespread peat accumulation was initiated at the Boreal-Atlantic transition about 5500 BC, when rainfall increased markedly and the presence of glacial boulder clay caused impeded drainage and waterlogging.

Accumulation of peat has continued at varying rates since the Atlantic period, with *Sphagnum*, *Eriophorum*, and *Calluna* as the chief peat-forming species. As its depth increases, blanket peat becomes increasingly unstable and erosion has probably occurred since the sub-Atlantic period. Massive erosion after heavy rainfall occasionally occurs (Crisp et al., 1964) but eroding gullies are frequent, probably initiated by channelling of surface drainage water. Wind and water continue the erosion, which is influenced by man's management (Bower, 1962). Eroding peat comprises 10–15% of the almost continuous peat cover on the eastern plateau-like area. Most of the peat which developed on the steeper western side has been lost through erosion.

Where blanket peat is absent, the soils developed on a given site are usually related to the presence or absence of glacial drift, the thickness of any such drift and the slope (Fig. 1). When drift is very thin (<20 cm) or absent, the underlying bedrock has influenced soil formation. Thus, as the thickness of drift increases on and around limestone exposures, the soils vary from rendzinas and brown calcareous soils, to acid brown earths and peaty gleyed podzols: soil complexes of this type occupy 4% of the Reserve but only 2% of the eastern escarpment. Apart from these complexes, brown earths are restricted to the very steep (>25°) valley sides of the escarpment, while peaty gleyed podzols tend to occupy moderately steep drift-covered slopes (10–20°) and peaty gleys the gently sloping or level areas. On the summit ridge, humus iron podzols have developed on sandy material derived from in situ disintegration of drift-free areas of sandstone bedrock. Small patches of alluvial soils have developed along the larger streams of the eastern slopes.

The mosaic of soils and vegetation which has developed in the past 10,000 years has probably been fairly stable during the last few hundred years; the main soil physical and chemical characteristics are given in Tables 1 and 2.

1.2.3.1 Characteristics of Peat Deposits

The deep layer of blanket peat covering 80–85% of the eastern part of the Reserve usually overlies a gleyed mineral soil. A horizon containing remains of birch scrub occurs at the base of the peat and minor climatic fluctuations have produced bands of peat of varying degrees of humification. On the better-drained slope and summit sites the peat is more humified than on lower, wet sites.

The water table, as indicated by the height of water in pits, is usually between 0 and 30 cm below the surface, depending on micro and local topography and drainage. Its position is influenced by climate, falling in the early stages of dry spells by as much as 2 cm d^{-1}. The rise in the water table after a dry spell is rapid—as much as 20 cm d^{-1} with a 4 mm rainfall (Forrest and Smith, 1975). Because of the high water-holding capacity of peat, however, the moisture content above and below the theoretical water table varies little within a site and is usually more than 500% dry weight. Diurnal variation in the moisture content of the surface litter can be two-fold and equals the seasonal variation (Springett et al., 1970).

The waterlogging is directly responsible for the development of reducing conditions deep in the profile, especially at wetter sites, and Urquhart and Gore (1973), while confirming earlier suggestions of seasonal variation in redox, emphasized the great variability of the redox profile between, and within, sites.

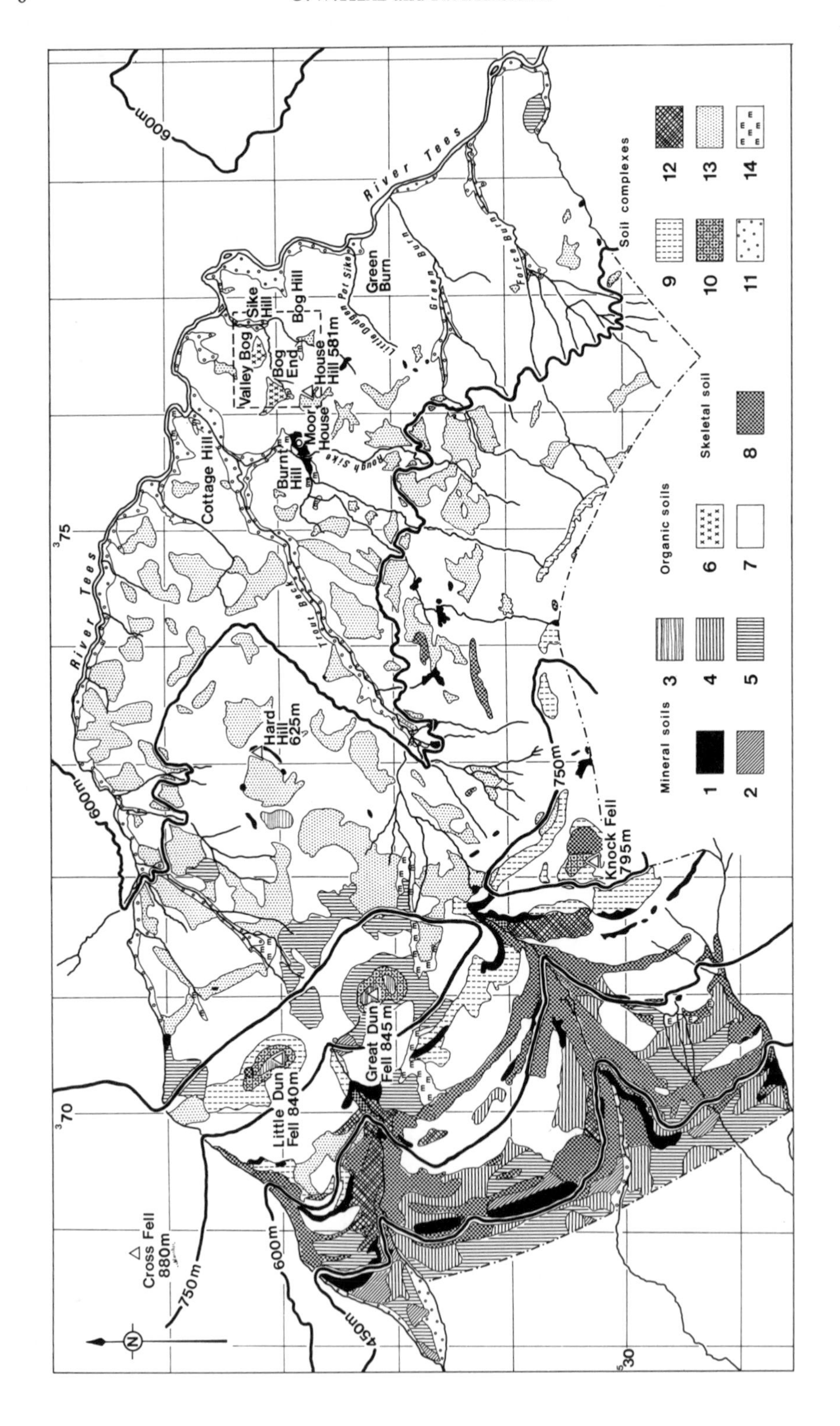
River Tees
600m
750m
450m
Cross Fell 880m
Little Dun Fell 840m
Great Dun Fell 845m
Knock Fell 795m
Hard Hill 625m
Cottage Hill
Burnt Hill
Moor House
House Hill 581m
Valley Bog
Bog End
Sike Hill
Bog Hill
Green Burn
Trout Beck
Rough Sike
Force Burn
N
375
370
530
Mineral soils
1
2
3
4
5
Organic soils
6
7
Skeletal soil
8
Soil complexes
9
10
11
12
13
14

There is a general decline in redox potential with depth, with values at Bog End decreasing from Eh 500 mV at the surface, to 336 ± 69 at 16.7 ± 6.1 cm depth (corrected to pH 4). Mean values over the whole profile are lower at the wetter Burnt Hill site, decreasing to Eh_4 112 ± 38 mV at 18.4 ± 4.0 cm. Although poorly defined, the mean depths of redox minima occur between 10 and 20 cm, approximating to the depth over which the water table fluctuates. Thus the pattern of redox potential within a site is complex, varying with temperature, which influences microbial activity, and with the water table, which varies in relation to climate and the moisture-holding and drainage characteristics of the peat.

The minimum redox values are associated, particularly at wetter sites, with sulphide deposition (Urquhart, 1966), and with maximum numbers of sulphate-reducing bacteria (Chap. 5). The profile of redox potential in the peat is probably responsible for the development of the coloured zones used to define sample positions for between-site comparison (Chaps. 5, 6).

A layer of fresh *Calluna* and *Eriophorum* litter, a few cm thick, overlies a layer of mainly unfragmented litter and *Sphagnum* which is dark brown in colour. At about the level of the redox dip (10–20 cm) the peat is often a greenish colour, darkening on exposure to air. Although the position of this green-brown horizon does not apparently correspond to the position of the water table at a particular time, its position and thickness seem to be directly related to site wetness. The colour may reflect the presence of ferrous iron or bleaching by hydrogen sulphide which is produced at this level. Conditions are normally reducing in this and the red-brown horizon below. The peat is increasingly humified with depth, but recognisable *Sphagnum* leaves, *Eriophorum vaginatum* leaf bases and roots occur throughout the profile.

In general, the blanket peat is very acid (pH in distilled water 3.0–4.2) with little seasonal variation and little between-site variation (Table 1). There is a tendency for pH to decline with depth in the Bog End profile: litter 3.57 ± 0.13, dark brown 3.50 ± 0.08, green-brown 3.36 ± 0.08 and red-brown 3.29 ± 0.11. Limited data from other sites generally support this, though at the Cottage Hill sites, where pH values in all zones are rather higher, there is a slight increase with depth, possibly as a result of flushing: litter 3.73, dark brown 4.06, green-brown 4.28 and red-brown 4.63.

◄

Fig. 1. Map of Moor House NNR showing major soil types and study sites (redrawn from Johnson and Dunham, 1963). The area marked ----- at Bog End is shown in detail in Figure 1 of Chapter 10

Key to the soil types and their equivalent names in the current classification of the Soil Survey of England and Wales:

1: Red-brown limestone soil ≡ soil complexes including rendzinas, brown calcareous soils, Brown Earths (sensu stricto), Stagnopodzols and Stagnohumic gley soils. *2:* Brown earths ≡ Brown Earths (sensu stricto). *3:* Fell top podzols ≡ Podzols (sensu stricto). *4:* Peaty podzols ≡ Stagnopodzols. *5:* Peaty gleys ≡ mainly Stagnohumic gley soils, some Humic gley soils. *6:* Valley bog and *7:* Blanket bog ≡ raw peat soils. *8:* Coarse scree ≡ raw skeletal soils. *9:* Peaty gley—peaty podzol soil complex. *10:* Solifluxion creep complex ≡ complex of Brown Earths and Brown Podzolic soils. *11:* Mixed bottom lands. *12:* Peaty Podzol—Brown Earth soil complex. *13:* Eroded blanket peat soil complex. *14:* Made ground (mine tips) ≡ man-made raw soils

Values of pH measured in peat homogenised with 1M KCl, which releases a larger proportion of the total acidity, are considerably lower than those obtained in distilled water: Bog End litter 3.0 ± 0.09, dark brown 2.91 ± 0.1, green-brown 2.75 ± 0.1, red-brown 2.73 ± 0.01.

The blanket peat, divorced from the underlying mineral substratum, is low in available nutrients (Table 1), and the only input of nutrients to the system is from the atmosphere. Bog plants depend on recirculation of nutrients from decaying remains, but release by decomposition is slow—a result of low pH, low temperature, lack of adequate aeration and waterlogging. Moreover, availability of many ions declines markedly below *c* pH 4.5; hence extraction by ordinary methods, at pH *c* 7.0, may not measure availability to the plant, much of what is extractable being biologically unavailable under these conditions. The capital is further depleted by losses in run-off, erosion and burning (Crisp, 1966). Dynamics of nutrient economy of the bog have been explored by Gore (1972) and Gore and Olson (1967).

Comparison of total nutrient concentrations in mineral and organic soils is difficult because of differences in availability and comparability of the depths of different layers (Table 2). Total concentrations of most nutrients are much higher in mineral soils than in peat, but a higher proportion of the nutrients is usually extractable from peat. Values for extractable amounts of most elements, on a weight basis, are not greatly different between the soil types, but comparison on a volume basis reveals the nutrient poverty of the peat, which is six times less dense than mineral soil; plant roots, therefore, have access to a considerably smaller nutrient supply. Moreover, the nutrients are strongly held in humic complexes in the highly organic peat.

Within the peat profile, the concentrations of iron, phosphorus and potassium are generally highest in the dark brown horizon. Iron may be concentrated there through oxidation of ferrous iron brought up with a rising water table, and its deposition as ferric iron, while high P and K levels are possibly due to root uptake of material leached from the litter layer following mineralisation. The concentration of calcium increases down the profile, a reflection of its strong binding in the peat, or a result of capilliary rise from the mineral soil beneath. Magnesium and sodium concentrations remain constant, and the decreasing C:N ratio reflects increasing decomposition with depth. Presence of H_2S in the green-brown horizon clearly demarcates this from the peat above, where it is absent. There are few between-site differences, although total phosphorus tends to be higher on drier sites and total magnesium the reverse.

1.2.4 Vegetation

Virtually all the IBP work was carried out on the blanket bog-dominated eastern plateau, at about 550 m altitude. This is above the present-day tree line and the plant cover, typically dwarf shrub, reflects the underlying soil type and, to a lesser extent, altitude and management. The *Calluneto-Eriophoretum* which characteristically dominates the acid, waterlogged peat is replaced by *Eriophoretum* on recently burned areas or where low summer temperatures at high altitude,

Table 1. Some physical and chemical characteristics of blanket peat at Moor House. Four depth zones are recognised by their colour:

	Litter	Dark brown	Green-brown	Red-brown
Colour (Munsell's chart)	—	5YR/2/2	5Y/2/2	5YR/3/2
Mean thickness (cm)	3.8	6.0	8.7	9.3
Density (g cm^{-3})	0.03	0.07	0.09	0.10
pH (H_2O)	3.52 ± 0.07	3.53 ± 0.14	3.49 ± 0.16	3.52 ± 0.16
Moisture (% dw)	1,241 ±96	1,199 ±142	1,275 ±130	1,209 ±316
Ash (% dw)	2.12 ± 0.19	3.56 ± 0.4	3.22 ± 0.28	3.10 ± 0.22
Total N (% dw)	1.04 ± 0.19	1.64 ± 0.12	1.50 ± 0.08	1.71 ± 0.08
Total C (% dw)	50.1 ± 0.8	50.9 ± 1.1	50.3 ± 0.7	51.3 ± 1.3
C:N	53.2 ±15.6	43.8 ± 5.3	44.8 ± 3.3	40.1 ± 4.5
Total K (% dw)	0.07 ± 0.01	0.08 ± 0.004	0.03 ± 0.003	0.02 ± 0.003
Extractable K (mg 100 g^{-1} dw)	nd	28.7 ± 6.2	8.2 ± 1.7	3.3 ± 0.48
Total P (% dw)	0.05 ± 0.005	0.08 ± 0.007	0.06 ± 0.004	0.04 ± 0.003
Extractable P (mg 100 g^{-1} dw)	nd	<0.4	<0.4	<0.4
Total Ca (% dw)	0.19 ± 0.02	0.24 ± 0.01	0.22 ± 0.02	0.28 ± 0.03
Extractable Ca (mg 100 g^{-1} dw)	nd	162 ± 14	152 ± 13	186 ± 18
Total Mg (% dw)	0.056 ± 0.007	0.072 ± 0.01	0.08 ± 0.01	0.065 ± 0.01
Extractable Mg (mg 100 g^{-1} dw)	nd	64.6 ± 3.5	62.1 ± 3.0	53.4 ± 3.7
Total Na (% dw)	0.02 ± 0.001	0.018 ± 0.01	0.016 ± 0.001	0.017 ± 0.005
Extractable Na (mg 100 g^{-1} dw)	nd	13.3 ± 0.8	10.6 ± 0.8	8.9 ± 0.4
Cation exchange capacity (me 100 g^{-1})	199	246	188	157
Total Fe (% dw)	0.14 ± 0.01	0.48 ± 0.05	0.15 ± 0.02	0.15 ± 0.01

Extractant ammonium acetate at pH 7.0. Mean ± se given where applicable. Data provided by V.G.Collins, B.T.D'Sylva, A.J.Holding, P.M.Latter, N.J.Martin and R.A.H.Smith.

or sheep grazing locally, restrict *Calluna* growth (Rawes and Welch, 1969). *Trichophoro-Eriophoretum* replaces the *Calluneto-Eriophoretum* on very wet, usually flat, areas of bog. The typical *Callunetum* comprises a *Calluna*-dominated community, *c* 30 cm high, with an understorey of species such as *Empetrum nigrum*, *Erica tetralix*, *Rubus chamaemorus*, *Eriophorum angustifolium*, *E. vaginatum*, and *Trichophorum cespitosum*, and a ground layer of *Sphagnum* spp. On the shallow peaty gleys and peaty podzols, species-poor *Nardetum subalpinum* and *Juncetum squarrosi subalpinum* occur and, on mineral-rich alluvial and limestone soils, relatively species-rich *Agrosto-Festucetum* (Eddy et al., 1969).

The IBP primary production studies (Chaps. 2, 3) were carried out on seven sites, chosen to represent the range of Moor House blanket bog vegetation, at similar altitudes (Appendix; Fig. 1). The two Cottage Hill sites were seral, having been burned in 1961, and showed the characteristic dominance of *Eriophorum* species, observed to follow a burn. The remaining sites had not been burnt for at least 30 years (M. Rawes, pers. comm.). Intensive studies on *Sphagnum* production were carried out at Burnt Hill.

Frequency values for higher plants at these sites show a clear distinction between the four drier sites on typical *Calluneto-Eriophoretum*: Sike Hill A and B, Bog Hill and Bog End, and the four wetter sites: Cottage Hill A and B on *Eriophoretum*, Green Burn on *Trichophoro-Eriophoretum* and Burnt Hill. A classification of the vegetation of the sites by normal information analysis giving a primary division on the basis of *E. angustifolium* confirms the distinction between the wet and dry sites. Burnt Hill, characterised by a wide range of habitats and consequently of dominant species, from *Sphagnum*-dominated pools, lawns and hummocks to typical *Calluneto-Eriophoretum*, shows the expected spectrum of variation (R. S. Clymo in Forrest and Smith, 1975).

The growth and distribution of *Sphagnum* species is closely related to moisture conditions and shade (Clymo, 1973), and cover on the drier *Calluneto-Eriophoretum* may be less than 20%, and the plants often unhealthy (Clymo and Reddaway, 1974). Nevertheless, *Sphagnum* species are important on all sites, particularly Green Burn, Burnt Hill, Cottage Hill B, Sike Hill B and Bog Hill. Indeed, inclusion of Green Burn among the wetter sites is largely due to the presence of *Sphagnum magellanicum*, its omission from the vegetation analysis revealing Green Burn's clear affinity with drier sites. *Erica tetralix* and *Vaccinium oxycoccum* were important additional species at this site.

The three main bog plants, *Calluna*, *Eriophorum vaginatum* and *Sphagnum* spp, have characteristic growth forms and dynamics. In areas which have not been burnt for 20 years or more, the age structure of the *Calluna* population is very wide, but the modal age of emergent stems is low—about eight years on the drier sites. The low age results from overgrowth of stems by *Sphagnum* and from wind and snow pressure which force older stems into a decumbent position in the litter layer, with the consequent emergence of younger upright stems. Thus a steady state is attained, the younger age classes being continually replaced, in contrast to the relatively even-aged stands, depending on the time since burning, in many other *Calluna*-dominated areas in Britain. In wetter areas, rapid overgrowth by *Sphagnum* produces an even younger modal age, but a similar stable, dynamic equilibrium.

Table 2. Summary of chemical composition of Moor House soils

Vegetation type	Soil type	pH	Loss on ignition (% dw)	Total N (% dw)	Extractable (mg 100 g^{-1} dw) K	Ca	Mg	P
Agrosto-Festucetum	Brown earth	4.7–6.3	12.5– 25.9	0.5–0.7	9–18	120–468	7–20	0.6–1.6
Nardus/Agrostis/Festuca	Podzol/gley	4.6–5.6	11.5– 35.5	0.3–0.7	8–26	86–130	16–22	0.8–1.4
Juncetum	Thin peat/peaty podzol/peaty gley	3.4–4.4	53 –113	1.2–2.0	12–70	20–140	20–34	—
Calluna/Eriophorum	Blanket peat	3.1–3.9	95 – 99	0.9–1.9	12–52	81–160	28–69	0.8–1.4
Sphagnum	Blanket peat	3.0–3.4	97 – 99	1.0–1.4	42–78	90–130	30–50	—

The range of concentrations is given for samples analysed by the Chemical Service at Merlewood Research Station 1960–1967. Sample depth usually 0–5 cm. Large variation within and between sites, and the use of different extractant procedures, allow only general interpretation.

Table 3. Chemical composition of vegetation from blanket bog

	Ash	N	P	K	Ca	Mg	Na
(A) Ten sites							
Calluna green shoots	2.60 ± 0.04	1.35 ± 0.02	0.13 ± 0.007[a]	0.57 ± 0.01[c]	0.32 ± 0.01	0.19 ± 0.004	0.06 ± 0.003
live wood	1.48 ± 0.04	0.59 ± 0.02[b]	0.05 ± 0.004[a]	0.27 ± 0.004	0.14 ± 0.01	0.07 ± 0.002	0.05 ± 0.004[b]
Eriophorum vaginatum green leaves	2.20 ± 0.09[c]	1.83 ± 0.06[b]	0.17 ± 0.01[a]	0.64 ± 0.04[a]	0.15 ± 0.01	0.16 ± 0.01[a]	0.01 ± 0.001
(B) Three sites							
Calluna dead wood	1.82 ± 0.07	0.85 ± 0.10	0.04 ± 0.002	0.14 ± 0.03	0.21 ± 0.005	0.07 ± 0.007	0.03 ± 0.003
Empetrum nigrum	2.03	0.74	0.08	0.26	0.45	0.15	0.01
Erica tetralix	2.43	0.86	0.05	0.43	0.38	0.10	0.10
Eriophorum vaginatum brown leaves	1.03 ± 0.70	1.01 ± 0.08	0.06 ± 0.04	0.26 ± 0.07	0.12 ± 0.09	0.07 ± 0.002	0.02 ± 0.001
Eriophorum angustifolium green leaves	1.94 ± 0.10	1.52 ± 0.04	0.10 ± 0.001	0.49 ± 0.03	0.16 ± 0.01	0.13 ± 0.001	0.07 ± 0.002
brown leaves	1.17 ± 0.10	0.88 ± 0.05	0.07 ± 0.03	0.10 ± 0.01	0.17 ± 0.10	0.05 ± 0.04	0.02 ± 0.001
Trichophorum cespitosum	3.13	1.53	0.11	1.02	0.13	0.18	0.22
Rubus chamaemorus	4.93	2.43	0.16	0.93	0.84	0.71	0.04
Narthecium ossifragum	4.78	1.95	0.62	1.33	0.61	0.20	0.13
Sphagnum papillosum	1.99 ± 0.04	0.86 ± 0.03	0.04 ± 0.003	0.35 ± 0.005	0.18 ± 0.07	0.08 ± 0.02	0.11 ± 0.005

(A) From 10 sites with significant differences between sites shown as [a] $P<0.001$; [b] $P<0.01$; [c] $P<0.05$. (B) From three sites (Burnt Hill, Hard Hill, Cottage Hill B). Values (± se where available) are % dw.

1.2.4.1 Chemical Composition

Nutrient concentrations in vegetation reflect a complex interaction of factors including the nutrient-supplying power of the soil (Tables 1, 2), temperature and moisture fluctuations, and manuring and defoliation by herbivores. Grazing, which increases the rate at which nutrients are cycled, is minimal on the blanket bog, and the low temperatures, high rainfall and high water table combine with the low availability of nutrients in the peat to select an ombrotrophic vegetation with low nutrient concentrations (Table 3). Concentrations, particularly of potassium, fall at the low end of the range found in other ecosystems (Rodin and Bazilevich, 1967) including the grasslands at Moor House (Rawes and Welch, 1969) and Snowdonia (Chap. 16). Nutrient levels in the vegetation of the bog are similar to those for *Calluna* heaths (Robertson and Davies, 1965; Chapman, 1967; Gimingham, 1972) and in many tundra ecosystems (Bliss and Wielgolaski, 1973; Rosswall and Heal, 1975).

Although there is considerable variation in nutrient concentrations between species at any one bog site, the plants with relatively high concentrations (*Rubus chamaemorus*, *Narthecium ossifragum*, and *Trichophorum cespitosum*) constitute a small proportion of the biomass and production. Only *T. cespitosum* at Cottage Hill A reaches dominant status. For a given plant species or component the concentration of a nutrient may also vary significantly between sites. Analysis of shoots and live wood of *Calluna* and green leaves of *E. vaginatum* on ten bog sites showed significant differences in P and, to a lesser extent, N and K (Table 3). Vegetation of the wet flush Cottage Hill sites and the drier Bog Hill tended to have high concentrations while the wet non-flushed sites at Burnt Hill and Green Burn were low in nutrients. The latter sites were also rich in *Sphagnum*.

Seasonal variation in nutrient concentration is not well documented for Moor House but in *R. chamaemorus* (Marks and Taylor, 1972), the aerial parts showed a marked decrease in concentration of N, P, and K between May and August, while Ca and Mg increased over the same period.

1.2.5 Climate

The general climate (Table 4) is cool, wet and windy, severe by British standards, oceanic and subarctic rather than temperate. The winters are cold and snowy, with an average snow cover of about 70 days, ground frost most nights from November to April, and February mean temperature of 1.0° C. The summers are cool and wet, the temperature exceeding 10° C only in July and August (August mean 10.8° C) and frost usually occurring in every month. The growing season, based on mean temperature >5.6° C, is about 180 days from May to October, with a sum of day degrees above 5.6° C of 621° and above 0° C of 2250°. The high rainfall, *c* 1900 mm yr^{-1}, is irregularly distributed throughout the year, precipitation occurring on an average of 250 days, with late autumn the wettest period. Windspeeds are high, particularly in February and March (*c* 8.0 m s^{-1}) when easterly winds are frequent. Calculation of potential evaporation from mean monthly temperatures and rainfall suggests that a water deficit would seldom occur, but estimates based on shorter periods indicate that it is a regular occur-

Table 4. Long-term climatic averages at Moor House (from Heal et al., 1975)

	J	F	M	A	M	J	J	A	S	O	N	D	Year
Solar radiation (kJ cm^{-2}) 1972–1974	4.4	10.9	24.6	34.5	41.7	47.1	42.4	38.0	24.4	13.6	6.8	3.5	291.9
Screen temperature (°C) 1953–1972													
Mean max	2.1	1.6	3.9	7.1	10.8	13.8	14.7	14.4	12.7	9.7	5.1	3.3	8.3
Mean min	−2.9	−3.6	−1.7	−0.1	2.5	5.4	7.2	7.3	5.9	3.9	0.1	−1.7	1.9
1/2 (max + min)	−0.4	−1.0	1.1	3.5	6.7	9.6	11.0	10.8	9.3	6.8	2.6	0.8	5.1
Lowest grass min temp (°C) 1957–1973	−13.9	−14.5	−10.9	−10.3	−7.7	−5.2	−2.9	−2.8	−5.6	−6.8	−11.5	−12.7	−9.5
Soil temp at 30 cm, 0900 GMT (°C) 1956–1972	1.6	1.4	2.1	3.7	7.0	9.9	11.5	11.7	10.4	7.9	4.4	2.7	6.2
Relative humidity at 0900 GMT (%) 1953–1967	94.0	94.7	90.3	87.1	82.2	82.8	84.9	86.7	88.7	91.9	94.3	94.4	89.3
Rainfall (mm) 1953–1972	177.3	150.1	133.7	119.9	127.9	113.3	144.1	169.9	160.4	179.9	200.4	206.1	1,883.0
Number of days with rain 1953–1972	22.5	20.8	20.9	19.9	19.4	18.3	20.0	20.8	19.2	20.8	22.6	23.2	248.4
Number of days with snow lying at 0900 GMT 1953–1972	16.5	17.4	11.6	4.0	0.6	0	0	0	0	0.2	6.0	10.4	66.7
Wind velocity (m s^{-1}) 1956–1972	8.1	8.0	8.0	6.7	6.3	5.9	5.9	5.9	6.4	7.4	7.5	8.2	7.0
Daily duration of bright sun (h) 1954–1972	1.0	1.7	2.6	4.0	5.2	5.8	4.6	4.2	3.4	2.5	1.3	0.9	3.1

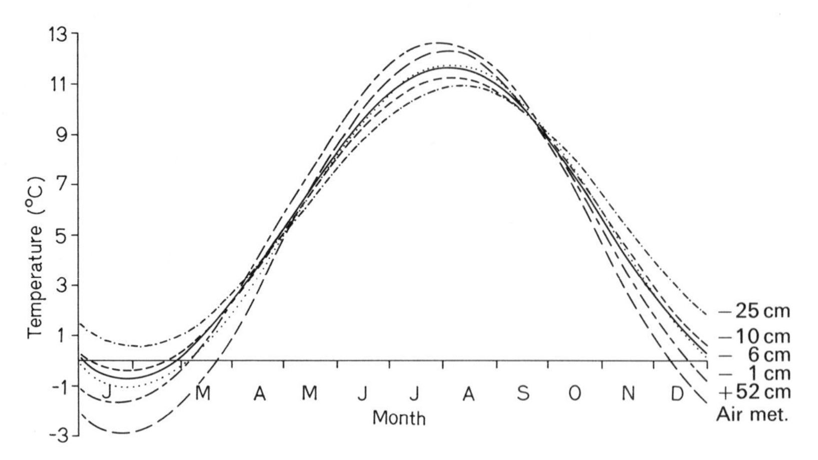

Fig. 2. Seasonal variation in temperature within a profile on blanket bog in 1969. The curves are calculated by Fourier analysis of the weekly mean temperatures at each position, using only the first harmonic. The first harmonic explains 85–92% of the variability of the data, the % increasing with depth

rence in summer. Periods of low humidity (RH < 10%) also occur occasionally and may have a more marked effect on the biota than does the general climate.

Maximum solar radiation occurs in June (5.8 h of bright sunshine d^{-1}) and the total annual radiation was 292 kJ cm^{-2} over the IBP study period. The important climatic characteristics of these years at Moor House are described in Chapter 2.

The above climatic regime refers to the area in the vicinity of the meteorological station at 558 m, but the Reserve covers an altitudinal range from 300–845 m. Limited meteorological data from the station at 845 m on Great Dun Fell show a mean annual temperature of about 3.5° C compared with 5.1° C at Moor House, indicating a lag of 0.53° C for 100 m rise in altitude. The decline in air temperature related to altitude is similar to that in soil temperatures. Coulson et al. (1976) recorded a difference of 0.48° C per 100 m rise on the western scarp at Moor House. The temperature gradient with altitude is not constant throughout the year, the decrease in temperature at 1 cm in the soil under mixed grasslands being greatest in autumn and least in winter (2.36 ± 0.1 and 1.52 ± 0.15° C per 500 m) (Coulson et al., 1976). The temperature differences within the Moor House area correspond to large regional differences; thus the mean soil temperature difference of 2.4° C between sites at 370 m and 847 m is equal to that between Plymouth and Edinburgh. Within the blanket bog seasonal variation in temperature shows the expected damping with depth (Fig. 2).

A peaty podzol with *J. squarrosi* about 100 m from the site of the bog temperature measurements showed a similar pattern with depth but mean annual temperatures were about 0.5° C higher under *Juncus* (Table 5). The litter microhabitats within the blanket bog also vary by *c* 0.5° C in mean annual temperature, *E. vaginatum* tussocks being warmest, *Sphagnum* lawn intermediate and *Calluna* litter lowest (Table 5; Standen, 1973). The differences probably relate to the degree of shading and to evaporative cooling from the wet *Sphagnum*. The mean

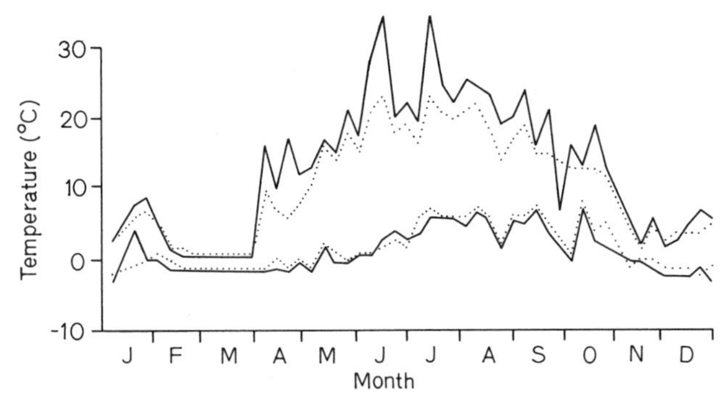

Fig. 3. Seasonal variation in weekly maximum and minimum temperatures in *Calluna* (........) and *Eriophorum* (——) litter on blanket bog in 1969

Table 5. Mean annual temperatures (°C) for 1969, calculated from hourly recording with thermisters, within profiles on blanket bog and *Juncetum squarrosi* at Sike Hill

Position (cm)		Blanket bog		*Juncus*		Difference
Within vegetation	(+ 4)	5.42	5.34	—	—	0.44[a]
Within vegetation	(+ 2)	—	—	5.71	5.93	
In litter	(− 1)	5.52	5.41	6.15	5.99	0.61[a]
	(− 3)	5.38	—	6.15	—	0.77
	(− 6)	5.44	—	5.98	—	0.54
	(−10)	5.44	—	5.86	—	0.42
	(−25)	5.75	—	6.07	—	0.32
E. vaginatum tussock litter		6.01	5.69			
Sphagnum lawn	(− 0.5)	5.83	5.83			
Sphagnum lawn	(− 5.0)	5.64	5.68			

Data for duplicate probes are given when available. [a] Calculated from means of duplicate probes.

temperatures do not adequately reflect the temperature differences between microhabitats, the weekly minima and maxima for *E. vaginatum* and *Calluna* litter showing that during the summer, differences of 5 to 10° C are frequent (Fig. 3).

1.2.6 Management

Man's interest in moorlands (Chap. 10) such as Moor House is associated particularly with two herbivores, grouse and sheep. Prior to its purchase by the Nature Conservancy in 1952, the area was a grouse moor and, although limited experimental burning is maintained, there is no shooting now. Grazing rights are retained by the local farmers and sheep range freely over the Reserve. At the lower altitudes there is potential for tree growth (Millar, 1964, 1965; Brown et al., 1964), but afforestation in this area of the northern Pennines is negligible. The eastern dip slope contains parts of the catchment for the River Tees and South Tyne, and the recently developed Cow Green reservoir borders the Reserve, emphasizing the importance of these moorlands in water resource management (Crisp et al., 1974).

Informal recreation, mainly walking, is increasing in the area, partly associated with the Pennine Way which crosses the Reserve. Despite man's interest in six aspects of the moorland (sheep, grouse, afforestation, nature conservation, water and recreation) his current impact on the ecology of the area is small. This contrasts with man's past activity in the Moor House area which can be seen in the relics of metalliferous mining, and which greatly influenced settlement in the area in the 19th century. The old mine tracks and drainage caused significant effects on the vegetation, and the contaminated waste tips support a distinctive and specialised flora.

Acknowledgments. The authors are grateful to M. Hornung for contributing to the soils section; to J. K. Adamson, A. D. Bailey, S. Carrick, P. Costeloe, C. Gill, G. Howson, B. Marsh, and J. M. Nelson for their help in collection and analysis of temperature, radiation and meteorological data; and to the Chemical Section at Merlewood Research Station for chemical analyses. M. Rawes was particularly helpful in a wide variety of ways throughout the project.

1.3 Appendix

The site names in the various chapters on Moor House refer to the following blanket bog sites. The main sites are shown in Chapter 1, Figure 1, and Chapter 10, Figure 1.

Name	Grid ref (NY)	Altitude (m)
Sike Hill A	769331	550
Sike Hill B	770331	547
Bog Hill[a]	767326	550
Bog Hill[b]	768327	550
Bog End	764329	556
Cottage Hill A	752335	561
Cottage Hill B	753335	558
Green Burn	774323	515
Burnt Hill	753329	572
Hard Hill	743330	594
Valley Bog	763331	550

[a] For primary production studies.
[b] For decomposition studies in sheep exclosure.

2. Field Estimates of Primary Production

R. A. H. SMITH and G. I. FORREST

From field studies, the production of the current blanket bog vegetation is estimated, with an assessment of the main factors causing variation in production.

2.1 Introduction

The blanket bog vegetation at Moor House has a *Calluna* canopy, an understorey dominated by *Eriophorum* species and dwarf shrubs, and a *Sphagnum* ground layer (Chap. 1). The different growth forms of these species necessitate methodological differences in the estimation of their primary production, which was carried out both by field cropping and by physiological techniques. Production estimates were derived directly from the field cropping data, and were predicted using modelling techniques, once relationships had been established between photosynthesis and environmental parameters (Chap. 3). The present chapter describes the field cropping procedures and results.

The main aim of the primary production studies at Moor House was to estimate production by species and community. In the present chapter, production of a range of blanket bog sites is compared to provide a measure of the variation in production and to determine whether this variation was related to the vegetation heterogeneity, which resulted from variation in microhabitat and management factors. Comparisons are also made with the production of other vegetation types on the Reserve, and possible relationships with soil conditions, floristics, microclimate and management factors are examined.

2.2 Definition of Steady State and Production

The blanket bog vegetation at Moor House is assumed to be in a steady state, except in those areas where burning has taken place over the last 20 years (Forrest, 1971). A steady state is defined as the situation in which there is no change in biomass and production from year to year over a long period of time. This assumption is mainly based on observation and on the population age structure characteristics, which show that there is continuous rejuvenation of *Calluna*. The stems of the *Calluna* bushes become decumbent, as a result of weight of snow and high winds, and are buried by upward growth of the *Sphagnum* carpet. The buried *Calluna* stems then root adventitiously into the *Sphagnum*, causing continuous rejuvenation of *Calluna* with increase of frequency in the younger age classes

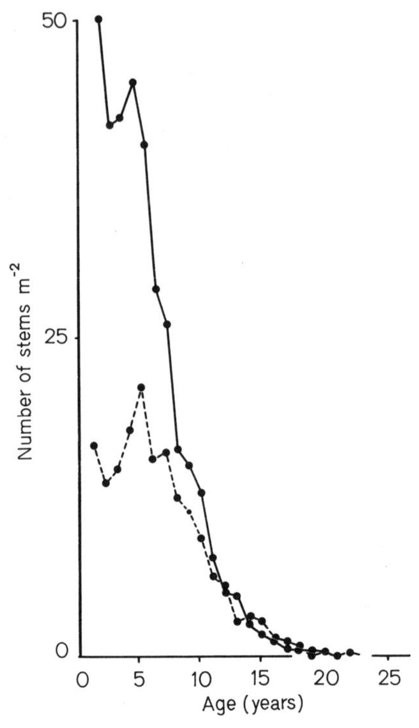

Fig. 1. *Calluna* age distribution for two blanket bog sites (after Forrest and Smith, 1975). ——— Green Burn - - - Sike Hill B

(Fig. 1). Although Sike Hill B has not been burnt for at least 30 years, the modal age of the *Calluna* population (as measured by stem growth rings at the level of the bog surface) is only five years. On a wetter site such as Green Burn the modal age is lower (< 1 year). An essential point is that the *Calluna* shows a mixed age distribution, in comparison with even-aged stands following burning in drier localities. There is also no evidence of succession from a *Calluna*-dominated community (Rawes, 1971), and the assumption of a steady state is supported by the fact that there is no measurable change in total biomass from year to year (Forrest and Smith, 1975). Although the blanket bog vegetation can be regarded as being in a steady state, this does not necessarily imply that the ecosystem as a whole can be similarly regarded (Chaps. 8, 9).

This assumption of a steady state necessitates precise definition of production terminology. There is, by definition of a steady state, no net accumulation of material over the year, above or below ground. There is, however, net production (P_n) over a period t as defined by Newbould (1967):

$$P_n = \Delta B + L + G$$

where ΔB is change in biomass over period t
L is loss by shedding and death during t
G is loss to consumers during t.

On the Moor House blanket bog, the amount of vegetation removed by herbivores is negligible, thus $G \simeq O$ (Chap. 4). Newbould probably intended this equation to refer to year-to-year changes in a seral situation. In the present context of a steady state, with t the length of the growing season, L represents losses from new growth only. In a steady state situation, there is no net annual increment, and P_n is a measure of the net annual amount of material assimilated. This assimilation is balanced by death and litterfall losses; indeed the measurement of these losses provides a further means of estimating net production, in addition to the measurement of biomass change over the growing season.

2.3 Sources of Error

Comparisons of production estimates from a range of sites, particularly when these have been obtained by different workers, are subject to many errors, both methodological and statistical. Quantitative expression of these errors is difficult because of the extrapolations often used to obtain production values.

The use of different techniques by each worker may result in important differences between production estimates. Also, in addition to normal random errors, systematic errors may be present, for example by summation, from a single harvest, of individual production estimates for species with peak biomasses at different times of the year. A further difficulty inherent in between-site and between-species comparisons of production estimates results from the differential effects of grazing and burning on the productivity of species and sites. Providing, however, that all these potential error sources are considered in their proper perspective, it is possible to draw limited conclusions from such comparisons.

2.4 Production Estimates at Moor House

Before the start of the International Biological Programme, several net production estimates had been obtained at Moor House, mainly as a result of work on the effects of grazing and burning (Rawes, 1961, 1963; Park et al., 1962; Rawes and Welch, 1964, 1966, 1969; Welch and Rawes, 1964, 1965, 1966) and on the recovery of blanket bog following clipping (Gore and Olson, 1967). Rawes and Welch (1969) obtained values for annual production of vascular species above ground of 154 g m^{-2} yr^{-1} for typical *Calluneto-Eriophoretum*, and of 69 g m^{-2} yr^{-1} for *Eriophoretum* derived from burning (Table 1). It was also estimated, by fitting a simple model to field results (Gore and Olson, 1967), that the net total primary production of *Calluneto-Eriophoretum* blanket bog attained a value of 420 g m^{-2} yr^{-1} following clipping. The "climax" net production value of 420 g m^{-2} yr^{-1} is similar to the 447 g m^{-2} yr^{-1} subsequently estimated on an adjacent site by direct measurement (Forrest and Smith, 1975; Table 1).

The more recent IBP-orientated primary production work included detailed investigations on single plant genera or species (Clymo, 1970; Clymo and Reddaway, 1972, 1974; Taylor and Marks, 1971; Marks and Taylor, 1972), and also measurements of component and total net production of a range of blanket bog sites (Forrest, 1971; Forrest and Smith, 1975).

Table 1. Summary of estimates of total primary production obtained on blanket bog sites at Moor House

Site	Vegetation type	Net production (g m^{-2} yr^{-1})				
		Above-ground		Total		Source of data
		−*Sphagnum*	+*Sphagnum*	−*Sphagnum*	+*Sphagnum*	
Sike Hill A	*Calluneto-Eriophoretum*	202	247	547	592	Forrest (1971)
Sike Hill B	*Calluneto-Eriophoretum*	298	454	589	745	Forrest and Smith (1975)
Bog Hill	*Calluneto-Eriophoretum*	238	355	564	681	Forrest and Smith (1975)
Bog End	*Calluneto-Eriophoretum*	259	308	447	496	Forrest and Smith (1975)
Bog End	*Calluneto-Eriophoretum*	nd	nd	420	nd	Gore and Olson (1967)
B2[a]	*Calluneto-Eriophoretum*	154	nd	nd	nd	Rawes and Welch (1969)
Green Burn	*Trichophoro-Eriophoretum*	105	318	278	491	Forrest and Smith (1975)
Burnt Hill	See text	nd	nd	nd	305	Clymo and Reddaway (1972)
Cottage Hill A	*Eriophoretum*	196	231	706	741	Forrest and Smith (1975)
Cottage Hill B	*Eriophoretum*	405	612	661	868	Forrest and Smith (1975)
E2[a]	*Eriophoretum*	69	nd	nd	nd	Rawes and Welch (1969)

nd: no data. [a] Sites on Cottage Hill.

2.4.1 Sphagnum Production

The investigation of a single genus was adopted by Clymo and Reddaway working on *Sphagnum*. Their main study area consisted of a mosaic of *Sphagnum*-dominated pools, lawns and hummocks, and general blanket bog vegetation at Burnt Hill, Moor House. The blanket bog vegetation was typical *Calluneto-Eriophoretum*, tending to the *Trichophoro-Eriophoretum* characteristic of wetter areas. Over five years, net primary production of *Sphagnum rubellum* on Burnt Hill averaged 80 g m^{-2} yr^{-1} (Table 2) (Clymo and Reddaway, 1974). There was a four-fold variation in results over two years between sites with healthy (200 g m^{-2} yr^{-1}) and unhealthy (50 g m^{-2} yr^{-1}) *Sphagnum* (Clymo and Reddaway, 1972). About 60% of *Sphagnum* production on this site occurred in pools, lawns and hummocks, although these habitat types occupy only 40% of the total area. The remaining *Sphagnum* production is associated with general blanket bog vegetation (Clymo and Reddaway, 1972). The results in Table 2 show that the production in 1968/69 was low compared with the two- and five-year averages which were themselves similar. Clymo and Reddaway (1974) suggest that this difference

Table 2. Mean net annual growth rate (g m^{-2} yr^{-1} ± se) of *Sphagnum rubellum* at three Moor House blanket bog sites over different periods (after Clymo and Reddaway, 1974)

Period	Sike Hill	Bog Hill	Burnt Hill
1 yr, 1968–69	86 ± 10 (276)	nd	75 ± 15 (71)
2 yr, 1968–70	111 ± 13 (118)	124 ± 14 (153)	96 ± 13 (57)
5 yr, 1968–73	109 ± 8 (41)	135 ± 6 (127)	80 ± 4 (94)

Number of observations given in parentheses.
nd: no data.

could be a result either of compensation of the 1968/69 figure by a higher than average production in 1969/70, or of bias in the 1968/69 results.

Comparisons of the production of *S. rubellum* on three sites (Table 2) showed few between-site differences; the growth rate of this species was greatest on Bog Hill, least on Burnt Hill, but there are difficulties of interpretation of these differences because of variation in *Sphagnum* and *Calluna* cover, water supply and grazing intensity. The average net production of *S. rubellum* on the three sites was 130 g m^{-2} yr^{-1}, with a between-site range varying by a factor of 1.7 (Clymo and Reddaway, 1974).

These production estimates were obtained by measurement of the growth in length of *Sphagnum* relative to the free end of cranked wires inserted vertically into the *Sphagnum* carpet. The growth figures were converted to production, using values for the mass per unit area of each species over a given carpet depth. Production was also estimated by measurement of the weight increase of the top 5 cm of *Sphagnum* plants, severed and then replaced or transplanted into the bog surface, Corrections were made for material originally present in the capitulum and carried into the "new growth" by internode extension. Both these methods give a figure for net production since the total increase of material over one year was measured (including standing dead production if any), and litterfall from current year's growth of this genus is probably insignificant.

The transplant method was used to investigate the actual and potential production of *Sphagnum* species in different habitats (Clymo and Reddaway, 1972). The results (Fig. 2) showed that each species had a greater net production than the others in its own habitat; this, however, was not necessarily the habitat in which each had the greatest net production. In the case of *S. rubellum*, *S. recurvum* and *S. cuspidatum*, net production increased with increasing wetness, but *S. papillosum* had a peak in its own habitat. With declining water table, *S. rubellum* can maintain a higher capitulum water content than *S. papillosum* and *S. cuspidatum* (Clymo, 1973) and this may explain some of the within- and between-site differences in species composition and production. However, preliminary microclimatic data (R.S. Clymo, pers. comm.) indicate that pools were often 2–3° C warmer than lawns or hummocks, and this factor, in addition to wetness, may account for the productivity maxima in pools for these three species. There are also pH differences which may influence production; in summer the pH was 3.3–3.5 in hummocks and 3.8–4.2 in pools, although during the remainder of the year a fairly uniform pH of 4.5 existed in all habitats (R.S. Clymo, pers. comm.).

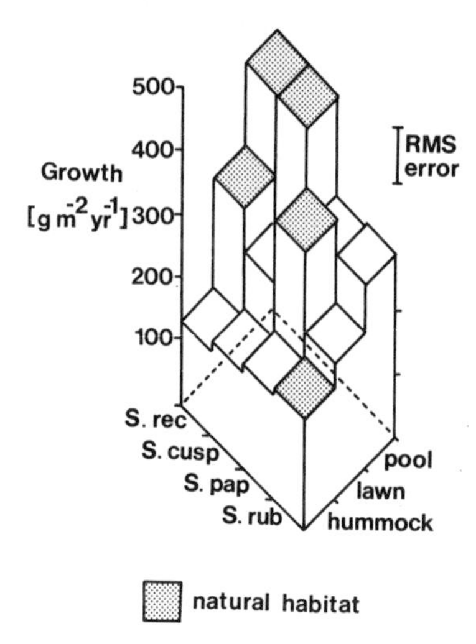

Fig. 2. Net productivity of four species of *Sphagnum* in three habitats on Burnt Hill (after Clymo and Reddaway, 1972)

2.4.2 Total and Component Net Production of Blanket Bog Vegetation

The total and component net production of seven blanket bog sites was measured (Forrest, 1971; Forrest and Smith, 1975). These sites included true *Calluneto-Eriophoretum* blanket bog, the floristically richer *Trichophoro-Eriophoretum*, and *Eriophoretum* derived from burning (Chap. 1).

Separate estimates of the production of each species were obtained in most cases both for above- and below-ground components, the above:below-ground boundary being defined as the top of the *Sphagnum* carpet. Current year's net green shoot production and component litterfall of woody species were measured directly by harvesting and by collection of material respectively. It was assumed that conversion of green shoots to standing dead was insignificant, and that no green shoots were lost to litter in the inter-harvest period (spring-summer). Figures of net wood production of woody species by radial increment were derived from regressions of various components on age; production was calculated as the difference between the calculated live wood biomass m^{-2} in successive years, assuming that litterfall losses were similar in the two years. An estimate of below-ground production of dwarf shrubs was obtained by assuming that production: biomass ratios were similar above and below ground, this assumption being reasonable as most of the below-ground biomass of woody plants comprised buried stems. However, the assumptions involved in deriving a figure for below-ground production are large and as no method was found to assess the magnitude of any losses, they were assumed to be equivalent to those above ground. The annual transfer from above to below ground by *Sphagnum* overgrowth was also estimated. Estimates of outputs to standing dead for woody species were obtained from the mortality rates, based on the age distribution curves (Fig. 1) and the regressions of wood dry weight on age.

Table 3. Net production of blanket bog sites at Moor House ($g\,m^{-2}\,yr^{-1} \pm se$ where calculable) (after Forrest and Smith, 1975)

Component		Sike Hill A	Sike Hill B	Bog Hill	Bog End	Green Burn	Cottage Hill A	Cottage Hill B
Calluna vulgaris	Green shoots	130±10	240±21	181±26	210±13	71±17	5±7	219±25
	Below-ground	183	151	180	64	110	nd	19
	Total live	313	391	361	274	181	nd	238
	Wood overgrown	—	12	29	—	8	nd	nd
	Losses–litter	108	95	90	122	73	nd	nd
	Losses–standing dead	60	61	92	80	30	nd	nd
Empetrum nigrum	Total above-ground	11	5	2	3	—	—	1
Erica tetralix	Total above-ground	—	—	—	—	11	0	—
Eriophorum vaginatum	Total	221	191	199	168	84	—	19
E. angustifolium	Total	—	0	—	—	2	120	399
Trichophorum cespitosum	Total	—	—	—	—	—	581	4
Other species	Total above-ground	2	2	2	2	0	—	—
Sphagnum	Total	45	156	117	49	213	35	207
Total all vascular species		547	589	564	447	278	706	661
Total all species		592	745	681	496	491	741	868

nd: no data

Production figures for *Eriophorum vaginatum* were obtained from weight measurements and detailed component sampling of individual tussocks. Estimates of production of other vascular species were derived mainly from the direct harvest figures, and production of *Sphagnum* spp from cover estimates and Clymo's growth values which included measurements on two of Forrest's sites. These production estimates obtained by derivation and extrapolation are probably less valid than those obtained by direct measurement. The production of only one site, Sike Hill A, was investigated to the detail specified above (Forrest, 1971); on the remaining sites some of the production parameters were extrapolated using the figures obtained at Sike Hill A, on the assumption that these figures were similar at all sites (Forrest and Smith, 1975).

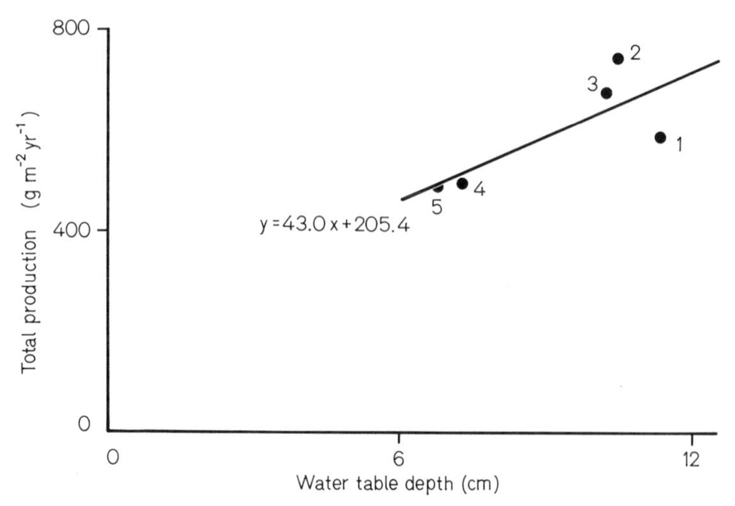

Fig. 3. Total production as a function of water table depth for blanket bog sites in steady state condition (after Forrest and Smith, 1975). *1:* Sike Hill A. *2:* Sike Hill B. *3:* Bog Hill. *4:* Bog End. *5:* Green Burn

The mean total net primary production of the seven blanket bog sites was 659 ± 53 g m^{-2} yr^{-1} (Forrest and Smith, 1975). This figure excludes *Calluna* radial increment, which was negative on two sites and thus obviously incorrect. Inclusion of the production of this component for the remaining sites gave a maximum increase in the total production figure of individual sites of less than 10%. The highest figure for total net primary production was 868 g m^{-2} yr^{-1} at Cottage Hill B, and it is likely that this was significantly greater than the lowest production figure, from Green Burn (Table 3). Probably no other differences were significant, but the production of the other Cottage Hill site was also high (741 g m^{-2} yr^{-1}). Cottage Hill had been burnt eight years earlier, resulting in a vegetation dominated by *Eriophorum angustifolium* and *Trichophorum cespitosum* which probably have higher relative growth rates than the dominants on the other sites. The high production of these sites may also reflect increased relative growth rates of some species related to juvenility due to burning. The importance of this juvenility cannot be assessed adequately since no comparable data are available on the production of the dominant species, *E. angustifolium* and *T. cespitosum*, on unburnt areas, although *Calluna* on these sites has a high production: biomass ratio above ground in comparison with other sites, probably implying a juvenile state. The dominance of species with high relative growth rates following burning compensates for the low absolute productivity of regenerating *Calluna* resulting from its low biomass. To what extent the dominance of these species at Cottage Hill is a result of the recent burn cannot be determined; from the present wetness of the area it seems likely that the vegetation was distinct from that of the other blanket bog sites prior to the burn.

The blanket bog sites other than Cottage Hill, assumed to be in a steady state, show a significant trend of decreasing production with increasing wetness, expressed in terms of water table depth measured in pits (Fig. 3). This trend is

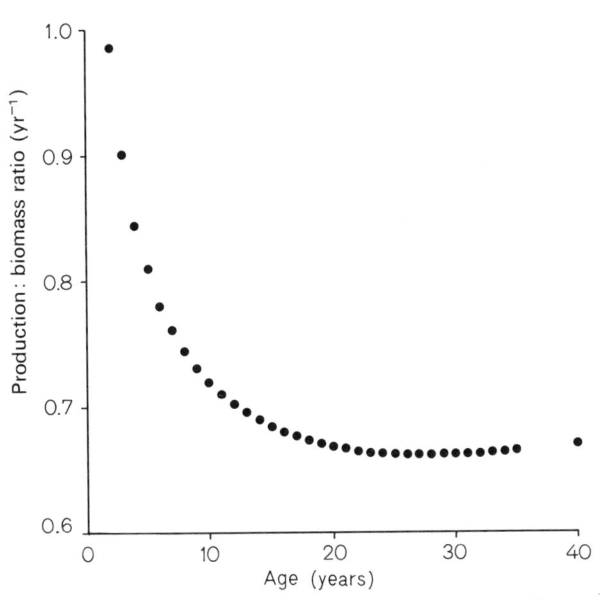

Fig. 4. Total production:biomass ratio as a function of age for *Calluna* at Sike Hill A

probably a result of a decreasing contribution of *Calluna* (190 g m^{-2} yr^{-1} to 70 g m^{-2} yr^{-1}) and *E. vaginatum* (195 g m^{-2} yr^{-1} to 84 g m^{-2} yr^{-1}) which is only partly replaced by increased *Sphagnum* production (92 g m^{-2} yr^{-1} to 213 g m^{-2} yr^{-1}) on the wetter Green Burn site and also on Burnt Hill, which had a total production of only 305 g m^{-2} yr^{-1} (Clymo and Reddaway, 1972). Burnt Hill is not included in Figure 3 because no comparable water table data are available; it is, however, at least as wet as Green Burn, as it is the only site with standing water present. On the *Sphagnum*-dominated areas of Burnt Hill the production of the ericaceous shrub category and *Eriophorum* spp was only 15–30% of that of *Sphagnum* (Clymo and Reddaway, 1972). In contrast, there were few differences in the net production of *S. rubellum*, a species of drier hummock habitats, on three sites of differing wetness (Table 2). *E. vaginatum* was shown to have a positive growth response to waterlogging under experimental conditions, although this may have been a response to nutrient release from *Sphagnum* killed by waterlogging, rather than a direct response to the water regime (Gore and Urquhart, 1966).

Although the increased *Sphagnum* production at the wetter sites does not wholly compensate for reduced vascular plant production, the total production of a *Calluna*/*Sphagnum* community with active *Sphagnum* growth may be greater than that of the two species growing separately. This could result from rejuvenation of *Calluna* by *Sphagnum* overgrowth followed by adventitious rooting, the ratio of total production (green shoots + radial increment): biomass decreasing with age up to 28–29 years after which there is a slight increase (Fig. 4). Since rejuvenation by *Sphagnum* does not reduce *Calluna* biomass, as is the case when rejuvenation occurs due to fire, there may be an enhancement of total production due to interaction of these two species. However, it would seem that at Green Burn at least this is not the dominant factor controlling *Calluna* production, since the ratio of green shoot production:biomass (i.e. live material) above ground, is lower at this site (0.172) than on six other bog sites which show a range from 0.176 to 0.625. It is likely that the wetness of Green Burn is the most important limitation to *Calluna* productivity. The presence of *Calluna* also influences *Sphagnum* production through shading. *Sphagnum* growth in both weight and length is reduced by increased shade, but the interactions of shade with water table and species make it difficult to relate the experimental response pattern to field distribution (Clymo, 1973).

There were few significant differences between years in the production of one site (Sike Hill B) (Forrest and Smith, 1975). Measurements were carried out over three years with similar climate characterised by unusually warm summers with July mean temperatures up to 1.7° C higher than the 10 year mean for the month (1971), August up to 1.6° C (1969) and September up to 0.9° C higher (1971) (Chap. 1). However in 1968/69 and 1969/70 cold winters probably caused severe damage to *Calluna*. This was evident in the spring harvest of 1970 when all the previous year's *Calluna* green shoots were browned. Despite the lack of initial photosynthetic capacity, *Calluna* green shoot production by the end of the 1970 growing season (319 ± 16 g m^{-2} yr^{-1}) was higher than in the other two years (240 ± 21 and 212 ± 13 g m^{-2} yr^{-1}). No production data are available from this site for a year characterised by a poor summer.

Thus, the variation in production between years is much less than the variation between sites resulting from species and habitat differences. The total community net production on the blanket bog shows a three-fold variation (300–900 g m^{-2} yr^{-1}) over the range of floristic composition and soil conditions. However, there are large between-site differences in the contributions made by each species to the total net production of the site.

2.5 Effects of Management on Production

The two herbivores of economic importance at Moor House are sheep and grouse, with densities, on the blanket bog, of about 0.02–0.2 and 1–2 individuals ha^{-1} respectively. At these densities they have little influence on total primary production on most of the bog, removing only a small fraction of the vegetation (Chap. 4). Where sheep grazing intensity is increased on blanket bog either experimentally (Rawes and Williams, 1973) or through long-term management (Welch and Rawes, 1966), the major changes are a decline in the standing crop of *Calluna* and an increase in *Eriophorum vaginatum*. The presence of *Juncus squarrosus* in heavily grazed bogs (0.75 individuals ha^{-1}) suggests that there may be a seral development towards a *Juncus*-dominant sward. Total primary production may not change greatly but the proportion of food suitable for sheep increases.

Burning has a marked effect both on the floristic composition and production of blanket bog vegetation. *E. vaginatum* recovers quickly after burning and its above-ground standing crop after five years is about 65% of the total higher plant community (Gore and Olson, 1967). *Calluna* regenerates more slowly, taking about 20 years to reach steady state when it contributes 70% to the above-ground standing crop (Forrest, 1971).

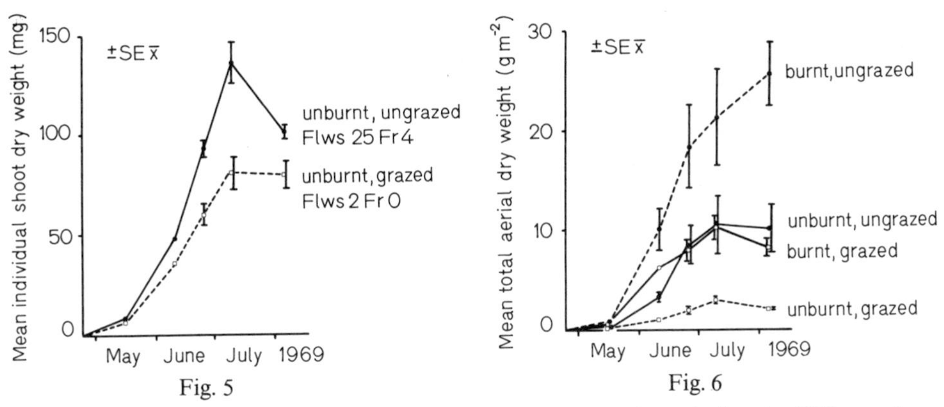

Fig. 5. *Rubus* individual shoot dry weight, flower and fruit numbers, during the 1969 growing season (after Taylor and Marks, 1971)

Fig. 6. *Rubus* total aerial dry matter during the 1969 growing season (after Taylor and Marks, 1971)

Grazing at an intensity of less than 0.1 sheep ha^{-1} can have a detectable effect on *Calluna* production during development after burning, shoot weight being reduced by 30–40% compared with growth in exclosures (Rawes and Williams, 1973).

Both sheep grazing and burning were found to affect significantly the production of *Rubus chamaemorus* (Taylor and Marks, 1971). The effect of grazing with a sheep density as low as one individual 40 ha^{-1} is apparently a result of a preference by the sheep for the leaves and shoots of this species. This is supported by the fact that the period of peak grazing intensity on the blanket bog corresponds with the time of maximum vegetative growth and flowering of *Rubus* (Taylor and Marks, 1971).

Taylor and Marks found that removal of grazing on unburnt plots resulted in greater above-ground standing crop with an increased shoot density, and larger shoots bearing many more flowers and fruits than the grazed plots (Fig. 5). Burning also increased the mean dry matter production of aerial parts (Fig. 6), giving greater shoot density and larger shoots. If burnt plots were grazed, this increase in above-ground standing crop was manifest as a large number of small shoots and virtually no flowers or fruits developed. Ungrazed plots which were burnt were characterised by maximum increases in all components, including those specified above and rhizome and root dry weight. The mean aerial production in this treatment was 25.8 g m^{-2} yr^{-1}, compared with only 3.0 g m^{-2} yr^{-1} in the control.

2.6 Effects of Nutrients on Production

There have been few investigations of the factors limiting the primary production of blanket bog under steady state conditions with negligible grazing pressure. However, some work has been carried out on possible nutrient limitation of the production of different species (Gore, 1961 a, b, 1963, 1972; Rawes, 1961, 1963,

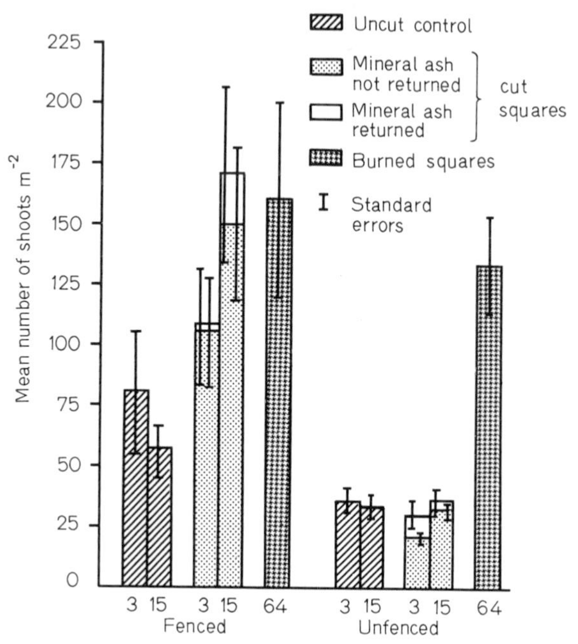

Fig. 7. The effects of canopy removal and ash addition on shoot density in *Rubus* (after Marks and Taylor, 1972)

1966; Gore and Urquhart, 1966; Brown and White, 1970; Marks and Taylor, 1972).

Marks and Taylor (1972) concluded that none of the nutrients investigated (nitrogen, phosphorus, potassium, calcium and magnesium) was limiting to the growth of *Rubus* under the burning and grazing regimes investigated. This conclusion was based on chemical analyses of the vegetation under different treatments which showed no major differences in nutrient concentrations, and on experimental additions of ash and removal of the vegetation canopy. The addition of ash did not cause any significant increase in shoot density of *Rubus*, but there was a pronounced response to removal of the *Calluna* canopy by cutting, the mean shoot density being almost doubled in this treatment by the end of the second growing season (Fig. 7). Hence the considerable effect of grazing and burning treatments on the production of *Rubus* probably results from reduction of above-ground biomass in the case of grazing, and from alteration of microclimate consequent upon canopy removal by burning. The main microclimatic change after canopy removal was probably in light regime, light intensity being 250% more in the burnt areas than the unburnt (T. C. Marks, pers. comm.). Temperature differences were apparently small, but may be important; the daily maximum temperatures 8 cm above the ground surface were higher on the burnt than the unburnt plot, presumably due to removal of the *Calluna* canopy, whereas the ground surface showed reduced temperature range and mean in the burnt plot, believed to reflect the insulating properties of the thick *E. vaginatum* mat on the burnt area. The physiological responses of *Rubus* to these microclimatic factors are discussed in Chapter 3.

It was also shown by fertiliser application to clipped blanket bog that calcium and phosphorus did not increase tiller dry weight of *E. vaginatum* (Gore, 1961a), but it was later stated that if all major nutrients were supplied, annual growth of native species increased substantially (Gore, 1972). In the case of *Molinia caerulea*, a species not indigenous to Moor House, none of the nutrients tested (Ca, P, and N) increased mean dry weight of transplants in the first year of growth, but there was a significant response to nitrogen in the second year (Gore, 1961b, 1963). Trials of *Pinus contorta* on blanket bog at Moor House showed that phosphorus and potassium, but not nitrogen, were limiting to its height growth (Brown and White, 1970).

On an *Agrosto-Festucetum* site fertilisation with N, P, K plus ground limestone doubled production in the first year (Rawes, 1963), but the residual effect of this treatment did not last beyond four years (Rawes, 1966). Rawes also found little response to fertiliser application on *Nardetum* (Rawes, 1961).

It therefore appears that, at least in the case of *Rubus* and possibly *E. vaginatum*, nutrients are not the main factors limiting primary production. Since available (ammonium acetate extractable) nutrients in the Moor House peat are present in low concentration (Gore and Allen, 1956; Chap. 1), it follows that the indigenous species are adapted to this soil infertility. The exceptionally deep rooting characteristics of species such as *E. vaginatum* (>30 cm; Forrest, 1971) may also enable tapping of nutrients from depths inaccessible in most soils, and allow stratified exploitation of nutrient resources (Boggie et al., 1958).

The evidence for or against nutrient limitation is inconclusive but the apparent lack of response to increased nutrients of at least some native species in the Moor House peat differs from conclusions reached for other similar sites. Phosphorus is the factor which has most frequently been found to be limiting, this being demonstrated for *M. caerulea*, *Myrica gale*, and also *P. contorta* on Irish peat (McEvoy, 1954), for *E. vaginatum* on a Swedish bog (Tamm, 1954), and for several bog species on Scottish blanket peats (McVean, 1959). Goodman and Perkins (1968a, b) also showed potassium, and to a lesser extent phosphorus, to be limiting to growth of *E. vaginatum* on two mire sites in South Wales. The adequacy of nutrient supply at Moor House in comparison to other similar sites may reflect higher nutrient availability on the very acid Moor House peat, and/or somewhat less deficient total levels than at other sites; Gore (1961b) showed that phosphorus contents at Moor House were higher than on a comparable lowland bog.

This tentative conclusion of lack of limitation by nutrients of the production of indigenous blanket bog species at Moor House suggests that the controlling factors, in addition to wetness, may be largely climatic. There may also be interactions between these various factors.

2.7 Effects of Climate on Production

The hypothesis of the importance of climatic factors in limiting production of vegetation at Moor House is supported by the physiological work on *Calluna* and *Rubus* (Chap. 3). The work on *Calluna*, however, did not investigate the possibility

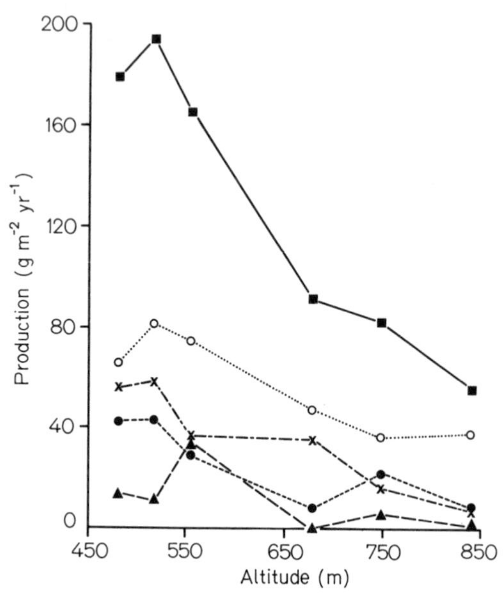

Fig. 8. Above-ground production under grazing of *Agrosto-Festucetum* and *Festucetum* at different altitudes (after Rawes and Welch, 1969). ○·····○ Fine-leaved grass. ●– – –● Broad-leaved grass. ▲– –▲ Flowering plants. × —-— × Dead. ■——■ Total

of an over-riding nutrient limitation in the field situation. Climatic factors would be expected to be particularly important in the case of species such as these, which are nearing the extremes of their distribution. *Rubus* is at the southern end of its range, whereas *Calluna* is a species near the altitudinal limit of its rather oceanic distribution. The sensitivity of *Calluna* production to climatic factors is additionally supported by the finding of a marked decline in *Calluna* standing crop from about 500 g m^{-2} at 560 m to only 16 g m^{-2} at 688 m (Rawes and Welch, 1969). Measurements of *Calluna* radial increment in successive years at two Moor House sites (Grace and Woolhouse, 1970) revealed parallel year-to-year fluctuations related to trends in late summer temperature (Grace, 1970; Grace and Woolhouse, 1974).

There is evidence for the effect of the severe climate of Moor House on other species, such as *M. caerulea* (Gore, 1961b, 1963). Data from Rawes and Welch (1969) show that above-ground production declines with altitude (Fig. 8). Regressions are significant for fine-leaved grasses ($r = 0.897^{**}$) broad-leaved grass ($r = 0.866^{*}$), dead material ($r = 0.954^{***}$) and total production ($r = 0.970^{***}$), but not for flowering plants ($r = 0.591$). The regressions indicate that, relative to production at 500 m, production declines by about 22% per 100 m. Over this altitudinal range, for each 100 m the mean annual soil temperature declines by about 0.5° C and the sum of day degrees above 0° C by about 200° C.

Monthly standing crop increments of three *Agrosto-Festucetum* sites were correlated with hours of bright sun and mean maximum temperature (Welch and Rawes, 1965) and on one of these sites a severe August frost greatly reduced production (Rawes and Welch, 1964). However, on the Moor House meadow there was no clear relationship between yield and climate in different years, but high incidence of frosts, especially early in the season, tended to reduce production (Rawes and Welch, 1969). Thus frost may affect growth of the species present in the meadow more than high temperatures in the first part of the season. A correlation between height growth and temperature was also demonstrated in *Betula pubescens* by Millar (1965).

It is likely that the effects of climatic factors on primary production are complex and differ for each species (Chap. 3). Climatic factors affect both the growing season length for a particular species and also its relative growth rate during the growing season (Pearsall, 1950). Further elucidation of the specific effect of each factor on particular species cannot be obtained without experimental studies, such as those detailed in Chapter 3.

2.8 Production of Other Vegetation Types at Moor House

The total above-ground net production of the blanket bog (361 g m^{-2} yr^{-1}) is similar to that of *Juncetum squarrosi* but about twice that of *Nardetum* and *Agrosto-Festucetum* at the same altitude (Table 4). Production by vascular plants is similar on all vegetation types because about one third of the total production on blanket bog and *Juncetum* is contributed by non-vascular plants which are negligible on the *Nardetum* and *Agrosto-Festucetum.*

No data are available for production below ground on sites other than the blanket bog, but the standing crop below ground on *J. squarrosi* and *Agrosto-Festucetum* is larger than that on the bog, above:below-ground ratios being 0.13, 0.11, and 2.05 respectively (Table 4). The high ratio for blanket bog is partly caused by the estimated *Sphagnum* standing crop being included in the above-ground fraction. The ratio for all vascular plants which is more closely comparable with other sites is about 1.20. On sheep-grazed montane *Agrosto-Festucetum* in Snowdonia, the above:below-ground standing crop ratio is about 0.19. This includes leaf bases in the above-ground fraction as in the estimates for Moor House. The total production is about 1482 g m^{-2} yr^{-1} with about 23% being

Table 4. Total above-ground net production and late summer standing crop on different vegetation types at Moor House

Vegetation type	Production	Standing crop	
		Above-ground	Below-ground
Blanket bog ($n=7$)	361 ± 50	$1,129 \pm 150$	550 ± 72
Nardetum	190[a]	—	—
Juncetum squarrosi	340[a]	207 ± 17[b]	$1,643 \pm 215$[b]
Agrosto-Festucetum	190[a]	201 ± 27[b]	$1,785 \pm 203$[b]

Results (g m^{-2} yr^{-1} $\pm$ se where available) are for areas grazed by sheep but the amount removed is included in the production estimate.

[a] Data from Rawes and Welch (1969) are for a number of sites.

[b] Estimates from five cores from a single site.

produced below ground (Chap. 16). The broad similarity between the Snowdonia and Moor House grasslands in composition, standing crop and aboveground production, using comparable methods, indicates that total production on the grasslands at Moor House may be above the range observed on the blanket bog (305–868 g m^{-2} yr^{-1}).

There is no obvious relationship between total above-ground net production and the number of plant species in the community. Using the number of species m^{-2} as an index of diversity (Chap. 10, Fig. 3), blanket bog is low in numbers of higher plants compared with *Agrosto-Festucetum* and, to a lesser extent, *J. squarrosi* and *Nardetum*. The difference in numbers is partly compensated by the relatively high numbers of lichen and liverwort species on the bog compared with the grasslands. The differences between vegetation types are shown in the species and chemical composition of the vegetation related to the soil nutrient status (Chaps. 1 and 10).

2.9 Turnover and Growth Rates

In order to examine the "strategy" (i.e. partitioning of dry matter between different parts of the plant) evolved by different species in the severe environment at Moor House, ratios of different components were compared. Because of limitations in the data, calculation of ratios of production: biomass (live material) above and below ground is possible for only two species at Forrest's intensive site (Sike Hill A), *E. vaginatum* having much higher values than *Calluna* (Table 5). For *E. vaginatum* the ratio of total components is higher even than that of above-ground parts, although most of the structural component in this species is below-ground (above-ground: below-ground biomass 0.81). The very high total production:

Table 5. Production : biomass ratios for *Calluna vulgaris* and *Eriophorum vaginatum* at Sike Hill A

Species	Above-ground			Total		
	Production (g m^{-2} yr^{-1})	Aug biomass (g m^{-2})	Production: biomass (g g^{-1} yr^{-1})	Production (g m^{-2} yr^{-1})	Aug biomass (g m^{-2})	Production: biomass (g g^{-1} yr^{-1})
C. vulgaris	130	740	0.18	351	1,399	0.25
E. vaginatum leaves + flowers[a]	59	82	0.72	221	183	1.21
E. vaginatum leaves + leaf bases[a]	175	82	2.13	221	183	1.21
E. vaginatum leaves + leaf bases[b]	52	82	0.63	130	183	0.71

For *E. vaginatum* the original production estimate [a] of Forrest (1971) and the revised estimate [b] by Jones and Gore (in press) are given (Sect. 2.9).

biomass ratio for *E. vaginatum*, which is greater than unity (i.e. production > biomass) partly results from an overestimate by Forrest (1971) of basal leaf production.

The estimate of basal leaf production and Forrest's other estimates of component transfers in this species have been revised by H. E. Jones (Jones and Gore, in press) using a computer simulation of the observed seasonal changes in the standing crops of the main components of *E. vaginatum*. The simulation incorporated concurrent transfers between the plant compartments and a set of simultaneous differential equations were solved over a year for weekly time steps.

From the flow diagram (Fig. 9b), the equations to describe input and output for each component were as follows:

$$\frac{dL}{dt} = A0 - (A1 + A3)\,L + (A4)\,S$$

$$\frac{dS}{dt} = (A3)\,L - (A4 + A6 + A7 + R1)\,S$$

$$\frac{dRt}{dt} = (A6)\,S - (A8 + R2)\,Rt$$

$$\frac{dD}{dt} = (A1)\,L + (A7)\,S + (A8)\,Rt - (A5)\,D$$

where L = leaves and leaf bases

S = rhizomes and shoot bases

Rt = roots

D = dead parts of all components

$A0$–$A8$ = the transfer coefficients between components

and $R1$–$R2$ = respiration losses from the live storage organs

The model format forced all the photosynthate via the leaves to redistribution in other plant parts as in Forrest's (1971) flow diagram for the same data on this species. Seasonal variation in the photosynthetic parts was provided by a six-month cosine wave on $A0$, six months being a reasonable estimate of the length of the growing season. Variation in other components was simulated by seasonal switches.

Forrest (1971) derived figures in his flow diagram of 221 g m^{-2} yr^{-1} for total net throughput and 175 g m^{-2} yr^{-1} for the transfer from live to dead leaves. By trial and error Jones and Gore (in press) assigned values to the transfers between components which provided over the year the nearest possible approximation to the observed standing crops. These values included a total net throughput of 130 g m^{-2} yr^{-1} and a loss from live to dead leaves of 52 g m^{-2} yr^{-1}. It was possible that the values for transfer coefficients estimated in this way represented only one of a number of possible sets of values which would also provide a good fit to the field data, so the procedure was repeated using the estimates of Forrest. The simulated results showed a large overproduction of live leaves because the net throughput was an overestimate, largely because of the overestimate of dead leaf production.

The dynamic simulation approach allowed a check to be made with the field data on the consistency of the output derived from estimated transfer coefficients. It was concluded that the estimates of 130 and 52 g m^{-2} yr^{-1} for total net throughput and dead leaf production respectively were more realistic than the alternative estimates of 221 g m^{-2} yr^{-1} and 175 g m^{-2} yr^{-1}.

The incorporation of the revised total throughput figure reduces the total *E. vaginatum* production: biomass figure considerably, to a value similar to the above-ground ratio which was little altered by the revision (Table 5).

The total production: biomass ratio for *E. vaginatum* demonstrates that this species, and probably the other cyperaceous species such as *E. angustifolium* and *T. cespitosum*, is characterised by a high production and low structural component (i.e. by high turnover). In contrast, *Calluna*, and doubtless other dwarf shrubs such as *Erica tetralix* and *Empetrum nigrum*, has low production and a high structural component (i.e. low turnover). This is supported by tentative estimates for the other species using limited data.

The strategy of the two species is summarised in Figure 9 which shows the standing crop and annual transfers derived from the computer simulation discussed above (Jones and Gore, in press). This shows that in *Calluna* about 76% of the annual production is converted into non-photosynthetic structural and storage tissue and 24% into production of new photosynthetic tissue. This compares with 60 and 40% respectively in *E. vaginatum*. Net production per unit standing crop is much lower in *Calluna* (0.25 g g^{-1} yr^{-1}) than *E. vaginatum* (0.7–1.2 g g^{-1} yr^{-1}). However, the proportion of photosynthetic tissue in the peak standing crop of *Calluna* (20%) is lower than in *E. vaginatum* (45%) and therefore the production per unit of photosynthetic tissue is only slightly lower in *Calluna* (1.17 g g^{-1} yr^{-1}) than *E. vaginatum* (1.59–2.70 g g^{-1} yr^{-1}). Thus the efficiency of the photosynthetic tissues in the two species may be similar and the main difference lie in the proportion of photosynthate converted into long-lived structural support tissue.

The production: biomass ratios of total vascular species above ground for different vegetation types considered as a whole (Table 6) naturally reflect the ratios of the dominant species. They are thus lowest for the typical *Calluneto-Eriophoretum* which has a high above-ground standing crop, and highest for *Eriophoretum* and *Agrosto-Festucetum* with low biomass above ground. Grazing on the *Agrosto-Festucetum* causes a great increase in the production: biomass ratio above ground, a result of decreased biomass under grazing with little reduction of production. Thus grazing increases the turnover of the system but not the efficiency of conversion of incoming solar radiation.

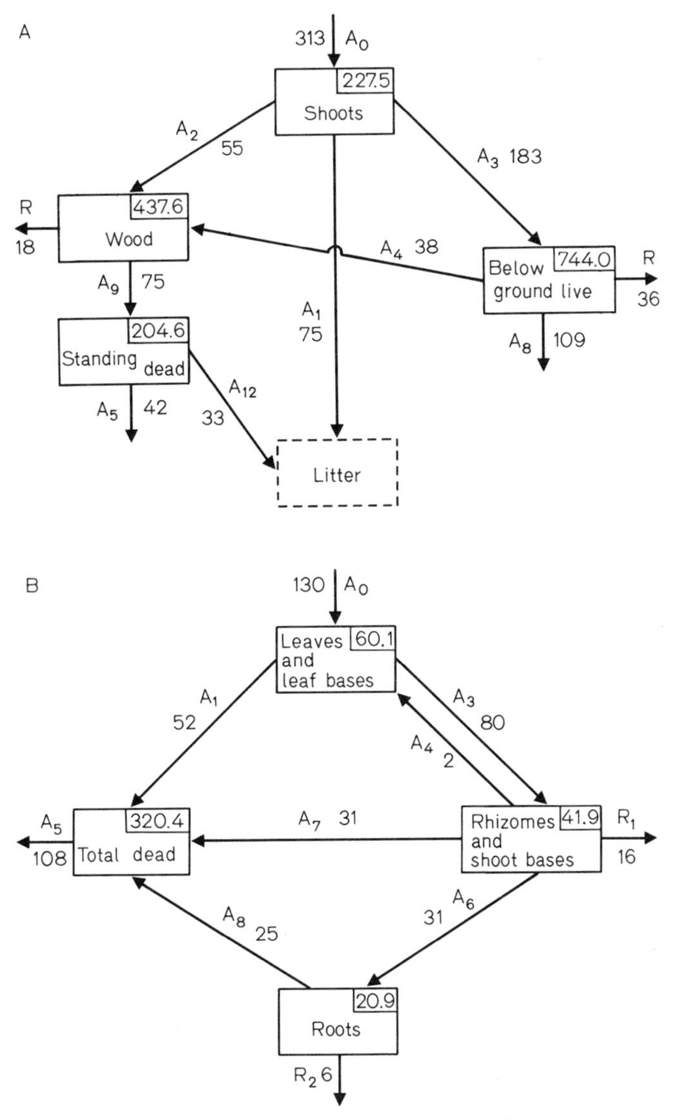

Fig. 9. Standing crops (g m^{-2}) and annual transfers (g m^{-2} yr^{-1}) in (A) *Calluna vulgaris* and (B) *Eriophorum vaginatum* at Sike Hill A (after Jones and Gore, in press)

A useful way to express production estimates for comparative purposes is in terms of production per growing season day (Table 7). The definition of the growing season length for a particular year is difficult, this parameter being variously and somewhat arbitrarily defined as the number of days with the mean temperature above 0° C, or above 5.6° C, or with the minimum temperature above 0° C. Although the normally accepted growth threshold is 5.6° C, physiological studies have shown that plants can adapt to cold conditions, and photosynthesis and respiration may occur below 0° C. However, net photosynthesis

Table 6. Mean production : biomass ratios of total vascular species above ground for different vegetation types (±se where available)

Vegetation	Grazing intensity (sheep ha^{-1})	Above-ground production ($g\,m^{-2}\,yr^{-1}$)	Above-ground Aug biomass ($g\,m^{-2}$)	Production: biomass ($g\,g^{-1}\,yr^{-1}$)	Source of data
Calluneto-Eriophoretum	0.02	249± 20	776± 53	0.33±0.04	Forrest (1971)
Eriophoretum	0.4	302±106	432±244	0.82±0.22	Forrest and Smith (1975)
Nardetum	1.7	190	250	0.76	
	0.0	264	305	0.87	
Juncetum squarrosi	1.4	343	275	1.25	Rawes and Welch (1969)
	0.0	367	325	1.13	
Agrosto-Festucetum	5.7	174	46	3.78	
	0.0	202	176	1.15	

Values for grazed vegetation include the amount removed by sheep. Data from Rawes and Welch (1969) are for specific sites at about 550 m.

Table 7. Mean (± se) growth and efficiency figures for blanket bog sites

	Vascular species above-ground		All species above- and below-ground	
	Calluneto-Eriophoretum	All blanket bog sites	*Calluneto-Eriophoretum*	All blanket bog sites
Production ($g\,m^{-2}\,yr^{-1}$)	249 ±20	243 ±36	628 ±54	659 ±53
Production/Aug biomass ($g\,g^{-1}\,yr^{-1}$)	0.33± 0.04	0.45± 0.11		
Production per growing season day ($g\,m^{-2}\,d^{-1}$)	1.44± 0.14	1.40± 0.20	3.64± 0.37	3.79± 0.31
Production/Aug biomass per growing season day ($g\,g^{-1}\,d^{-1}$)	1.90×10^{-3} $\pm0.25\times10^{-3}$	2.60×10^{-3} $\pm0.61\times10^{-3}$		
Efficiency (phot. active radiation) (% yr^{-1})			0.90± 0.08	0.93± 0.07
Efficiency (phot. active radiation) (% growing $season^{-1}$)			1.27± 0.10	1.31± 0.10

only occurs in *Calluna* above 7.2° C (Chap. 3). In Table 7 the threshold of mean temperature above 5.6° C is used to enable comparison with other work. On this basis, the average growing season length at Moor House over the years 1968–1971 was 175 days. Mean total primary production on the blanket bog was 3.79±0.31 g m^{-2} per growing season day, and mean efficiency of conversion of photosynthetically active radiation over the growing season was 1.31±0.10%.

2.10 Conclusion

Production values from a range of *Calluneto-Eriophoretum* blanket bog sites at Moor House in the north Pennines were compared. The vegetation of these sites was assumed to be in a steady state, rejuvenation of *Calluna* occurring by *Sphagnum* overgrowth.

The average net production of *S. rubellum* at three sites was 130 g m^{-2} yr^{-1}. The total net production of the blanket bog showed a threefold variation, from about 300–900 g m^{-2} yr^{-1}, with a mean of 659 ± 53 g m^{-2} yr^{-1}. The highest values were on recently burnt sites in a seral condition, and the remaining variation was largely related to water table and species composition, the wettest *Sphagnum*-dominated sites having a lower production than the drier sites. Nutrient and microclimatic variation is probably unimportant in affecting the total production of the blanket bog vegetation, but the physiological studies and other work indicated that the production of individual species on a particular site is probably largely limited by climatic factors and by interaction between species.

The "strategy" of the two main bog species, *Calluna vulgaris* and *E. vaginatum*, was compared. They showed similar efficiency of production per unit of photosynthetic tissue but differed in the proportion of photosynthate converted into long-lived structural support tissue (76 and 60% respectively).

The growing season at Moor House is about 175 days, and within that period the mean total primary production on blanket bog was 3.79 ± 0.31 g m^{-2} d^{-1}. About 1.3 ± 0.1% of the photosynthetically active solar radiation in the growing season was converted to plant production.

Limited data from other vegetation types indicated that, although there were major differences in species composition, the total standing crop and annual production of *J. squarrosi* and *Agrosto-Festucetum* were within, or possibly above, the range observed on the blanket bog.

Acknowledgements. We wish to thank Dr. R.S.Clymo, Mr. T.C.Marks and Mr. M.Rawes for making their unpublished data so freely available, and thus making possible the compilation of this paper. We are also grateful for all the helpful advice and criticism we received from them and from Dr. O.W.Heal, Dr. H.E.Jones, Mr. A.J.P.Gore and Dr. J.Grace.

3. Physiological Aspects of Bog Production at Moor House

J. GRACE and T. C. MARKS

Analysis of the rates of photosynthesis and respiration of plant species in controlled environments provides a quantitative assessment of factors controlling production. Predictions of production, based on these results and on the climate in the vegetation, are compared with the observed seasonal changes in standing crop described in the previous chapter.

3.1 Introduction

Physiological studies can contribute much to our understanding of primary production. By isolating plants in the laboratory, away from the rapidly fluctuating and complex natural environment, we may study some of the processes involved in growth and development of constituent species. In controlled environment situations we can exploit the speed and ease with which photosynthesis can nowadays be measured, to determine response curves in relation to selected environmental variables. In doing this we are trying to assess the sensitivity of plant processes, in this case photosynthesis, to changes in environmental variables. If we know how each plant process contributes to overall production, then we are placed in a stronger position to make predictions regarding the likely effect on production of natural or man-made changes of variables in the field situation.

Our ideas on how such plant processes interact and contribute to the whole may be put to the test if we are prepared to embark on mathematical modelling, which enables us to explore the numerical consequences of making various assumptions. A wrong assumption, for example, frequently leads to negative plant weight, impossibly enormous plant weight, or results which are otherwise not in accord with our general experience, or with field observations.

There are a number of pitfalls in this approach. First, we must exercise skill in choosing the variables for laboratory experimentation, recognising the danger of overlooking key variables whose effect might be so large as to swamp the effect of the variables actually chosen. In this connection it is certainly useful if we can carry out supporting physiological studies in the natural situation. Second, we must exercise skill in choosing the plant processes to be studied, our choice being dependent on current ideas of how plants function. Third, we must recognise the penalty associated with the convenience of controlled environment situations; growth room and assimilation chamber environments are artificial and different in so many ways from natural conditions. We can never be sure, therefore, of the extent to which plant performance in the laboratory reflects that in the field.

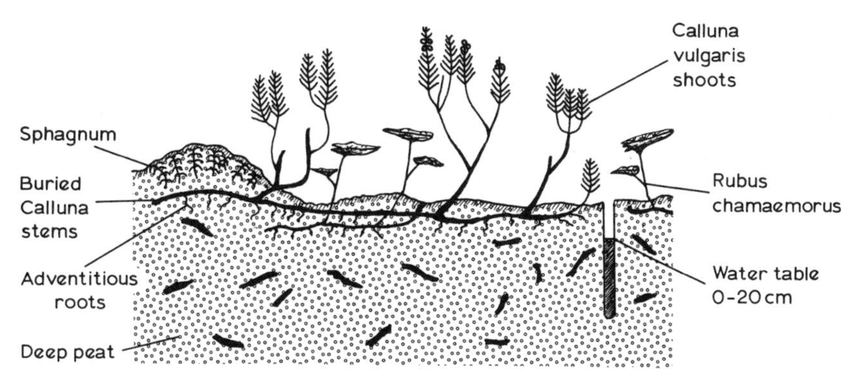

Fig. 1. Spatial arrangement of *Calluna*, *Rubus* and *Sphagnum* in the blanket bog

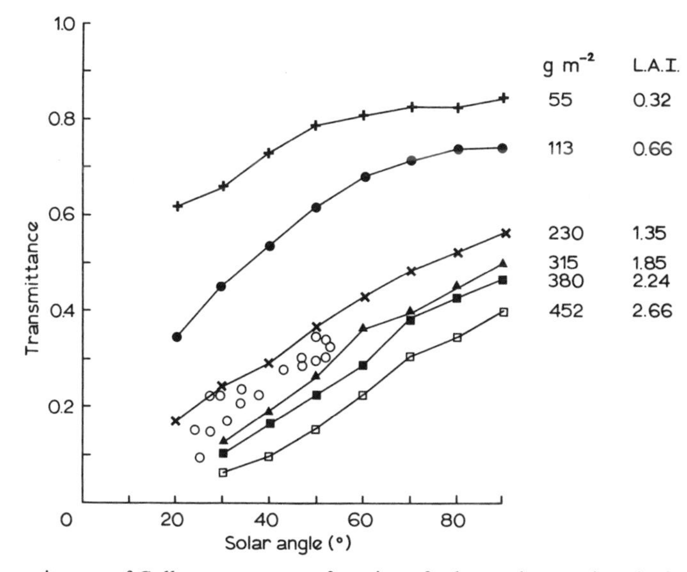

Fig. 2. Transmittance of *Calluna* canopy as a function of solar angle at various leaf area indices. Field data (○), other points from laboratory reconstruction of the canopy (after Grace and Woolhouse, 1973a)

The present paper considers the physiological work relevant to the IBP production studies at Moor House, and attempts to assess its contribution to our understanding of the control of primary production in upland Britain, this southerly piece of the Tundra Biome. Available information is concerned largely with *Calluna vulgaris* and *Rubus chamaemorus* with limited observations on *Sphagnum rubellum*.

3.2 Light Interception and Microclimate

Figure 1 shows the spatial arrangement of the plant species for which information is available. Leaf laminae of *Rubus chamaemorus* are held in a horizontal plane and occur both within and between *Calluna* plants. Leaves of the *Calluna*

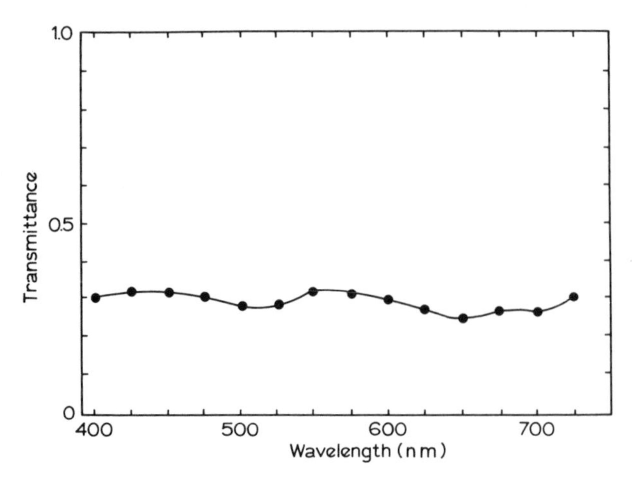

Fig. 3. Transmittance of *Calluna* canopy as a function of wavelength (after Grace, 1973)

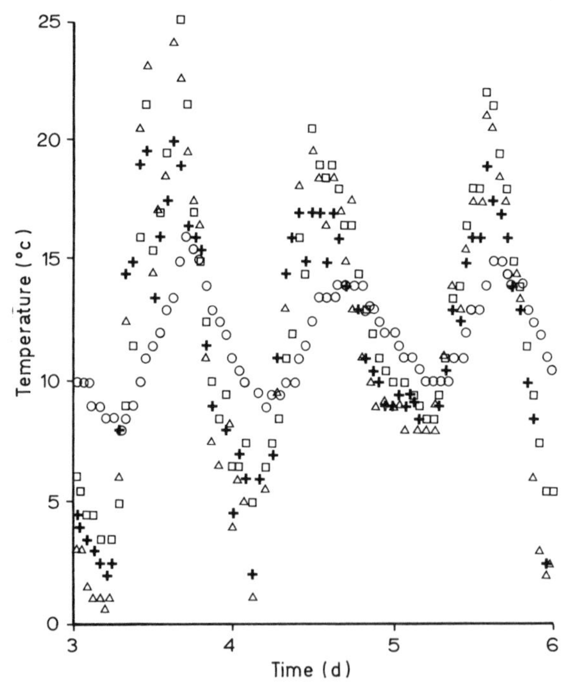

Fig. 4. Temperature fluctuations for three summer days: above *Calluna* canopy (+), within canopy (△), 0.5 cm below bog surface (□) and 5.0 cm below bog surface (○)

canopy are by no means randomly dispersed, there being concentrations of leaves on individual shoots and clusters of such shoots associated with individual stems. In studying light interception by *Calluna* conventional theory was, therefore, discarded in favour of an empirical approach (Grace, 1973; Grace and Woolhouse, 1973a).

The result, Figure 2, shows how the insolation available for *Sphagnum* and *Rubus* is a function of the thickness of the *Calluna* leaf canopy, as well as of the solar angle. The leaf area index at many of the Moor House sites lies between 1 and 2, small in relation to that of most crop canopies. The spectral energy distribution beneath *Calluna* is little different from that above, the canopy acting as an almost neutral filter (Grace, 1973; Fig. 3).

Figure 4 indicates some of the microclimatological differences between the environments within and beneath the *Calluna* at Sike Hill (Grace, 1970; see also Delany, 1953; Gimingham, 1972; Marks, 1974). From this figure it can be seen that the temperatures of the bog surface and the air within the leaf canopy may rise several degrees above the ambient air temperature.

In nature there are rapid and complex fluctuations of temperature and light. In the laboratory situation the aim is to produce a simplified environment, varying only in steps at the will of the experimenter.

3.3 Net Photosynthesis

The photosynthetic rates of plants collected from Moor House were measured in the laboratory using an infra-red gas analyser to sense the difference in the CO_2 concentration of the airstream before and after it had passed through the chamber containing a single plant or leaf (Grace and Woolhouse, 1970; Marks, 1974). The light source was a Xenon arc, chosen because of the similarity of its spectral energy distribution to that of the sun. *Calluna* and *Sphagnum* plants were taken from the Moor House Sike Hill site at intervals during the year and stored for several days in a controlled environment room before measurements were made (Grace and Woolhouse, 1970).

Turves containing *Rubus chamaemorus* were collected from the House Hill site and placed in controlled environment cabinets under temperature and light intensity conditions approximating to those at Moor House in early June.

Photosynthetic rates of *Rubus* determined by Marks (1974) are shown in Figure 5. The points are calculated from a regression equation, used as a convenient way of summarising a mass of laboratory data collected under a variety of conditions. The maximum rates are equal to those one might expect from crop plants growing in near-optimal conditions (Gaastra, 1959; Hesketh and Moss, 1963). Similar rates have been obtained for other native species under arctic tundra conditions at Barrow, Alaska, by Tieszen (1972).

Light saturation for *Rubus* occurs at low intensities and the temperature optimum is 12–18° C. Temperatures in excess of 25° C may cause a long-term decline in photosynthetic rate which persists for several hours after the return to a lower temperature.

Photosynthetic rates of *Calluna* determined by Grace and Woolhouse (1970), originally calculated on a leaf weight basis, have been recalculated for comparative purposes on a leaf area basis (Fig. 6). These rates are low compared with those for *Rubus* but similar to those obtained for arctic-alpine species by Hadley and Bliss (1964) and Scott and Billings (1964). Rather higher rates have been obtained for *Calluna* using glasshouse raised plants (McKerron, 1971) and plants obtained

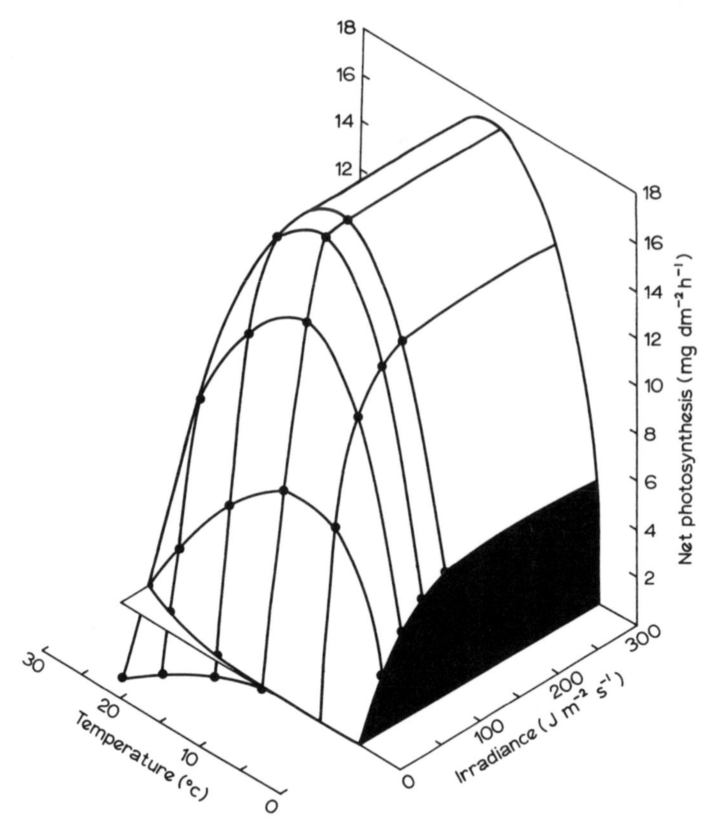

Fig. 5. Photosynthesis of *Rubus chamaemorus*, aged 11 weeks, as a function of temperature and irradiance (380–720 nm)

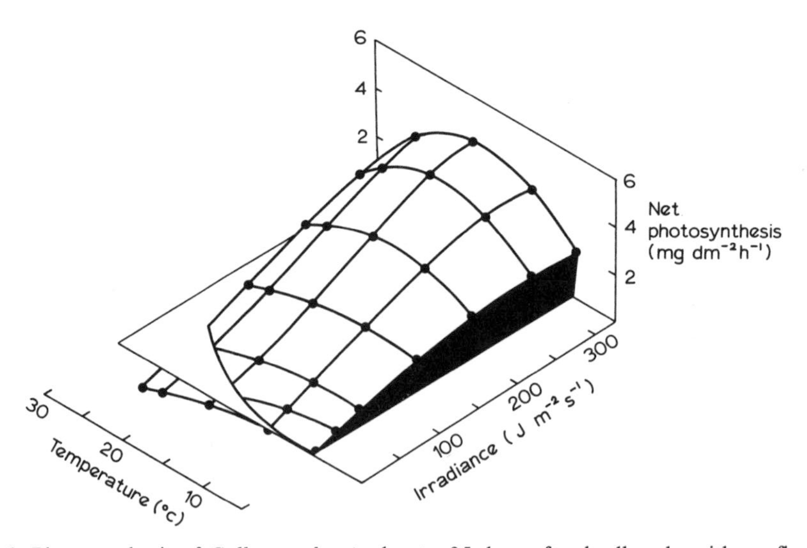

Fig. 6. Photosynthesis of *Calluna vulgaris* shoots, 35 days after budbreak, with no flowers, and with a temperature treatment of 10° C

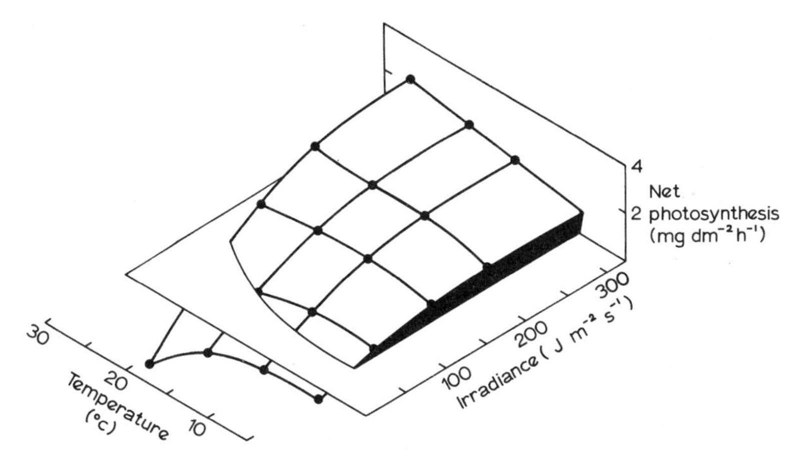

Fig. 7. Photosynthesis of *Sphagmum rubellum* as a function of temperature and irradiance (380–720 nm)

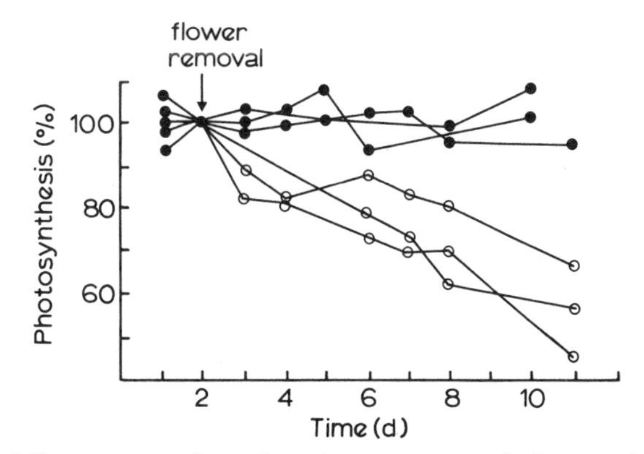

Fig. 8. Effect of flower removal on the subsequent rate of photosynthesis in *Calluna* (○) and controls (●) (after Grace and Woolhouse, 1970)

from Scottish mountains (Summers, 1972; Chap. 12). The temperature optimum for *Calluna* at Moor House is 18° C and the effect of temperature is less pronounced than for *Rubus*. *Calluna* is, however, sensitive to previous temperature treatment (Grace and Woolhouse, 1970), capacity to photosynthesise depending on the temperatures to which the plants have been exposed in the preceding days. Light saturation does not occur until the highest intensities of light have been reached, and light compensation points are high.

Photosynthetic rates of *Sphagnum rubellum* (Fig. 7), determined by Grace (1970), are only one ninth of those of *Rubus*, but comparison between species should be made with care, taking into account the widely differing amount of photosynthetic tissue present per unit area of leaf.

Analyses of the crop contents of Red grouse have shown that *Calluna* flowers are often eaten by the birds (Leslie and Shipley, 1912). When flowers were removed from shoots in a laboratory situation (Fig. 8) the rate of photosynthesis fell

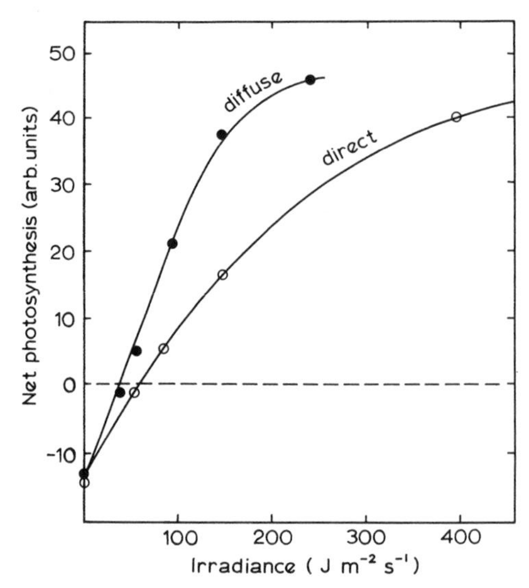

Fig. 9. Effect of diffuse (●) and direct (○) light fields on net photosynthesis of *Calluna* (after Grace and Woolhouse, 1973a)

abruptly (Grace and Woolhouse, 1970). Therefore, grouse may significantly reduce photosynthesis of *Calluna* under field conditions.

Photosynthesis of *Calluna* is also sensitive to light direction (Grace and Woolhouse, 1973a), the photosynthetic response being much modified when an experiment is repeated with a diffusing shield fitted in the top of the chamber (Fig. 9). Light saturation then occurs at lower intensities because of the improved efficiency of light interception by *Calluna* foliage when receiving light from all directions.

The differing responses of *Calluna* and *Rubus* to various external and internal variables (Table 1), may lead to a rather different photosynthetic performance over the course of each day. The daily course of photosynthesis for the two species has been plotted (Fig. 10) on the basis of their established responses in the labora-

Table 1. The effect of external and internal variables on net photosynthetic rates in *Calluna vulgaris*, *Rubus chamaemorus* and *Sphagnum rubellum*

	Calluna	*Rubus*	*Sphagnum*
Irradiance	**	**	**
Light direction	**	?	?
Temperature	*	**	*
Previous temperature	**	*	0
Reproductive status	*	?	?
Leaf or shoot age	*	*	?
Time of day	*	0	0
Water content	?	?	**

** and *: extent of effect; 0: no effect; ?: effect not investigated

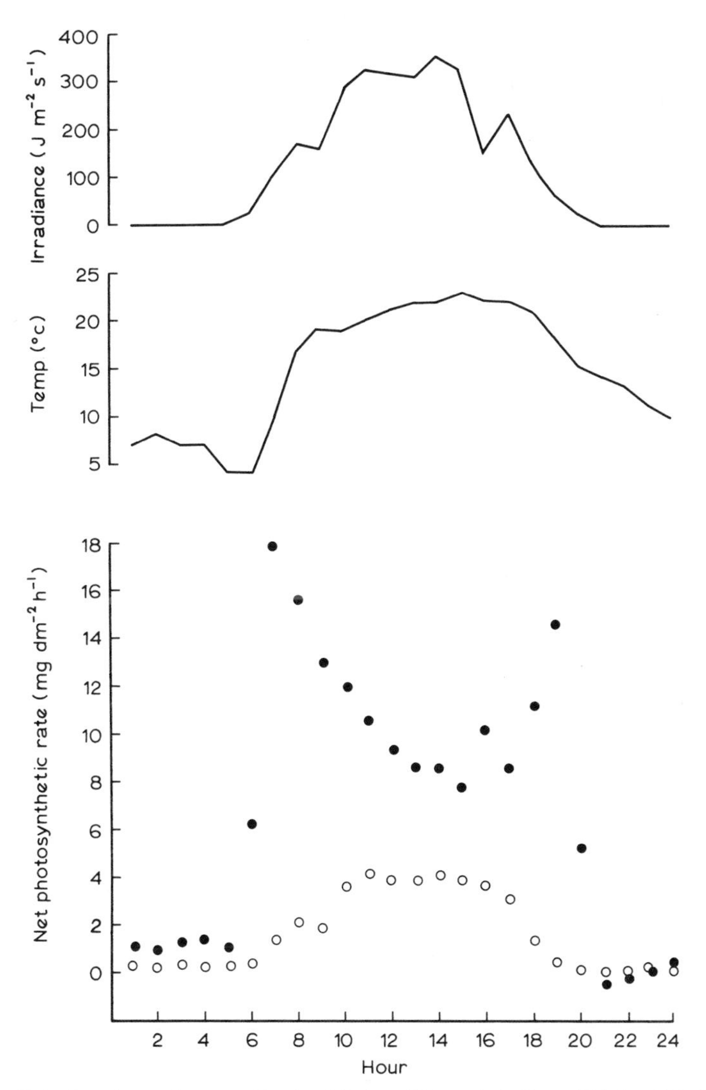

Fig. 10. Calculated net photosynthesis of *Rubus* (●) and *Calluna* (○) for 25 July 1969

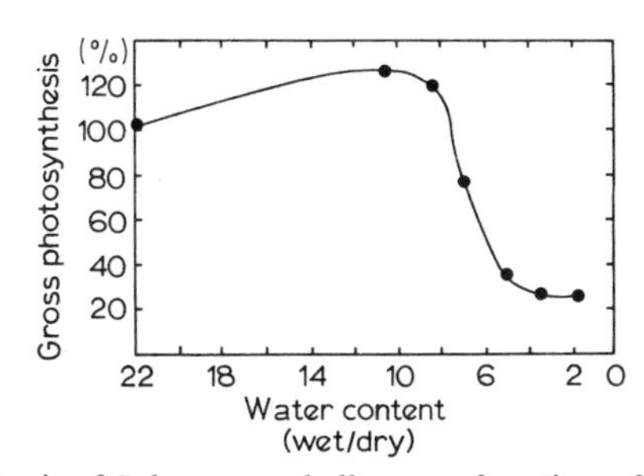

Fig. 11. Gross photosynthesis of *Sphagnum rubellum* as a function of capitulum water content. Irradiance 400 J m^{-2} s^{-1}, temperature 22° C

tory, using climatological data from a day when both air temperatures and insolation were very high. *Rubus* demonstrates a clear midday depression resulting from supra-optimal temperatures at high irradiance.

Photosynthesis of *S. rubellum* is sensitive to water content of the capitulum (Grace, 1970; see also Rudolph, 1968). The maximum rate of photosynthesis does not occur at the maximum water content, presumably because of a reduced CO_2 diffusion rate through the water film (Fig. 11). This effect may go some way to account for the differing production of *S. rubellum* on sites of varying wetness (Clymo and Reddaway, 1972; Chap. 2, Fig. 2).

3.4 Distribution of Assimilate

Plant growth cannot be estimated from rate of photosynthesis alone since growth rate depends so much on the relative rates of utilisation by the various parts of the plant (Blackman, 1919; Boysen Jensen, 1949). It is a feature of trees, and woody plants like *Calluna*, that a proportion of the assimilate which might otherwise be used to build extra leaf area must be utilised as wood, and this feature has prompted the hypothesis that woody plants may have inherently low growth rates (Jarvis and Jarvis, 1964). If we are to understand the contribution of photosynthesis to growth rate we must therefore also know how the plant utilises photosynthate; how much is destined for wood and how much for the development of new leaf material. In *Calluna* for example a third of the season's photosynthate appears to be used to create woody stems and branches (Grace and Woolhouse, 1973b), the pattern of use varying during the year (Fig. 12).

Other workers (Mooney and Billings, 1960; Hadley and Bliss, 1964; Warren-Wilson, 1966; Tieszen, 1972) have emphasised the role of stored carbohydrate reserves in the life strategies of arctic and alpine plants, serving to ensure the rapid development of leaf area with the onset of the short growing season. Available data for *Calluna* (Grace and Woolhouse, 1970, 1973b) and *Rubus* (Marks, 1974)

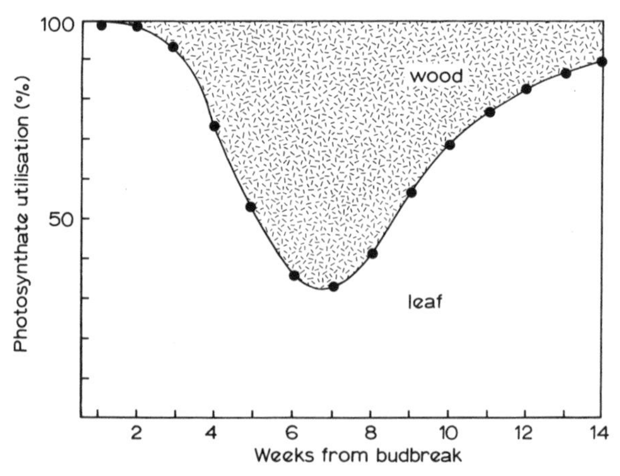

Fig. 12. Photosynthate utilisation by wood and leaf in *Calluna*, estimated from field measurements (after Grace and Woolhouse, 1973b)

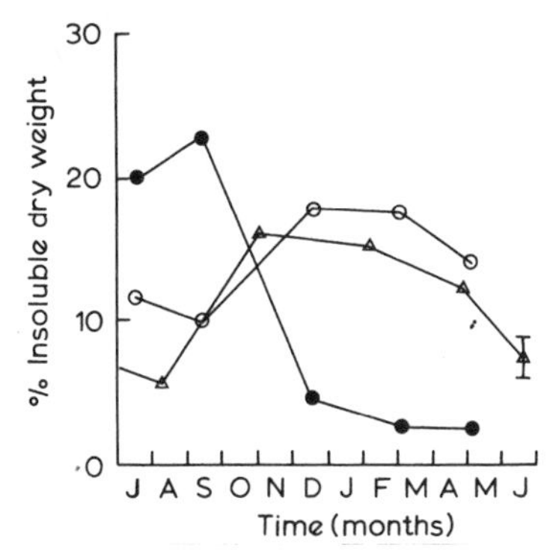

Fig. 13. Seasonal changes in carbohydrate content of *Rubus* rhizomes and *Calluna* green shoots. Starch in *Rubus* (●), sugars in *Rubus* (○), sugars in *Calluna* (△)

Table 2. The effect on levels of leaf sugar of bringing *Calluna* plants into a warm glasshouse during the early spring

Time (days)	1	2	3	5	7	8	10
Sugar (% dry weight)	6.9	7.9	8.0	7.4	0.2	0.2	0.2

Budbreak was observed at day 5

are summarised in Figure 13. The main storage carbohydrates in *Calluna* are mono- and disaccharide sugars stored in over-wintering leaves and stems. When budbreak occurs the level of sugar in these leaves may fall abruptly (Table 2). The main storage carbohydrate of *Rubus* rhizomes during summer and autumn is starch, but in winter much of this appears to be converted to sugars (Marks, 1974).

3.5 Modelling

In an attempt to explore some of the consequences of their data for *Calluna* and *Rubus*, Grace and Woolhouse (1974) and Marks (1974) have proposed empirical models in which hourly estimates of photosynthesis are integrated over a whole season. The *Calluna* model, described in detail in Grace and Woolhouse (1974), is summarised in Figure 14 using standard Systems Dynamics notation. The input data are climatological records collected at Moor House during 1968 and 1969. To test this model the resulting growth curve was compared (Fig. 15) with observed values of the standing crop in the same year for the same site (Forrest, 1971).

Encouraged by the apparent good agreement between Forrest's observations and the model's output, the sensitivity of the model to small changes in temperature and solar radiation was examined (Grace and Woolhouse, 1974). In these tests, changes in solar radiation had little effect on model performance but temperature increases, especially when made in the second half of the growing season, led to marked increases in plant production. Using *Calluna* dendrochronological

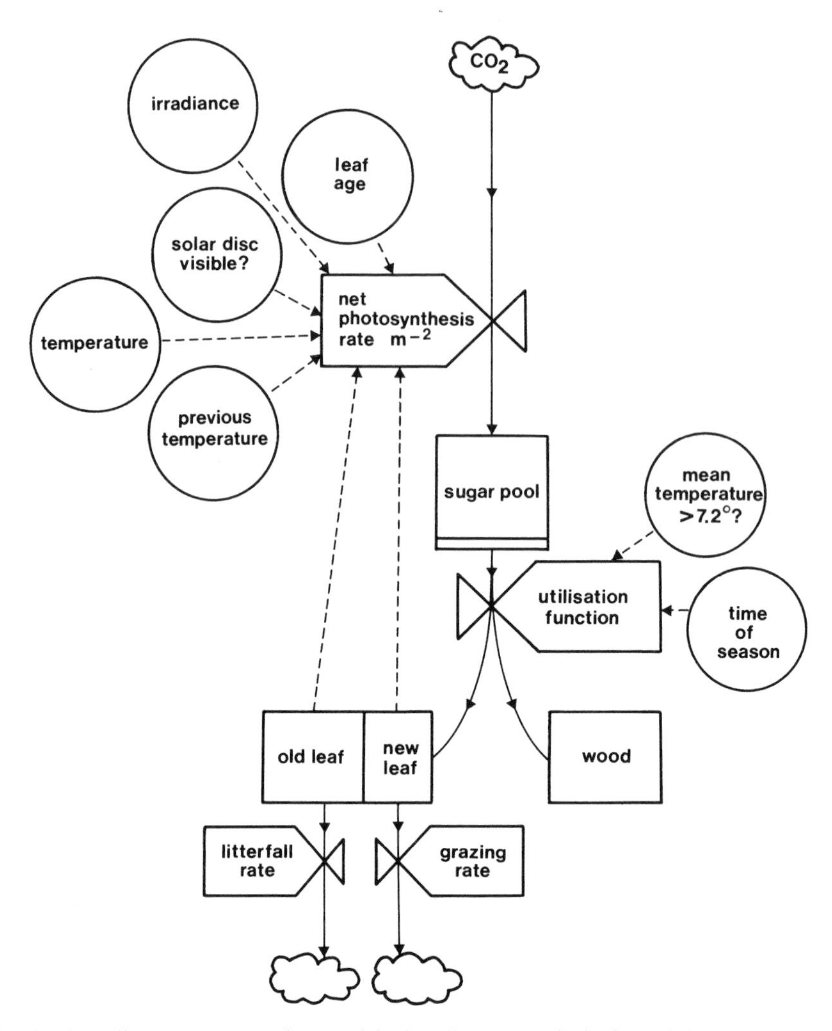

Fig. 14. Flow diagram representing model of *Calluna* growth, in formal Systems Dynamics notation. Measurable physical quantities are shown by rectangles, rates influencing those levels by valves. Delays are indicated by sections within rectangles. Flows of carbon are denoted by solid arrows and causal relationships by broken lines. Circles represent variables influencing the rate equations and clouds are sources or sinks that are unimportant to model behaviour (after Grace and Woolhouse, 1974)

data it has been demonstrated that real plants, like the model, are highly sensitive to temperature changes made in the second half of the growing season.

The *Rubus* model was devised to determine the effects of microclimate changes on carbon assimilation (Marks, 1974). To improve the accuracy of the model and reduce the number of assumptions made, field biomass and leaf area data were used as information input. Consequently, the model does not provide an independent estimate of productivity but it does give an insight into the relationship between climate and *Rubus* growth.

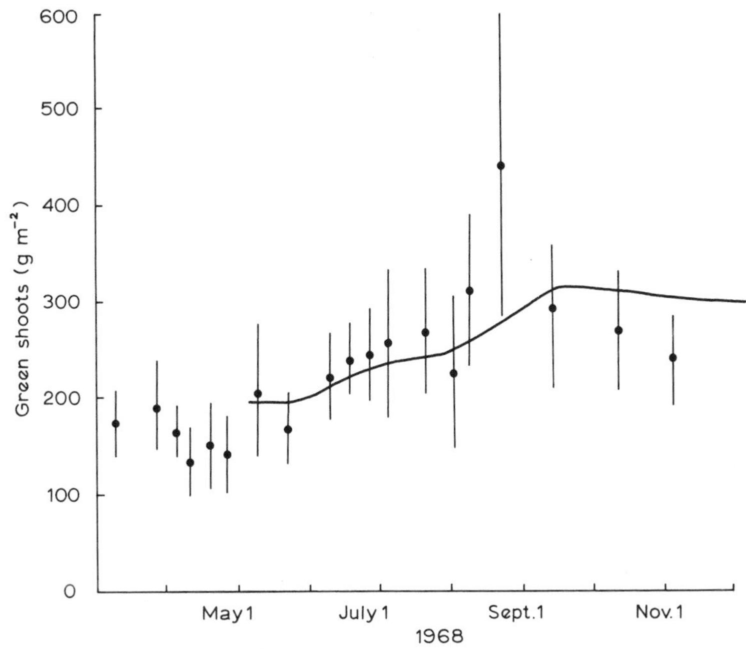

Fig. 15. Total green *Calluna* shoots at Sike Hill, Moor House, 1968 as measured by Forrest (1971) and as calculated by the model. Forrest's data denote 95% confidence limits (after Grace and Woolhouse, 1974)

The presence of a *Calluna* canopy and the consequent reduction in solar radiation available to *Rubus* leaves to 40% of full daylight (Fig. 2) is shown by the model to cut seasonal carbon assimilation per unit area of bog surface to 33% of that where *Calluna* is absent. Under these conditions the seasonal carbon assimilation per unit leaf area is reduced to 78% of normal. This is a fairly large reduction considering the low light saturation level of *Rubus* and points to the fact that light intensities are below saturation level for a high proportion of daylight hours when a *Calluna* canopy is present.

The response of *Rubus* in the model to overall change in air temperature regime was slight but the direction of response was surprising. An overall reduction of 2° C above ground during the growing season increased carbon assimilation by 5%. The growing season of *Rubus chamaemorus*, unlike that of *Calluna*, does not appear to be extended by warm late-season conditions. Leaf senescence occurs in early September and high temperatures during the growing season merely reduce net photosynthesis. During the 1969 growing season 20% of the above-ground, daytime temperatures were above the optimum for net photosynthetic rate of *Rubus*.

The *Rubus* model, therefore, gives a useful insight into the relationship between the production of *Rubus*, its microclimate, and the influence of *Calluna*; suggesting how a *Calluna* moorland might be managed as a cloudberry-producing amenity. Earlier work (Marks and Taylor, 1972) showed that factors other than mineral nutrient availability, such as microclimate, were of primary importance in regulating *Rubus* growth at Moor House.

3.6 Discussion

The general conclusion that temperature exerts an over-riding influence on *Calluna* performance in this upland situation is of great interest. Manley (1945, 1952) compared the annual pattern of temperature at Moor House with that at lowland sites and, on the assumption that plant growth begins at 5.6° C and increases with increasing temperature, he demonstrated how plants in the British uplands may be expected to grow at a reduced rate and over a shorter period than those in lowland places. Gore (1963) compared growth rates of *Molinia caerulea*, a moorland grass, at Moor House and at an edaphically similar, but warmer, site near sea level in Lancashire. Both relative growth rate and the length of the growing season were much reduced at Moor House, providing evidence for Manley's speculative hypothesis. Similarly Millar (1965) found significant correlations between height growth of *Betula pubescens* planted at Moor House and temperature during the year, although this does not exclude the possibility of weight changes occurring independently of temperature.

Calluna seems to conform well to Manley's simple model, growth beginning at a certain temperature (although 7.2° C, not 5.6° C), and thereafter increasing with increasing temperature (Grace and Woolhouse, 1970, 1973b, 1974). *Rubus* does not conform to this pattern, and undergoes leaf senescence at a time when air temperatures are still quite high (Marks, 1974), whilst high temperatures during the growing season result in reduction in net photosynthesis.

The differing responses of *Calluna* and *Rubus* are remarkable, and seem to reflect the species' position at Moor House relative to their geographic and climatic ranges. *Calluna* is centred on NW Europe (Gimingham, 1960) and seems to thrive best in an oceanic situation, so that any amelioration of the severe Moor House climate leads to an increase in growth. On the other hand, *Rubus* is here at the southern edge of its circumpolar distribution (Taylor, 1971) and temperatures in excess of those found at Moor House further retard its performance.

3.7 Conclusion

The dominant species at many Moor House sites is *Calluna vulgaris*, the foliage of which transmits 10–40% of the incident light, depending on the solar angle and the leaf area index. This leaf canopy markedly modifies the microclimate of those species growing beneath, such as *R. chamaemorus* and *Sphagnum* spp. Net photosynthesis of these constituent species was measured in controlled environment conditions. The highest rates were achieved by *Rubus*, in which the light saturation curve was very steep and maximum rate was 18 mg dm^{-2} h^{-1}. The effect of temperature at high irradiance was marked and maximum rate was achieved at 12–18° C. In *Calluna* net photosynthesis depended not only on irradiance, present and previous temperature and leaf age, as in *Rubus*, but also on light direction and whether or not the plant was flowering. Maximum rates were about quarter of those of *Rubus*. In *Sphagnum*, water content, as well as irradiance and temperature was an important determinant of photosynthetic rate. *Rubus* and *Calluna* show quite different strategies of assimilate distribution. The main storage component in *Rubus* was starch, whilst in *Calluna* mono- and disaccharide sugars fluctuated seasonally and appeared to be the main storage products, starch being practically absent. Using climatological data as input, computer models of *Calluna* and *Rubus* were proposed to investigate the sensitivity of plant performance to changes in climate or microclimate. In these tests the two species responded to small changes in temperature in opposite senses, *Calluna* growth being much promoted by higher temperatures whilst *Rubus* was slightly retarded. These results are discussed in relation to the geography of the species, and to our knowledge of plant growth in the British uplands.

Acknowledgements. The authors wish to thank all those people who have contributed in various ways to the work described here, especially Professor H.W. Woolhouse, Dr. K. Taylor, and members of the Botany Department of Sheffield University and of the Department of Botany and Microbiology, University College London. We wish also to thank those members of the Nature Conservancy who have collaborated to collect data, encouraged us, and helped to provide us with a working environment which has been stimulating both intellectually and physically. We wish to thank Mr. I. Stenhouse who has skilfully prepared our diagrams for publication. Much of the work has been sponsored by the Natural Environment Research Council.

4. Ecology of Moorland Animals

J. C. COULSON and J. B. WHITTAKER

A comprehensive review of the composition, distribution and abundance of the fauna, with analyses of the population dynamics of individual species, and of the trophic structure in the major habitats of the moorland, based on research over more than 20 years.

4.1 Introduction

The great majority of information on moorland animals has been acquired over the last twenty years. As recently as 1950, Pearsall in his excellent book *Mountains and Moorlands* wrote, "one of the most noticeable features of upland life is the apparent smallness of the animal population, and this is perhaps particularly the case on moorlands ...". At that time much of the available information concerned vertebrates, since those are often of financial and aesthetic interest to man. The invertebrates, however, had been neglected because in many instances they are inconspicuous, and are difficult to study quantitatively and to identify accurately. Accordingly, it is an almost impossible task to work on the whole fauna of a particular type of moorland and it has proved much more successful to have each worker studying the ecology of one of the most abundant groups, thus ensuring accurate and reliable taxonomic treatment of the material. As far as invertebrates are concerned, most of our knowledge of British moorlands comes from Moor House and neighbouring areas in the northern Pennines of England which are dominated by blanket bog. A full description of the area is given in Chapter 1.

Almost all of the work on moorland animals has been in the form of three-year studies and the majority of the groups have now been investigated. It is becoming possible to make an integrated picture of the faunal composition of moorland communities and to evaluate the energy and nutrient turn-over of moorland fauna.

Two approaches have been made in this paper. First, an account is given of certain aspects of the biology of the fauna, particularly those which appear to characterise the moorland system, with emphasis on species composition, habitat selection and the reduction in predators and parasites. Second, an attempt is made to examine the role of animals in the moorland ecosystem. Because of the ease with which most animals move from one area to another, the differences between the fauna on neighbouring peat and mineral soils reflect the choice of the animals and give valuable leads as to the cause of variations in the ecological processes of energy and nutrient flow which occur in markedly different ways on these two soil types. Further, the soil and vegetational mosaic, which characterises areas dominated by blanket bog, is shown to be important to the distribution and abundance of several animal species. This takes the zoological investigation beyond the scope of the Moor House IBP Programme which focused attention on the blanket bog only.

In both these approaches, it is necessary to bear in mind that all of the investigations were not made simultaneously and it is sometimes necessary to assume that population figures obtained over, say, three years, are representative of a much longer period of time (see Appendix 4.11). These studies have been spread over a twenty-year period and there is much observational and semi-quantitative evidence to suggest that they are typical of this longer period and form representative estimates of the animal populations in time and for vegetation types.

The detailed studies on particular groups have been carried out by workers specialising in those groups who have tested alternative sampling methods and used the best ones for the particular conditions. These are described in the appropriate publications and generally conform to the IBP guidelines (Petrusewicz and Macfadyen, 1970; Phillipson, 1971).

4.2 Composition of Moorland Fauna

Most investigations of the Moor House fauna have emphasised its affinity with subarctic and arctic situations. The point is discussed by Cragg (1961) in his review of work at Moor House and this tendency has been confirmed in more recent studies. Nelson (1971) recognised about fifty species of invertebrates at Moor House which have a montane distribution in the British Isles. By comparison with lowland Britain the fauna is sparse in number of species and superficially, at least, in biomass. Approximately 1100 species of animals have been recorded from the whole of the Moor House Reserve (4000 ha) compared with over 4000 species from Wytham Woods, Berkshire on about 1500 ha (data from Wytham Ecological Survey). This reduction is not equally shared among the animal groups (Table 1). For example, amongst the insects, Diptera are exceptionally well represented at Moor House. Thus Nelson (1965) recorded that 96% of all the invertebrates captured on sticky traps were Diptera, whilst they comprised 63% of the catch in liquid-filled pitfall traps, and *c* 60% of the total species taken. The dominance of the Diptera is typical of arctic and subarctic faunas. Aquatic insects, particularly stream living groups such as Ephemeroptera, Plecoptera and Trichoptera, but not Odonata, are particularly well represented.

There are some marked differences between the Moor House fauna and that of lowland heaths in Britain. For example, Richards (1926) recorded a much more diverse insect fauna associated with *Calluna* heaths than is found at Moor House, with both the Coleoptera and the Lepidoptera outnumbering the Diptera in number of species.

Some of these differences are almost certainly connected with the relative lack of flowers on high altitude moorland and consequent reduction in pollen- and nectar-feeding species. Others are probably associated with the wet conditions

Table 1. Approximate number of species in the main groups recorded at Moor House, Wytham Woods and in Britain

	Moor House	Wytham	British list	Ratio[a] Wytham/Moor House		
				High	Medium	Low
Myriapoda	6	28	90		4.7	
Pseudoscorpionidae	1	8	26	8.0		
Phalangidae	3	17	20		5.7	
Araneida	71	219	580		3.1	
Acarina	107	260	1,700		2.4	
Thysanura	1	3	23		3.0	
Protura	0	1	17	∞		
Collembola	56	56	305			1.0
Orthoptera	1	9	38	9.0		
Dermaptera	0	1	9	∞		
Plecoptera	25	6	32			0.2
Psocoptera	0	21	68	∞		
Ephemeroptera	12	5	46			0.4
Odonata	2	19	42	9.5		
Thysanoptera	+	17	133			
Hemiptera	71	273	1,411		3.8	
Neuroptera	3	21	60	7.0		
Mecoptera	1	2	4		2.0	
Trichoptera	31	23	188			0.7
Lepidoptera	70	532	2,187	7.6		
Coleoptera	115	1,074	3,690	9.3		
Hymenoptera	85	774	6,191	9.1		
Diptera	453	638	5,199			1.4
Siphonaptera	5	5	47			1.0
Mollusca	8	65	188	8.1		
Fish (freshwater)	2	20	45	10.0		
Amphibia	2	4	6		2.0	
Reptiles	1	2	6		2.0	
Mammals (excl. marine)	9	24	47		2.7	
Birds (breeding)	24	73	226		3.0	
Total	1,165	4,200	22,624			

[a] Gives an indication of which groups are poorly represented at Moor House (ratio high), of average occurrence (ratio medium) and very well represented (ratio low).
* Present. Number of species not known.

and the permanently high water table which makes pupation in the soil difficult without special modifications such as those found in some Diptera. The success of the Diptera undoubtedly relates to their preference for and adaptation to wet or semi-aquatic environments, whilst their ability to be active at low temperatures is also important. Mosquitoes are poorly represented. No larvae were found over ten years and only three specimens were taken: two *Culex pipiens* L. and one female *Theobaldia annulata* (Schrank). The contrast with many arctic and subarctic areas is unexplained, but may result from the dull, wet summers at Moor House.

Table 2. Number of species in the main insect groups recorded at Moor House and three other tundra sites (Maclean, 1975)

	Moor House 54°65′N	Hardangervidda Norway 60°36′N	Barrow Alaska 71°21′N	Lake Hazen Canada 81°49′N
Collembola	56	33	32	14
Ephemeroptera	12	3	0	0
Odonata	2	0	0	0
Orthoptera	1	1	0	0
Hemiptera	71	12	2	4
Neuroptera	3	1	0	0
Coleoptera	115	74+	12	4
Trichoptera	31	3	5	1
Lepidoptera	70	48	8	19
Diptera	453	84	120	142
Hymenoptera	85	35+	30	57

The high rainfall and low temperatures ensure that the soils rarely dry to an appreciable extent. Accordingly, the risk of desiccation, a major threat to small organisms, is much reduced and this is reflected in an abundant and rich soil fauna. Hale (1966a) recorded 56 species of Collembola, comparable to numbers on lowland sites (Wallwork, 1970). Similarly, the 78 species of mites recorded by Block (1965) indicate a fairly rich fauna. Enchytraeid worms and nematodes are exceedingly abundant and Tipulidae occur in most soils, 68 species out of a British list of some 300 having been found on the Reserve.

Only four vertebrates are abundant on the moor. Sheep have been introduced by man, whilst the Common frog (*Rana temporaria* L.), Red grouse [*Lagopus scoticus* (Lath.)] and Meadow pipit [*Anthus pratensis* (L.)] are the only species regularly encountered on the blanket bog. Perhaps this habitat offers adequate cover but not enough food at certain times to support a wider range of vertebrates (Sect. 4.6).

Many of these features of the Moor House fauna are found in the more extreme climatic conditions of the tundra (Maclean, 1975). Acarina, Collembola, are well represented on the arctic fauna, whilst Diptera are particularly prominent. In general, saprovores predominate and herbivores are scarce (Table 2). The main way on which the broad characteristics of the Moor House fauna differ from those of the arctic fauna is in the reduction of parasitic Diptera and Hymenoptera at Moor House (compared with lowland Britain). This point is discussed further in Section 4.7.

4.3 Distribution of the Fauna Within the Study Area

Blanket bog covers nearly 80% of the study area but the presence of other habitat types is important in determining the faunal composition of the whole moorland. The heterogeneity introduced into the system by the dissection of the blanket bog by streams and the presence of mineral soils along the stream banks and on limestone areas is particularly important. The streams themselves support an extraordinarily high proportion of the British Plecoptera and Ephemeroptera species (Table 1) as well as a high density of individuals. The interchange of animal material between banks and streams (Sect. 4.6.1.2) seems to be very important relative to the area occupied by the streams themselves (Crisp, 1966).

Even within the area covered by peat, considerable variations occur in the dominant vegetation, usually testifying to differences in the depth or nature of the peat. *Juncus squarrosus* L. becomes dominant on shallow peat, supporting higher densities and a larger biomass of animals than the blanket bog area (Sect. 4.4.1) whilst flushes offer a variety of additional habitats for animals, particularly when the water has flowed over limestone, although forming a small proportion of the total area. Such high-density pockets of animals can play an important role in supplying food to predators or acting as a population reservoir in the infrequent dry periods.

Many of the animal grazers and decomposers, in contrast to predators, are confined to particular vegetation types. They can be divided into three main groups: those living on soils with a high organic content (peats); those in a predominantly mineral soil, usually associated with stream sides and outcrops of rocks, particularly limestone; and the aquatic species (Sect. 4.6.2). While there is a good deal of overlap between these categories, the soils, and hence parameters such as water holding ability, pH, mineral content and vegetation, are radically different (Chap. 1).

4.4 Soil Fauna

4.4.1 The Fauna of Peat Soils

Since the distribution of acid peats is markedly northern, it is not surprising that, in most peat fauna groups studied, strong affinities have been found with northern Scandinavia, and also with the species-poor fauna of Iceland. Only in a few cases is there an affinity with the central and southern European upland fauna. In contrast the fauna on mineral soils has a major component of the species which are found in lowlands in Britain, and which have a central European distribution (Sect. 4.10).

Although the taxonomy of nematodes is extremely difficult, Banage (1962) demonstrated that peat soils have characteristic groups and that the Dorylaimoidea and Mononchidae, important in lowland soils, are uncommon at Moor House. The density of nematodes is higher in the *Juncus squarrosus* (3.9×10^6 m^{-2}) than the mineral soil (3.3×10^6 m^{-2}) and considerably lower in the wetter blanket bog soil (1.9×10^6 m^{-2}).

Banage (1963) divided the nematodes according to their feeding habits and showed that on the peat soils plant feeders predominated (83%), though this category included many fungal feeders. Microbial feeders constituted 12% but again it was not established that they feed exclusively on bacteria.

The density of earthworms on peat soils is very low, ranging from one worm 100 m^{-2} on areas with a high proportion of *Eriophorum* to 13 worms 100 m^{-2} on *Calluna*-dominated areas (Svendsen, 1957a). No lumbricid species are restricted to peat, whereas one enchytraeid species, *Cernosvitoviella briganta* Springett, first described from Moor House (Springett, 1969), occurs only in peat soils. Peachey (1963) found enchytraeid worms in all the moorland soils, where they contribute by far the greatest biomass. Even bare, eroded peat supports *c* 20,000 worms m^{-2} whilst Springett (1970) recorded densities of over 300,000 m^{-2} on *J. squarrosus* moor.

Collembola do not show a marked species preference between mineral and peat soils (Hale, 1966a). Of the sixteen species in which more than 50 individuals were taken during quantitative sampling, only six were on the *J. squarrosus* moor, one (*Folsomia brevicauda* Agrell) was confined to peat soils and seven species were not found on the blanket bog at all. Feeding habits are highly varied (Christiansen, 1964); most of the species abundant at Moor House seem to be saprovores or bacterial or fungal feeders. Known herbivores (Sminthuridae) account for only a very small proportion, perhaps 1–6%, while there is no evidence of much predatory activity.

Like the Collembola, very few of the mite species are restricted to one of the two main soil types, though Cryptostigmata tend to represent a higher proportion (63% of the biomass) of the total on peat soils than on limestone grassland whilst Mesostigmata are relatively uncommon (37% of the biomass) on the peat soils (Block, 1966). Cryptostigmata are mainly saprovore and fungal feeders plus some bacterial and algal feeders, whilst the Mesostigmata are mainly predaceous, feeding on a variety of animals (Wallwork, 1970).

By contrast, none of the 66 species of Tipulidae on the Reserve has larvae which inhabit both peat and mineral soils (Coulson, 1959). Only one large species (*Tipula subnodicornis* Zett.) is common, and two small species [*Molophilus ater* Meigen and *Pedicia immaculata* (Meigen)] occur abundantly on the peat. Final instar densities of *T. subnodicornis* of >100 individuals m^{-2}, and >3000 larvae m^{-2} of *M. ater* have been found. The highest larval densities of both species are almost invariably on areas dominated by *Juncus squarrosus*.

Five invertebrate groups (Nematoda, Enchytraeidae, Acari, Collembola, Tipulidae) together make up well over 90% of the total biomass of the blanket bog fauna (Table 3). Approximately 75% is from three species, *T. subnodicornis*, *M. ater* and *Cognettia sphagnetorum* (Vejdovsky) which are also the dominant animals contributing to the biomass on *J. squarrosus*, forming about 50% of the total weight. On mineral soils, the three major species, two earthworms and *Tipula paludosa*, contribute less than 30% of the total biomass, mainly because more species are present. The *Juncus* area is thus intermediate between the blanket bog and the mineral soils, except for its almost complete lack of Lumbricidae.

4.4.2 The Fauna of Mineral Soils

The limestone grassland areas, dominated by *Festuca* and *Agrostis*, have a typical brown earth soil and, although they form only about 4% of the total area at Moor House, are greatly used by herbivores, particularly sheep. The more abundant alluvial grasslands have a relatively species poor vegetation and vary according to the nature of the mineral soil and proportion of peat added from neighbouring bog areas (Chap. 1).

Table 3: The mean population density (numbers m^{-2} in summer) and mean standing crop ($g\ m^{-2}$ dw) of animals on the four main habitat types at Moor House

	Mineral soils				Peat soils			
	Limestone grassland		Alluvial grassland		*Juncus squarrosus*		Blanket bog	
	Density ($n\ m^{-2}$)	Standing crop ($g\ m^{-2}$)	Density ($n\ m^{-2}$)	Standing crop ($g\ m^{-2}$)	Density ($n\ m^{-2}$)	Standing crop ($g\ m^{-2}$)	Density ($n\ m^{-2}$)	Standing crop ($g\ m^{-2}$)
Lumbricidae	390	23.2	390	23.22	4	0.04	0.04	neg
Enchytraeidae	80,000	4.1	120,000	5.15	200,000	4.60	80,000	2.16
Nematoda	3,300,000	0.14	3,300,000[a]	0.14[a]	3,900,000	0.18	1,400,000	0.07[b]
Collembola	46,000	0.15	40,000	0.12	23,000	0.05	33,000[b]	0.10
Acarina	33,000	0.35	36,000	0.66	45,000	0.32	60,000[b]	0.40
Tipulidae	120	4.1	87	6.1	2,500	1.96	700	0.58
Coleoptera								
Carabidae	3	0.007	3[a]	0.007[a]	2	0.004	1	0.002
Araneida	40	0.003	370	0.047	300	0.039	130	0.017
Lepidoptera	<1	neg	<1	neg	<1	neg	<1	neg
Hemiptera	100	0.010	50	0.007	250	0.029	3,500	0.025
Grouse	0	0.0	0	0.0	0	0.0	0.00011	0.023
Small rodents	0	neg	0.00001	neg	0.0001	neg	0.00001	neg
Sheep	0.00056	14.0	0.00026	6.5	0.00013	3.2	0.000014	0.35
Total		46.0		42.0		10.4		3.73

neg: negligible

[a] No direct measure, assumed to be equal to limestone grassland.

[b] Note: The figures quoted in Heal, Jones and Whittaker (1975) are incorrect.

The summary of the densities and biomass of the mineral soil fauna (Table 3) shows that Lumbricidae, generally assumed not to be important on moorland, form by far the greatest biomass of the animals of the mineral soils (Svendsen, 1957a). Densities (400–500 worms m^{-2}) are of the same order as those obtained by similar methods on lowland pastures. *Lumbricus rubellus* Hoff and *Dendrobaena rubida* Savigny are the dominant worms of the alluvial grassland. All the species recorded at Moor House occur at low altitudes and have probably penetrated from the lowlands through the mineral soils along the stream sides.

Enchytraeidae are next only to lumbricids in biomass on the mineral soils. At Moor House species of the genus *Achaeta* are only found in soils with an appreciable mineral content, whilst members of the genus *Fridericia* are almost exclusively restricted to limestone grassland (Springett, 1970).

The density of nematodes on mineral soils is estimated to lie between that on blanket bog and on the *Juncus squarrosus* areas (Banage, 1960; Cragg, 1961). As on the other sites, most were found to be plant and fungal feeders, but these comprised only 48% of the nematode population on the limestone grassland site compared with 83% on the mixed moor (Banage, 1963).

Hale (1966b) found average populations of *c* 46,000 m^{-2} Collembola on limestone grassland and 40,000 m^{-2} on alluvial grassland, higher than on many lowland soils.

On mineral soils, three species of tipulid occur at typical densities of between 30 and 100 final instar larvae m^{-2}. These enter their final instar at different times: *Tipula varipennis* Meigen emerges in late May, followed by *T. paludosa* in late July and early August and *Tipula pagana* in early October. In contrast to the situation in peat soils, the small short-palped craneflies are almost completely absent.

During the two years of his quantitative study, Block (1966) found that densities of mites were not so high on the mineral soils as on the mixed moor and other peat sites. In 1961 and 1962 densities of 29,000 and 45,000 mites m^{-2} were found on the limestone grassland. In contrast to the peat soils, here predaceous Mesostigmata predominate (69% of biomass) and the Cryptostigmata form the majority of the remaining mites (31%).

4.4.3 Vertical Distribution of Soil Fauna

In most soils, the majority of the soil fauna occurs in the surface zone. While the main groups of soil invertebrates are largely restricted to the top 3 cm in both peat and mineral soils at Moor House (Table 4), in all groups where comparisons are possible, the proportion of the animals in the surface zone is greater in peat soils. This difference is almost certainly related to the anaerobic conditions near the surface of peat soils caused by waterlogging.

There are measurable vertical migrations of the fauna with a tendency for tipulids (Hadley, 1971), enchytraeids (Springett, 1970) and Collembola (Hale, 1966b) to move into deeper layers in winter when the surface is liable to regular and prolonged freezing. This is possible because the food and oxygen requirements of the animals are low at low temperatures. There is also evidence (Springett et al., 1970) of a daily vertical migration in *Cognettia sphagnetorum*. The worms come closer to the surface at night, and move downwards, apparently in

Table 4. Percentage of fauna in the top 3 cm of soil

	Mineral soils	Peat soils
Enchytraeidae	84	91
Nematoda	52	76
Collembola	74	100
Acarina	91	95
Molophilus ater	absent	73

response to risk of drying out, during the higher day temperatures. Although the saturation deficits are usually small at Moor House, enchytraeid worms are singularly poorly adapted to withstand water loss.

4.5 Herbivores

The casual observer at Moor House could quite easily gain the impression that there are only two herbivores of any consequence: namely the sheep, which graze mainly on the grasslands, and Red grouse grazing on the blanket bog. The biology of the Red grouse has been the subject of detailed study in Scotland by the Nature Conservancy (Chap. 13) and has been given only limited investigation at Moor House (Chap. 10). Other vertebrate herbivores are certainly scarce. The vole [*Microtus agrestis* (L.)] occurs, but is limited in its distribution and does not attain high densities. Its distribution appears to follow the edges of peat and alluvial areas where the vole uses the blanket bog area for cover but feeds mainly, at least during the winter, on the alluvial vegetation. In late summer some individuals spread onto the blanket bog but their density probably does not exceed 1 ha^{-1} (D. Evans and P.R. Evans, pers. comm.). The water vole [*Arvicola amphibius* (L.)] is even more restricted and occurs in only one or two areas.

The sheep are removed from the Reserve each autumn and returned in the late spring, an appreciable proportion of their growth thus taking place off the Reserve. For comparison with the invertebrate herbivores on particular sites, sheep density m^{-2}, rather than the more conventional numbers ha^{-1}, is given in Table 3, together with the standing crop biomass of the principal herbivores.

Homoptera are shown by trapping and quantitative sampling to be the only abundant above-ground invertebrate herbivores. Although 70 species of Lepidoptera occur on the Reserve (J. Heath, pers. comm.), they have been taken infrequently by the sampling methods used. Figure 1 illustrates data obtained by comparable methods of the relative abundance of the main animal groups feeding on the *Calluna* canopy at a lowland heath site in Dorset (Chap. 11) and at Moor House. The low densities of Lepidoptera on the bog, considerably $<0.1\ m^{-2}$, have persisted throughout the study. For example, Coulson trapped only five individual macro-Lepidoptera on twelve sticky traps on blanket bog in two years and only two caterpillars have been taken in a series of twenty pitfall traps in a

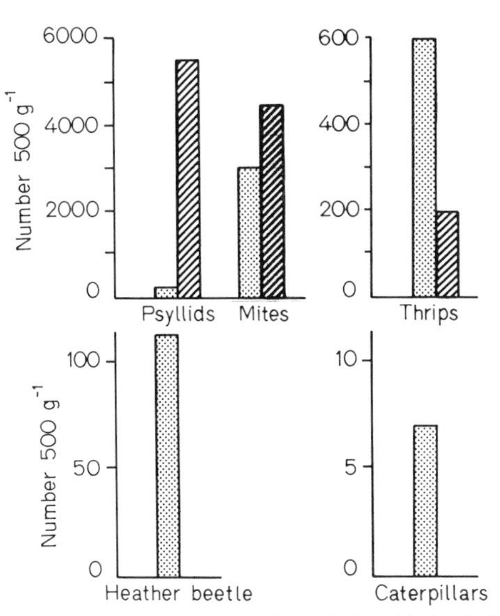

Fig. 1. Mean number of invertebrates extracted from 500 g (dw) of *Calluna* shoots in Arne, Dorset (▦) and Moor House (▨)

similar period. However, Nelson (pers. comm.) considers the moth *Entephria caesiata* Dennis and Schuffermüller to be abundant on the bog in some years.

The heather beetle, *Lochmaea suturalis* Thomson, one of the most important herbivores of *Calluna* is absent from Moor House. Psyllidae, mites and thrips are the most abundant invertebrate herbivores in the *Calluna* canopy (Fig. 1), the mites being present in large numbers for only brief periods (Block, 1966) and presumably returning to the litter zone. On blanket bog the psyllids, virtually absent from the Dorset heath, occur at densities averaging 2000 m^{-2} and may exceed 11,000 m^{-2} (Hodkinson, 1973). Living throughout the year in the leaf axils of *Calluna*, from which they extract sap with their stylet mouth-parts, they probably do not feed in winter, and certainly do not moult or even suffer much mortality then.

The counterparts of the psyllids on the grasslands are the froghoppers and leafhoppers (Homoptera, Auchenorrhyncha) at much lower densities (<1500 m^{-2}) but a far greater biomass (Whittaker, 1963, 1965a, b).

These are the dominant invertebrate herbivores, in the absence of Orthoptera, of aerial parts of plants (Table 1). All the grassland species at Moor House overwinter as eggs which hatch in late May or June, and complete their life cycle by September. They are considered important because not only are they numerically abundant but, like the Psyllidae, their effect on primary production may be many times greater than suggested by the amount of leaf material they consume since they selectively remove amino acids. Moreover, very large quantities of plant sap are passed through the gut and excreted as honey-dew or "spittle".

Prevailing microclimatic conditions may greatly influence the distribution of insects. On a sunny day, when there was a temperature gradient of 4° C in the 10 cm vegetation zone, 20 ± 8 Homoptera individuals m^{-2} were found in the top

5 cm of ungrazed limestone grassland and none in the bottom 5 cm. When the temperature gradient was absent, the numbers in the top and bottom layers were 14 ± 3.0 and 18 ± 4.5 respectively.

The moorland fauna is undoubtedly poorer because of the lack of trees or tall shrubs, which would add an appreciable third dimension to the habitat. Even on blanket bog with the dwarf shrub *Calluna vulgaris*, animals are restricted to about 40 cm above the ground. Within the last 20 years, several experimental plantations of trees have been established at Moor House. Although the nearest trees, except for an occasional rowan, are about six miles away, the sawflies [*Croeses latipes* (Villaret) and *Scolioneura betuleti* (Klug)] have abundantly colonised birch whilst *Neopridion sertifer* (Geoffroy) now also occurs commonly on pines (Nelson, 1971). Similarly Ibbotson (pers. comm.) has shown that the Frit fly (*Oscinella frit* L.) readily attacks oats transplanted to Moor House although there is no suitable food plant for many miles. There are many examples of insects ranging in size from thrips and aphids to *Sirex* and butterflies, which frequently drift onto the area from the neighbouring Eden valley although they have no suitable environment or food supply at Moor House. Only in the minority of cases can the absence of a species be attributed to the difficulty of reaching the area.

4.5.1 Effects of Herbivores on Vegetation and on Other Animals

Sheep densities on peat soils are low and their impact on the vegetation slight, but grazing tends to encourage *Juncus squarrosus* and associated plants at the expense of ericaceous plants. In contrast, sheep graze limestone grassland intensively, ingesting an estimated 60–110 g m^{-2} yr^{-1} (Rawes and Welch, 1969) and producing an extremely short turf. Experimental sheep exclosures cause marked changes in the vegetation, including the occurrence of taller plants, different species composition, and more flowers (Welch and Rawes, 1964). The related effect on the fauna has been little studied but there is clearly an increase in those insects which visit flowers, whilst spiders, whose numbers are relatively low on grazed limestone grassland, double in numbers (Cherrett, 1964). Although twenty eight species of insects were restricted to enclosed areas, they were mostly records of single occurrences, and the main effect of exclosures seems to be quantitative, rather than qualitative (Nelson, 1971). Whittaker (1963) recorded a four-fold increase in the biomass of Homoptera on the tall vegetation around limestone outcrops within exclosures, compared with heavily grazed areas, and he attributed this as much to the very different microclimates, and perhaps vegetation structure, as to floristic differences.

The deposition of sheep dung, an attractive microhabitat to many insects and earthworms, varies greatly between years (Rawes and Welch, 1969). White's (1960a) estimates of 119 cm^3 m^{-2} on grassland, 78 cm^3 m^{-2} on *J. squarrosus* moor and 18 cm^3 m^{-2} yr^{-1} on blanket bog indicate the approximate quantities available to dung-living animals. The dung is attractive to flies and beetles only for a few days. Its subsequent destruction by Lumbricidae and weathering is much more rapid on grasslands (40 d) than on *J. squarrosus* moor (64 d) or blanket bog (143 d). Svendsen (1957b) found that four surface-active lumbricids aggregated so habitually in sheep dung that he could use it as a means of trapping them and estimating population size.

Many animal species are solely associated with dung. White (1960b) recorded 16 species of *Aphodius* beetles associated with dung at Moor House, but failed to find any *Geotrupes*. Gibbons (pers. comm.) found three species of the dung-fly, *Scopeuma*, whilst Nelson (1971) lists 34 species of *Sphaeroceridae* (Diptera), probably mainly from this microhabitat, and a further 13 species of Diptera whose larvae feed on dung.

Dung provides relatively high concentrations of available Ca, N, P and K on the sward, locally exceeding amounts of minerals removed by sheep grazing (Rawes and Welch, 1969), and on all sites it results in a more rapid cycling of nutrients.

Because of the very high densities of the psyllid *Strophingia ericae* Curtis, recorded on *Calluna*, it was chosen for a study of the way in which a sap-sucking insect affected its host plant (Hodkinson, 1973). In a laboratory experiment designed to investigate the effects of psyllid feeding on the rates of respiration and photosynthesis of *Calluna vulgaris*, no significant effect could be measured even with psyllid densities $c \times 10$ those observed in the field. Under field conditions normal densities of psyllids have no measurable effect on shoot length, dry matter accumulation, flower production or mineral content of *Calluna*. Apparently, *Calluna* completely compensates for the materials removed by the psyllids and actually becomes more productive. At typical densities of about $2000\ m^{-2}$, the psyllids remove $c\,25\ kJ\ m^{-2}\ yr^{-1}$ from the total production of *Calluna*, a quantity usually ignored when determining primary production by measuring annual increment.

4.6 Predators

Although certain predators are common on moorland, others which might be expected to occur are either uncommon or completely absent. Predators of insects and other invertebrates are frequently encountered but the top predators are generally absent.

4.6.1 Predators of Invertebrates

Spiders (Araneida), harvestmen (Phalangidae), Ground beetles (Carabidae), Rove beetles (Staphylinidae) and mites form the major invertebrate predators at Moor House and frogs, Meadow pipits and shrews (both the Common, *Sorex araneus* L., and Pygmy, *Sorex minutus* L.) are the most frequently encountered vertebrate predators.

Seventy one species of spiders have been recorded from the Reserve, including 52 species of Linyphiidae, seven species of Lycosidae, one Tetragnathidae and three Argiopidae (Cherrett, 1964). The spiders show a marked distribution of species according to habitat, the Argiopidae being restricted to peat hags which dissect the blanket bog cover and provide the right site architecture. Only 14% of the Linyphiidae were found on all of the four main vegetation types, and each vegetation type has species peculiar to it.

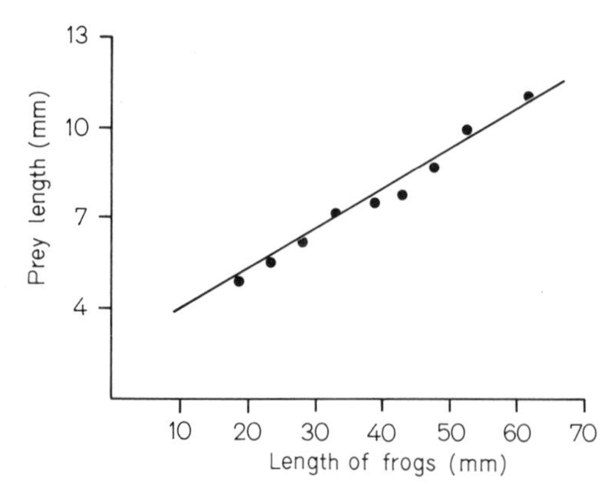

Fig. 2. The mean size of prey eaten by 359 *Rana temporaria* of different size groups (from Houston, 1973)

Vertical zonation, restricted by the low moorland canopy, is shown most clearly by the spiders, each species spinning its webs at a characteristic height in the *Calluna* canopy. This seems to be a direct reflection of the structure of the habitat rather than the microclimate (Cherrett, 1964). Structure is also important in determining their horizontal distribution.

Houston (1971) recorded 46 species of Carabidae and studied their distribution using pitfall traps. Densities of two to four adult carabids m^{-2} seem typical on blanket bog corresponding to a capture of about seven beetles per trap-year. A similar density (*c* 3 individuals m^{-2}) of sub-adult and adult harvestmen occurs, only *Mitopus morio* (Fab.) being common on the blanket bog.

Relatively little is known of the detailed feeding biology of most invertebrate predators compared with that of the vertebrates. A detailed study showed that the Common frog eats a wide range of food animals, closely reflecting the availability of invertebrates of suitable size in its habitat (Houston, 1973), except for the complete lack of mayflies in several thousands of food items examined. This unexpected absence is explained by the immobility of resting mayflies and the necessity for the frog to be attracted by moving prey. Both the average size and variation in size of food increased with the size of the frog (Fig. 2). For example, mites and Collembola are taken by frogs of all sizes but form a smaller proportion of the food of the larger individuals as other larger prey are taken.

The Pygmy shrew is about ten times as abundant at Moor House as the Common shrew (Houston, pers. comm.). Both are found on the blanket bog but, like the Meadow pipit and *Microtus*, they are most frequently found along the mineral/peat intersection, taking advantage of the food available on both areas and the cover on the blanket bog. In some winters, the peat at Moor House is frozen to a depth of 20 cm for several weeks, and how the shrews survive, and find sufficient food when their potential prey are encased in ice, remains a mystery.

Conditions for predators differ markedly within the soil and above ground. While there is an abundance and relative stability of numbers of potential prey such as Collembola, mites and enchytraeid worms throughout the year below ground, the small soil cavities limit the size and type of predators able to use them. Only predatory mites, some dipterous larvae and certain staphylinid beetles are common predators below ground. Above ground there is no restriction of the size of predators or their movement. The food supply, consisting largely of the emergent adults of soil animals which spend most of the year as larvae beneath the surface, fluctuates greatly. There is no emergence in winter and few insects on the surface. Poikilothermic predators cease activity during the winter, until the spring rise in temperature brings an increase in adult insects. Homoiothermic animals such as Meadow pipit, wheatear and wagtail, unable to decrease their food needs, leave the area.

4.6.2 Seasonal Availability of Food for Insect-Feeding Predators

In a detailed study of food availability for predators hunting on the surface of the moorland, Coulson (1959) measured the seasonal distribution of insects on a *Juncus squarrosus*-dominated peat area and an alluvial grassland by use of adhesive traps placed at ground level. The traps were used to evaluate the potential prey for the Meadow pipit, but the results are applicable to other ground-insect predators. They show a marked difference in both the abundance and seasonal distribution of insects in the two areas (Fig. 3). The period of abundance (captures exceeding 2 g dw) was limited to a month, from mid May to mid June, in the

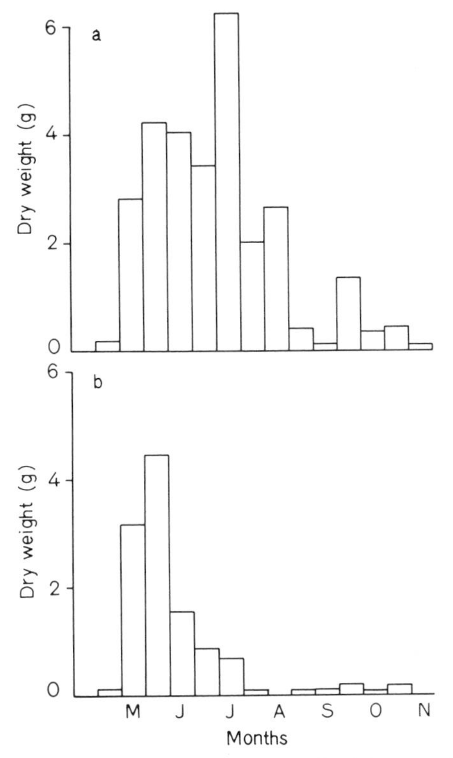

Fig. 3a and b. The dry weight of tipulids captured on 12 sticky traps in 1955 (from Coulson, 1956a). (a) On alluvial grassland (total 28 g), (b) on *J. squarrosus* (total 12 g)

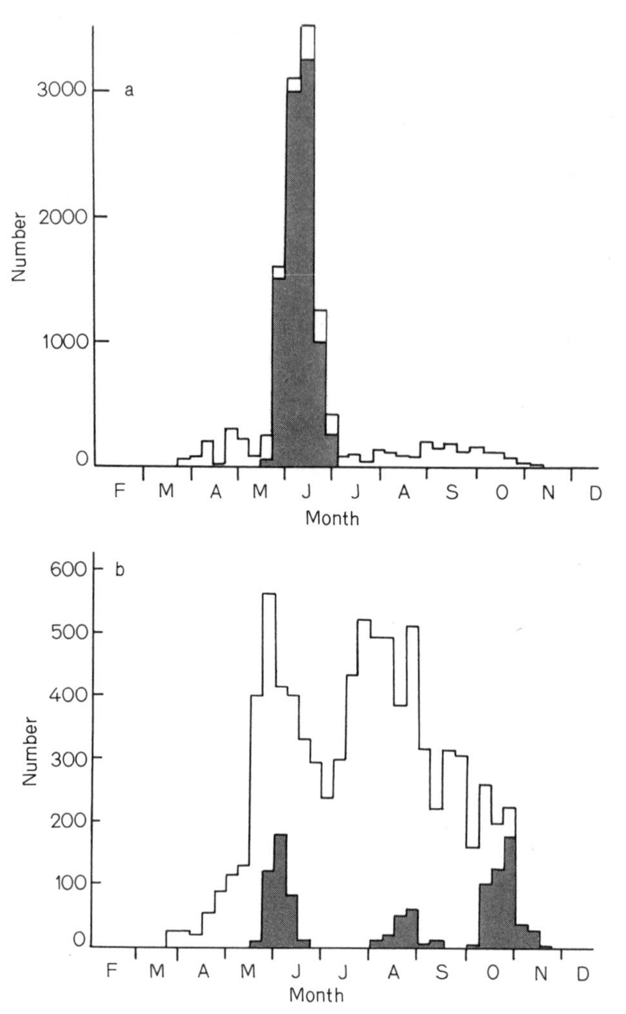

Fig. 4a and b. Numbers of invertebrates trapped by 10 water traps at weekly intervals in 1965. The shaded areas show tipulids only (from Nelson, pers. comm.) (a) on blanket bog, (b) on limestone grassland

Juncus area, representative of peat soils. On the grassland area the period of abundance was three times longer, extending from mid May to the end of August. Other data for *Calluna*-dominated blanket bog give similar results to the *Juncus* sites but the captures are roughly halved, exceeding the equivalent of 2 g only in the first two weeks in June.

Nelson (1971), expanding these investigations with the use of water traps, has made available data for most of the 637 species listed in his paper. These are summarised in Figure 4 and converted to biomass (Fig. 5) using dry weight data (Nelson, 1965; Whittaker pers. comm.). Again on the blanket bog there is a vast spring emergence in late May and June, >80% of the numbers and an even higher proportion of the biomass appearing and disappearing in six weeks. The

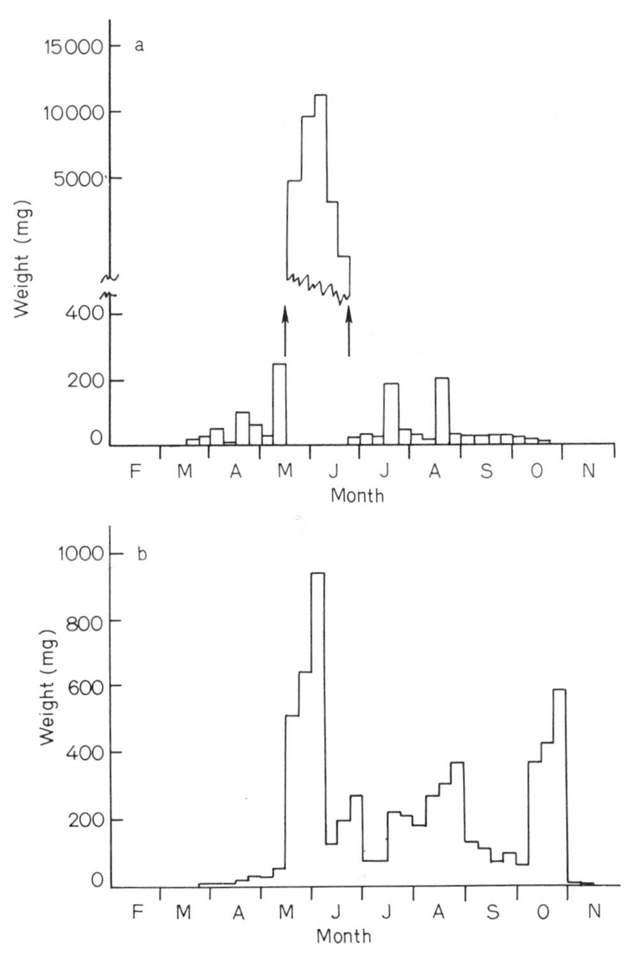

Fig. 5a and b. The dry weight of invertebrates trapped by 10 water traps at weekly intervals in 1965 (from Nelson, pers. comm.) (a) On blanket bog, (b) on limestone grassland

greater part of this peak is caused by the emergence of two craneflies, *Tipula subnodicornis* and *Molophilus ater*. Both have a highly synchronised emergence at the edge of the blanket bog and on the *Juncus squarrosus*, 95% emergence of *M. ater* occurring in eleven days (Hadley, 1969), about half as long as in *T. subnodicornis* (Coulson, 1962). While the peak biomass on the grassland is only about a tenth of that on the bog, the numbers and biomass are more evenly spread over six months. The three peaks in biomass correspond to the emergence of three craneflies, *Tipula varipennis*, *T. paludosa* and *T. pagana*.

Thus the situation for an insect predator on the blanket bog is one of superabundant food for a short period of time followed by a prolonged period of scarcity. To survive, predators have to be either specialist feeders or highly mobile so that they can utilise the spring peak on the peat and then resort to the mineral areas or the stream sides for food during the remainder of the summer and

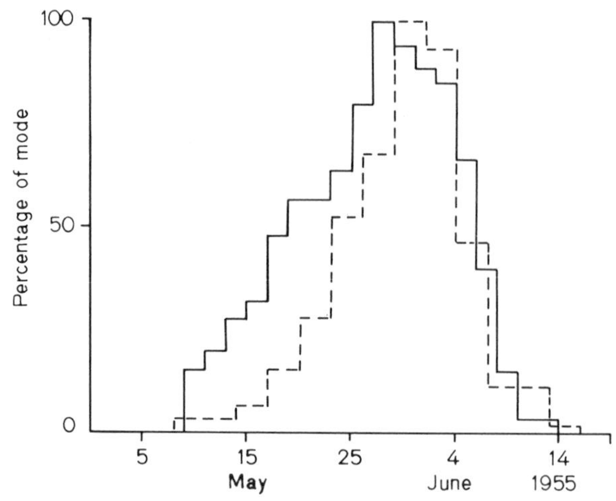

Fig. 6. The percentage emergence of 588 *Tipula subnodicornis* (—) and the occurrence of young in 15 broods of Meadow pipits (----) in 1955. Results are expressed as percentage of the mode (from Coulson, 1956a)

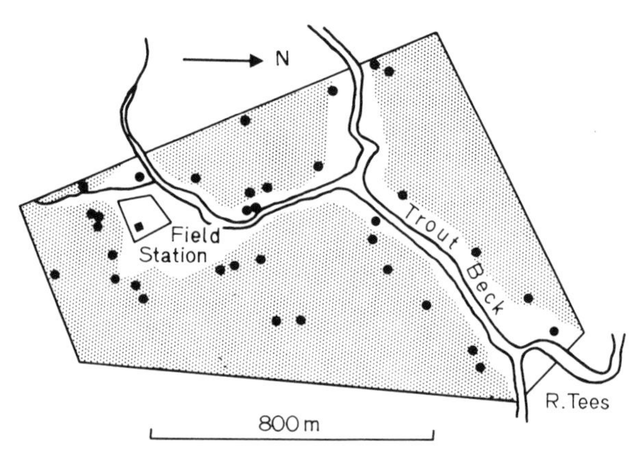

Fig. 7. Sketch map of the area near to the Field Station at Moor House showing positions of nest sites of the Meadow pipit. *Shaded areas:* blanket bog; *unshaded areas:* mineral soils

autumn. We believe that the difference in the availability of insect food between alluvial areas and blanket bog is crucial in determining the types of predators found in the area. For example, breeding in the Meadow pipit is closely synchronised with emergence of *T. subnodicornis* (Fig. 6). Although the birds lay their eggs three weeks before the peak of the craneflies' emergence, synchronisation results from response to the same stimulus which is responsible for the insects' pupation, viz the marked rise in temperature in late April and May. Whilst the first brood of the Meadow pipit is fed almost exclusively on adult *T. subnodicornis*, the second brood in July is fed extensively from stream sides and mineral grasslands (Table 5). This pattern of feeding is reflected in the distribution of breeding pairs of Meadow pipit. Few suitable nest sites exist on the grassland areas, but nests are concentrated on blanket bog in proximity with grassland (Fig. 7). Houston (1970)

Table 5. Food taken to 11 first and five second broods of the Meadow pipit

	Food	Percentage of visits
First brood (216 visits by parents)	*Tipula subnodicornis*	83
	Other Tipulidae	2
	Other adult Diptera	4
	Stream insects	11
Second brood (122 visits by parents)	*Tipula paludosa*	40
	Other Tipulidae	1
	Other adult Diptera	11
	Diptera larvae	5
	Moths	10
	Stream insects	33

Table 6. Proportions of predatory invertebrate species in different habitats (species data from Nelson, 1971)

Habitat	Total number of species	Number of predators	Percentage of predators
(A) Species occurring in only one habitat			
Meadow	48	14	29
Calluna moor	36	3	8
Limestone grassland	9	3	33
Alluvial grassland	30	11	37
Total	123	31	25
(B) Species occurring in at least eight habitat types (as defined by Nelson)			
	63	35	56

found that five species of carabid beetle, and also the Common frog, exhibit a similar migration from the bog areas to the grasslands, presumably in search of food.

Elton (1966) suggested that predators occur over a greater habitat range than herbivores or detritus feeders. This so-called inverse pyramid of habitats has been examined for 637 invertebrate species recorded by Nelson (1971) on Moor House blanket bog. The analysis shows that the proportion of predators is higher among the widespread species at Moor House (Fig. 8), and the same is true of other vegetation types (Table 6). This reflects not only the advantage gained from the greater mobility of predators in being able to follow the food supply (Sect. 4.6.1.2) but also their lack of the need to associate closely with plants in the same way as do herbivores.

Perhaps alone of the invertebrate predators, the spiders show specialised and restrictive habitat requirements. While their density on the blanket bog is, in fact, lower than on any other habitat studied by Cherrett (1964), apart from limestone

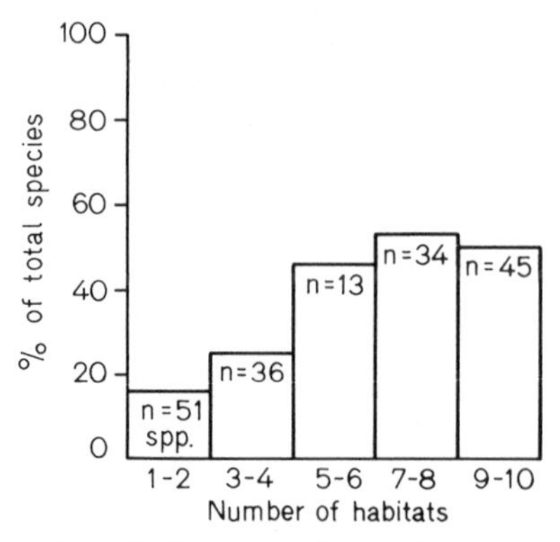

Fig. 8. The percentage of total species known to be predators which are restricted to one or two, three or four—nine or ten of the habitat types of Nelson (1971). The number of species is shown in each block

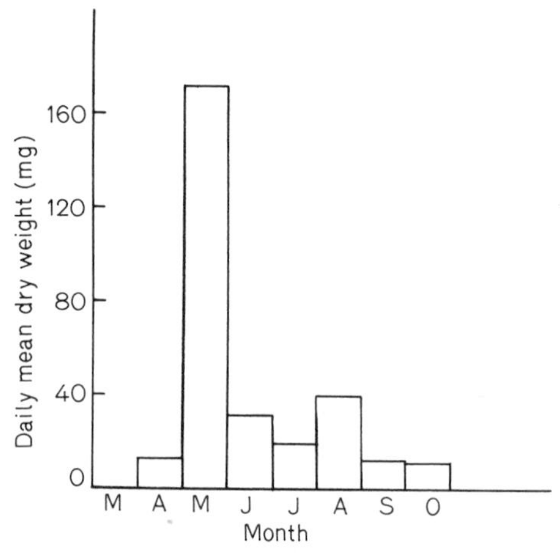

Fig. 9. Daily mean dry weight of potential prey of *Araneus cornutus* caught by artificial webs on blanket bog (from Cherrett, 1961)

grassland where grazing is the limiting factor, the structural diversity of the bog doubtless partially accounts for their relative abundance and species diversity. Cherrett (1961) measured seasonal availability of prey for web-spinning spiders by using artificial webs on blanket bog hags alongside a small stream (Fig. 9). Although the major peak is composed mainly of empids as opposed to tipulids, possibly because the flightless *M. ater* and female *T. subnodicornis* were not available for capture on the webs, the general similarity to Figure 3 is not accidental. However, a study of the prey remains in web retreats of *Araneus cornutus* Clerck on blanket bog (Table 7) showed that the most frequent prey were tipulids, followed by empids and caddis flies (Trichoptera). The large number of the latter is of special interest in that it indicates that part of the prey not only originates away from the bog, but has aquatic origin, emphasising the considerable interconnection between terrestrial and aquatic habitats (Chap. 10).

Table 7. Prey in retreats of *Araneus cornutus*

	Hymen-optera	Trich-optera	Tipu-lidae	Lepid-optera	Empi-didae	Other Diptera	Neur-optera	Cole-optera
Numbers	5	27	97	20	37	3	3	8
%	2.5	13.5	48.5	10	18.5	1.5	1.5	4

It is evident that, unlike many predators, the majority of spiders remain on the blanket bog throughout the year. Clearly the quantity of available food varies (Figs. 4 and 5) and they must rely upon their ability to survive for long periods without food (Cherrett, 1961) and the continuous presence of the small soil arthropods such as mites and Collembola.

4.6.3 Predators of Vertebrates

Large carnivores are almost completely absent from Moor House and the neighbouring areas. A few foxes (*Vulpes vulpes* L.) occur, largely on the well-drained western scarp slope, probably feeding predominantly in the Eden Valley on farm land. Mustelids are clearly of little importance in the ecosystem, observations by several fieldworkers of the stoat (*Mustela erminea* L.) and weasel (*Mustela nivalis* L.) averaging one a year. In two cases they are recorded as feeding on frogs and, while other vertebrates are no doubt taken, the lack of small mammals undoubtedly accounts for their scarcity, as well as that of most birds of prey.

In many years, no Short-eared owls (*Asio flammeus* Pont.) nest on the Reserve and in the remaining years only one pair has been present, tending to hunt at lower altitudes. Buzzards [*Buteo buteo* (L.)] do not feed over the greater part of the Reserve and avoid the blanket bog. Like the fox, they tend to restrict themselves to the western part of the Reserve. Similarly kestrels (*Falco tinnunculus* L.) are more frequent on the western scarp slope but infrequently hunt over the blanket bog areas.

The two bird-eating falcons, the peregrine (*Falco perigrinus* Tunst.) and merlin (*Falco columbarius* Tunst.) are rare and the former hunts almost exclusively outside the Reserve. To our knowledge (and surprise) the merlin has not nested on the Reserve between 1952 and 1972 but in one year it bred just over the boundary. During three years of study on the Meadow pipit at Moor House, Coulson did not encounter a single merlin and found no evidence that the pipits were being preyed upon at all. Investigation of the British Trust for Ornithology nest record cards suggests, however, that the merlin does not normally breed above 600 m (2000 ft).

We believe that the almost complete lack of top predators is due to the density of suitable prey on the Reserve and neighbouring moorland being too low to support them, rather than to keepering. While old records, showing that a large number of predators were shot annually, appear to be incompatible with this view, they represent a time when predator numbers over the whole of the north of England were higher than they are now. Moreover, many of the predators were killed on the western side of the Reserve where blanket bog is absent and they probably consisted of wandering, immature individuals from more typical habitats, searching for a potential breeding area, rather than representing production from Moor House. There is no evidence that there have ever been sufficient small mammals, such as voles, to support predators. Thus Moor House is an area where there is clearly a very flat pyramid of numbers and members of the top trophic levels are, to all intents and purposes, not represented.

4.7 Parasites

Several investigators have commented on the scarcity of hymenopterous and dipterous parasites. The majority of those taken are small braconids and chalcids (Nelson, 1971), and parasitised insects and other invertebrates are only infrequently recorded. In an examination of a large sample of cercopids (frog-hoppers) Whittaker (1971) failed to find any parasitism, although he recorded rates of up to 63% in lowland situations. No hymenopterous parasites have been found in several hundred tipulid larvae examined; and larvae of the parasitic dipteran, *Siphona*, have been found on only four occasions. The Diptera have been well studied at Moor House, apart from the Chironomidae, and the species list is fairly complete. Nevertheless only 1.3% of the 453 species are parasitic, compared with 6.7% in the British list of 5200 species (Kloet and Hincks, 1945). While *Mermis* and *Gordius* larvae have occasionally been found, no insect parasites have been found in carabid beetles (Houston, 1970). The larvae of the rush moth *Coleophora alticolella* Zell. are found on the seeds of *Juncus* plants from sea level to almost 600 m in the north of England. Over most of this range they are heavily parasitised by hymenopterous parasites of several species, but in the upper part of their range, parasitism decreases with altitude and finally disappears leaving a zone where the moth is completely free of parasitism (Jordan, 1962; Reay, 1964) (Fig. 10). The cuckoo (*Cuculus canorus* L.), an important nest parasite of the

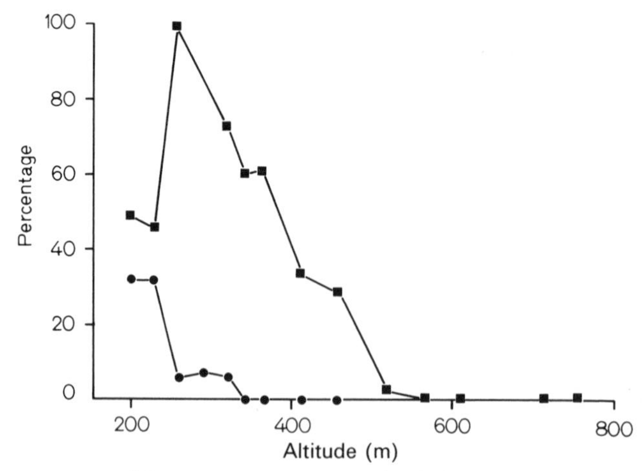

Fig. 10. The percentage of *Juncus squarrosus* capsules infested by *Coleophora* (■) and the percentage of *Coleophora* parasitised (●) by Hymenoptera on the west transect at Moor House 1953 (from Jordan, 1962)

Table 8. Nest mortality rate in Meadow pipit in relation to altitude. 27% of all cases of mortality were due to cuckoo laying in the nest

		Altitude			
	m	0–15	15– 76	76–229	>229
	ft	0–50	50–250	250–750	>750
Number of nests examined on at least two occasions		71	63	36	32
Number predated		25	17	5	1
Nests with contents taken (%)		35	27	14	3

Meadow pipit, does not penetrate above 300 m although its host is both widespread and abundant above this altitude. This lack of cuckoos at higher altitudes largely explains the decrease in nest mortality of the Meadow pipit with altitude (Table 8) (Coulson, 1956 b).

It appears that parasitic species play a less important role in upland moorland ecosystems than at lower altitudes. This finding is not expected from a comparison with other tundra sites (Table 2) where parasitic insects are usually well represented. However, Burnett (1951, 1954) has shown that the rate of attack by a parasitic insect is halved when the temperature falls 4° C in the laboratory or field. It may be that in spite of good representation in the arctic, the ecological significance of parasites is not so great as elsewhere. In several species at Moor House, the hosts have a much greater altitude range than the associated parasites, perhaps because the parasites' search for hosts is inefficient at low temperatures or under the wet, misty conditions which frequently typify moorland. This is clearly a field of investigation which will repay further study.

4.8 Aspects of the Biology of Moorland Animals

4.8.1 Structural Changes Related to Altitude

Mani (1968) considers that the most obvious features of montane insects are reduced wing size and mean body length and a tendency towards melanism. These features are found in the Moor House fauna.

In those species of Auchenorryncha (Homoptera) which exhibit both macropterous and brachypterous adults, the former were taken only in the meadow, a sheltered, ungrazed area on limestone and at the lowest altitudes on the western scarp slopes of the Reserve. Several other species, e.g. *Diplocolenus bensoni* (China) and *Streptanus marginatus* (Kirsch), while macropterous at lower altitudes on the Reserve, had shorter wings at higher altitudes (Whittaker, 1963). Many species of stoneflies (Plecoptera) are also represented by brachypterous forms at Moor House and this general subject is discussed by Brinck (1949) for this group in the Scandinavian fauna. Little is known of the biology of the scorpion-fly (Mecoptera) *Boreus hyemalis* (L.), apterous in both sexes, which has been taken on the Reserve.

Varying degrees of wing reduction, accompanied by shortening and strengthening of the legs for walking, have been found in the Tipulidae at Moor House. In some species, only the female exhibits brachyptery, e.g. *Tipula subnodicornis*, *T. gimmerthali* Lacks., *T. pagana*, *Limnophila pulchella* (Meigen), but in *Molophilus ater* and *Dicranota robusta* Lunds. the wings are extensively reduced in both sexes.

Within the Carabidae, smaller species occur at higher altitudes: Houston (1970) has shown that at about 450 m the mean body length was 9.75 mm (37 spp), at 550 m it was 9.23 mm (40 spp), at 750 m it was 7.18 mm (20 spp), while at 850 m there were only six species, and the mean body length had decreased to 6.16 mm. He also showed a significant decrease in mean body length of *Patrobus assimilis* Chaudoir between 550 m and 850 m, males decreasing from 7.05 mm to 6.65 mm and females from 7.66 mm to 7.12 mm.

However, a detailed study by Butterfield (1973) of the factors responsible for size variation in *T. subnodicornis* failed to show any direct or indirect influence of altitude. The size of adults of both sexes was affected by the larval density, the year and the site. Successive generations showed a marked tendency to be consistently either large or small at a particular site, presumably indicating different nutritional qualities of the sites. Similarly, the fecundity and size of *M. ater* adults appear to be independent of altitude.

Observations on melanism in the Moor House fauna are subjective, but Houston (1970) states that many species of Carabidae are darker than at lowland sites, whilst Whittaker (1965a) says that there is a greater than usual tendency towards the production of dark pigments in the Homoptera.

4.8.2 Reactions to Low Temperatures

Those stages of the life cycle of an animal which overwinter at Moor House are often subject to severe climatic conditions. Invertebrate activity ceases for several months and there is a greater tendency than at lowland situations to overwinter as eggs. For example, while in lowland situations species of Homoptera frequently overwinter as nymphs or adults (Whittaker, 1965a), only two species do so at Moor House. The nymphs of one of these, *Strophingia ericae*, studied by Hodkinson (1973) live in the leaf axils of *Calluna*, where the temperature not infrequently falls below −10° C. In spite of frequent exposure to low temperatures, they do not appear to produce the polyhydric alcohols which act as an anti-freeze in other insect groups.

The soil fauna is generally subject to less extreme temperatures, being insulated by the soil and by snow. In exposed areas, however, where strong winds prevent snow from settling, and in years of low snowfall, peat may freeze to a depth of 20 cm. In view of the long periods at very low temperatures involved in overwintering, it was surprising to find that many larvae of *Tipula paludosa* and *T. subnodicornis*, taken from the field in early autumn, died on exposure to −4° C (Horobin, 1971; Butterfield, 1973). A period of acclimatisation, during which the temperature was progressively lowered, however, resulted in a high survival rate at −4° C.

4.8.3 Life History Patterns

The vast majority of invertebrate species at Moor House appear to have no more than a single generation each year. Nelson (1971), discussing pit-fall captures of insects and a few other groups says: "As a rule there appears to be a single generation of adults annually ...". None of the 34 species of Homoptera studied by Whittaker (1965a) had more than one generation a year although this is frequent in lowland Britain. Hale (1966a) showed that most species of Collembola have an annual life cycle, each species having a characteristic period of egg laying, usually in the spring or autumn. Only in two species, *Dicyrtomina minuta* (Fab.) and *Dicyrtoma fusca* (Lucas) (Sminthuridae) was any evidence obtained of two life cycles per year under Moor House conditions. Although in most mite species studied the life cycle was less synchronised and some nymphs were found in most months of the year, Block (1966) nevertheless showed that there were two main types of life cycle, eggs being laid predominantly in the autumn and spring respectively. These patterns give rise to quite well marked annual peaks in the populations of Collembola and mites (Figs. 15, 16).

Most Tipulidae are univoltine at Moor House, including small species where the same or related members often have two generations a year at low altitudes. *Tipula rufina* also has two generations a year in lowland Britain but one at Moor House and in Iceland. It is probable that *Pedicia rivosa* L. takes two years to complete its life cycle at Moor House, there being two distinct sizes of larva present throughout the year and a single, synchronised emergence of adults.

The parthenogenetic enchytraeid, *Cognettia sphagnetorum*, dominant on the moor, reproduces throughout the year, though at a lower rate in winter, by fragmentation. A detailed account of its population dynamics is given by Standen (1973), while studies by Springett (1970) on other moorland Enchytraeidae suggest that *C. cognettii* (Issel), which took 321 days from hatching to maturity at 10° C, has a life cycle of over a year at Moor House (mean annual temperature 5° C). There is evidence of two types of development, unrelated to environment, in *Marionina clavata* N. and C. since at 10° C with identical conditions and excess food, six worms took 223 days and eighteen worms took 65 ± 3 days to reach maturity.

Some of the larger Carabidae species have life cycles taking two, three or even four years (Houston, 1970). In some *Carabus* species, e.g. *C. glabratus* Paykull, the larval stages last almost two years, whilst the adult beetle takes a further year to reach sexual maturity. From a study of mandible wear, Houston was able to show that three year groups of adult *Carabus* were present at one time, the oldest cohort consisting only of individuals which had completed breeding. The presence of a biennial life cycle at Moor House and an annual one below 250 m (820 ft) in the minute psyllid *Strophingia ericae* is due to the occurrence of two different physiological races (Parkinson and Whittaker, 1975).

Evidently some moorland invertebrates survive and succeed in the cold Moor House climate by extending their generation time. This is relatively simple in species whose lowland forms have two or more generations a year, but requires elaborate adaptations in the development of a biennial life cycle, with two over-

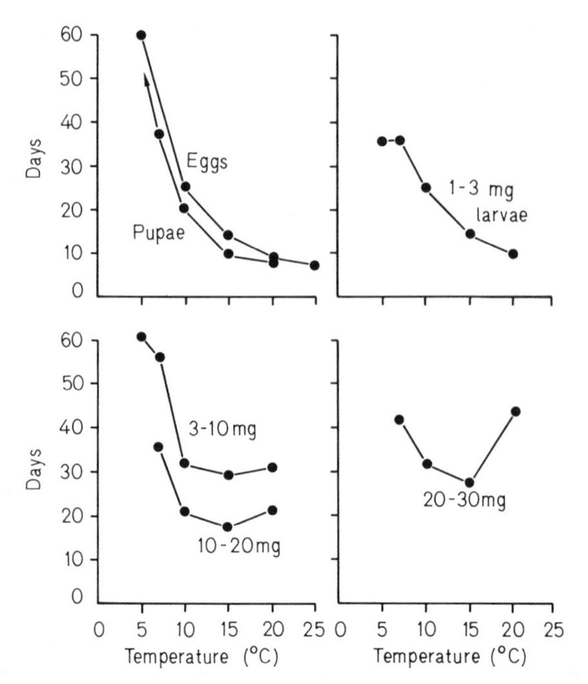

Fig. 11. Rates of development of *Tipula subnodicornis* at different constant temperatures in the laboratory (from Butterfield, 1976)

wintering stages, or a more rapid development relative to temperature, in species with an annual life cycle in warmer regions.

Growth and development of some carabid larvae and adults are greatly influenced by the photoperiod, short day length inhibiting moult and maturation of gonads (Houston, 1970). Thus, in a colder environment, delayed development results in a postponement of maturity, though not necessarily growth, until the following summer. Species which are autumn breeders at low altitudes may extend the life cycle and become spring or summer breeders at high altitude. This allows the vulnerable eggs to hatch before winter in both localities. Egg development time for species of Collembola at Moor House shows the typical inverse linear relationship with temperature (Hale, 1965). The developmental zero obtained by extrapolation is close to 0° C, lower than the 4° C recorded for lowland species. While there are dangers in such extrapolations, Hale concludes that a physiological mechanism probably allows eggs of moorland species to develop faster at low temperatures than those of lowland species.

In a detailed study on the physiology of development of *Tipula subnodicornis*, Butterfield (1976) showed that it is univoltine from high moorland to lowland bogs (e.g. New Forest). The life cycle and environmental temperatures indicate that the development time at Moor House would be twice that in the New Forest. At different, but constant, temperatures only eggs and pupae (Fig. 11) show the classic relationship between temperature and development rate. For larvae, the curve is flatter than expected and the optimal temperature for development decreases in larger larvae. As a result, the difference in the development time at temperatures

Table 9. April population densities of *Molophilus ater* at the peaty podzol site at Moor House

Year	1965	1966	1967	1968	1969	1970	1971	1972
Number m^{-2}	1,900	2,320	2,350	2,230	1,610	2,230	1,523	557

representing those of the New Forest and Moor House is less than expected. The development at Moor House further converges on the New Forest population through a shortday photoperiod effect which inhibits pupation. Thus the lowland larvae wait much longer for the extended day length in late spring.

4.8.4 Population Regulation at Moor House

Only four studies of animals at Moor House are of sufficient duration to permit analyses of population changes. Jordan (1962), Reay (1964), and Welch (1965) studied the moth *Coleophora alticolella* from 1952 to 1958 and again from 1961 to 1964. Whittaker (1971) compared the population dynamics of *Neophilaenus lineatus* (L.) at Moor House and at a lowland site at Wytham Woods, Berkshire, over a period of nine years. Coulson and his co-workers have studied population changes in *T. subnodicornis* and *M. ater* from 1965 to 1972 (Table 9).

C. alticolella and *N. lineatus*, widespread at low altitudes, are not well adapted to an upland life. The former has an upper altitudinal limit of little more than 600 m (2000 ft), determined by the upper limit of seed setting of the host plant, *Juncus squarrosus*, and the moth's limited mobility (Jordan, 1962). The correlation between the numbers of larvae and the ripening capsules (roughly two consumed per larva during its development) is close at high altitudes (Fig. 12), but less close at lower altitudes, e.g. 289 m (940 ft). At the lower altitudes larval numbers are less closely related to food availability and could well be regulated by parasitoids which are absent at higher altitudes. Whether or not the population at low altitudes is well regulated has not been demonstrated but in the upper part of its range it is certainly not, and is subject to wide fluctuations and frequent local extinction as seed supplies vary. The upper limit varied over 13 years between

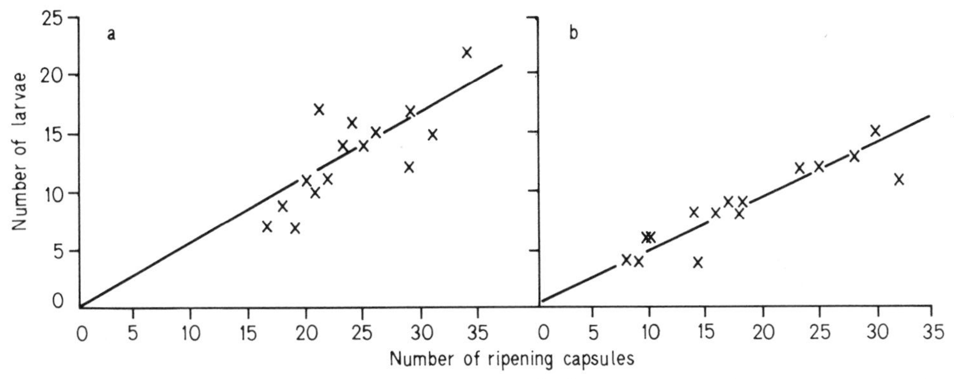

Fig. 12a and b. Relation between the food available (number of ripening capsules of *J. squarrosus*) and the number of larvae of *Coleophora alticolella* (a) at 290 m (950 ft), (b) at 365 m (1200 ft) (from Jordan, 1962)

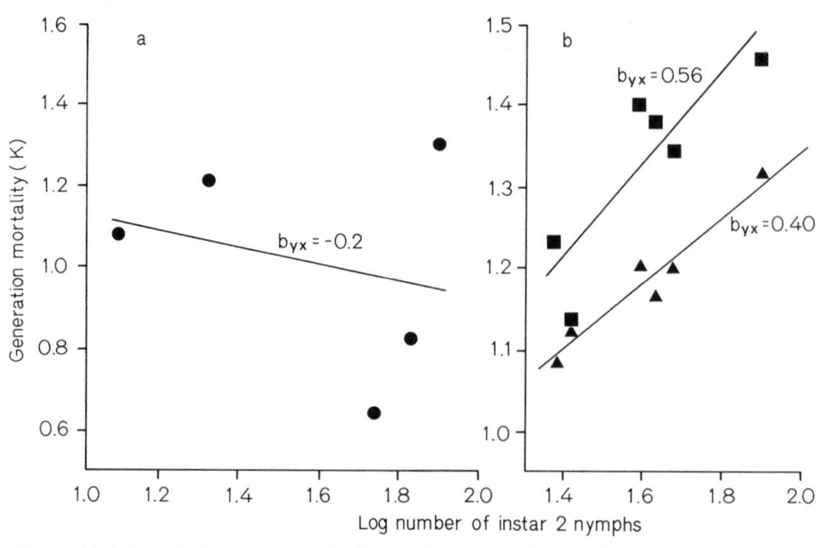

Fig. 13a and b. The relation between the logarithm of numbers of instar 2 in each generation and the subsequent "generation" mortality (K) in *Neophilaenus lineatus* (from Whittaker, 1971), (a) at the *Juncus* site, Moor House, (b) at Upper Seeds, Wytham Woods. ■: including parasitoid; ▲: excluding parasitoid

660 m and 510 m on one transect (Welch, 1965). Low temperatures were thought by Welch to contribute to the local extinctions, by delaying emergence of adults until too late to oviposit on *J. squarrosus* florets.

N. lineatus, a lowland species which can tolerate conditions up to 832 m, may similarly suffer local extinction at places where low temperatures so delay maturity that females are unable to mature and lay eggs before dying (Whittaker, 1965b). K-factor analysis shows that although it could be well regulated at low altitudes where it was heavily parasitised, there is no regulation at 550 m at Moor House (Fig. 13) and it suffers local extinction if climatic conditions prevent the life cycle from being completed. A positive slope significantly different from zero indicates regulation in these graphs. The coefficient of variation (100 x sd/arith mean) shows the relative instability of *N. lineatus* at Moor House (91.5) compared with Wytham Woods, Berkshire (47.5). Experimental evidence for the lack of regulation at Moor House was obtained by halving the density of one population and comparing its subsequent performance with that of a control. Although these two populations had been similar for the seven years previous to the reduction, they remained significantly different four years later.

Coulson and his co-workers recorded densities of eight stages in the life cycle of *Molophilus ater* from 1964 to 1972. The generation mortality (K) in each year on a podzol site at 550 m is plotted against log egg density in the same generation in Figure 14. The slope of 2.07 differs significantly from a slope of zero ($P<0.01$) indicating that during the period of the observations the population was well regulated. This is borne out by the low coefficient of variation of the population at all stages on this site, while on a gley site the population was subject to much wider fluctuations (Table 10) and regulation, if present, was less precise.

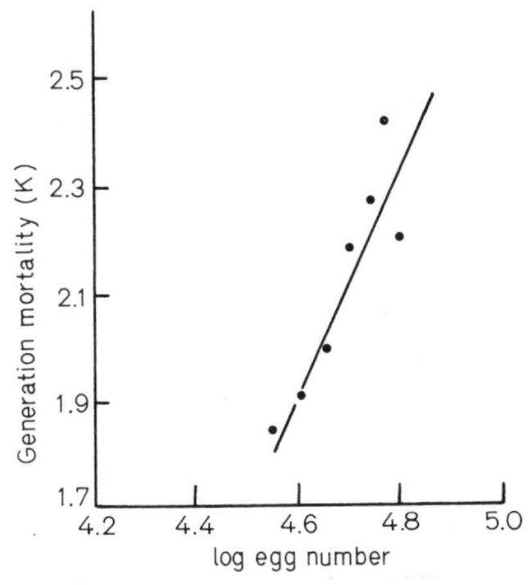

Fig. 14. The relation between the logarithm of numbers of eggs and the subsequent "generation" mortality (K) in *Molophilus ater* at Moor House (from Horobin, 1971)

Marked fluctuations in population density occur in *Tipula subnodicornis*. Apparently as a result of exceptionally dry conditions in June and July, 1955, causing a high egg mortality and an unusually high mortality in newly hatched larvae, the population crashed over most of the peat areas (Coulson, 1962). However, the species survived the drought in small, isolated pockets where the water content remained fairly high, e.g. in flush areas. These areas do not normally support a high population, but are clearly important in species survival, acting as reservoirs from which it can recolonise its typical habitat (Table 11). A similar drastic decline, but apparently not due to climatic conditions, in the population of *T. subnodicornis* on Knock Fell occurred between 1 and 24 March, 1972 (Butterfield, 1973) with a 45% mortality. The dead larvae showed dark markings on the cuticle and spiracular disc. Live larvae with the same markings died rapidly, and it is probable that the condition resulted from injury by other larvae. Despite these occasional vast changes in abundance, evidence suggests that density-dependent processes operate on the populations. Butterfield (1973) altered the population size of emerging adults, creating adjacent plots with different egg densities, comparable with the range found in the field (Table 12) and found that survival was associated with density. Similar experiments using larvae in culture chambers gave comparable results, survival decreasing from 25 to 0.2% with a ten-fold increase in numbers per chamber.

Density-dependent regulation also appears to occur in the overwinter mortality and in the fecundity of *T. subnodicornis*, so there are at least three stages in the life cycle where some degree of regulation can occur. There is no indication of a large reduction in numbers of eggs or larvae by parasites or predators. Adult emergence is so synchronised (Fig. 6) that their numbers swamp any predators, and Coulson (1956a) estimated that the Meadow pipit, the chief predator of adults, takes $<1\%$ of those available.

It is tempting to speculate that lowland populations, at the edge of their ranges at Moor House, are not well regulated, with frequent local extinctions and

Table 10. *Molophilus ater* population data from two Moor House sites in seven successive generations, 1964–1971

	Podzol			Gley		
	Mean density m^{-2}	standard deviation	$\frac{sd \times 100}{M}$	Mean density m^{-2}	standard deviation	$\frac{sd \times 100}{M}$
Egg	51,190	7,901	15.4	33,414	15,729	45.8
Instar 2	3,267	327	10.0	2,279	599	26.3
Instar 3	2,468	305	12.3	1,694	674	39.8
Instar 4 (autumn)	2,299	279	12.1	1,533	632	41.3
Instar 4 (spring)	2,023	320	15.8	1,257	660	52.5
Adults	1,917	587	30.6	1,125	615	54.6
Adult ♀	576	190	33.0	327	254	77.6
Calculated adult ♀[a]	406	102	25.0	262	143	54.8

Note: The ratio of the mean and the standard deviation for eggs and adult ♀ should, theoretically, be the same but the egg data include one earlier year than the female data.
[a] Assuming that all adult females produce 120 eggs (mean fecundity).

Table 11. The change in numbers m^{-2} of 4th instar larvae of *T. subnodicornis* between 1954 and 1955

	Site				
	A	B wet (valley bog)	C	D wet (flush)	E
1954	150	80	110	25	45
1955	0	4	0	12	0
Reduction	100%	95%	100%	48%	100%

Sites A, C and E were blanket bog areas with different vegetation types.

Table 12. The survival of eggs and larvae from experimental field plots with added adult *T. subnodicornis* (Butterfield, 1973)

Adults m^{-2}	Eggs m^{-2}	First instar m^{-2}	Survival rate (%)
6	1,910	780	41
40	12,760	1,360	11
320	102,000	7,300	7

marked density fluctuations. By contrast, those species which are well adapted to upland life may be much less affected by the harsh conditions and are better regulated. The idea is supported by circumstantial evidence. Hodkinson (1971) found some evidence of regulation in the upland psyllid *Strophingia ericae* and Whittaker (1965c) suggests that of three Homoptera on a *Juncus squarrosus* site at 550 m, the only species showing any density related mortality is *Macrosteles alpinus* Zett., the only upland species.

Table 13. The chemical composition (% dw) of moorland invertebrates and certain vegetation (data provided by Chemical Section, ITE, Merlewood)

A. Taken from specific habitats

Element	Enchytraeidae		Tipulidae larvae			Psyllidae
	Cognettia sphagnetorum		*Tipula subnodicornis*		*Molophilus ater*	*Strophingia ericae*
	Blanket bog	*J. squarrosus*	Blanket bog	*J. squarrosus*	Blanket bog	Blanket bog
C	46.3	48.1	47.2	49.0	51.2	—
N	11.4	11.9	11.5	10.5	10.6	9.9
P	1.06	1.12	0.90	0.88	0.26	1.75
Na	0.20	0.15	0.50	0.45	0.11	—
K	0.52	0.55	0.90	0.85	0.12	0.73
Ca	0.15	0.09	0.12	0.13	0.08	—
Mg	0.10	0.11	0.14	0.15	0.05	—

B. Taken from vicinity of streams. Insects collected by D.T. Crisp

Element	Adult Trichoptera (mixture of species)	Adult Ephemeroptera (mixture of species)	Adult *Tipula subnodicornis*	Adult small Empididae	Large adult Diptera excl. Tipulidae
C	—	—	—	—	—
N	10.6	9.9	9.4	9.0	9.3
P	0.84	0.86	0.77	1.00	0.81
Na	0.81	0.27	0.25	0.22	0.18
K	0.52	0.89	0.76	0.78	0.75
Ca	0.13	0.15	0.07	0.24	0.14
Mg	—	—	—	—	—

4.8.5 Mineral Requirements of Moorland Invertebrates

The limited information available on the chemical composition of moorland animals is given in Table 13. The figures are of interest in the light of the low levels of ions and cations in the rain, peat and moorland plants (Chap. 1). Further, several elements, although fairly abundant in peat, are not available to most organisms. This is particularly true for nitrogen which is trapped in stable organic compounds in the peat, and most plants may depend on the N in the rain or animal faeces.

While the chemical composition of many plants, e.g. *Calluna* (Gimingham, 1972) varies with the mineral content of the soil, moorland animals, with the possible exception of *Molophilus ater*, have concentrations typical of invertebrates generally (e.g. Uvarov, 1966). Considerably higher mineral concentrations occur in vegetation and peat on *J. squarrosus* areas than on blanket bog (Chap. 1) but larvae of *Cognettia sphagnetorum* and *Tipula subnodicornis*, collected from these habitats, show no such differences. The near constant chemical composition, and the high concentrations of P, K, and N (*c* ten times those in the vegetation and peat) evidently present in animals (Table 13) are probably necessary for physiological processes specific to animals, such as nerve impulse transmission and muscle function. The high concentrations of certain ions, together with the loss of N in particular in the excreta of herbivores and decomposers, indicate their need for a high degree of selectivity in order to obtain adequate amounts of the necessary ions. This aspect is at present under investigation.

Moorland animals and their excreta have high nutrient concentrations and decay rates, suggesting that they may be an important source of minerals to plants, concentrations being high and decomposition rapid. For example, the synchronised emergence and rapid death after one or two days of *T. subnodicornis* and *M. ater* adults puts back into circulation quantities of minerals in a very short period. Thus, 77% of all the production by *M. ater* is returned in two weeks in early June, and 33% of the production of *T. subnodicornis* in late May–early June, equivalent to 0.11 g N m^{-2} and 0.019 g P m^{-2} on blanket bog and 0.44 g N m^{-2} and 0.04 g P m^{-2} on *Juncus*. Large quantities of minerals are also recycled in September–October, when large numbers of tipulid larvae die.

4.9 Productivity

Since 1952, most of the numerically important groups have been studied at Moor House, and while the studies have not all been directed primarily towards productivity and energy flow, basic data on population size, habitat and life cycle allow energy flow and production figures to be calculated. Population size obviously varies between generations; the values in Table 3, considered typical for each group, are based on data of several years, though not all in the same period of time.

Productivity terms follow the definitions in Petrusewicz and Macfadyen (1970) and calculation of production has been made from detailed life tables wherever possible. In other cases it was determined from the mean standing crop and estimated mean population turnover rate. Respiration rates have been determined in the laboratory by appropriate methods: for mites, enchytraeids and small tipulid larvae in Cartesian divers, for psyllids by an electrolytic respirometer and for large tipulid larvae in a Warburg respirometer.

In the psyllids and enchytraeids the annual field respiration by the population was calculated from daily temperature measurements, the stage of development and a regression equation describing the change in respiration rate with temperature (Hodkinson, 1971; Standen, 1973). In both studies, this procedure was compared with the simpler method of using the mean annual population and mean annual temperature from field records (5.0° C) and the respiration rate at that temperature. This underestimated the enchytraeid population respiration rate by 14% and the psyllid rate by 20%. Population respiration of other animals has been estimated either from the summer and winter population sizes and the respiration rate at 5° C or from the regressions of respiration and production presented by McNeill and Lawton (1970).

A comparison of typical population densities and standing crops of animals on the four selected sites shows that earthworms and sheep dominate on mineral soils and avoid the bog (Table 3). Respiration and productivity data (Table 14)

Table 14. Respiration (R) and productivity (P) in kJ m^{-2} yr^{-1} on four main habitat types at Moor House

	Mineral soils				Peat soils			
	Limestone grassland		Alluvial grassland		*Juncus squarrosus*		Blanket bog	
	R	*P*	*R*	*P*	*R*	*P*	*R*	*P*
Lumbricidae	1,130	242	1,130	243	12.6	2.5	0.4	0.04
Enchytraeidae	456	113	577	142	515	130	243	63.0
Nematoda	25.1	14.6	25.1	14.7	29.7	17.2	13.0	8.4
Collembola	7.9	6.7	8.8	7.5	3.8	2.9	7.1	5.9
Acarina (herbivore and decomposer)	6.3	3.3	11.7	6.7	8.8	5.4	14.2	8.4
Tipulidae	335	385	498	573	244	159	66.1	48.1
Hemiptera	2.1	1.3	1.7	0.8	7.1	4.2	3.3	1.7
Lepidoptera	neg		neg		neg		neg	
Small rodents	neg		neg		neg		neg	
Grouse	0	0	0	0	0	0	23.4	1.3
Sheep	1,004	52.3	464	24.3	229	12.1	25.1	1.3
Subtotal herbivores plus decomposers	2,966	819	2,717	1,012	1,050	333	396	138
Araneida	0.8	0.4	10.0	7.5	8.4	6.3	3.8	2.5
Coleoptera	0.4	0.4	0.4	0.4	0.4	0.4	0.2	0.1
Acarina (pred)	13.4	7.9	25.9	15.0	9.2	5.4	8.4	5.0
Meadow pipit	0.8	0.04	0.8	0.04	0.8	0.04	0.8	0.04
Subtotal predators	15.5	8.8	37.2	23.0	18.8	12.1	13.1	7.6
Grand total	2,982	828	2,754	1,035	1,069	345	409	146
Assimilation	3,810		3,790		1,414		555	

neg: negligible.

Table 15. Energy flow m^{-2} through primary production, herbivores, decomposers and predators on four vegetation types at Moor House

	Mineral soils		Peat soils	
	Limestone grassland	Alluvial grassland	*Juncus squarrosus*	Blanket bog
1. Primary production ($g\ m^{-2}\ yr^{-1}$)	174 (above ground)		367 (above ground)	629
2. Corrected primary production ($g\ m^{-2}\ yr^{-1}$)	348	348	734	629
3. Energy equivalent of 2. ($kJ\ m^{-2}\ yr^{-1}$)	6,552	6,552	13,820	11,841
4. Corrected total primary production (kJ) (see text)	9,280	9,790	14,962	12,351
5. Herbivore and decomposer assimilation (kJ) $(R+P)=A$				
(a) total	3,785	3,729	1,383	534
(b) excluding sheep	2,728	3,240	1,142	482
6. Proportion of primary production assimilated	41% (58%)[a]	38% (57%)[a]	9.2% (10.0%)[a]	4.3% (4.5%)[a]
7. Predator assimilation $(R+P)=A$ (kJ)	24.3	60.2	29.9	20.8
8. Herbivore plus decomposer production (kJ)	819	1,012	333	138
9. Proportion of herbivore and decomposer production assimilated	2.9%	5.9%	9.0%	15.0%

[a] The figures in parenthesis ( ) are calculated from line 3.

indicate a progressively smaller total animal respiration from limestone grassland, through alluvial grassland and *Juncus squarrosus*, to blanket bog, where it is only one seventh of that on limestone grassland. There is a clear difference between mineral and peat soils in secondary production which, due mainly to a larger enchytraeid and tipulid population, is 25% higher on alluvial than limestone grassland, seven times as high as on blanket bog.

Estimates of the above-ground primary production on limestone grassland and *J. squarrosus* areas, and above-ground plus below-ground primary production on the blanket bog, are available (Rawes and Welch, 1969; Forrest, 1971; Chap. 2). To estimate total production in the limestone grassland and the *Juncus* areas, below-ground production is assumed to be equal to that above ground. It is evident from incomplete studies that the primary production on alluvial grassland is similar to that on limestone grassland and, in the absence of further detailed studies, the limestone grassland value has been applied to both sites (Table 15).

While primary production is considerably higher on peat than on mineral soils, secondary production is appreciably less. The proportion of the primary production assimilated by animals is five to ten times greater on the mineral soils than the peat soils (Table 15). Sheep show a marked preference for the limestone grassland and graze it to a very short turf. Nevertheless they assimilate only about one third of the total assimilation by animals, viz about 16% of the total primary production. However, this value is equivalent to ingestion of *c* 40% of the total and 80% of the above-ground production.

At this stage it is worthwhile considering the methods used to calculate primary production. On grassland the vegetation is cut at frequent intervals within mobile exclosures which keep out sheep and other large herbivores, but not invertebrate grazers such as Homoptera and tipulids. These consume a part of the primary production not accounted for by summing the increments harvested throughout a season. Thus the energy value of the plant material removed by invertebrate herbivores must be added to the value estimated by harvesting, to obtain the true primary production. Similarly, on blanket bog, where sheep and grouse are not excluded, their consumption must be added to harvested production to obtain the primary production.

Whilst measurements of ingestion by all herbivores are not available, the assimilation/ingestion ratio has been determined at 30% for tipulid larvae, the main herbivore after sheep. By assuming a similar value for plant feeding mites and Collembola, it is possible to estimate the shortfall in primary production values. These are small (4%) on the blanket bog and the *J. squarrosus* (10%), but considerable on the alluvial and the limestone grassland (34 and 51% respectively) (line 4, Table 15). The effect of this adjustment is to reduce the proportion of primary production assimilated by the fauna to 41 and 38% on the limestone and alluvial grasslands respectively, but no appreciable change is caused on the two peat sites (9.2% on *Juncus* and 4.1% on blanket bog). Despite these corrections the effect of the fauna on the grasslands remains very high.

Whilst there is no clear distinction between herbivores and decomposers, lumbricids and enchytraeids are evidently the most important decomposers, followed by some of the nematodes. Apart from the decomposer *Molophilus ater*, most common tipulids are herbivores, and non-predatory mites and Collembola are both herbivores and decomposers, in unknown proportions. Thus on the mineral soil, assimilation by herbivores is about matched by the decomposers although, since sheep and tipulids consume so much of the primary production, much of the decomposer substrate must be in the form of animal faeces rather than uningested plant remains. Tentative calculations indicate that decomposers have about 60% of the organic matter available in the form of herbivores' faeces, mainly sheep dung. This conclusion agrees closely with the known importance of dung as food for earthworms (Svendsen, 1957b) and with the speed of decomposition of dung by animal activity (White, 1960a).

On the peat soils earthworms are much less important and are replaced by enchytraeids, sheep and tipulids and, to a lesser extent, nematodes as the main animals groups utilising the primary production. Partly because *M. ater* is primarily a decomposer and accounts for *c* 30% of the total tipulid assimilation, the proportion of assimilation due to decomposers increases from 50% on the grasslands to 60% on *Juncus* and 75% on *Calluna*.

The predator component is small (Tables 14, 15) and the proportion of the herbivore and decomposer production assimilated is low, especially on mineral soils. Predation of the few frogs and shrews is, as far as can be determined from the limited studies, negligible. In view of the lower density of potential prey on blanket bog, it is surprising that a higher proportion of its herbivore and decomposer production is assimilated by predators.

Lindeman's efficiency between the second and third trophic levels, i.e. grazer to predator transfer, ranges between 0.5 to 1.5% on the Moor House mineral soils and 2.5 to 4.0% on the peat soils, low values compared with other published studies. The efficiency of exchange between successive predator trophic levels would appear to be even lower, probably well below 1%. Growth efficiency is also rather low, varying between 22 and 27% for herbivores on the four vegetation types and between 38 and 39% for predators.

The utilisation of primary production is obviously very different on peat and mineral soils (Table 15). If the high utilisation of the primary production on mineral soils, particularly the proportion consumed by herbivores, also occurred on peat, the nature of the areas now dominated by blanket bog would be markedly different, and the accumulation of organic matter would not occur. Possible reasons for the low utilisation of blanket bog production are discussed in Section 4.10.

4.10 Conclusion

The general picture of the moorland fauna is one of scarcity of vertebrates other than sheep and grouse, generally low numbers of species and individuals of above-ground herbivores, and few parasites and top predators. On the other hand, the soil and litter fauna has at least as high a biomass as that of many lowland soils and some of the groups are rich in species. Below-ground herbivores (root feeders) and decomposers are the most important fractions of the fauna.

Blanket bog and its associated vegetation does not form a uniform cover over the underlying mineral soils. Peat varies in depth and in a number of places, such as along the sides of streams, it is absent and the mineral soil has a typical grassland vegetation. This and other sources of variation result in a soil and vegetation mosaic which is reflected in the distribution of the animals. The difference between the fauna on peat and that on mineral soils is due largely to the different origins of the two groups. The former is derived mainly from subarctic regions where peat soils are more abundant, and the latter is a poor lowland grassland fauna, further restricted by the inability of some species to cope with the wet and cold climate at high altitudes.

Adult insects, which form the main source of food for many predatory species, have a very different pattern of emergence on the two soil types. The short, active season of the subarctic results in the majority of insect species using the spring rise in temperature or lengthening photoperiod to initiate pupation. A similar situation on the peat areas at Moor House results in a vast "spring" emergence with little additional emergence throughout the rest of the year. In contrast, the insects associated with mineral soils show a much more even distribution of emergence during the six months which form the Moor House "spring", "summer" and "autumn". Above-ground predators on the blanket bog need to be either specialised in their feeding methods or mobile, moving off the peat areas after the spring "bloom" of food and feeding on the mineral soils during the remainder of the year. There is evidence that predators are more abundant along mineral/peat soil margins, probably because this situation offers a more stable food supply. Thus whilst the proportion of predators to herbivores is low at Moor House (whether this is measured in number of species, biomass or production), it is evident that it would be even lower were it not for the mosaic of soil and vegetation which adds variety and perhaps more stability to the system.

The low proportion of predators and the small extent of parasitism at Moor House is probably the result of the colder climate and the considerable daily fluctuations in the weather. The duration of precipitation at Moor House is much greater than would be expected from the annual rainfall since, compared with lowland areas, much falls as mist and drizzle. Such conditions are unsuitable for the intensive searching necessary for parasitic insects to find sufficient numbers of hosts to ensure the next generation. The dry, warm conditions necessary for activity in the Hymenoptera occur too infrequently and too irregularly and this almost certainly accounts for their virtual absence from the Moor House fauna. Since most of the dominant animal groups suffer relatively little predation or parasitisation, their numbers must be regulated by other means.

The extension of the zoological investigations to three other soil and vegetation types in addition to blanket bog has been informative. It is evident that the distribution of herbivores and decomposers does not parallel the primary production, which is as large, or even larger on the bog than on neighbouring mineral soils. On the mineral soils sheep play an important part in consuming the aboveground primary production, which is almost completely utilised by animals, leaving very short turf. Similarly, the extensive soil fauna accounts for the ingestion of much of the below-ground primary production. Over a third of the primary production is assimilated by the herbivores and the remainder by the decomposers, including the microorganisms.

On the blanket bog the assimilation is lower and much of the vegetation is not ingested. It is evident from the smaller numbers of herbivores and decomposers that many animals find the vegetation on the blanket bog unattractive or unsuitable as a source of energy and minerals. For example, sheep clearly prefer the mineral soils (Rawes and Welch, 1964). Blanket bog vegetation is comparable in energy content with that on mineral soils, but has lower digestibility and mineral content. The bog plants grow on peat which is incessantly leached by rain, is low in available nutrients, and is generally too deep to allow root penetration to the underlying mineral soil.

An examination of the chemical composition of animals living on a peat area compared with those living on mineral soils at high and low altitudes shows no consistent differences in their chemical composition. Thus the animals feeding on blanket bog have to deal with a much greater mineral "gradient" to meet their physiological needs. Highly selective feeding and the rejection of much material with a suitable calorific content but low in minerals, particularly Ca, K, P, and N, is known in sheep (Rawes and Welch, 1964). The Red grouse has been shown to select *Calluna* with greater mineral content (Moss, 1972). It is not impossible that similar selection is carried out by other animals, resulting in an appreciable proportion of the primary production remaining uningested and unsuitable for both animals and microorganisms.

Acknowledgements. Credit for the conception and development of much of this zoological study belongs to Professor J. B. Cragg. He initiated biological studies at Moor House in 1951 and played a major part in advising and directing many of the studies which form this paper.

The role of the authors has been to collate and synthesise the many studies which have been made at Moor House. It is a pleasure to acknowledge the enthusiasm and willing assistance given by all of the persons approached for information; in few other cooperative studies could so much good will exist. However, we accept full responsibility for the interpretation presented in this paper and for many of the calculations.

We gratefully acknowledge the assistance of and the opportunity to refer to work by: W. B. Banage, W. C. Block, V. M. Brown, J. E. L. Butterfield, J. B. Cragg, J. M. Cherrett, D. T. Crisp, L. Davies, D. Evans, P. R. Evans, G. I. Forrest, W. G. Hale, M. J. Hadley, O. W. Heal, J. C. Horobin, I. D. Hodkinson, W. W. K. Houston, A. M. Jordan, P. M. Latter, J. M. Nelson, J. E. Peachey, M. Rawes, R. C. Reay, G. R. J. Smith, J. A. Springett, V. Standen, J. A. Svendsen, D. Welch and E. White.

4.11 Appendix: Population Fluctuations and Accuracy of Estimates

One of the major problems facing investigators of community phenomena is that the component populations are not stable and change seasonally and from year to year. Moreover, it is often not feasible to study all the component organisms simultaneously. In this study, the earliest quantitative work was carried out in 1953 and the latest in 1972. In some cases we have frequent estimates over long periods of time and in others we have to rely on a single estimate of population or biomass. What we do not know is whether the populations fluctuate randomly with respect to each other. The best we can do, therefore, is to use average population figures as if these are additive and to indicate the extent to which the populations studied have varied, and hence the validity of the mean value used. We must recognise that it is conceivable, but unlikely, that all the population maxima or minima may correspond so that the average figures used could be an over- or underestimate of reality. From the evidence available, there are many species in which maxima and minima did not correspond in the years studied. The types of data for the most abundant groups or members of a group are summarised here.

Collembola population estimates were made from monthly samples using a Macfadyen high gradient extractor over the period March 1960 to November 1961 (Hale, 1966b). Although individual species fluctuated a great deal, total Collembola counts remained remarkably constant (Fig. 15). On the four main sites the maximum biomass was usually only four times the minimum: limestone grassland 0.45–0.13 g m^{-2}, alluvial grassland 0.54–0.19 g m^{-2}, *Calluna* litter 0.47–0.10 g m^{-2} and *Juncus squarrosus* 0.24–0.06 g m^{-2}.

Mite samples were taken by W. C. Block at monthly intervals between January 1961 and December 1962 and extracted using the same equipment as for the Collembola (Block, 1966). Figure 16 shows the seasonal fluctuations in total mite numbers and indicates the degree of variation from one year to another. Figure 17 shows the seasonal fluctuations in biomass of mites on limestone grassland and blanket bog. They vary by a factor of less than three from minimum to maximum on the grassland and by a factor of about five on the bog. Production and respiration by mites and Collembola are even more constant than these figures would suggest because most production and respiration takes place in the summer months when densities and biomass were comparable in the two years.

Enchytraeidae have been studied in detail by Peachey (1963), Springett (1970) and Standen (1973) over the periods 1956–57, 1962–63 and 1968–70 respectively. Figure 18 shows the fluctuations of the most numerous species on *Juncus* moor, *Nardus* grassland and limestone grassland. Errors on sample estimates are low. On blanket bog, *Nardus* grassland and *Juncus* moor, the numbers of the commonest species, *Cognettia sphagnetorum*, were relatively constant over the two year study period. Production of *C. sphagnetorum* was calculated by Standen (1973) over a two year period using a mathematical model which allows incorporation into the estimate of fragmentation and regeneration of this species. Population respiration was obtained by relating laboratory Cartesian diver results at different temperatures to

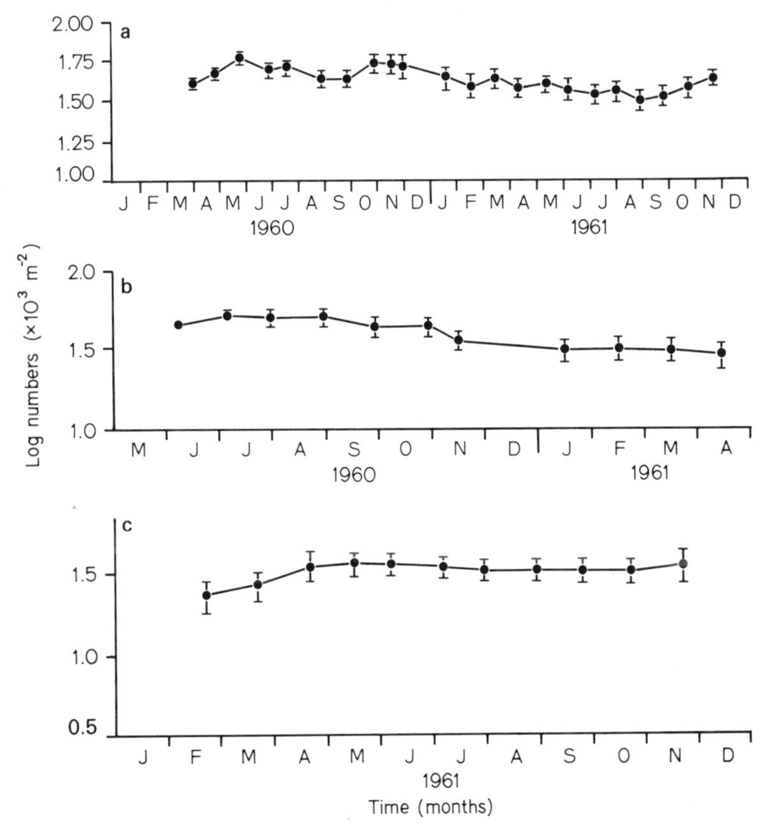

Fig. 15a–c. Three monthly running means and standard errors of the means of total Collembola numbers at Moor House (from Hale, 1966b). (a) On limestone grassland, 1960–1961, (b) on alluvial grassland, 1960–1961, (c) on *Calluna* litter, 1961

the field population counts over the two year period and the mean soil temperature between sample dates. Standen also calculated production for a sexually reproducing enchytraeid, *C. glandulosa* (Michaelson).

Data on nematode populations (Fig. 19) were collected by Banage (1960) over a two-year period. He gives sufficient information for these to be converted to biomass.

Tipulidae have been studied extensively by Coulson and his co-workers. Three species are found commonly on mineral soils. *Tipula paludosa* occurred at final instar densities of 57 m^{-2} in 1965 and 70 m^{-2} in 1971. Less extensive information indicates that densities of this order occur in most years, both on the limestone and the alluvial grassland. In contrast *T. pagana* fluctuates markedly in numbers, with densities up to 300 ± 15 m^{-2} recorded in 1967, dropping to 35 ± 2 m^{-2} by 1971. Less extensive samples at other sites suggest that the pre-adult population is usually below 100 m^{-2}. *T. varipennis* has a patchy distribution, generally more abundant on alluvial than limestone grassland, with local high densities. An overall density of 56 ± 1 larvae m^{-2} was estimated from 125 samples over an extensive area in 1971.

Two species of Tipulidae, *T. subnodicornis* and *Molophilus ater*, are the dominant species on blanket bog; other species, such as *Ormosia pseudosimilis* Lundstroem, *Limnophila pulchella*, *Pedicia immaculata* and *Limnophila meigeni* Verrall are relatively unimportant.

In addition to the data on *Molophilus ater* given in Tables 9 and 10 extensive information has been collected on its abundance on other sites and the densities used in Table 4 can be regarded as typical.

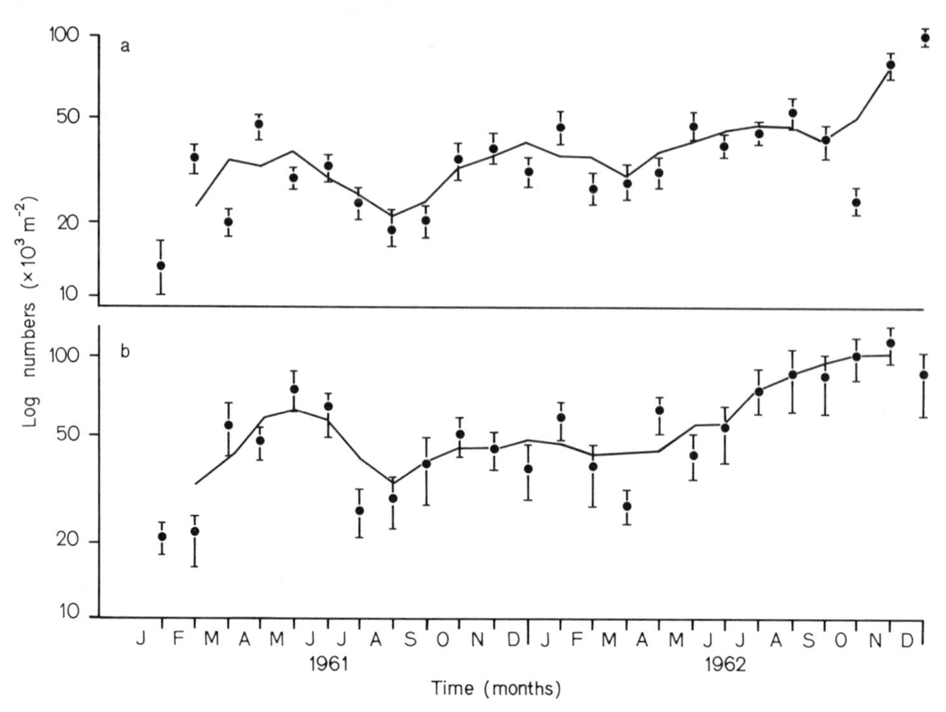

Fig. 16a and b. Seasonal fluctuations in the numbers (mean ± se) of total soil Acarina at Moor House. Three-point running mean values are indicated by the lines (from Block, 1966). (a) On limestone grassland, 1961–1962, (b) on blanket bog, 1961–1962

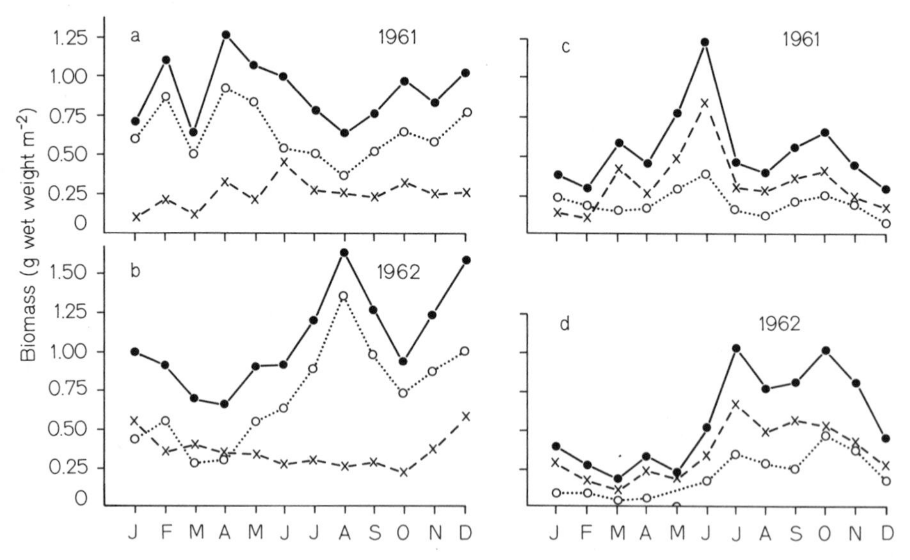

Fig. 17. Seasonal fluctuations in the biomass of mites on limestone grassland and blanket bog, 1961–1962 (from Block, 1966). × : Cryptostigmata; ○: Mesostigmata; ●: total Acarina

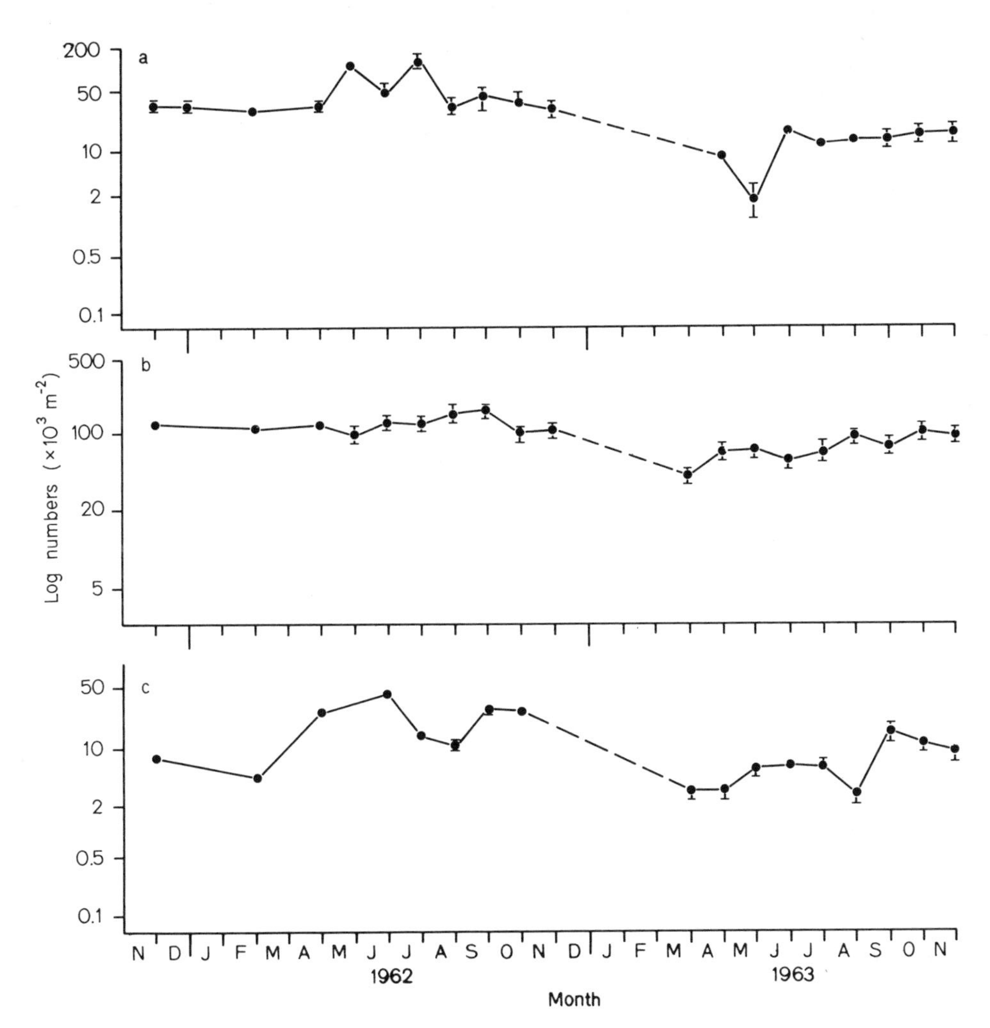

Fig. 18a–c. The seasonal variation in mean density (±se) of some enchytraeid species at Moor House (from Springett, 1970). (a) *Marionina clavata* on *Juncus squarrosus* moor; (b) *Cognettia sphagnetorum* on *Nardus stricta* grassland; (c) *Fridericia* species on *Festuca-Agrostis* grassland

The population size of *T. subnodicornis* has been discussed by Coulson (1962) and more recently by Butterfield (1973). In general, the population levels on *Juncus squarrosus* sites have not returned to those found prior to 1956, although on some sites densities as much as three times the 1954–55 densities have been found. The densities of *T. subnodicornis* on blanket bog have remained consistently lower than those on *Juncus* sites.

Smith (1973) has calculated production and respiration for *Molophilus ater* using the population data of Coulson (1956a), Hadley (1969, 1971) and Horobin (1971) and laboratory studies. Production of *T. subnodicornis* has been calculated for a complete life cycle in 1954–55.

Grassland Homoptera were sampled at approximately weekly intervals from 1961 to 1963 (Whittaker, 1963). Studies continued for a further seven years in the case of some species (Whittaker, 1971). Densities are rather variable from year to year (e.g. Fig. 20) and the average

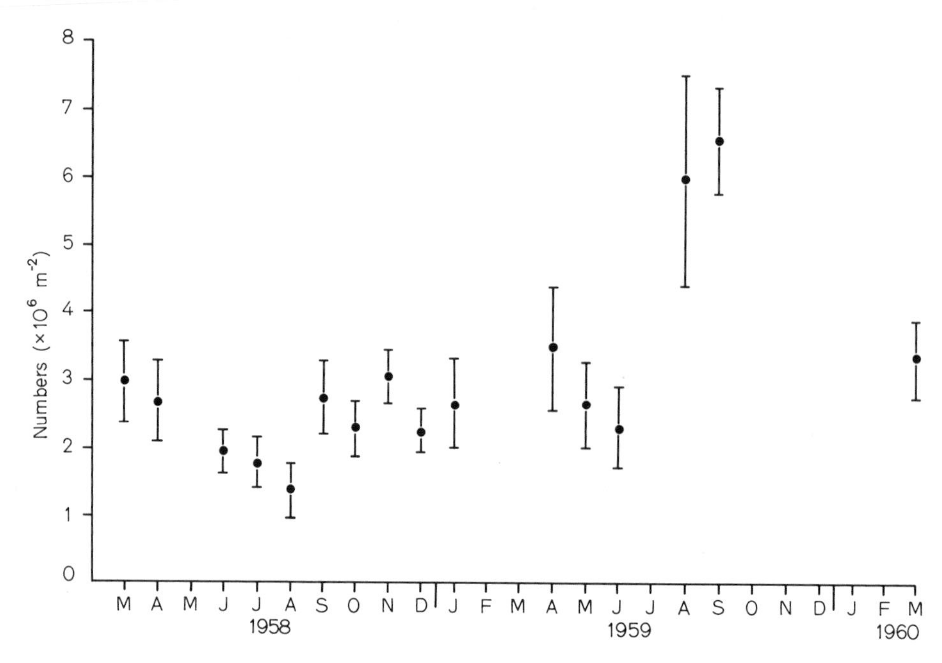

Fig. 19. Seasonal variation in mean numbers (±se) of Nematoda (from Banage, 1960)

Fig. 20. Numbers m^{-2} of second instars of *Neophilaenus lineatus* at Moor House from 1964 to 1969 (from Whittaker, 1971). ○: *Juncus* site; ●: *Nardus* grassland

figures for population density, production and respiration tend to have large errors (Tables 3 and 14). Hodkinson (1973) made a detailed population and production study of *Strophingia ericae* from 1969 to 1971. He showed that the density could vary by a factor of at least eight from one year to another although individual population estimates had errors <20% of the mean. The population density was recorded on three sites at Moor House by samples taken at approximately monthly intervals and sufficient data are available for the calculation of energy budgets for two years on the main study site (Sike Hill) and for one year on two supplementary sites.

Earthworms were studied in 1952 (Svendsen, 1957a) by digging and hand-sorting soil cores (0.04 and 0.028 m^2 × 20 cm deep), by examining dung and by searches of the soil surface and litter at night. Fifty three sites along the banks of three streams were sampled twice in the summer. The estimated density from this technique agrees well with an independent estimate using a different technique on alluvial areas in Teesdale (Crisp, pers. comm.).

Spider densities given by Cherrett (1964) were obtained by visual counts of webs and retreats, pitfall traps and by extraction from 0.025 m^2 turfs using a lateral heat extraction apparatus.

Coleoptera densities are estimated from the records of Houston (1970) and Nelson (1971) mainly by trapping out known areas.

These are the data from which our estimates of mean density and production have been made. In most cases the standard errors of the individual means of population density are quite low (see Figures), but the fluctuations in some of the populations from year to year introduce a large error, particularly since the studies are spread over about 15 years and we do not know how these fluctuations were phased with respect to one another. Fortunately, however, the most abundant populations (in the soil and litter) fluctuate least (Figs. 15–19).

For these reasons we have made no attempt to put errors on our final estimates of mean density, production and respiration in Tables 3, 14, and 15.

5. Microbial Populations in Peat

V. G. Collins, B. T. D'Sylva and P. M. Latter

A survey of the bacterial and fungal populations provides a basis for understanding the dynamics of decomposition within the blanket bog ecosystem.

5.1 Introduction

Preliminary studies of the microbiology of some Moor House sites (Latter et al., 1967) revealed the restricted range of taxonomic groups and low numbers and biomass in acid waterlogged peat compared with more base-rich mineral soils. These features were investigated more extensively in a series of IBP projects, to determine the distribution of bacteria and fungi in Moor House peats, the potential activity of selected groups in relation to nutrient cycling and the taxonomy of those groups participating in the nitrogen and sulphur cycles, or in the hydrolysis of some naturally occurring organic compounds. A description of the numbers and composition of the microbial population, and its variation in space and time, is given in this chapter.

Aerobes, facultative anaerobes and obligate anaerobes were studied as components of the heterotrophic bacterial population. Analyses of aerobes, obligate anaerobes and Actinomycetes were by A. J. Holding and N. J. Martin, and of facultative anaerobes, viz those bacteria capable of growth and activity under the reduced oxygen regime known to exist in the peat profile, by Collins and D'Sylva. Studies on selected biochemical groups of bacteria were carried out by Collins and D'Sylva, and of fungi by Latter. Identification of isolates and laboratory investigations of response to environment were carried out by all authors.

Between-site variation in the microbial population in relation to site, depth in the profile, vegetation cover and season, and results of the laboratory studies, were used to interpret observed between- and within-site variations in activity.

5.2 Methods

5.2.1 Field Sampling Procedures

Samples of peat were taken from the Bog End, Sike Hill, Cottage Hill A and B, Green Burn and Burnt Hill sites at Moor House (Chap. 1). Unless stated otherwise, methods and procedures apply to samples from Bog End, the most intensively studied site.

Peat cores (50 cm × *c* 0.01 m^2 cross section) were extruded from the corer in the field, wrapped in polythene and taken to the laboratory; some were stored at 4° C. The depths of the four colour horizons (Chap. 1, Table 1) were measured, and the cores sampled aseptically at measured depth intervals. Peat subsamples were weighed, diluted and inoculated into the appropriate bacteriological and mycological media. Numbers of cores sampled are given in Table 3.

5.2.2 Heterotrophic Bacteria

Aerobic bacteria were counted from mechanically macerated peat samples serially diluted with 0.01% peptone, and inoculated into pour plates of Casein Peptone Starch (CPS) medium (Collins and Willoughby, 1962) as modified for peat studies by Martin (1971).

Facultative anaerobes and microaerophilic organisms were counted from peat samples hand-shaken in dilution tubes and inoculated into pour plates of Tryptone Soya Agar (TSA) and CPS with an overlay of 1.5% plain agar. Anaerobes and facultative anaerobes were estimated from macerated peat serially diluted with phosphate peptone cysteine HCl and 25% soil extract (Siwasin, 1971), incubated in anaerobic jars. All plates were incubated for 14 days at 22° C. Direct counts of bacteria were made on four soil films (Jones and Mollison, 1948) prepared from each of one to three samples of litter or peat. Six microscope fields were examined on each film using a × 100 oil immersion objective.

5.2.3 Dilution Counts of Selected Physiological Groups of Bacteria

Proteolytic bacteria, i.e. those capable of liquefying gelatin, were counted in a gelatin medium, nitrifying bacteria in a medium containing NH_4^+ salts and nitrate-reducing and denitrifying bacteria in the liquid-medium of Stanier et al. (1966) containing NO_3^- as the main nitrogen and energy source.

Nitrogen-fixing bacteria were counted at 25° C on the nitrogen-free medium of Norris (1959) or the mineral salts nitrogen-limited medium of Biggins and Postgate (1969). Counts were made aerobically, or anaerobically with the air in the incubator replaced by CO_2.

Thiosulphate-oxidising bacteria were isolated from a liquid medium of Starkey (1934) for aerobes and of Baalsrud and Baalsrud (1954) for anaerobes. After incubation at 30° C for 15 days, thiosulphate utilisation was detected by titration of 10 ml of the medium against N/10 iodine solution (BDH standard); the pH of the medium was also recorded. Sulphate-reducing bacteria were isolated from the liquid medium and the lactate agar medium B of Starkey (1938). After eight weeks' incubation at 30° C, sulphate reduction was detected by the odour of H_2S and the production of a black ferrous sulphide precipitate. Starch-hydrolysing bacteria were isolated on plates of nutrient agar containing 0.2% (w/v) soluble starch. Starch hydrolysis was recognised by the occurrence of cleared zones around colonies after flooding the plates with iodine.

Cellulolytic bacteria were counted on enriched Cellvibrio medium (Manual of Microbiological Methods, 1957) and later transferred to mineral salts agar containing finely precipitated cellulose (Skerman, 1967). Finally, chitinolytic bacteria were estimated on chitin agar media (Campbell and Williams, 1951; Willoughby, 1968).

5.2.4 Actinomycetes, Yeasts and Fungi

Actinomycetes were counted on dilution plates of chitin agar (Campbell and Williams, 1951) and yeasts on Sabouraud Dextrose Agar.

The frequency of occurrence of fungi was estimated by recording colonisation of washed peat particles of *c* 0.3 mm diameter, plated on Czapeks-Dox and Malt agar. Direct counts of stained and unstained fungal material were made using a × 40 bright-field objective by the Jones and Mollison (1948) technique described for bacteria in Section 5.2.2. The method is described by Latter and Cragg (1967).

5.2.5 Response to Environmental Factors

The effect of different temperatures, oxygen and carbon dioxide concentrations and pH on bacterial and fungal growth has been examined in a number of studies. The physiological characteristics of the heterotrophic bacterial population were investigated on a random sample of 240 of the 1500 pure culture isolates obtained from the primary isolation plates of peat from various depths of the soil profile (see Sect. 5.2.2). Growth of bacteria at low temperatures was assessed visually on streak agar plates of TSA, incubated at 0–2°, 5°, 7–8°, 12°, 16°, 20°, and 25° C. The experiments used to investigate the effect of low temperature on fungal growth are described by Heal et al. (1967) and Latter and Heal (1971).

Bacterial growth in, and tolerance of, varying oxygen tensions was tested by inoculation into deep agar stabs of TSA with 0.9% agar incubated at 25° C, and also in liquid CPS medium in a continuous culture chemostat. Bacteria were counted on spread plates of CPS agar incubated at 10° C for 15 days.

Response of the predominant fungi to varying proportions of oxygen and carbon dioxide was tested by screening in deep Czapeks-Dox agar (0.8%) tube cultures and also by measuring the dry weight of isolates from liquid CDA cultures incubated in the gas mixtures for 10 days at 25° C. In order to examine fungal response to pH, isolates used in the temperature study were grown on Czapeks-Dox agar at pH 4.5, 5.5, and 6.5, adjusted with sterile HCl after sterilisation of the medium.

5.3 Results and Discussion

5.3.1 Field Sampling of Bacteria, Replication, Credibility and Error

To minimise the possibility of contamination of lower horizons with material from above, samples for analysis were taken from the centres of cores; the distribution of numbers in relation to depth (Tables 1–3) is, therefore, probably realistic (see Chap. 1 for description of horizons). A number of other points must be taken into consideration when interpreting the data, however. The method of sample preparation and the selectivity of the medium affect actual numbers of colonies

Table 1. The numbers ($\times 10^5$ g^{-1} $\pm$ se) of aerobic and anaerobic heterotrophic bacteria in the different horizons of peat from Bog End

Horizon	Aerobic	Anaerobic	Ratio Aerobic/anaerobic
Litter	260 ± 66	5.9	44
Dark brown	110 ± 15	9.3	12
Green-brown	76 ± 19	6.3	12
Red-brown	15 ± 4	0.5	30

rather than their distribution pattern (Table 8), but to reduce the possibility of obtaining misleading results, a wide range of selective media and incubation procedures was used. Further, the growth of organisms under laboratory conditions may merely reflect their ability to survive the adverse conditions e.g. anaerobiosis of the horizon from which the sample was taken. Table 3 provides a general description of the distribution and size of the Moor House blanket peat bacterial population and the data for potential nitrogen fixers show the precision obtained with the chosen sampling regime and indicate the degree of credibility in support of the interpretation.

5.3.2 Distribution and Taxonomy of Heterotrophic Bacteria

The distribution of bacteria in the peat (Table 1) shows the aerobic population to be much greater than the anaerobic in all horizons. The counts were very variable throughout the profile, numbers in the green-brown horizon sometimes exceeding those from the litter, but in general, while the aerobic population decreased with depth, the anaerobes showed no marked reduction until the red-brown horizon. The ratio of aerobic:anaerobic populations decreased abruptly below the litter. Numbers of facultative anaerobes recorded on the double layer plates increased down the profile, with highest numbers in the green-brown and red-brown zones (Table 3). Had only aerobic counts been carried out, this facultative to microaerophilic population would have been missed—an indication of the importance of methodology in peat soils.

5.3.2.1 Aerobes

The dominant bacteria in the upper two horizons were Gram-negative rods and in the lower horizons *Bacillus* spp, other groups clearly representing only a small variable fraction of the population (Table 2A). The majority of the Gram-negative rods in any horizon belong to the aerobic genera: *Pseudomonas*, *Alçaligenes* and *Acinetobacter-Moraxella* complex, with only an occasional facultative anaerobe, but the oxygen requirements of the *Bacillus* population appear to vary with depth.

In the litter 75%, and in the dark brown 68% of strains tested were strict aerobes, while all those isolated from the green-brown were facultative anaerobes. Strangely, in the red-brown, only 18% of strains were facultative anaerobes. A

Table 2. The numbers of different groups of heterotrophic bacteria, expressed as a percentage of the total isolates, from the different horizons of peat

A. Aerobes	*Bacillus* spp	Gram-negative rods	Gram-positive cocci	*Arthrobacter-Nocardia* spp
Litter	13 (100)[a]	84	0	3
Dark brown	46 (24)	42	13	0
Green-brown	84 (30)	10	6	0
Red-brown	80 (44)	18	0	0

B. Anaerobes	Facultative	Obligate	"Unknown"
Litter	43	24	33
Dark brown	23	24	53
Green-brown	27	39	35
Red-brown	38	10	52

C. Obligate anaerobes	*Actinomyces*-like	*Clostridium* spp	"Vibrioid"
Litter	84	3	13
Dark brown	74	21	5
Green-brown	87	13	0
Red-brown	75	25	0

For *Bacillus* spp the percentage of the *Bacillus* population present as spores is given in brackets
[a] No simultaneous count for vegetative cells and spores. Almost certainly an overestimate.

high proportion of the *Bacillus* isolates from aerobic plate counts appeared to belong to the *Bacillus* circulans complex (i.e. Group II of Smith et al., 1952), mainly Gram-variable and with swollen sporangia. They could be divided into four groups on the basis of sporangial size, spore morphology and colonial growth characteristics on nutrient agar. The *Bacillus* isolates in this group showed slow growth and activity in biochemical response to various tests and enrichment procedures in culture, possibly indicating that their activity in peat is restricted, perhaps by lack of suitable growth factors. It is of interest to note that the *Bacillus* circulans complex is also a dominant type in arable soils.

The proportion of heat-resistant *Bacillus* endospores, investigated by pasteurising a soil sample at 75° C for 15 min, was surprisingly high in the litter zone, where active decomposition would be assumed to lead to a higher proportion of vegetative cells than in lower horizons. The limited data, however, preclude the forming of firm conclusions.

It is also interesting to note that activation of superdormant spores by Ca-dipicolinate (Busta and Ordal, 1964) led to an increase in the endospore count equal to about 55% of the *Bacillus* spp which grow with normal plating procedures, but only the green-brown and red-brown horizons were tested by this procedure.

5.3.2.2 Anaerobes

Additional data on the distribution of facultative and obligate anaerobes (Siwasin, 1971) are given in Table 2B.

It could be assumed that colonies of bacteria which could not be successfully subcultured from the original isolation plates must be largely anaerobes, since other bacteria, including facultative anaerobes, could be subcultured. It follows that the facultative anaerobe component decreased with depth, except in the red-brown horizon. *Staphylococcus* spp and *Bacillus* spp appear to occur in roughly equal proportions, though no accurate estimates were made. Since growth of any strains in cultures at below pH 5.0 could not be demonstrated, it is likely that the staphylococci may form an inactive component of the Moor House blanket bog population. The dominance of staphylococci appears from previous reports to be restricted to acid and anaerobic soil conditions.

On the basis of physiological characteristics, the staphylococci were classified into biotypes 1, 3, 4, and 5 of *Staphylococcus epidermidis* according to Baird-Parker (1965a, b), in roughly equal numbers. No strains with the characteristics of biotype 2 or of *Staphylococcus aureus* were obtained.

Table 2C gives an analysis of those obligate anaerobes successfully subcultured from the original isolation plates. Unexpectedly, *Clostridium* did not dominate any horizon; a Gram-positive organism, apparently closely related to *Actinomyces* and forming small clusters of cocci or coccal rods, was dominant in all horizons. Vibrioid bacteria, in almost semi-circular shapes, also occurred frequently; they did not correspond to any previously described obligate anaerobe.

The *Actinomyces*-like bacteria were non acid-fast, grew best at 37° C and in the absence of oxygen, and gave a catalase-positive reaction—unusual in anaerobic or microaerophilic bacteria. When cultured in shake tubes, visible growth began in most cases 0.5 cm from the surface and, while many strains grew throughout the medium, some grew only in a narrow band about 1 cm from the surface, indicating a preference for microaerophilic conditions. The organisms would not grow in a medium containing less than 1% peptone and 0.3% yeast extract; they utilised a narrow range of sugars and related substances, decomposing only glucose, glycerol and mannose out of 12 such compounds; and they did not hydrolyse casein, gelatin or starch nor produce indole or H_2S. The occurrence as the anaerobic dominants in Moor House peat of these hitherto apparently undescribed organisms warrants further investigation.

The vibrioid organisms, found only in the upper two horizons of the peat profile, appear not to be proteolytic and do not reduce sulphate, but ferment a wide range of sugars. The only other catalase-negative curved rods reported from soils are *Desulphovibrio* spp.

On the basis of morphological, saccharolytic, proteolytic and certain other properties, the *Clostridium* spp were considered to be closely related to five previously described species. *Clostridium thermosaccharolyticum* was dominant and represented more than 75% of the strains investigated, with *Clostridium difficile* and *Clostridium sartogoformum* occurring rarely. *Clostridium bifermentans* was the most numerous proteolytic type, *Clostridium cylindrosporum* occasional.

5.3.2.3 *Microaerophilic Bacteria*

The species composition of the microaerophilic component of the heterotrophic population was determined from those bacteria which grew on the original double layer isolation plates (Sect. 5.2.2), 57% of which, including *Aeromonas* spp, *Vibrio* spp and Enterobacteriaceae, demonstrated a fermentative metabolic pathway with sugars. *Bacillus* spp accounted for 15%, *Pseudomonas* 7%, *Arthrobacter* and coryneform types 5% and micrococci 8% while, due to difficulty in maintaining cultures, 12% were not identified.

Bergey (1948) lists 77 genera, including both heterotrophs and autotrophs, known to occur in soils. The result of the present studies on heterotrophs shows that the spectrum of genera present in peat at Moor House is restricted.

5.3.3 Selected Physiological Groups of Bacteria

5.3.3.1 *Proteolytic Bacteria*

The number of proteolytic bacteria, i.e. those capable of gelatin hydrolysis, declined markedly with depth in two peat cores (Table 3), from 2×10^4 g^{-1} in the litter to complete absence from the red-brown horizon. Results of gelatin hydrolysis tests on the 1500 pure culture isolates indicated that about 17% of Gram-negative bacteria and 50% of bacilli cultures (at least 3×10^6 g^{-1} out of a total bacterial population of 10×10^6 g^{-1}) are proteolytic.

This is important in relation to the nitrogen cycle in peat (Fig. 1 and Chap. 6) the first stage of which is the breakdown of proteinaceous material to ammonia compounds. The response of diluted peat samples on direct inoculation into gelatin-containing media, and the large numbers of Gram-negative bacteria in the litter zone in peat, provide additional evidence that this horizon contains the largest number of proteolytic bacteria, which are evidently an active component of the heterotrophic population. The fact that the ammonia compounds released on proteolysis are a potential nitrogen source for uptake by higher plants may explain the concentration of roots in the litter region. It is almost certain that fungi also participate in proteolysis, particularly in acid soils (Alexander, 1961).

5.3.3.2 *Nitrifying Bacteria*

The ammonia salts medium used for nitrifying bacteria would favour the growth of autotrophs but after five to six weeks incubation neither nitrite nor nitrate was detected from any of the four horizons of two cores, indicating that if organisms capable of the autotrophic utilisation of ammonia salts are present, they must be few and not very active. Peat samples from four horizons in another core, inoculated into autotrophic media used for detection of *Nitrosomonas* and *Nitrobacter*, also gave negative results, while control samples from lake muds gave positive evidence of nitrification within two to three weeks in the same media. There is, however, evidence in the literature that in certain environments heterotrophic nitrification by bacteria and fungi is possible, and though this was not tested in the absence of autotrophic nitrification, it is a possible cause of nitrification in this stage of the nitrogen cycle (Fig. 1).

Table 3. The distribution of bacterial groups with depth in peat

			Litter	Dark-brown	Green-brown	Red-brown	No. of cores
Autotrophic bacterial groups							
thiosulphate-oxidising			5×10^3	3×10^2	2×10^1	2×10^2	3
sulphate-reducing (range)			$0.09–7\times10^3$	$0.3–9\times10^3$	$4–7\times10^3$	$3–14\times10^1$	1
nitrifying			0	0	0	0	2
Heterotrophic bacterial groups							
proteolytic			2×10^4	1×10^4	2×10^1	0	2
nitrate-reducing			2×10^4	2×10^4	1×10^3	1×10^3	3
denitrifying (anaer)			2×10^4	2×10^4	1×10^3	0	3
starch-hydrolysing			5×10^6	not tested	not tested	not tested	
cellulolytic			2×10^3	3×10^2	8×10^2	8×10^1	1
chitinolytic		aer	1×10^4	1×10^4	2×10^4	1×10^4	
		anaer	17×10^4	3×10^4	3×10^4	2×10^4	4
aerobes (range)			9–260	6–150	11–76	0.7–42	
facultative to microaerobic (range)			3–77	23–138	10–380	27–210	
anaerobes (range)			6–173	9–62	6–116	0.5–52	
Potential N-fixing							
	Feb.	aer	153 ± 14	150 ± 38	60 ± 10	42 ± 14	3
		anaer	173 ± 27	62 ± 16	48 ± 13	$52\pm$—	
	Apr.	aer	61 ± 6	47 ± 14	30 ± 11	33 ± 3	
		anaer	70 ± 39	66 ± 22	44 ± 23	10 ± 7	
	June	aer	44 ± 15	25 ± 2	11 ± 3	10 ± 3	
		anaer	39 ± 15	17 ± 3	15 ± 1	10 ± 1	
	Sept.	aer	9 ± 4	91 ± 11	13 ± 4	$2\pm$—	
		anaer	26 ± 3	44 ± 5	19 ± 2	7 ± 3	
	Oct.	aer	48 ± 3	18 ± 7	52 ± 3	20 ± 9	
		anaer	68 ± 11	24 ± 11	116 ± 20	25 ± 25	
Cottage Hill A	Apr.	aer	106 ± 25	78 ± 18	34 ± 11	5 ± 3	
		anaer	144 ± 14	65 ± 20	27 ± 7	10 ± 5	
Green Burn	Apr.	aer	47 ± 17	102 ± 29	62 ± 22	22 ± 8	
		anaer	81 ± 19	111 ± 30	52 ± 19	19 ± 8	
Sike Hill	Sept.	aer	33 ± 8	34 ± 9	15 ± 6	20 ± 8	
		anaer	15 ± 6	25 ± 16	14 ± 4	13 ± 9	
Cottage Hill B	Sept.	aer	471 ± 210	76 ± 25	74 ± 15	22 ± 12	
		anaer	447 ± 190	78 ± 10	36 ± 10	15 ± 11	

Counts $\times 10^5$ g^{-1}, unless otherwise stated. All data are for the Bog End site unless stated. aer = aerobic; anaer = anaerobic.

5.3.3.3 Nitrate-Reducing and Denitrifying Bacteria

Since it is evident that little nitrate is supplied in these peats by microbial nitrification, it follows that the bulk of it comes from rain (Fig. 1 and Chap. 6). Results from samples from four horizons of three cores (Table 3) indicate the presence of 1×10^3–2×10^4 g^{-1} potential nitrate-reducing organisms, with highest numbers in the litter zone, and decreasing with depth in all cores. Similarly, organisms capable of denitrification by anaerobic respiration and conversion of nitrites to nitrogen, ammonia and other gases, were most abundant in the litter and dark brown horizons and decreased with depth. As shown in Figure 1 the conversion of nitrites to N_2 may be a chemical, as well as a microbial activity.

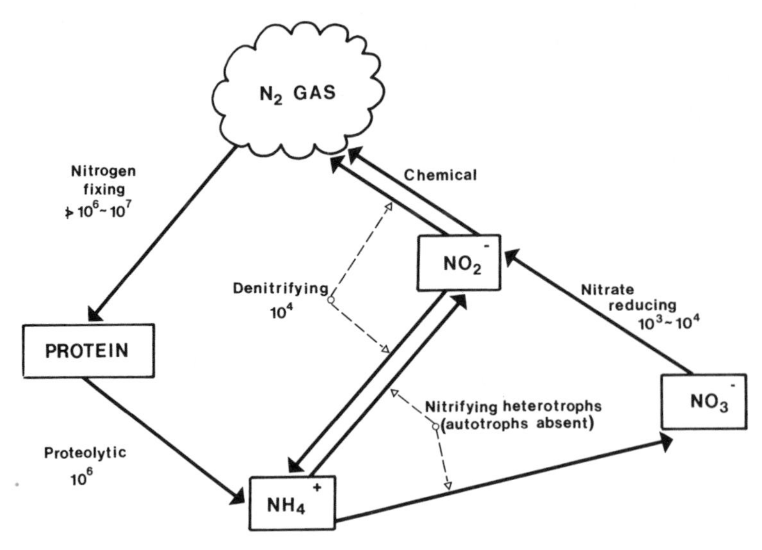

Fig. 1. Numbers of bacteria (g^{-1}dw) in the main groups associated with transformation of nitrogen

5.3.3.4 Nitrogen-Fixing Organisms

Limited seasonal sampling, at a number of sites, of the organisms responsible for the last stage of the nitrogen cycle, i.e. the fixation of atmospheric nitrogen into protein-bearing compounds, indicated that, under aerobic or anaerobic conditions of incubation, numbers decrease down the profile and, while there is no marked seasonal pattern, are highest for the Bog End sites in February. Similar results were obtained from other sites, except that both aerobic and anaerobic incubation gave higher counts in the litter zone at Cottage Hill B in September. The ability to live on a nitrogen-limited medium is, however, not proof of nitrogen fixation.

A random sample of 30 isolates from the original isolation plates from three cores was repeatedly subcultured on the nitrogen-limited medium to confirm their ability to grow actively. Twelve successful cultures were obtained from both the litter and dark brown and three each from the green-brown and red-brown horizons. When half the isolates from the litter were tested for their ability to reduce acetylene to ethylene at 25° C, using a modification of the methods of Hardy et al. (1968), the amount of nitrogen fixed by 10^5 organisms ml^{-1} solution ranged from 0.2296 ng to 23.71 ng N_2 h^{-1}. Similar tests on the isolates from the red-brown gave negative results, and no tests were carried out on dark brown or green-brown isolates. The limited evidence makes it clear that nitrogen-fixing bacteria are present and probably active, at least in the litter zone.

Some fungi, mainly segmented and other sporing types, which grew on the nitrogen-limited medium, were subcultured onto the same medium; 33% of the subcultures from the litter, 38% from the dark brown, 10% from the green-brown and 19% from the red-brown showed active growth with negligible available nitrogen.

5.3.3.5 *Thiosulphate-Oxidising Bacteria*

Whilst two of the three cores, tested in May and November, showed decreasing activity of thiosulphate-oxidising bacteria with depth, in the third, also taken in May, activity increased slightly down the profile. As many of the pure culture isolates from the original cultures proved to be strictly aerobic, and the oxygen concentration decreases with depth in the peat, the distribution of these bacteria could result from contamination of the corer as it passed through the litter zone where their activity is greatest. With this reservation in mind, the most probable number of the thiosulphate-oxidising organisms in litter is 5×10^3 g^{-1}, dark brown 3×10^2 g^{-1}, green-brown 2×10^1 g^{-1} and red-brown 2×10^2 g^{-1}.

Biochemical tests of a random collection of 43 isolates subcultured from the original isolation flasks showed *Thiobacillus trautweinii* to be the predominant organism (67% of the isolates). The remainder (eight cultures from the litter, four from the dark brown and two from the red-brown) were *Thiobacillus denitrificans*, which can oxidise thiosulphate anaerobically and, using nitrate as a hydrogen acceptor, reduce nitrate to nitrogen. This group of organisms can therefore participate in both the nitrogen and sulphur cycles in peat in the absence of oxygen.

5.3.3.6 *Sulphate-Reducing Organisms*

Intensive sampling of a single core (Table 3) showed that between 5 cm and 20 cm depth in the peat, the number of sulphate-reducing organisms increased, from 9×10^1–7×10^3 g^{-1} in the litter to 4–7×10^3 g^{-1} in the green-brown, decreasing again in the red-brown (3–14×10^1 g^{-1}) to a depth of 32 cm below which they were not detected. This distribution, despite the limitations of the study, corresponds well with the redox potentials (Urquhart and Gore, 1973) which show very low values, indicating low oxygen tension, in the 5–20 cm zone. Such an environment could be considered as a selective ecological niche which these organisms are able to exploit by reducing sulphates under anaerobic conditions. Identification of the organisms from 80 random pure culture isolates from the original blackened and H_2S-producing cultures showed the dominant organism to be *Desulphovibrio desulphuricans*, already known to be able to reduce sulphates anaerobically.

5.3.3.7 *Starch-Hydrolysing, Cellulolytic and Chitinolytic Bacteria*

About half (5×10^6 g^{-1}) of the estimated bacterial population of the litter zone appears to be capable of hydrolysing starch to sugar-containing compounds. Of the bacteria isolated from aerobic plate counts, especially of the *Bacillus* circulans complex, 90% were starch hydrolysers and this, together with the observed increase in numbers of bacilli with depth (Table 2), indicates an increase in the proportion of organisms capable of starch hydrolysis down the profile.

In the single core sampled, numbers of cellulose-decomposing bacteria decreased from 2×10^3 g^{-1} in the litter to 8×10^1 g^{-1} in the red-brown zone. Identification of pure cultures, though not yet completed to species level, shows *Cellvibrio* sp, *Cellulomonas* sp and *Pseudomonas* sp to be included in the main genera.

No zonation in numbers of chitinolytic bacteria ($1–17 \times 10^4$ g^{-1}) was shown under aerobic or anaerobic conditions (Table 3); few of those which grew on chitin agar showed active chitin hydrolysis by clearing of the medium.

5.3.4 Actinomycetes, Yeasts and Fungi

Neither actinomycete nor yeast colonies were observed visually or microscopically on subsamples from dilution plates of the original peat from four horizons of two cores, and yeasts formed only 1% of identified strains of the heterotrophic bacteria isolated, and rarely occurred on plates used for isolation of fungi and bacteria, though they were occasionally found in *Juncus* peat, particularly on slide traps placed in the litter zone in winter (Latter and Cragg, 1967). It appears that yeasts, if present at the Bog End site, occur in numbers less than 10^1 g^{-1}. Limitations of the method used, however, may account for the negative actinomycete results, as an unidentified *Actinomyces*-like form is the dominant anaerobe throughout the peat (Sect. 5.3.2.2).

Though estimates of the percentage colonisation (Table 4) of peat particles by fungi were made on only one occasion, the genera are similar to those recorded

Table 4. The percentage of washed particles colonised by fungi

A. Mean % for three cores, 300 particles for each horizon

	Horizon			
	Litter	Dark brown	Green-brown	Red-brown
Fast growing	12	2	4	0
Slow growing	49	80	32	2
Total	61	82	36	2

B. Mean % for each core, 400 particles for each core

	Calluna	*Eriophorum*	*Sphagnum*
Fast growing	10	3	1
Slow growing	58	41	23
Total	68	44	24

C. Frequency of major genera, mean % for three cores, 300 particles for each horizon

	Litter	Dark brown	Green-brown	Red-brown
Segmented fungi	17	39	29	1
Sterile dark spp	23	30	13	0
Mortierella spp	14	4	4	0.3
Basidiomycete	0.3	6	2	1
Sporing imperfecti[a]	16	3	4	0

[a] Includes *Penicillium, Verticillium, Trichoderma, Oidiodendron, Cladosporium* and *Tolypocladium* species.

One core was sampled under each plant type. From each horizon of each core, 100 washed particles were plated onto Czapeks-dox and malt agar.

from other sampling procedures (Latter et al., 1967). Sterile dark and segmented forms constitute the major part of the population, slow growing fungi predominating in all horizons of cores under *Calluna*, *Eriophorum*, and *Sphagnum* (Table 4 B). There was a decrease in number of genera and percentage colonisation down the profile, more marked for fast than for slow growing forms and for sporing than sterile dark and segmented types, and very few particles in the red-brown were colonised. The *Calluna* core showed the highest colonisation and *Sphagnum* the lowest, particularly of the sporing imperfecti, i.e. *Cladosporium*, *Oidiodendron*, *Penicillium*, and *Trichoderma*, but *Tolypocladium* spp occurred on dilution plates. Cellulolytic activity has been demonstrated for *Cladosporium* spp, *Oidiodendron tenuis*, *Penicillium spinulosum*, *Trichoderma viride* and *Tolypocladium geodes* and for a few isolates of basidiomycete mycelium, sterile dark and sterile white fungi, but not for any isolates of segmented fungi that were tested. Limited information on the biochemical characteristics of the fungi is given in Flanagan and Scarborough (1974).

5.3.5 Direct Counts and Biomass Estimates of Bacteria and Fungi

Estimates for bacterial numbers obtained by direct counts are approximately 1000 times higher than those based on dilution methods, ranging from 4 to 21×10^9 g^{-1} in the litter with a maximum count in the green-brown on a number of occasions.

Stained mycelium was present in largest amounts in the litter horizon, and unstained in the dark brown horizon, both decreasing in the red-brown zone (Table 5). The proportion of total mycelium which was stained decreased to the black-brown and green-brown horizons. While unstained mycelium is almost certainly dead, the proportion of stained mycelium which is live is unknown (Frankland, 1975).

Biomass estimates (mg g^{-1} dw) based on direct counts (Table 6) show a marked decline between the green-brown and red-brown horizons for both bac-

Table 5. The quantity of bacteria ($\times 10^9$ g^{-1}) and stained and total fungal mycelium (m g^{-1}) estimated by direct counts

Horizon	Bacteria	Fungal mycelium		Number of cores
		Stained	Unstained	
Litter[a]				
1–2 yr	21	2,760	6,050	3
2–3 yr	17	2,190	9,110	3
3 cm deep	18	2,390	10,050	3
Dark brown[b]	18	1,030	9,760	8
Green-brown[b]	26	750	4,980	8
Red-brown	12	200	790	1

[a] Compounded from data for individual litters (*Sphagnum*, *Calluna*, *Eriophorum*, *Rubus*), adjusted for the proportion of each present at each stage of decomposition (Chap. 8). All samples taken 7. 5. 1970.
[b] Mean of eight seasonal samples, 5. 4. 1962–12. 3. 1963.

Table 6. Estimates for the blanket bog of bacterial and fungal biomass (mg g^{-1} and g m^{-2}) based on dilution and direct counts of bacteria and direct estimates of mycelium length

Horizon	Depth (cm)	Bulk density (g cm^{-3})	Biomass (mg g^{-1})			Biomass (g m^{-2})			
			Bacteria	Fungi		Bacteria		Fungi	
			Direct	Stained	Total	Dilution	Direct	Stained	Total
Litter									
1–2 year	1.0		1.36	1.43	3.15				
2–3 year	1.0	0.03	1.11	1.14	4.74	0.109	3.3	3.6	14.4
older	4.6		1.13	1.24	5.22				
Dark brown	2.9	0.07	1.15	0.54	5.08	0.026	2.3	1.1	10.3
Green-brown	12.9	0.09	1.67	0.39	2.59	0.132	19.4	4.5	30.1
Red-brown	10.0	0.10	0.78	0.10	0.41	0.028	7.8	1.0	4.1
Total	32.0					0.295 (0.018)[a]	32.8	10.2	58.9

Conversion factors: Bacteria, dilution count: based on the proportions of groups shown in Table 2A and weights in g × 10^{-12} per bacterium, of *Bacillus* vegetative cells 0.58, *Bacillus* spores 4.2, Gram-negative rods 0.2, cocci 0.44, *Arthrobacter* 0.1 with density of 1.1 and moisture contents of 80% and 25% fresh weight for vegetative cells and spores respectively. Bacteria, direct count 6.4×10^{-14} g per bacterium based on mean size of 0.29 μ^3, a density of 1.1 and moisture content of 80% fresh weight. Fungi 5.2×10^{-7} g m^{-1} of mycelium based on a mean width of 2 μ, a density of 1.1 and moisture content of 80% fresh weight.

[a] Estimate based on dilution count but using the conversion factor for direct counts.

teria and fungi and indicate that, while fungal biomass exceeds that of bacteria in the litter zone of peat, the reverse is true in deeper layers, probably reflecting their differing responses to redox potential and pH. Thus in the dark brown and green-brown zones, bacterial biomass exceeded that of stained, but not total, fungi, while in the red-brown zone it was greater than the total fungi count; in winter direct counts of bacteria decreased relative to fungal counts (Latter et al., 1967). Results for colonisation of washed peat particles support this trend, colonisation of the litter (61%) and dark brown (82%) horizons by fungi being higher than by bacteria (56 and 39%), while bacterial colonisation of the green-brown (40%) and red-brown (20%) zones exceeded that by fungi (36 and 2% respectively).

5.3.6 The Influence of Specific Environmental Parameters on the Microbial Population

5.3.6.1 Temperature

Summarising the results of tests on 240 pure cultures of bacteria, 40% grew at 0–2°, 81% at 5°, 98% at 7–8°, 99% at 12° and 100% at 20° and 25° C. Similar results were obtained by Rosswall and Clarholm (1974) in an analysis of 90 isolates of aerobic bacteria from Moor House peat. The trend for good growth between 7° and 25° C may reflect the original isolation temperature, viz 22° C. Isolates from the litter zone produced more vigorous growth at low temperatures than those from deeper layers.

Fungi isolated from the field under winter conditions were more cold tolerant than dominant summer isolates. Of seventeen types tested, 53% grew moderately well at 1° C (Latter and Heal, 1971).

Although the data on seasonal variation are limited, they indicate that bacteria and fungi are potentially capable of functioning throughout the year under the temperature regime of Moor House blanket bog peat (Chap. 1).

5.3.6.2 Oxygen and Carbon Dioxide Concentration

On inoculating the same 240 pure isolates into deep semisolid TSA stabs, 93% grew well under microaerophilic conditions, i.e. they were incapable of growth under fully aerobic or anaerobic conditions but were able to grow well under reduced oxygen tension. Only 7% were strictly aerobic as indicated by surface growth in the stabs.

Two hundred of these, chosen for their ability to grow under microaerophilic conditions in the TSA stabs, and representing each of the four peat horizons, were selected for continuous culture experiments. Although populations decreased if oxygen concentrations fell below 3%, all the cultures, regardless of their oxygen preference in the stabs and their origins in the peat profile, grew well in the range 3–110% O_2 at 10–15° C; the criterion for growth was a colony count at 10° C. Fresh medium was continuously added as the O_2 tension decreased due to microbial activity, so neither temperature nor low O_2 tensions restricted growth in these cultures. However, when the nutrient concentration was reduced by 75%, only 4% of the cultures survived below 50% oxygen saturation.

While these studies have their limitations, applying as they do to pure cultures, they indicate that although the majority of the isolates are aerobic, they are potentially capable of growth and activity at low temperatures and oxygen concentrations, provided the nutrient level is not also reduced.

A number of predominant fungi were screened for growth in deep agar tube cultures (Taylor, 1970), and while only one isolate, a segmented form, grew better below the agar surface, several grew as a surface mat and also within the agar, while *Mortierella* isolates grew throughout the agar. Because it has been suggested that the microaerophilic type of growth could be the result of a changing CO_2 regime in tube cultures, four of these fungi, demonstrated as being capable of some anaerobic growth (two segmented forms, a *Verticillium* sp and a *Mortierella* sp) were grown as liquid cultures subjected to concentrations of oxygen ranging from 0 to 5% and of carbon dioxide from 0 to 10%. A wide range of tolerance was found, all the fungi growing to some extent under all the conditions tested, but all grew best with 5% CO_2 and 5% O_2. In the absence of oxygen *Mortierella* showed best growth, but it and the segmented forms grew poorly in the absence of CO_2.

It is evident that a high proportion of the bacteria and at least some of the fungi are capable of growth under microaerobic conditions. Although oxygen concentrations were not measured in the field, the work of Urquhart and Gore (1973) on redox potentials showed decreasing oxygen levels with increasing depth in Moor House peats. At Bog End the lowest redox potential observed was at *c* 10 cm depth (Chap. 1).

It is obvious that in general microbial populations and redox potentials follow the same trend of decreasing levels with depth. The seasonal decline in redox potential, increasingly correlated with temperature as depth increases (Urquhart and Gore, 1973) indicates that the reducing conditions, partly due to the influence of water table on gaseous diffusion rates, are to some extent the result of a depletion of oxygen by microbial activity. Although a high proportion of the microflora is capable of growth under reduced oxygen conditions, the decline in numbers and in decomposition rates with depth (Chap. 7) indicates a retarding of their activity by the low redox potential which they have helped create. A few selected groups of organisms are, on the contrary, stimulated by the low redox potential, e.g. the sulphate-reducing bacteria can utilise substances released by the oxidation and reduction of the chemically resistant substances present in this region.

Urquhart and Gore (1973) state that "Redox is essentially qualitative, but its relative behaviour can give an indication of microbiological activity and also of the depths to which the roots of higher plants can penetrate", and Urquhart (1966) demonstrated an increase in redox potential values associated with a high proportion of aerenchymatous tissue. Since roots of *Eriophorum vaginatum* extend well into the anaerobic zone in peat soils, occupying an estimated 1% of the volume of total peat to a depth of 32 cm (Table 7), they could thus provide, through exudates and at death, fresh substrates for microbial activity or inhibitory substances not otherwise present, as well as aerobic microsites which could influence the microbial population, particularly in relation to its decomposing activities.

Table 7. Comparison of bacterial numbers ($\times 10^5$ g^{-1} wet weight) on *Eriophorum* roots and in the peat beneath *Eriophorum* tussocks

Date	Horizon	Condition of roots	Number of cores	Mean numbers		Root: peat ratio
				Root	Peat	
April	Dark brown	50% white 50% yellow	3	9.3	12.4	0.75
	Green-brown	10% white 90% yellow	3	10.1	4.1	2.5
	Red-brown	all brown or yellow	3	0.15	0.5	0.30
June	Dark brown	all white	3	90	15	4.4
	Green-brown	all white	3	7.9	5.5	1.2
	Red-brown	all white	1	0.4	1.3	0.36
October	Roots only in tussocks	all white	2	9	45	0.20

However, Martin (1971), studying the effect on the bacterial population of the microhabitat associated with the rhizosphere, estimated that the numbers of aerobic heterotrophic bacteria on and below roots of *E. vaginatum* tussocks (Table 7) were not significantly different from those in the surrounding peat. While this may be the result of methodology in that the whole root was sampled, it suggests that the rhizosphere effect is small or insignificant, which may help to explain the low rate of decay of *E. vaginatum* roots (Chap. 7).

5.3.6.3 pH

The pH (in distilled water) ranged from 3.0 to 4.6 in the sites studied; the values measured in 1 M KCl were lower (2.4–3.9) due to release of the exchangeable acidity in peat by the KCl. The values are typical of bog peat but there are microhabitats within the peat and litter where the pH may be as much as a unit higher. Latter and Howson (1977) found a pH range from 3.5 *(Sphagnum)* to 4.4 but no significant correlations occurred with microbial numbers over this range.

In cultures, bacteria grew best at or above pH 6.0, lower values restricting their ability to metabolise nutrients in the growth medium. Rosswall and Clarholm (1974), investigating the influence of a pH range from 5.0 to 9.0 on pure cultures of bacteria from Moor House blanket peat, found that of 90 aerobic isolates, only 1% grew at pH 5 compared with more than 80% at pH $>$ 5.5. Growth in culture below pH 5.0 could not be demonstrated in staphylococci, a component part of the facultative anaerobic population in peat, and known to be restricted to acid anaerobic conditions. It is, therefore, unlikely that they are active under field conditions of low pH.

The thiosulphate-oxidising bacteria, particularly the anaerobic *Thiobacillus denitrificans*, on the other hand, produce acid conditions while oxidising sulphur-bearing compounds, and continue activity as the pH of the medium falls from 6.8 to 2.0; in fact they are largely responsible for the production and maintenance of acid conditions in peat soils. This group is therefore the only one able to tolerate the low pH conditions, because other main species of the group found in the peat are probably restricted in their activity by low nitrate availability owing to the absence of nitrification processes. Thus though the peat soils may contain groups of organisms capable of exploiting one environmental parameter, e.g. low pH, the combined effects of other parameters, e.g. low oxygen tension and nutrient availability, can restrict their in situ activity. The population estimates obtained for all other bacterial groups may reflect their growth in isolated pockets of higher pH in the peat, or their ability to survive under low pH and resume activity when the pH rises above 6.0.

No marked pH optima were obtained for fungi within a pH range 4.5–6.5 and fungi tested all grew well at pH 4.5. During culture, however, the pH of the agar usually increased, particularly with nitrate rather than peptone as a nitrogen source, and with segmented forms it reached pH 7.0–9.0. With a few isolates, pH fell below 4.5. This environmental parameter undoubtedly restricts the activity of most groups of organisms, particularly bacteria, now known to be present in peat soils.

5.3.6.4 Quantity and Quality of Substrate

Since the precentage of organic matter varies little within the profile, from $95.8 \pm 0.5\%$ in the dark brown to $97.0 \pm 0.3\%$ in the green-brown, the total quantity of substrate can hardly account for the observed drop in microbial numbers down the profile. However, the organic matter consists of increasingly resistant fractions in successive layers, and hence substrate quality could explain the presence of the greatest microbial numbers in the litter zone. Evidence for this effect on the blanket bog is limited and often indirect; core samples under different vegetation show that bacterial and, to a lesser extent, fungal numbers are generally higher under *Calluna* than under *Sphagnum* (Table 8), while fungal colonisation of peat particles was also higher under *Calluna* than under *Sphagnum* (Table 4B).

This effect may be related to microclimate differences caused by the plant cover and/or to chemical and physical differences in substrate. Latter and Howson (1977) found bacterial numbers to be correlated positively with nitrogen ($r = 0.915^{***}$) and negatively with crude fibre ($r = 0.718^{*}$) content of the substrate, and the quantity of mycelium was greater on the nutrient-rich *Rubus* leaves than on other litter though, possibly because of the limited data set, no significant correlations between mycelium and substrate composition were obtained. The influence of nutrients and carbon source on microbial activity is shown in Chapters 6 and 7. Thus variation in substrate quality is undoubtedly responsible for some of the within-site variation in population, but cannot entirely explain the great drop in numbers with depth. The marked decline in decomposition rates of the same substrate placed at different depths in the peat (Chap. 7) must be attributed to environmental factors rather than substrate quality. It is suggested that there is a broad positive relationship between microbial population size and rate of decomposition and therefore substrate quality is a contributing, but not necessarily the major, factor responsible for the decline in population with depth.

Table 8. The distribution of bacteria ($\times 10^6\ g^{-1}$) and fungi ($\times 10^5\ g^{-1}$) under two types of vegetation

Bacteria	*Calluna*				*Sphagnum*			
	TSA		CPS		TSA		CPS	
	a	b	a	b	a	b	a	b
Horizon								
Litter	4.0	0.5	9.0	2.0	2.0	1.0	6.0	4.0
Dark brown	20.0	1.0	10.0	0.9	2.0	0.9	3.0	1.0
Green-brown	6.0	3.0	10.0	3.0	6.0	0.8	5.0	0.8
Fungi	*Calluna*				*Sphagnum*			
	TSA		CPS		TSA		CPS	
	a	b	a	b	a	b	a	b
Horizon								
Litter	17.0	3.0	18.0	7.0	13.0	5.0	16.0	3.0
Dark brown	25.0	5.0	18.0	6.0	6.0	3.0	16.0	4.0
Green-brown	2.0	1.0	3.0	2.0	5.0	0.5	8.0	2.0

Data are mean values for three cores. Using Tryptone Soya Agar (TSA) and Casein Peptone Starch Agar (CPS) with an agar overlay for facultative to microaerophilic counts. Two methods of preparation: dilution tubes (a) shaken by hand (b) macerated.

5.4 Summary and Conclusions

The results indicate that the fungal biomass exceeds that of bacteria in the litter zone of the peat, but that the situation is reversed with increasing depth, and though bacterial numbers decrease down the profile, the decline is greater with aerobic than with facultative and anaerobic types. The number of genera of bacteria and fungi is limited, the greatest diversity of types being present in the litter zone of the peat, though certain bacterial genera are present in greater numbers within deeper zones of the peat soil profile than in the litter zone.

Twelve selective bacterial groups, as well as actinomycetes, yeasts and fungi are shown to be present, and while some of these occur in much lower numbers than in arable soils, suggesting a restriction of their potential activity, numbers of aerobic and facultatively anaerobic heterotrophs are similar to those in arable soils. It is likely that, as most of the microbial population can grow between 0° and 25° C in tests, some microbial activity occurs throughout the year under the Moor House temperature regime. Similarly, evidence suggests tolerance of a wide range of oxygen concentrations, low redox potential at certain depths in the peat selectively encouraging certain groups at the expense of others.

A low pH regime, below 5.5, restricted metabolic activity of pure cultures of bacteria, and although isolated pockets of relatively high pH may occur in the peat, the general level of pH 3.0–4.6 existing at all depths must be a major factor affecting numbers and activity of microbes.

Thus, the restricted numbers and activity of most groups in peat at Moor House is likely to be due to a combination of the low pH and redox potential, and the low substrate quality available at any depth within the peat soils studied.

Acknowledgements. We gratefully acknowledge the financial assistance of the Natural Environment Research Council through the period of these studies, 1966–1972, the Nature Conservancy for laboratory facilities both at Merlewood Research Station and at Moor House, Mr. S. van Zeller for skilled technical assistance and Dr. O. W. Heal who was responsible for coordinating these studies and gave a great deal of his time for discussion.

6. Nutrient Availability and Other Factors Limiting Microbial Activity in the Blanket Peat

N. J. MARTIN and A. J. HOLDING

An experimental approach indicates the influence of nutrient and substrate availability in the respiration and nutrient cycling by microbial populations within the blanket bog.

6.1 Introduction

Peat accumulation is generally attributed to low rates of organic matter decomposition through the adverse conditions of both the vegetation and climate. Whilst there is little published information concerning the influence of microbial nutrient availability, studies suggest that inorganic nutrients have little effect on microbe numbers or activity in peat soils (Waksman and Stevens, 1929; Waksman and Purvis, 1932; Knowles, 1957; Clymo, 1965) although peptone, a combined carbon and nitrogen source, has been shown to stimulate microbial numbers (Holding et al., 1965). Because of the lack of data, studies at Moor House investigated the relative effects of inorganic and organic nutrients on microbial activity, and whether or not enhanced activity improves mineralisation of essential nutrients.

Laboratory investigations using a non-cyclic percolation system through a peat/sand mixture provided information about nutrient influence on microbial numbers and the release and immobilisation of nutrients. Respiration measurements were made on homogenised and unhomogenised peats. Results of these experiments, and the changes produced in the microbial population by nutrient addition in the field, are discussed in relation to other chemical and physical soil characteristics, and to studies on nitrogen fixation carried out by B. T. D'Sylva and V. G. Collins.

6.2 General Methods

6.2.1 Sources of Samples for Laboratory Studies

Peat samples from the dark brown horizon, which started at 3–10 cm depth and was 2–4 cm thick (Chap. 1) were taken from 10 cm diameter cores from the Bog End site, stored at 4° C in polythene.

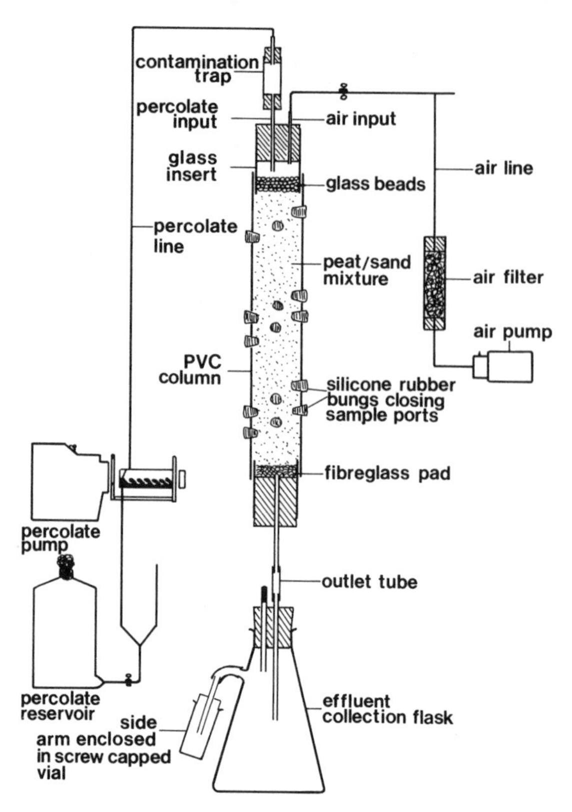

Fig. 1. Apparatus for continuous non-cyclic percolation of peat/sand mixtures

6.2.2 Microbial Counting Procedures

The samples (10 g for field samples, 0.3 g from percolation experiments) were blended for 1 min with 90 ml distilled H_2O in an "Osterizer" homogeniser unit, and serial decimal dilutions prepared in 0.01% peptone water. Using a pour plate procedure with 1 ml aliquots of suitable dilutions in Casein Peptone Starch (CPS) medium (Collins and Willoughby, 1962, modified by Martin, 1971) colony counts were made after 14 days' incubation at 22° C. Percolation samples (Sect. 6.2.3) were plated on nutrient agar containing 1:500,000 crystal violet as a general Gram-negative selective medium (Holding, 1960). Only a proportion of the peat Gram-negative bacterial populations grew on it, however, and these also showed better growth and activity than the remainder in laboratory test media (Martin, 1971). Bacterial endospores were counted on CPS plated with 1 ml samples from pasteurised dilution tubes (15 min at 75° C).

6.2.3 Methods to Evaluate the Influence of Nutrients on Microbial Activity

The non-cyclic percolation system (Fig. 1) used in this study consisted of five pairs of PVC columns, packed with a mixture of 15% peat ground up with 85%

acid-washed sand, through which selections of nutrients were percolated from five separate reservoirs at the rate of 35 ml d^{-1}. Samples were removed aseptically after 0, 2, 4, 7, 15, and 30 days' percolation at 22° C, and effluents from the columns were collected at intervals to determine K^+, Ca^{2+}, Mg^{2+}, NO_3^-, and NH_4^+ nitrogen, and phosphate contents.

6.2.4 Field Treatments

The influence of six selected nutrients on microbial numbers in the field was investigated by ten 1 ml injections of the nutrient, or distilled water as a control, equidistantly round the circumference of the dark brown horizon of peat cores, each treatment being replicated six times. The depth of injection was marked with sterile plastic pegs and the core replaced in situ. After four weeks the cores were lifted and a 2 cm thick slice centred on the injection zone removed from each. The outer 1 cm ring of the disc was removed and the remainder mixed and subsampled to determine microbial populations.

6.2.5 Laboratory Respiration Measurements

A Warburg manometer was used to measure the respiration of 2.9 ml hand mixed peat amended with 0.1 ml nutrient salt solution, or 0.1 ml distilled water as a control. Alternatively a Clark-type oxygen electrode was used and 25 g of peat homogenised with 100 ml distilled water, amended with one of the selected salt solutions. The pH, reduced by exchange acidity, was restored to that of the controls by addition of 1 N NaOH, and the homogenates incubated statically or shaken on an orbital incubator at 29° C. After incubation, 3 ml aliquots were equilibrated for 10 min in the sample cuvettes (Yellow Springs Instrument Company model 53 Biological Oxygen Monitor) and the oxygen electrode inserted. Oxygen uptake rate was measured from the slope of the recorder trace.

6.2.6 Determination of Nitrogen Fixation Rates

B.T.D'Sylva and V.G.Collins used a modified acetylene reduction technique (Hardy et al., 1968) to measure nitrogen fixation rates under field conditions at Bog End. Peat samples, flushed first with Ar, Ar + O_2 or Ar + CO_2, and then injected with acetylene, were sealed and incubated at field temperature (11° C) for 1 h. Using gas chromatography, the quantity of ethylene produced, corrected for abiological ethylene production, was measured in the litter, dark brown, green-brown, and red-brown horizons from each of three cores in April and June 1971. The annual fixation rate was calculated from the sum of the mean monthly rates m^{-2}, corrected for temperature, using the mean monthly temperatures at 10 cm depth (Chap. 1) and a Q_{10} of 3 (Granhall and Sellander, 1973). Determinations were also made at Cottage Hill A and Green Burn in April (incubation temperature 15.5° C) and at Cottage Hill B and Sike Hill in September 1971 (16° C and 18° C), and were corrected for temperature.

6.3 Results and Discussion

6.3.1 The Effect of Inorganic and Organic Nutrients on Numbers, Types and Activity of Bacteria

6.3.1.1 Percolation Experiments

The nutrient effect on microbial numbers in each of five experiments (Table 1) was investigated by a two-factor (treatment and time) analysis of variance, using a $\log_{10}$ transformation of all except the first count, which was made before percolation commenced. If the mean count for a nutrient differed significantly from that of its control (distilled water or glucose), the nutrient was said to have elicited a response. Thus amendment with Na glutamate, NH_4Cl, $NaNO_3$, $SrCl_2$, $CaCl_2$, $MgCl_2$, $NaHPO_4$, all with glucose, and $CaCl_2$ without glucose, produced a significant response, while $NaNO_3$ and KCl alone did not. In experiment 1 the counts for 15 and 30 days, but not the overall counts, were significantly greater for phosphate + glucose than for glucose alone.

In all treatments bacterial numbers changed significantly over the experimental period. The statistical analysis also showed significant interactions ($P<0.05$) between the treatment and the time factors showing that the bacterial growth curve differed for different treatments.

All the divalent cations tested produced a large response, but this was relatively slow with $CaCl_2$, with or without glucose, and $SrCl_2$, in contrast with other nutrients. In general there was a rapid increase in bacterial numbers, even in distilled water columns, in the first two days' percolation, assumed to be primarily a response to disturbance of the peat during preparation of the peat/sand mixture (Clark, 1967). The initial increase in microbial numbers varied with different treatments, with mean generation times of 7.3–10 h in those nutrients which produced the largest increases in the count, i.e. NH_4Cl, Na glutamate and $NaNO_3$, all with glucose, compared with 24 h in distilled water and 16 h in glucose. Although the latter produced a distinct response after 15 and 30 days' percolation, it had little effect on the maximum count achieved.

Respiration data (Sect. 6.3.1.3) suggest that the stimulatory effect of mixing the peat may be attributed to the consequent increased availability of organic substrate. While the increased rate of growth and the longer period when maximum counts are obtained in the glucose percolated columns support this, the lack of a marked increase in the maximum count indicates that organic substrate is not the only, or most important, limiting factor. The maximum count was much higher, and the mean generation time shorter, when a combined nitrogen and carbon source was provided, though nitrate in the absence of glucose produced little response. It seems that, as in the neutralised acid soil studied by Lowe and Gray (1973), both C and N must be supplied to elicit a large response. The small reserves in homogenised peat, while adequate to permit a certain increase in microbial growth in distilled water percolated columns, are such that neither C nor N is substantially more available, and the addition of either alone produces only limited additional growth due to inadequate availability of the other.

Table 1. Bacterial populations in peat/sand mixtures percolated with various nutrient solutions

Experiment number	Treatment	Time (d)							Log mean	Log LSD ($P<0.05$)
		0	2	4	7	15	30	mean		
1	Control (dist H_2O)	0.91	4	5	—	3.3	3.6	4	0.599	0.13
	glucose	0.91	8	8	—	3.4	6.2	7	0.795[a]	
	glucose + $NaNO_3$	0.91	25	39	—	29	24	31	1.453[b]	
	glucose + Na glutamate	0.91	30	33	—	51	37	36	1.535[b]	
	glucose + NaH_2PO_4	0.91	4	7	—	6.6	11	8	0.874[a]	
2	Control (dist H_2O)	nd	25	22	—	21	17	21	1.321	0.102
	$NaNO_3$	nd	22	21	—	18	28	20	1.307	
	glucose + $NaNO_3$	nd	36	35	—	27	23	30	1.475[b]	
	glucose + NH_4Cl	nd	81	110	—	30	14	39	1.593[b]	
3	Control (glucose)	0.11	—	25	—	15	5	11	1.053	0.135
	glucose + $CaCl_2$	0.11	—	28	—	43	67	44	1.634[b]	
	glucose + $SrCl_2$	0.11	—	30	—	44	94	51	1.707[b]	
	glucose + $MgCl_2$	0.11	—	41	—	24	21	26	1.416[b]	
4	Control (dist H_2O)	0.08	—	18	—	17	5.2	11	1.057	0.077
	glucose	0.08	—	21	—	15	11	15	1.178[b]	
	glucose + NaH_2PO_4	0.08	—	22	—	26	20	22	1.349[b]	
5	Control (dist H_2O)	0.84	—	22	4	8	—	7	0.852	0.244
	KCl	0.84	—	16	8	10	—	9	0.972	
	$NaNO_3$	0.84	—	14	9	24	—	13	1.096	
	$CaCl_2$	0.84	—	15	14	68	—	39	1.316[b]	

All counts are $\times 10^8 g^{-1}$ dw peat. Nutrient solutions were added at 0.002 M concentrations except glucose (0.003 M) and $NaNO_3$ (0.004 M) in experiment 5. nd: no data; LSD: least significance difference. Treatments differing from controls are shown: [a] $P<0.05$; [b] $P<0.01$.

The delayed response to phosphate suggests that the supplies of available phosphorus, while initially adequate, became limiting later due to leaching and immobilisation in the columns.

The response to the divalent cations is more complex. Unlike other nutrients, including $MgCl_2$, calcium and strontium elicited a steady increase in microbial numbers over the experimental period. Direct nutritional effect of calcium seems improbable since the level of calcium in the peat, well above that required by microorganisms (Nicholas, 1963), is unlikely to be limiting, and strontium elicited a slightly greater response. A pH effect may be excluded as calcium amendment produced a slight decrease in pH by exchange acidity. It is likely that the calcium may act on the organic matter, possibly overcoming inhibitory effects on extracellular enzymes, and thereby increasing the availability of nitrogen or other substrates. Ladd et al. (1968) and Ladd and Butler (1969a, b, 1970) found that 10^{-3} M $CaCl_2$ overcame the inhibitory effects of humic acids, abundant in peat, on proteolytic enzymes. It is interesting to note that the calcium content of plant litters was one of the factors positively correlated with their rate of decomposition (Chap. 7).

The contrast between the stimulatory effect of calcium salts on microbial numbers and their lack of effect on respiration rates (Sect. 6.3.1.3) reflects the two different parameters being measured. In organic soils, the relationship between viable bacterial counts and soil respiration is indirect, the latter being a measure of the rate of energy utilisation by the total soil microflora, which is determined by a range of factors such as nutrient and substrate availability, soil conditions and interactions between organisms. However, amendment with calcium did increase the proportion of the yellow pigmented bacteria, mentioned below, in the homogenates used for the O_2 probe measurements.

The level of magnesium in the peat, 600 ppm (Latter, pers. comm.), 70% of which is available (Gore and Allen, 1956) is much higher than that required for optimal bacterial growth (Nicholas, 1963), and it is not clear why percolation with $MgCl_2$ stimulates microbial numbers in the same way as does percolation with a nitrogen source plus glucose. Possibly the high Mg levels help organisms to tolerate other inhibitory effects, e.g. high hydrogen ion concentrations (Nicholas, 1963).

In bacterial population studies, Gram-negative rods comprised 94–99% and spore-forming bacteria only 2% of the organisms isolated from the treatments and numbers of bacterial endospores did not increase in the distilled water-percolated columns over the first 15 days. Amendment with a nitrogen source + glucose increased the proportion of bacteria able to grow on crystal violet nutrient agar especially in the early and final stages of the percolation run.

Percolation with Na glutamate produced large numbers of facultatively anaerobic acid- and gas-producing bacteria, probably of the coliform group which, using the techniques described in this paper, have been isolated in significant numbers at Moor House only from the rhizosphere of *Eriophorum* spp. This is an interesting observation in view of the known secretion of amino acids by many plant roots.

Gram-negative organisms producing a yellow non-diffusible pigment formed 10–20% of colonies on dilution plates from columns percolated with $MgCl_2$,

Table 2. The effect of nutrient amendments on bacterial populations in the dark brown horizon in the field

Bacterial population	Dist H_2O control	Glu-cose	NaCl	$CaCl_2$	KCl	NaH_2PO_4	$NaNO_3$	Na gluta-mate
Total viable count	8.6	4.9	8.3	5.0	6.0	6.4	6.4	124[a]
% crystal violet tolerant bacteria	9.7	18.0	9.5	9.0	8.5	14.0	6.5	27[a]

Data are given as numbers $\times 10^6$ g^{-1} dw, after four weeks.
[a] Significantly higher than control $P<0.05$.

$CaCl_2$ or $SrCl_2$, but not more than 1–2% of colonies with other treatments. The significance of this observation is not known.

6.3.1.2 Field Experiments

To relate laboratory observations to field effects, 0.2 M solutions of NaCl, KCl, $CaCl_2$, NaH_2PO_4, $NaNO_3$, Na glutamate or glucose were injected into peat cores in the field. Na glutamate produced a 10-fold increase in bacterial numbers (Table 2), and the significantly ($P<0.05$) higher proportion of crystal violet-tolerant bacteria shows that it also altered the composition of the bacterial population. There was no significant response ($P>0.05$) to the other nutrients tested, including glucose or Na NO_3. The response to glutamate, which may serve as both a carbon and a nitrogen source, supports the laboratory observation that a combined carbon and nitrogen source in necessary to promote microbial activity. The absence of any response to carbon or nitrogen added separately or to other nutrients such as calcium or phosphate, which stimulated microbial numbers in the laboratory experiments, probably reflects reduced demand for those nutrients because of reduced microbial activity under the less favourable conditions in the field.

6.3.1.3 Respiration in Laboratory Experiments

The influence of salt solutions on the respiration of hand-mixed peat samples was measured in the Warburg manometer by adding singly 0.1 ml of a solution of the following salts (final molarities in the peat) 0.07 M NH_4Cl, 0.025 M NaH_2-PO_4, $CaCl_2$, KCl or a mixture of 0.07 M NH_4Cl and 0.025 M NaH_2PO_4. A treatment with 0.025 M NaCl was included to compensate for any non-specific inhibitory salt effects.

Declining respiration with time, as is indicated in Table 3, has been shown by Chase and Gray (1957) to produce a straight line if plotted logarithmically. The regressions for log uptake against log time were therefore calculated for all treatments and tested for parallelism (gradient m', using the terminology of Chase and Gray, 1957) and coincidence (intercept of y axis f')by the method of Eisenhart and Wilson (1943). The symbols used are those used by Chase and Gray (1957) who expressed the relationship as log uptake rate $= \log f' - m' \log$ time.

Table 3. The influence of nutrient treatments on microbial respiration rates in peat measured in the Warburg manometer

Treatment	Time (d)					Total uptake	Regression log rate/log time (see text)	
	2	3	4	7	18		gradient $-m'$	intercept f'
Dist H_2O	32	34	37	24	19	9,106	−0.323	2.1004[a]
0.02 M NaCl	26	21	25	14	14	5,910	−0.333	1.9541
0.07 M NH_4Cl	46	34	26	21	11	6,941	−0.643[a]	2.647[a]
0.025 M NaH_2PO_4	31	26	29	22	15	6,368	−0.339	2.061[a]
0.02 M KCl	26	21	26	18	14	6,402	−0.345	1.996
0.015 M $CaCl_2$	41	22	24	22	13	7,732	−0.449	2.265
0.07 M NH_4Cl+ 0.025 M NaH_2PO_4	52	26	26	17	10	6,483	−0.56[a]	2.522[a]

Data are expressed as $\mu l\ O_2\ g^{-1}\ h^{-1}$ dry peat at 27° C. Total uptake expressed as $\mu l\ O_2\ g^{-1}$ over the experimental period.
[a] Significantly different from the NaCl control ($P<0.05$).

In the early stages of the experiment, there was a large response to NH_4Cl and a weak response to Na phosphate, but absence of a significant difference ($P>0.05$) between the regressions for NH_4Cl or NH_4Cl + NaH_2PO_4 showed there was no synergistic effect between these nutrients. Despite the initially high respiration rate in $CaCl_2$-amended peat, its regression line was not significantly different from the salt control. This contrasts with the report by Küster (1963) that $CaCl_2$ stimulated respiration, which reached a maximum two to three days after amendment, declining thereafter. However, the initially high rate in the Warburg studies may represent a similar effect, but one which is too short-lived to appear significant by the statistical analysis used. If the effect is real, it is in marked contrast to the slowly developing, large and prolonged stimulation of microbial numbers by $CaCl_2$ amendment.

For experiments using the oxygen electrode, replicate peat homogenates were amended with 0.01 M $NaNO_3$, 0.008 M NaH_2PO_4, 0.005 M $NaHSO_4$, 0.005 M KCl, 0.005 M $CaCl_2$, 0.003 M $Ca(OH)_2$, or 0.003 M Na glutamate. Although these concentrations are lower than those used in Warburg manometers, the concentrations g^{-1} peat are little altered. A two-factor analysis of variance, without a log transformation of the data, was used to assess the significance of the results obtained.

Only Na glutamate significantly ($P<0.01$) stimulated the oxygen uptake rate of the homogenates (Tables 4 and 5). There was no significant response ($P>0.05$) to any of the other nutrients tested including Na nitrate. The respiration rates of the control homogenates were about three-fold higher than those observed for intact peat in the Warburg manometers.

The oxygen uptake rate of peat homogenates incubated statically for three days or longer was significantly ($P<0.01$) higher than for those incubated under well-aerated conditions in an orbital incubator (Table 6). A large part of this increase appeared to be caused by non-biological oxygen uptake since, compared

Table 4. The effect of treatment with various nutrients on the oxygen uptake rates (µl O_2 g^{-1} h^{-1}) of peat homogenates incubated on an orbital shaker at 29° C and measured with a Clark type O_2 electrode

Treatment	Time (d)			Mean rate	LSD
	2	6	10		
None	149	98	160	136	50
0.005 M $CaCl_2$	nd	63	149	106	
0.01 M $NaNO_3$	114	114	206	145	
0.005 M $NaHSO_4$	126	100	98	108	
0.008 M NaH_2PO_4	137	160	80	125	
0.005 M KCl	nd	114	114	114	

nd: no data; LSD: least significant difference at 5% level.

Table 5. The effect of $Ca(OH)_2$ and Na glutamate on the oxygen uptake rates (µl O_2 g^{-1} h^{-1}) of peat homogenates

Treatment	Time (d)			Mean rate	LSD	Total uptake
	3	4	7			
Shaken[a]						
Macerate	114	103	125	120	60	18,300
Macerate + $Ca(OH)_2$	114	nd	160	137		nd
Macerate + Na glutamate	149	515[c]	366[c]	257[c]		49,300[c]
Static[b]						
Macerate	538	629	274	480	200	nd
Macerate + $Ca(OH)_2$	172[c]	252[c]	172[c]	199[c]		nd
Macerate + Na glutamate	596	985[c]	870[c]	817[c]		nd

Nutrient solutions were added at 0.003 M concentration. nd: no data; LSD: least significant difference at 5% level.
[a] Homogenates previously incubated on an orbital incubator at 29° C. [b] Homogenates incubated statically at 2[illegible] C. [c] Significantly different from control, $P<0.05$.

Table 6. The effect of various treatments on oxygen uptake (µl O_2 g^{-1} h^{-1} at 29° C) by peat homogenates (20% w/v in distilled water)

Treatment	Time (d)			
	0	1	3	6
Shaken	217	148	137	148
Shaken + azide	68	57	41	38
Static	217	240	606	686
Static + azide	68	125	469	549
Static + γ radiation sterilised	240	nd	23	23

Homogenates were incubated on an orbital incubator (shaken) or static. Na azide was added in some treatments to give a concentration of 0.001 M, just before inserting the O_2 probe. nd: no data.

with the aerated homogenates, there was a relatively small decrease in the oxygen uptake rate when biological activity was inhibited by the addition of Na azide immediately before measurement. Non-biological oxygen uptake can be eliminated with calcium amendment (Table 5) and, since it was absent from γ radiation-sterilised homogenates and increased over the first three days' incubation, microbial production of reducing conditions may result in the formation of a pool of readily oxidisable reduced compounds. Flocculation of peat in the homogenates shows that calcium affects peat organic matter and may inhibit non-biological oxygen uptake by stabilising either the reduced or oxidised organic compounds.

The decline in respiration rate with time in the Warburg studies (Table 3) suggests depletion of a small pool of organic matter (Chase and Gray, 1957) made available by the disturbance of the peat in sample preparation (Clark, 1967). The absence of such a pool in undisturbed peat may explain the lack of response to inorganic nitrogen in the field, while homogenising the peat promotes a far greater increase in organic matter availability, resulting in a higher respiration rate, maintained for at least seven days (Tables 4 and 5). Waksman and Stevens (1929) and Waksman and Purvis (1932) found an increase in availability of organic matter in peat from which lignin and hemicellulose had been extracted by alkali, or waxes by toluene. This suggests that otherwise available substrates may be masked by resistant materials such as lignin or protein-tannin complexes (Handley, 1954; Benoit et al., 1968; Benoit and Starkey, 1968a, b). In support of this, an inverse relationship was found between rate of decomposition and tannin and lignin contents in various litters (Chap. 7).

Response to nitrogenous compounds was also affected by homogenisation, homogenised peat showing no response to inorganic nitrogen amendment while there was an initial increase in rate of oxygen uptake in relatively undisturbed peat. The absence of any increase in total oxygen consumption in the latter reflects the smallness of the available organic matter reserves. The different forms of nitrogen used may partly explain the differing response to nitrogen amendment, since the NO_3^--N added to the homogenates is less available to the microflora than the NH_4^+-N used in the Warburg experiments. In view of the responses to both forms in the percolation studies, however, such an effect is likely to be much less important than that due to the greater nutrient stress in intact peat.

While homogenising peat relieves the nutrient stress so that neither nitrogen nor carbon limits organic matter decomposition, addition of a combined carbon and nitrogen source, e.g. Na glutamate, exerts a priming effect on respiration, as it does in the percolation system (Broadbent, 1947). Nitrogen plus glucose, but not nitrogen alone, stimulated decomposition of a muck soil (Stotzky and Mortensen, 1957); similarly, the C:N ratio in a peat soil determines the effect of added nitrogen. Thus addition of large amounts of nitrogen has an unfavourable effect if the available carbon reserve is small, but activity improves if the reserve is increased, e.g. by treatment with organic solvents (Küster, 1963).

The microbial processes which lead to the formation of reducing compounds in the homogenates are presumably related to those occurring in nature which result in the formation of reducing conditions in the blanket bog. The reducing compounds themselves may be ecologically important in stabilising redox potential in environments subjected to fluctuating water levels, and may, by competing with the microflora for the oxygen released, account for the small rhizosphere effect around *Eriophorum* roots (Martin, 1971; Chap. 5).

6.3.2 Nitrogen Fixation Rates

The rates of ethylene production $cm^{-3} h^{-1}$ (Collins and D'Sylva, pers. comm.) at Bog End were not significantly different ($P > 0.05$) for the three sample treatments, the two sampling times or the four horizons investigated (Table 7). The data for each horizon at Bog End were pooled and the mean annual nitrogen fixation rates m^{-2}, corrected for seasonal temperature changes, were calculated for each horizon and for the peat to a depth of 30 cm (Table 8).

The very approximate figure, $3.2 g m^{-2} yr^{-1}$, for the total quantity of nitrogen fixed at Bog End, based on an extrapolation from a few highly variable samples taken on only two occasions, is nevertheless similar to that found for a range of soil types (Chang and Knowles, 1965; Knowles, 1965) though higher than that observed in most tundra soils (Alexander, 1974). In contrast to the general trend of decreasing activity with depth (Chaps. 5, 7) the rate of N fixation cm^{-3} is similar in all horizons, possibly because of the participation of both aerobes and anaerobes although the numbers of potential nitrogen fixers decrease down the profile (Chap. 5). The large amount of N fixed in the green-brown horizon is a reflection of the greater depth of this horizon, rather than a greater rate.

The nitrogen fixation rates at the other sites (Tables 7, 8), based on a smaller number of samples obtained by Collins and D'Sylva, are much lower than those for Bog End, emphasising the need for more extensive sampling to provide a realistic estimate of the annual input of nitrogen through fixation.

6.3.3 The Release of Nutrients from Peat

To evaluate the role of microorganisms in nutrient cycling and the effect of added nutrients on the release of other nutrients from the peat, the effluents from the percolation columns were analysed for K^+, Ca^{2+}, Mg^{2+}, NO_3^--N, NH_4^+-N and phosphate.

The rates of release of nutrients from separate pairs of columns percolated with distilled H_2O, glucose or $CaCl_2$ respectively are plotted against time (Fig. 2) and total quantities released (Table 9) are indicated by the areas of the blocks. The kinetics of the process indicate whether physical processes, such as leaching or ion exchange (an initially rapid then declining rate) or biological processes (a sustained or increasing rate of release) are in operation.

Potassium and phosphorus were released by leaching alone and calcium and magnesium by leaching and ion exchange processes, e.g. the release of Mg from $CaCl_2$ percolated columns. The total amount of Ca and Mg released from glucose percolated columns exceeded that from controls and the rate increased with time. Since there was no evidence from pH measurements of stimulation by H^+ ions, microbial action, perhaps by production of a chelating agent (Louw and Webley, 1959), is assumed to be involved.

Table 7. Acetylene reduction at five Moor House sites

Horizon	Bog End						Sike Hill			Cottage Hill A			Cottage Hill B			Green Burn		
	April 1971			June 1971			Sept 1971			April 1971			Sept 1971			April 1971		
	1	2	3	1	2	3	1	2	3	1	2	3	1	2	3	1	2	3
Litter	0.161	0.119	0.564	0.162	0.093	0.124	nd	0.076	0.035	nd	0.089	0.119	nd	nd	0.061	0.081	0.024	0.005
Dark brown	0.414	1.199	0.768	0.059	0.463	0.228	nd	nd	0.052	nd	nd	0.272	nd	nd	0.073	0.018	0.017	0.162
Green-brown	0.182	0.013	0.904	0.638	0.280	0.069	nd	0.114	0.152	nd	nd	nd	nd	nd	nd	nd	1.813	nd
Red-brown	0.284	0.084	0.338	0.638	0.160	nd	nd	nd	nd	nd	nd	nd	nd	nd	nd	0.028	1.157	nd
Incubation temperature (°C)	11			11			18			15.5			16			15.5		

Results (μ mol ethylene $cm^{-3} h^{-1}$) are means of three subsamples from three cores. 1: Ar; 2: $Ar+O_2$; 3: $Ar+O_2+CO_2$; nd: no data.

Table 8. The annual rate of nitrogen fixation ±95% confidence limits where available (g N m^{-2} yr^{-1})

Site/horizon	Fixation rate
Bog end	
Litter	0.46±0.6
Dark brown	0.52±0.3
Green-brown	1.53±1.2
Red-brown	0.63±0.3
Total	3.20±2.6
Sike Hill[a]	0.06
Cottage Hill A[a]	0.12
Cottage Hill B[a]	0.05
Green Burn[a]	0.054

[a] Results quoted from Alexander (1974) corrected for annual temperature variation.

Table 9. Quantities (μ mol g^{-1} dry peat) of nutrients released from peat/sand columns, percolated for 30 days with various inorganic salts, compared with quantities in the original peat (Latter, pers. comm.)

Percolate	K	Ca	Mg	N(NO_3+NH_4 or total)	P
Dist H_2O	8.8 ±2.1	7.2 ±2	4.7 ±1	5.4 ±1.7	3.2 ±1.5
Glucose	8.7 ±2.7	10.4 ±7.2	10.2 ±2.3	4.1 ±2.2	1.9 ±1.2
$CaCl_2$+glucose	7.8 ±1.3	−67 ±2	18.1 ±0.3	4.7 ±2.1	1.5 ±1
$MgCl_2$+glucose[a]	5.3	52	−164	3.3	2.6
NaH_2PO_4+glucose	8.6 ±1.3	29 ±1	15.1 ±2.7	6.2 ±1.6	−44 ±11
$NaNO_3$+glucose[a]	12.4	7.2	7.5	−396	3.9
Na glutamate +glucose[a]	11.7	14.9	9.7	322	3.2
Original peat	20	51	25	870	33

A negative value denotes fixation, or loss, presumably by denitrification, in the case of NO_3^-. Data, ± mean difference for each replicate pair from the mean, are averages from four to six columns except those marked[a], which are readings from two columns only. Nutrient solutions were added at 0.002 M concentrations, except in the case of glucose (0.003 M) and $NaNO_3$ (0.004 M).

A small amount of nitrate was leached from the columns but the absence of detectable nitrate production, except possibly in the $CaCl_2$ percolated columns, indicates a lack of nitrification, even with NH_4Cl or Na glutamate. This is expected from the absence of nitrifying bacteria in the peat (Chap. 5).

The considerable and steady loss of nitrate from columns percolated with nitrate alone (220 μ mol g^{-1} dry peat) and especially with nitrate + glucose (280 μ mol), however, implies active denitrification by the large numbers of nitrate reducing and denitrifying bacteria in the peat (Latter et al., 1967; Chap. 5).

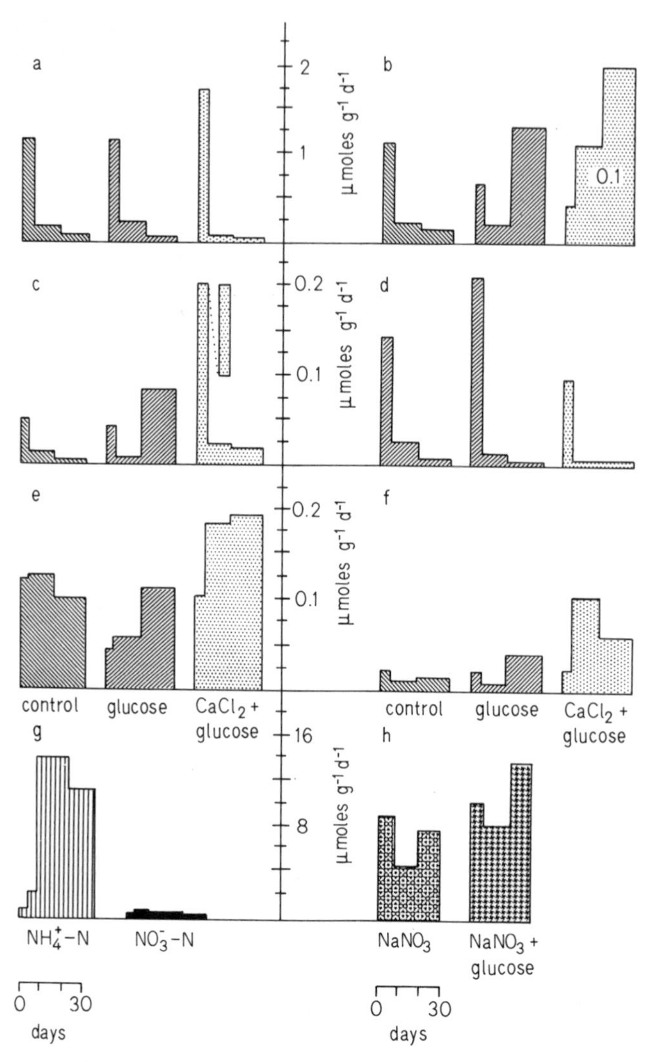

Fig. 2 a–h. Changes with time in the rate of release of inorganic nutrients from peat/sand mixtures percolated with distilled H_2O (control), glucose and $CaCl_2$ + glucose. (a) K^+; (b) Ca^{2+} (note rate shown for $CaCl_2$ + glucose treatment is 0.1 × true value); (c) Mg^{2+}; (d) PO^{3-}; (e) NH_4^+; (f) NO_3^-; (g) comparison of the rate of release of NH_4^+ and NO_3^- from peat/sand mixtures percolated with Na glutamate; (h) comparison of the rate of loss of NO_3^- from mixtures percolated with $NaNO_3$ alone and $NaNO_3$ + glucose. In all experiments the concentration of nutrients in the input solutions was 0.002 M except for glucose which was 0.003 M

A steady, even increasing, rate of NH_4^+ release implies biological ammonification (mineralisation of organic N, a crucial step in the N cycle in peat) but this is difficult to quantify owing to release by leaching and possibly extensive immobilisation. However, active deamination of glutamic acid was shown to occur (Fig. 2).

An inverse relationship between maximum bacterial count and the quantity of phosphorus released for a given treatment indicates immobilisation of phosphorus (Martin, 1971). With an increase in viable bacterial count of 1×10^9 g^{-1} relative to the corresponding control, 0.2 μmol of phosphorus was immobilised, close to the value calculated on the basis of a live weight cell mass of $1 \cdot 1 \times 10^{-12}$ g (small rods were the predominant organisms on the plates) and a phosphorus content of 0.5% (Luria, 1960).

6.3.4 Immobilisation of Nutrients

Assimilation of mineral nutrients such as nitrogen, phosphorus, potassium, and magnesium by microorganisms removes them from circulation for the cell's lifespan. Because such immobilisation may be important in the ecology of a nutrient poor system such as blanket peat, the amounts were roughly estimated; most parameters are only approximate and the estimate is thus significant in order of magnitude only.

On the basis of a total bacterial count of 2×10^{10} g^{-1} dw peat, a biomass of 1.1 and 11×10^{-12} g^{-1} for small and large bacterial cells respectively and a moisture content of 80% (Luria, 1960) the bacterial biomass was estimated to be 7.4 mg g^{-1} peat, or 190 g m^{-2}. Cell nutrient contents, based on data from A. F. Harrison (pers. comm.) for mean concentrations in cells of two species grown on soil extract agar and TSA, varied considerably with the different growth media, but the nitrogen content of bacteria isolated directly from the peat (P. M. Latter, pers. comm.) is close to that used to calculate the values of nutrient immobilisation in distilled water percolated columns (Table 10).

The significance of microbial immobilisation of any nutrient depends on the proportion of the potentially available pool (as opposed to the total pool) rendered unavailable to other organisms. Although, with the natural turnover of the microbial population, the immobilised nutrient is released, its total amount will vary only if the total bacterial biomass changes. The available pool is calculated as the quantity immobilised in organisms plus the quantity available after immobilisation [i.e. the maximum quantity leached from any of the treatments

Table 10. Estimates of the amounts of nutrients (mg g^{-1} peat) in peat and the proportions contained in bacterial cells i.e. immobilised, and leached in peat/sand mixtures

Nutrient	Content in peat	I Immobilised in bacteria	L Leached	$\frac{I}{I+L} \times 100$
Potassium	0.77	0.08 (11)	0.34 (44)	20
Magnesium	0.59	0.04 (7)	0.43[a] (72)	9
Phosphorus	0.98	0.08 (8)	0.10 (10)	44
Nitrogen	12.20	0.44 (4)	0.09[b] (1)	87

Results are from columns percolated for 30 days with distilled water or with $CaCl_2$[a] or NaH_2PO_4 plus glucose[b]. Values in parentheses show amounts as a % of that in peat. Nutrient concentrations in bacterial cells are calculated using data from A.F. Harrison (pers. comm.) K 1.12; Mg 0.56; N 5.91; P 1.07% dw.

(Table 10) which, on the basis of Gore and Allen's (1956) data, probably underestimates potassium by 50%].

Whilst microbial immobilisation of almost 90% of the biologically available nitrogen pool occurs in the columns (Table 10), calculations based on the lower microbial biomass (43.9 g m^{-2}, excluding unstained fungal hyphae) in the blanket peat (Chap. 5) suggest only 25% immobilisation in the field. Forty four percent of the biologically available phosphorus is immobilised in the columns. The corresponding value in the field is 50% if the extractable phosphorus [2% of the total in peat (Allen, 1964)] represents the net biological availability under field conditions, or only 10% if the proportion leached from the columns (10%, Table 10) is nearer the true estimate for availability.

Only a relatively small proportion of the biologically available pool of magnesium (9%) and potassium (20%) is immobilised in the laboratory, while corresponding figures for the field are 2% and 4% respectively. As these cations are much more available than those of nitrogen and phosphorus (Gore and Allen, 1956), microbial immobilisation is unlikely to be important.

6.3.5 The Nitrogen Cycle in the Blanket Bog

A broad outline of the nitrogen budget for the blanket bog at Moor House is given in Heal and Perkins (1976) and Chapter 10 but microbial studies on nutrient circulation provide more detailed information on transfers within the ecosystem (Fig. 3). Despite gaps in the data, which necessitate extrapolation, the main pathways and rates are indicated, and areas for further research indicated.

Nitrogen input is in precipitation and by fixation. Of the annual input in precipitation of *c* 0.8 g m^{-2} yr^{-1}, *c* 0.7 g is inorganic N (Gore, 1968) and, on the basis of data for rainwater (Allen, 1964), consists of 0.6 g NH_4^+ and 0.1 g NO_3^-. Preliminary estimates for N fixation range from 0.05 to 3.2 g m^{-2} yr^{-1} on the bog (Table 8).

Whilst the total amount of N in peat to 2–3 m depth is of the order of 2000–3000 g m^{-2}, the biologically active upper 30 cm has a standing crop of around 300–400 g m^{-2} (Gore, 1972), distributed between four major components: live plants *c* 10–15 g, surface litter *c* 20 g, microflora *c* 2 g and a massive peat component of *c* 340 g. From amounts of N immobilised and released as NH_4^+ from control columns (Table 9) minus nitrogen initially present as NH_4^+ and that estimated to be present initially in the microflora, it appears that the peat nitrogen consists of only a small (10 g) pool of actively cycling, readily available N, the remainder forming a large passive pool.

In addition there are two small pools of inorganic nitrogen which are subject to considerable flux through mineralisation of organic nitrogen by microflora and uptake by plants. Extrapolation to 30 cm depth of Allen's (1964) data for the 0–18 cm region, estimates the NH_4^+ pool to be *c* 0.1 g m^{-2} 30 cm^{-1} and that of $NO_3^- < 1$ mg m^{-2}.

It is of interest to compare the nitrogen levels at Moor House with those in a wet meadow tundra soil in Alaska (Flint and Gersper, 1974) which have a similar quantity of nitrogen in the upper 30 cm, divided into a large passive and a small

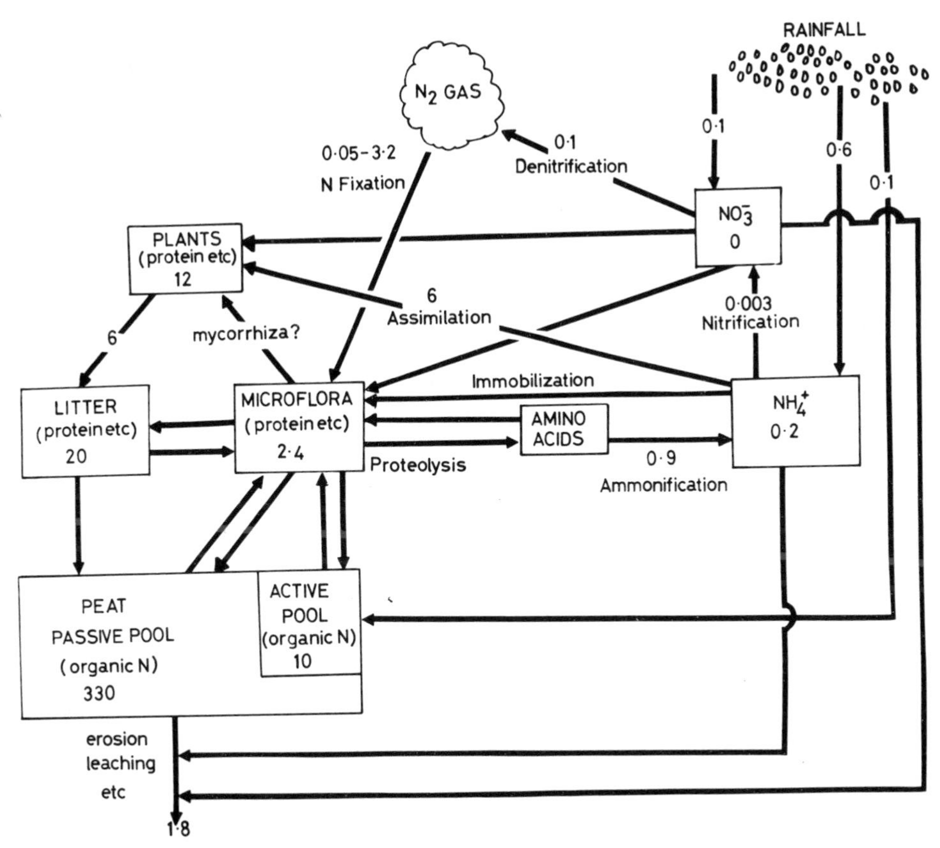

Fig. 3. A preliminary summary of the nitrogen cycle in blanket peat showing the participation of microorganisms. Values *in boxes* represent the steady state quantity (g m^{-2}) of nitrogen in that particular pool to a depth of 30 cm. Values *beside arrows* represent the calculated rate of transfer (g m^{-2} yr^{-1}) between pools. Where no value is given no estimate of rate has been possible

active pool, and a similar standing crop of microbially immobilised N. In the tundra soil, however, the available N seems to be largely in the form of NH_4^+-N, increasing seasonally from 2 g to 6 g m^{-2}, with a smaller amount of organic N. This contrasts with the situation at Moor House, where the bulk of the available N, *c* 10 g m^{-2}, is organic; the data do not permit conclusions on seasonal variation.

The total amount of nitrogen in circulation annually, *c* 40–50 g m^{-2}, represents the transfer between litter, the active pool, the microflora, NH_4^+ and plants. The time span for the cycle is probably many years, as indicated by the nitrogen changes in decomposing litters. Leaves of *Rubus chamaemorus* show slight mineralisation of nitrogen during the first few years of decomposition with C:N ratios reaching 16 after three years (Dowding, 1974; Chap. 7). However, in other, more abundant species, a longer period elapses before C:N ratios decline to a level at which nitrogen may show net mineralisation.

Before net mineralisation occurs, the organic nitrogen in the decomposing litter undergoes either chemical reactions with tannins and polyphenols or microbial transformation to form decay-resistant compounds which are incorporated into the large, passive N pool. This net loss from the actively circulating fraction is probably balanced by a very slow release from the passive pool, and by inputs in precipitation and fixation. Although some N is mineralised within the litter, it is rapidly reabsorbed by the microflora which, with a biomass of 6–9 g m^{-2} (Chap. 5) and a N content of 6% (Table 10), contains *c* 2% of the total nitrogen in the litter, i.e. most of the actively circulating fraction.

The mineralisation of litter protein and other nitrogenous compounds involves both fungi and bacteria. Besides their role as decomposers, the fungi may aid bacterial proteolysis, e.g. mycorrhizal fungi in ericads decompose tannins and other inhibitory materials and produce microhabitats at more favourable pH values (Burgeff, 1961). They also produce unidentified substances which stimulate bacterial proteolysis in anaerobic horizons of the peat. Free-living individuals of these and other fungi probably produce similar effects in litter not penetrated by roots, thus accelerating the nitrogen cycle in their immediate environment. Because of the high C:N ratio of plant remains, most of the mineralised nitrogen is reabsorbed immediately by other microorganisms in the litter. The nitrogen re-entering the main cycle is provided by mineralisation of microbial cells (Jannson, 1958; Chu and Knowles, 1966; Knowles and Chu, 1969), presumably when the ratio of available carbon to available nitrogen declines to 20–30:1 (Alexander, 1961).

Litter and microbial protein undergoes proteolysis to amino acids which are further broken down by ammonification, liberating NH_4^+. Mineralisation of other nitrogenous compounds e.g. nucleotides, amino sugars etc., also liberates NH_4^+ (Alexander, 1961). The rate of ammonification (0.9 g N m^{-2} yr^{-1}) is estimated from the mean rate of release of NH_4^+ from the distilled water percolated columns in the laboratory (Table 9) multiplied by the ratio of bacterial plate count in the field to that in the columns to correct for reduced bacterial activity in the field. A Q_{10} value of 2 is used to correct for lower field temperatures (Flanagan and Veum, 1974). These corrections applied to this and other extrapolations from laboratory data may give significant underestimates of rates since no allowance is made for a possible increase in fungal activity, or for differences in the proportion of the active microflora not enumerated by plate counts, under field conditions. Nitrogen transformation in the peat may also be limited by the low N availability rather than temperature.

The small pool of NH_4^+ from mineralisation and rain can undergo assimilation by plants or microorganisms, nitrification or leaching.

The mean value for throughput of vegetation dry matter is about 650 g m^{-2} yr^{-1}, with a mean N content of 0.9% at death. It is assumed that this is equal to the amount assimilated, i.e. 6 g N m^{-2} yr^{-1}, mostly from the NH_4^+ pool. The magnitude of the microbially assimilated to microbially immobilised nitrogen transfer for NH_4^+ is difficult to estimate since results discussed earlier suggest that microorganisms also use amino acids as a nitrogen source. The total nitrogen assimilated by the litter microflora can, however, be estimated, assuming, by the method of Babiuk and Paul (1970) 30 generations per annum. The value of *c*

0.4 g N immobilised in the litter microflora, equivalent to the assimilation of *c* 13 g N yr^{-1}, suggests that on average an atom of combined nitrogen from the litter input is recycled twice through the microflora before becoming mineralised and available to the vegetation. This assumes that most mineralisation occurs in the litter horizon, an assumption supported by the distribution of proteolytic organisms (Chap. 5) and by decomposition data (Chap. 7).

Mycorrhizal fungi are known to play an important role in the nitrogen and phosphorus nutrition of many plants. Their significance in the cycling of both these nutrients in the blanket peat will depend on their incidence and activity. Vesicular-arbuscular mycorrhizae resulting from *Endogone* infection are of little significance in the wet blanket peat (B. Mosse and D. S. Hayman, pers. comm.) However, *Calluna* has well-developed mycorrhizae (Pearson and Read, 1973a) which almost certainly plays an important role in the nitrogen nutrition of the plant. Read and Stribley (1973) and Stribley and Read (1974) showed the growth and nitrogen content of mycorrhizal *Vaccinium macrocarpon* plants were greater than in non-mycorrhizal plants and that the former had access to a soil nitrogen fraction inaccessible to non-mycorrhizal plants. This links interestingly with Burgeff's (1961) observation that symbionts from ericads decomposed a wide range of often quite resistant substrates, some not utilised by other peat organisms, and in addition produced some factor which stimulated bacterial proteolysis.

If *Calluna* behaves similarly to *Vaccinium* the mycorrhizal fungi could contribute roughly 0.8 g N m^{-2} yr^{-1} of the total 6 g N m^{-2} yr^{-1} assimilated by the primary producers. Since phosphorus may be translocated from the mycorrhizal fungi to the host plant (Pearson and Read, 1973b) they may also contribute to phosphorus cycling in the blanket bog ecosystem.

Nitrification (NH_4^+ to NO_3^-) is probably a very limited process in acid waterlogged peats, no nitrification of added NH_4^+ being observed in peat/sand columns. Autotrophic nitrifiers could not be isolated from fresh peat (Chap. 5) and though slow heterotrophic nitrification may occur, the maximum rate, estimated from the quantity of nitrate in the effluent from unamended peat/sand mixtures, amounts to an insignificant 0.003 g N m^{-2} yr^{-1}.

Although 2 g NH_4^+ m^{-2} were leached from distilled water percolated columns, this third possible transfer from the NH_4^+ pool seems unlikely under field conditions, where NH_4^+ release is much slower and more intensively competed for by assimilation, and the total depth of peat much greater.

The extremely small nitrate pool, with a small input, mainly from rain, is unlikely to contribute significantly to nitrogen assimilation by plants or microflora and, with the possible exception of nitrate falling in heavy rain, it is doubtful if any escapes to be leached. The bacterial populations capable of denitrification (Latter et al., 1967; Chap. 5) are presumably concerned with any denitrification of nitrate in rain. The potential rate for denitrification, 0.1 g NO_3^--N lost as N_2 m^{-2} yr^{-1}, is estimated from nitrate loss from $NaNO_3$ percolated columns, corrected as for ammonification and using figures for denitrifier numbers given in Chap. 5. There are insufficient data to calculate the actual field rate, which depends on the balance between assimilation and denitrifier competition for the small amount of available nitrate.

Against the inputs and internal transfers in the bog (Fig. 3) can be set estimates of loss in runoff, both in solution (average over the catchment area *c* 0.3 g m^{-2} yr^{-1}) and as peat (Crisp, 1966). The source and form of the soluble N loss are unknown, but much is probably in the form of colloids or phenolic substances. Although peat particle losses are *c* 5 × as high, they represent loss from the large fraction of organic N and, probably being restricted to areas of eroding peat, are unrelated to nutrient transfers within the vegetated parts of the blanket bog.

An additional, potentially large, transfer of nitrogen occurs during burning. Allen (1964) estimated that approximately 4.5 g N m^{-2} could be lost from the standing crop of vegetation, i.e. the actively circulating fraction, in one burn, in gaseous constituents and smoke. The annual loss from burning on the long rotation of about 15 years practised at Moor House, is of the order of 0.3 g N m^{-2} yr^{-1}, and a high proportion of this may be redeposited on the adjacent bog surface and retained in the litter and peat rather than leached (Allen et al., 1969; Evans and Allen, 1971).

These estimates of annual transfer are clearly subject to large errors and will vary with annual and, particularly in the case of N fixation, local conditions. This gives rise to some uncertainty as to the actual contribution of N fixation, potentially the most important N input to the blanket peat, to the nitrogen cycle in the bog as a whole.

In conclusion, the blanket peat is a nutrient deficient soil, only 10% of its nitrogen content (0.6–1.8%) actively cycling and this passing through an ammonia rather than nitrate pathway as a result of environmental conditions. About 100 g m^{-2} yr^{-1} of organic matter, containing about 1.5 g N (25% of that passing through the vegetation annually), are lost from the biologically active layers to the deeper peat layers, where decomposition rates are extremely slow or negligible. The loss is apparently offset by nitrogen input in rainfall and fixation, but the evidence, in spite of uncertainties in the estimates, suggests a delicately balanced budget for the blanket bog ecosystem. Refinement of these estimates would require a massive research effort.

6.4 General Discussion of Factors Affecting Microbial Activity in Peat

Nutrient availability is only one of an interacting series of factors which influence microbial activity and result in the observed rates of decomposition and nutrient cycling and the amount and chemical composition of the organic matter.

The factors most likely to determine the microbial population of the peat are summarised (Chap. 5) as low oxygen tension, low pH and the low quality and availability of the organic matter.

The oxygen status of the peat, probably best expressed by the redox potential, declines down the peat profile, becoming reducing, i.e. anaerobic from a microbial viewpoint, at a depth of 14–16 cm (Urquhart, 1969; Urquhart and Gore, 1973), roughly corresponding with the green-brown horizon and the first obvious occurrence of H_2S. This decline in redox potential would be expected to be matched by a corresponding decline in microbial decomposing activity. Decomposition rates for a number of substrates decline with depth down the profile (Chap. 7) and Clymo (1965) found a much reduced decomposition rate for *Sphagnum* and filter paper under the anaerobic conditions prevailing below the level at which H_2S was permanently present.

The decline in redox potential not only affects microorganisms, but also results from their reducing activity. Because of the high water content and table in the blanket peat, oxygen diffusion is restricted and microbial consumption of oxygen produces anaerobic conditions.

Waterlogging is also partly responsible for the possible restriction of microbial activity in the peat by lower soil temperatures, though Latter and Heal (1971) showed the fungal microflora from Moor House to be better adapted to low temperatures in terms of growth at 1° C than that from the warmer Roudsea Wood, and winter isolates showed greater tolerance than summer isolates. However, only 40% of bacterial isolates from Moor House were capable of growing at 0–2° C (Chap. 5), and Martin (1971) found a plate count at 4° C for the green-brown horizon (10–20 cm), where *Bacillus* spp predominated, only 29% of that at 26° C, compared with 62% for the litter and dark brown horizons where Gram-negative rods predominated. Thus the adaption to cold does not apply to all genera or all horizons, and low temperatures will reduce microbial activity in general. Clymo (1965) found the rate of decomposition of *Sphagnum* was less at Moor House than at the warmer Thursley Bog in southern England. Although the reduction in activity does not necessarily lead to organic matter accumulation, waterlogged soils heat relatively slowly, leading to substantial above: below-ground temperature differences, which tend to favour plant over microbial growth, so contributing to the imbalance between production and decomposition, and the accumulation of peat (Latter and Heal, 1971).

It is difficult to assess the effective pH value in an environment as complex as peat. Raising the pH of a peat macerate (20% peat in distilled water) from pH 3.6 to pH 4.2 (dist H_2O) produced a three-fold increase in the microbial respiration rate (from 120 to 320 μl O_2 g^{-1} h^{-1}) while reducing the pH to 3.0 almost halved the rate (from 129 to 67 μl O_2 g^{-1} h^{-1}) (Martin, 1971). Yet the pH of the microhabitats in which the microorganisms are growing may be very different from the bulk pH of the soil; indeed none of the bacteria isolated from the peat grew in laboratory media at the measured pH of the blanket bog. Collins, D'Sylva and Latter (Chap. 5) found a degree of acid tolerance in a representative collection of fungi from the peat which showed little difference in growth over a pH range 4.5–7.0. They may play an important role in the ecosystem: Williams and Mayfield (1971) and Lowe and Gray (1973) have shown that fungal growth on organic matter added to an acid soil produced localised areas of raised pH, so providing favourable microhabitats for bacterial growth.

Experimental data (Sect. 6.3.1.3) indicate that microbial access to organic matter is restricted by physical factors, partly removed by homogenising, resulting from the reduced physical breakdown of the plant litter through low faunal activity. Only half the annual litter production passes through the soil fauna and some major litter macerating organisms, such as earthworms, are virtually absent (Chap. 4).

The overall restricting effect of the environment is seen in the lower microbial numbers in the blanket peat (11×10^6 g^{-1} in the dark brown horizon) compared to a laboratory model such as the percolation system (17×10^8 g^{-1} in distilled water columns). Respiration data show the effect even more clearly. Assuming an RQ of 1 and 50% C in the peat organic matter, the total oxygen uptake in unamended peat in the Warburg manometers was equivalent to the oxidation of 20% of the peat per year, and in the oxygen probe 100%. These values, similar to those reported by Latter et al. (1967), are much higher than those observed in the field, where comparable rates (5–40%) are found only in decomposing litter (Chap. 7).

Using data in Clymo and Reddaway (1972) for the combined production of CO_2 and CH_4, the net rate of breakdown of *Sphagnum* peat on the Burnt Hill area at Moor House is 0.35–0.63 g C dm^{-2} yr^{-1}. With an average peat depth of 2.7 m and a dry matter content of 10%, the mean output rate rate is 0.027–0.051 μl CO_2 g^{-1} h^{-1}, about 0.1–0.2% of the Warburg rate. Although the rate may be somewhat higher in the upper peat layers, environmental factors which are altered during laboratory measurements, particularly temperature and oxygen availability, clearly retard potential microbial activity.

Although nutrient amendments in the field demonstrated a combined carbon and nitrogen limitation, Clymo (1965) found no significant stimulation of decomposition of air dried *Sphagnum* in litter bags at the surface of Thursley bog by peptone or other nutrients. While this suggests that other factors are more important than nutrient limitations to microbial activity, Clymo found decay rates of more nutrient rich *Sphagnum* capitula nearly three times as great as those of mature *Sphagnum*, and it is possible that less decay-resistant materials may demonstrate nutrient limitation.

Thus microbial activity appears to be limited primarily by low availability of both organic matter and nitrogen, and perhaps other nutrients, in the native peat. Other factors may be secondary or complementary. Low availability of organic matter, and probably nitrogen and phosphorus, results from many factors, some of which are interrelated: physico-chemical barriers to microbial access, lack of oxygen, low pH and the formation of decay-(especially anaerobic) resistant complexes with lignin and tannins.

6.5 Conclusion

An attempt has been made to determine the nutritional and other factors limiting microbial activity in the blanket peat and to evaluate the role of microorganisms in nutrient cycling. Peat bacterial populations, mainly the Gram-negative component, increased when certain nutrient solutions were percolated through a peat/sand mixture. The greatest increases in numbers (relative to NH_4Cl as 100) were produced by NH_4Cl (100), Na glutamate (53), $NaNO_3$ (34), all with glucose, and $MgCl_2$ (17) and $CaCl_2$ (70) alone. Smaller effects occurred with NaH_2PO_4 + glucose (10) and glucose alone (4) and none with $NaNO_3$ alone. Only Na glutamate promoted an increase in the population when the nutrients were inoculated into cores in the field.

Homogenised, as opposed to unhomogenised, peat incubated under aerated and non-aerated conditions showed increased oxygen uptake, attributable to microbially produced reduced substances. Respiration rate was further increased by addition of Na glutamate, but not $CaCl_2$, NaH_2PO_4, or $NaNO_3$, while respiration in unhomogenised peat was increased by NH_4 Cl addition.

Chemical analyses of the effluents from percolated peat/sand mixtures, with or without glucose, showed that NH_4^+ was released from both added and native organic nitrogen compounds. No nitrification was detected but denitrification of added NO_3^- occurred. Published estimates of field nitrogen fixation rates are discussed. No detectable microbial release of Ca^{2+}, Mg^{2+}, K^+ or PO_4^{3-} was observed during percolation with distilled water, but various percolates plus glucose promoted microbial release of Ca^{2+} and Mg^{2+}. Significant microbial immobilisation of both phosphorus and nitrogen compounds appeared to occur.

The annual rate of nitrogen fixation, measured by the acetylene reduction technique, estimated as 3.2 g m^{-2} at one site, was often an order of magnitude lower at other sites. Calculations indicate that the blanket peat (to 30 cm depth) contains 45 g m^{-2} of actively cycling nitrogen and 330 g m^{-2} of relatively inert nitrogen. Microbially mediated processes make about 5 g m^{-2} yr^{-1} available to green plants. Contributions from nitrogen fixation and rainfall more than compensate for nitrogen losses and produce an apparent net nitrogen gain of about 2 g m^{-2} yr^{-1}.

Microbial activity in Moor House blanket peat is limited by the combined low availability of carbon and nitrogen, by environmental factors, particularly lack of oxygen, low pH and low temperature, by the physical structure of peat and to a lesser extent by the availability of essential nutrients other than carbon and nitrogen.

Acknowledgements. The research described in this paper was financed by a grant from the Natural Environment Research Council. We also gratefully acknowledge the assistance and provision of unpublished data by Mr. S. E. Allen, Dr. Vera G. Collins, Miss Barbara T. D'Sylva, Mr. A. J. P. Gore, Dr. A. F. Harrison, Mr. D. S. Hayman, Miss Pamela M. Latter and Dr. Barbara Mosse, and technical assistance by Miss Catherine McKenzie and Miss Jean Kaye. In addition we would like to acknowledge the invaluable help and discussion provided by Dr. O. W. Heal in the preparation of this chapter.

7. A Study of the Rates of Decomposition of Organic Matter

O. W. HEAL, P. M. LATTER and G. HOWSON

The previous chapters analyse the faunal and microbial populations in the bog ecosystem. The combined activities of these populations result in the decomposition of the plant remains and, under the influence of environmental factors, result in variation in the rate of decomposition in time and space.

7.1 Introduction

Peat is a dominant feature of Moor House and of much of upland Britain. Accumulations of up to 4 m, amounting to 400 kg m^{-2}, have occurred over 5000–10,000 years reflecting the imbalance between rates of primary production and decomposition. Many of the biological characteristics of the Moor House bog ecosystem result from the presence of peat and the present chapter describes a series of experiments designed to measure the rates of decomposition of plant remains and define the influence of environmental factors on these rates. Information is derived from unpublished studies by the authors and from published work (Clymo, 1965; Latter and Cragg, 1967; Latter et al., 1967; Clymo and Reddaway, 1972; Heal et al., 1974; Standen, in press), as well as from papers in the synthesis of decomposition studies in the IBP Tundra Biome (Holding et al., 1974).

7.1.1 The Decomposer System of the Bog

The major components and transfers within the decomposer system of the blanket bog, formalized in Figure 1, are derived from three plants (*Calluna vulgaris*, *Eriophorum vaginatum*, and *Sphagnum* spp) whose markedly different growth forms affect their input to the decomposer system. *Calluna* input consists of woody stems and shoots which remain in the canopy for a number of years subject to large temperature fluctuations and to a low moisture regime before falling as litter. In the litter layer temperature variation is reduced and moisture is higher and more constant than in the canopy. *Eriophorum* leaves die while still attached to the plant and become overlain by dead leaves in successive years to form a tussock. Temperatures near the surface of *Eriophorum* tussocks are higher than those in *Calluna* litter (Chap. 1, Table 5) and moisture level increases in the lower layers of the litter. *Sphagnum* dies from its base; its remains are thus immediately in a wet environment with a limited temperature fluctuation.

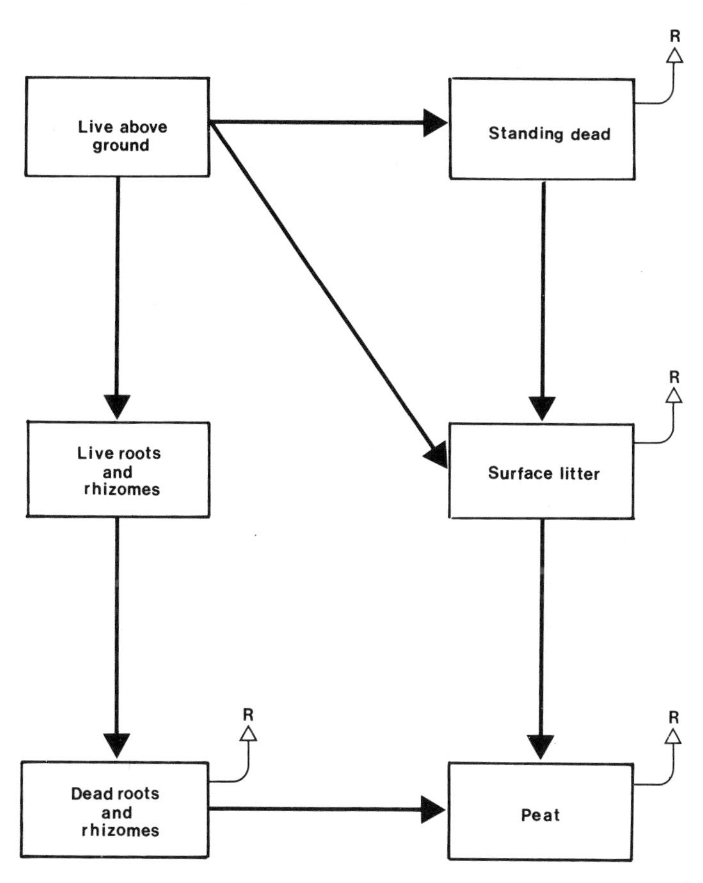

Fig. 1. The main components and transfers within the decomposer system of the blanket bog. *R*: respiration

Below the surface, underground stems and roots of *Calluna* and leaf bases and rhizomes of *Eriophorum* die in the upper 5–10 cm where moisture levels are high but aerobic conditions are normal. Roots of *Eriophorum* grow to about 1 m and, on death, are in a predominantly waterlogged anaerobic environment. The pH of the peat (3.3–3.9) is lower than that of the surface litter horizon (3.9–4.4).

A number of other plant species *(Rubus chamaemorus*, *Erica tetralix*, *Empetrum nigrum)* form minor components on most of the blanket bog, while *Eriophorum angustifolium* and *Scirpus cespitosus* are common on some areas. These add some variety to the type of plant remains contributed to the decomposer system, but the input is predominantly of dwarf shrubs, Cyperaceae and mosses and only *Rubus* provides an input of herb litter. The differing chemical and physical composition of these plant litters provides a range of potential rates of decomposition. The growth form of the plants results in their remains entering the decomposer system in a range of temperature, moisture and oxygen conditions, but with a generally occurring low pH.

7.2 Methods

The decomposition studies were done largely under field conditions and used weight loss from plant litter as the main measure of decomposition rate. Plant remains which were recently dead were collected from the field, air-dried, made up as samples and weighed before replacing in the field. The samples had to be recognizable and retrievable after a number of years and were placed in the litter layer to simulate the natural material with respect to position, microenvironment, contact with adjacent litter, subsequent incorporation into the profile and availability to soil fauna. Some of these aims are conflicting, so the methods used were a compromise and differed with the various types of litter. All litters were attached by a 45 cm piece of terylene line to a numbered plastic label. Nylon hair nets (1 cm mesh) were commonly used to enclose litters and fallout was reduced by rolling the sample loosely in the net so that it was covered by several layers of mesh. Because of their small leaflets, *Calluna* shoots were enclosed in fine mesh (1 mm) terylene bags. Samples of *Calluna* stems were unconfined but tied directly to the terylene line. Sample sizes were small enough (0.15–2.0 g air dry weight) to minimise artificial effects but sufficient to obtain realistic weights.

On retrieval from the field obvious extraneous material, e.g. roots or moss, was removed from the samples which were then weighed, oven dried and reweighed to provide an estimate of moisture content and, by comparison with the initial weight, of weight loss. The results are expressed as % weight loss or as the constant fraction loss rate defined by the regression coefficient of $\log_e W_t/W_o$ as a linear function of time in years, where W_t is the final weight and W_o the initial weight (Chap. 8).

The loss in weight of a litter sample during a period of time integrates loss due to respiration, leaching of soluble organic and inorganic matter and the removal of particulate matter by soil fauna or by physical factors. Respiration is a direct measure of the loss of carbon dioxide by catabolic processes in the litter and was used to estimate the proportion of weight loss resulting from microbial respiration and as a measure of that process in different litters and at different times. Oxygen uptake by litters retrieved from field experiments was measured at 10° C and field moisture in Gilson or Dixon respirometers. Carbon dioxide and methane output from the bog surface was measured directly by Clymo and Reddaway (1972).

Cellulose, in the form of filter paper (Clymo, 1965), strips of Borregaard cellulose board placed horizontally on the litter (Rosswall, 1974) and strips of unbleached cotton inserted vertically into the bog (Heal et al., 1974; Latter and Howson, 1977) were used as standard substrates for comparison of decay rates in different micro and macrohabitats. Dry weight, cellulose content and tensile strength respectively were the parameters measured.

Details of individual experiments are given with the results to avoid confusion through the variety of methods and experiments involved. A series of field experiments provides the basis for this chapter examining variation in the rate of decomposition with time, years, substrates, microhabitats, sites and depth.

7.3 Decomposition of Litter at the Bog Surface

7.3.1 Decomposition of Sphagnum

Sphagnum remains constitute a major part of peat and Clymo (1965) estimated the rate of decomposition of various species and parts of *Sphagnum* at Moor House and at Thursley Bog in Surrey using samples of known weight retained in fine mesh bags. The mean annual weight loss at the surface at Moor House was 12% with *S. acutifolium* and *S. cuspidatum* decomposing at about twice the rate of *S. papillosum*. Losses were higher from the capitulum (0–2 cm) than the mature section (2–6 cm) of the plant but these were live plants which were killed by drying. *Sphagnum* which dies naturally will probably have lower rates, as indicated by an experiment at Thursley in which *S. papillosum* with most of the leaves still attached was collected from 60 cm and replaced at the bog surface. The loss was about half that of comparable mature, killed material.

Further estimates of *Sphagnum* weight loss have been attempted using brown, apparently dead, material from the base of plants. Some samples gained weight because the *Sphagnum* grew but from samples which showed weight losses, Jones and Gore (Chap. 8) estimated a constant fraction loss rate of 0.023 g g^{-1} yr^{-1} for *S. acutifolium* compared to 0.078 for *S. acutifolium* and *S. papillosum* combined (data in Clymo, 1965). These loss rates are suspect because of methodological problems but independent estimates for *Sphagnum magellanicum* from carbon dioxide flux and from profile concentrations of Al, Ti and Mg give rates of the same order, 0.032 and 0.062 respectively (Chap. 9).

7.3.2 Weight Loss and Respiration of Calluna, Eriophorum, and Rubus

The results of Clymo (1965) and Latter and Cragg (1967) provided a guide to the expected rate of weight loss of plant remains on the bog, for an experiment initiated in autumn 1966, designed to define the decay rates, over a number of years, of other major types of litter. *Calluna* stems and shoots and *E. vaginatum* leaves were placed in a randomised block design on Bog Hill. *Rubus* leaves were added to the experiment in 1967. Although *Rubus* is not a major component of the bog vegetation it is the only common herb and, having a broad deciduous leaf with a higher nutrient concentration than the other bog litters, it was expected to provide an estimate of the maximum decay rate among the indigenous vegetation.

The results, summarised in Table 1, (and Chap. 9, Fig. 1) show a consistent weight loss over a number of years. The linear regression of weight remaining against time shows an annual loss rate of 0.04 for *Calluna* stems, 0.08 for *Calluna* shoots, 0.10 for *Eriophorum* leaves and 0.15 for *Rubus* leaves (Table 2). These rates are lower than the observed first year loss in all cases, and the regression constant of less than 1.0 indicates that there is an initial loss which increases in relation to the general loss, with *Calluna* stems lowest and *Rubus* leaves highest. This initial loss may result from high microbial respiration of soluble organic compounds or leaching or handling losses.

Table 1. A summary of the mean (± se) weight loss (%), respiration (μl O_2 g^{-1} h^{-1} at 10 °C) and moisture content (% dw) of four litter types on blanket bog (Bog Hill) over periods of 4–5 years

Weeks in field	Weight loss	Respiration	Moisture	Weeks in field	Weight loss	Respiration	Moisture
A. *Calluna* stems				B. *Calluna* shoots			
9	2.9±0.2	14.2± 1.9	92± 9	8	5.9±0.8	36.8± 6.1	237±15
27	4.3±0.3	10.0± 0.8	104± 3	26	10.4±0.7	28.7± 2.4	221±25
56	7.6±0.6	11.1± 2.2	76±19	55	14.8±1.4	32.7± 4.0	222±39
108	15.2±1.9	10.3± 2.9	94±14	107	29.3±2.5	39.0± 3.3	312±24
160	16.1±1.0	7.0± 1.4	104±11	159	34.5±2.2	29.3± 7.3	260±19
212	24.1±1.3	7.3± 0.9	167± 9	211	40.2±1.4	27.0± 2.4	360±62
264	23.5±1.1	7.5± 1.6	158±20	263	43.9±3.7	15.0± 5.8	438±37
C. *Eriophorum vaginatum* leaves				D. *Rubus chamaemorus* leaves			
5	8.5±1.6	62.6± 6.0	411±68	0	5.3±0.5	86.0±10.2	106±12
23	13.0±0.7	46.9± 7.3	432±48	11	21.1±0.8	83.6± 7.1	421±34
51	26.4±2.1	49.8±11.2	304±42	24	24.3±1.0	56.9±14.3	221±52
104	36.5±2.4	42.5± 8.5	582±99	39	33.3±1.3	37.7±11.9	76±23
156	46.7±2.1	24.8± 8.4	358±64	52	38.1±1.6	66.3± 4.3	389±24
208	65.1±5.1	nd	438±37	82	46.3±0.9	44.3± 5.4	390±54
260	54.7±3.2	nd	nd	104	53.8±2.0	38.5± 4.1	370±44
				130	53.7±1.9	44.3± 2.0	342±28
				157	62.3±3.9	31.5± 5.9	334± 8
				209	72.9±3.4	30.3± 8.6	116±76

Table 2. Regressions of weight remaining with time (t in years) for four litters, calculated from data over five years (*Calluna* and *Eriophorum*) and four years (*Rubus*)

Litter	Untransformed	$\log_e$ transformed
Calluna stems	$y=0.965-0.045\,t$ $r=-0.880$	$y=-0.029-0.052\,t$ $r=-0.867$
Calluna shoots	$y=0.917-0.077\,t$ $r=-0.877$	$y=-0.071-0.109\,t$ $r=-0.859$
Eriophorum leaves	$y=0.867-0.105\,t$ $r=-0.851$	$y=-0.107-0.185\,t$ $r=-0.787$
Rubus leaves	$y=0.820-0.147\,t$ $r=-0.903$	$y=-0.146-0.290\,t$ $r=-0.873$

A $\log_e$ transformation provides a similar, but slightly lower, correlation than the untransformed data (Table 2) but this may reflect only the variability in the measurements and the limited time scale. A linear loss with time implies that a constant amount of material is lost irrespective of the amount present. A negative

Table 3. Analysis of the relationship between respiration rate in $\mu l\ O_2\ g^{-1}\ h^{-1}$ at 10° C (y) and % weight loss (x)

	Regression equation	r	n
A. Individual samples			
Calluna stems	$y=11-0.127x$	0.282	28
Calluna shoots	$y=36-0.246x$	0.315	28
Eriophorum leaves	$y=64-0.680x$	0.577[b]	20
Rubus leaves	$y=88-0.928x$	0.526[c]	58
B. Means at each sampling			
Calluna stems	$y=13-0.247x$	0.828[b]	7
Calluna shoots	$y=38-0.316x$	0.609	7
Eriophorum leaves	$y=65-0.765x$	0.889[a]	5
Rubus leaves	$y=87-0.858x$	0.865[c]	10

[a] $P<0.05$; [b] $P<0.01$; [c] $P<0.001$.

exponential relationship is more probable with a constant fraction of material lost and the absolute amount lost directly related to the amount present (Chap. 8, 9).[1]

The contribution of the various processes which comprise weight loss, i.e. respiration, leaching and comminution, may vary at different stages of decomposition. Thus weight loss may be too insensitive to distinguish a change in decay rate, but the oxygen uptake is a direct measure of the rate of catabolic processes. The respiration rate, like the weight loss, was highest in *Rubus* leaves and lowest in *Calluna* stems (Table 1) and declined with time, particularly in *Rubus* and *Eriophorum* leaves which had the highest initial respiration rates. The percentage of weight lost was used to provide a measure of the state of decomposition for comparison between litters and the decline in respiration rate as weight loss proceeds is expressed in the regression of respiration against weight loss (Table 3A). The decline in respiration rate is most apparent in *Rubus* and *Eriophorum* litters with oxygen uptake being about 50% of its initial rate when the litter had lost 50% of the weight. These litters contain a high proportion of readily degradable compounds such as soluble carbohydrates and a low proportion of resistant compounds such as lignin. The decomposition of *Calluna* stems is dominated by the high proportion of lignin and cellulose, thus the composition does not change markedly during decomposition and, therefore, the respiration shows no significant decline. The regressions in Table 3A are based on individual sample data and some of the variation can be explained by differences in moisture con-

[1] Further samples have been taken at six years for *Rubus* and at seven and ten years for the other litters. Regressions on all mean data, including these samples, gave higher correlations for the $\log_e$ transformation. With the exception of *Calluna* stems, the regression coefficients were slightly lower than those given in Table 2 for the data at five years. Plotting the data showed the *Calluna* stems were continuing to lose weight, but the loss from other samples had virtually stopped and an asymptotic regression gave the best fit. The mean weight remaining at six years for *Rubus* was 27% and at ten years for the other samples was *Eriophorum* 34%, *Calluna* shoots 51% and stems 56%, but still with great variation between samples for any one litter.

tent between samples. However, although analysis of mean values at each sampling from Table 1 shows a similar pattern to regressions of individual data for *Calluna* shoots and *Eriophorum* and *Rubus* leaves, *Calluna* stems now show a significant decline in respiration (Table 3B).

The conclusion from the respiratory data is that, during the early years of decomposition, at least one of the component processes of decomposition, i.e. respiration, shows a declining rate with time. Whilst a negative exponential is a convenient and moderately realistic description of the weight loss curve up to five years, the hypothesis of a constant decay rate does not correspond with the recognition that readily degradable compounds are lost by microbial respiration and leaching, leaving the more resistant original compounds and thus reducing the decay rate as decomposition proceeds (Minderman, 1968). In addition, the production of resistant secondary compounds through chemical and microbial processes is also expected to reduce the decay rate (Paul, 1970). The possibility of distinguishing two components in the decomposition curve, representing fast and slow decomposition compounds and described by two or more negative exponentials, has been explored (Jenkinson, 1965; Paul, 1970; Bunnell et al., 1975). There is a need to develop explicit biological theory rather than to obtain the best mathematical description of the data (Bunnell and Tait, 1974).

7.3.3 The Contribution of Respiration to Weight Loss

Loss of carbon dioxide during microbial respiration contributes to weight loss of the litter, other losses being through leaching and detachment of particles from the litter. The contribution of microbial respiration to weight loss was estimated from the measured oxygen uptake over the sampling period and corrected for field temperature in the litter. No adjustment was made for moisture because the respiration rate was measured at field moisture levels. A respiratory quotient of 0.8 was assumed and weight loss was taken to be twice the carbon loss.

The relationship of respiration to temperature was examined experimentally and showed detectable activity down to $-5.0°$ C, with a Q_{10} of about 4.0 over the normally occurring field temperatures. A Q_{10} of 2.0 was also used in the calculations to test the influence of this factor. For the four litters the weight loss predicted from respiration was 55–100% of the observed weight loss using a Q_{10} of 4.0 and 74–136% with the lower Q_{10} (Table 4) indicating that most, and possibly all, of the weight loss results from microbial respiration.

Table 4. Weight loss of litters predicted from respiration rate compared with observed weight loss (± se)

Litter type	Time in field (weeks)	Predicted weight loss (%)		Observed weight loss (%)
		$Q_{10}=2.0$	$Q_{10}=4.0$	
Calluna stems	264	21.7	16.0	23.5 ± 1.1
Calluna shoots	263	60.0	43.8	43.9 ± 3.7
Eriophorum leaves	260	63.5	46.3	54.7 ± 3.2
Rubus leaves	209	54.3	39.7	72.9 ± 3.4

In a parallel analysis of litter respiration data from a number of Tundra Biome sites, Bunnell and Scoullar (in press) using a more sophisticated model which included effects of moisture, predicted respiratory losses for *Calluna* shoots and stems and *Rubus* leaves at Moor House which were between 65 and 81% of the observed weight losses. Making a number of assumptions on the process of leaching they estimated that leaching accounted for 8 to 34% of the observed weight loss. Water soluble compounds comprise between about 3 and 20% of the initial weight of the litter (Table 7) and the analysis suggests that, despite the high rainfall at Moor House, leaching is not a major process in decomposition, presumably because the soluble compounds are readily assimilated by the microflora.

7.3.4 The Influence of Moisture on Litter Respiration

Some of the variation in decomposition rate may be attributable to varying moisture content of the litter (Table 1). Weight loss is an integrated measure over time but respiration is an instantaneous measure and therefore likely to show a response to the immediate moisture conditions. Although the respiration was not significantly related to moisture, the correlation coefficients of respiration against weight loss (Table 3) increased when the moisture term was added (Table 5). The most comprehensive data set is for *Rubus* litter and, using the regression of respiration against weight loss, the respiration rates of individual samples were adjusted to respiration at time 0. The graph of corrected respiration against moisture content shows that microbial activity is inhibited by moisture levels below about 100%, but is apparently independent of moisture above that level (Fig. 2). There is no indication in these data of the decline in respiration at high moisture levels (>300%) observed in other tundra results (Flanagan and Veum, 1974). Whether the difference results from experimental technique, better aeration of Moor House litters or adaptation of the microflora is uncertain. Of the 58 *Rubus* samples examined, 11 had moisture levels below 100% and, as the sampling covers most periods of the year (Table 1), the data indicate that even with the high rainfall and waterlogged peat, low moisture inhibits respiration of surface litter for about 20% of the time.

7.3.5 Nutrient Changes During Litter Decomposition

It is generally recognised that the concentration of nutrients in bog vegetation is low and that because plant roots cannot usually reach the mineral soil, they are

Table 5. Regressions of respiration rate ($\mu l\ O_2\ g^{-1}\ h^{-1}$ at 10° C) against % weight loss (x_1) and % moisture content (x_2) for individual samples

	Regression equation	R	n
Calluna stems	$y=11-0.130\,x_1+0.001\,x_2$	0.282	28
Calluna shoots	$y=26-0.316\,x_1+0.049\,x_2$	0.501[a]	28
Eriophorum leaves	$y=49-0.705\,x_1+0.035\,x_2$	0.638[b]	20
Rubus leaves	$y=74-1.284\,x_1+0.097\,x_2$	0.734[c]	58

[a] $P<0.05$; [b] $P<0.01$; [c] $P<0.001$.

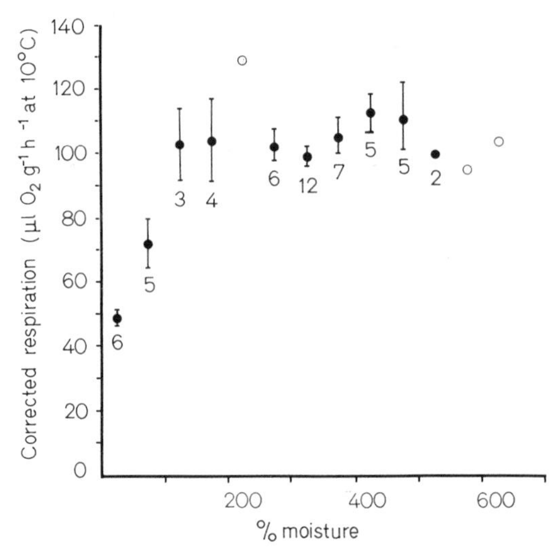

Fig. 2. Respiration of leaf litter of *Rubus chamaemorus* in relation to its moisture content at sampling. ●: mean respiration (± se) for the number of samples indicated, ○: single samples. Respiration rates are corrected to respiration at time 0 using the regression of respiration against time

directly dependent on nutrient release from decomposing organic matter or from rainfall for their supply of essential elements. During decomposition, nutrients may be transferred from complex to simple compounds or to elemental form, removed from the litter by leaching or as gas, taken up by plant roots or synthesised in microbial and faunal tissues until death initiates recycling (Fig. 3 in Chap. 6). The net result of these processes is reflected in the changes in concentration of elements during decomposition of the four litters described earlier but analysis of the transformation pathways has not been included in the decomposition studies.

Concentrations of carbon, nitrogen, phosphorus and potassium were determined for each litter sample, where sufficient material was available (Table 6). There were no significant changes in carbon concentration during the sampling period and the total range between all the litters (140 samples) was 46.3–57.7%. The final concentrations of nitrogen had increased by 50–127% of the initial concentrations, phosphorus increased by 15–70% but potassium varied from a 5% increase in *Calluna* shoots to a decrease of 84% in *Rubus* litter (Table 6). Comparison of the final concentration with those of the adjacent horizons indicates the stage of nutrient change reached by the litters. The concentrations of nitrogen and phosphorus in *Eriophorum* leaves and *Calluna* shoots indicate that there will be little further change in concentration, but *Calluna* stems are still below the concentrations in peat, a reflection of their low decay rate. The nitrogen concentration in *Rubus* leaves increased to a level higher than that in peat and this species may continue to form pockets of high nitrogen for some years. Potassium concentrations in litters were all higher than or similar to peat levels and further losses are expected before peat levels are reached.

Table 6. Initial and final concentrations (% dw) of total nitrogen, phosphorus and potassium in four litters

	Nitrogen		Phosphorus		Potassium		Weight loss (%)	Time (weeks)
	Initial	Final	Initial	Final	Initial	Final		
Calluna stems	0.44	0.66±0.04	0.026	0.030±0.002	0.050	0.028±0.002	24.1±1.3	212
Calluna shoots	1.14	1.77±0.04	0.060	0.088±0.002	0.080	0.084±0.002	40.2±1.4	211
Eriophorum leaves	1.0	1.96 —	0.057	0.097 —	0.115	0.089 —	65.1±5.1	208
Rubus leaves	1.33	3.02±0.07	0.072	0.104±0.003	0.55	0.088±0.007	62.3±3.9	157
Dark brown horizon	2.00		0.097		0.050			
Green-brown horizon	1.80		0.084		0.034			

Standard errors are given where individual samples were analysed. Weight loss, time period between initial and final samplings, and nutrient concentrations in the 5–10 and 10–20 cm horizons of the peat are given.

The increased concentration of nitrogen and phosphorus probably results from loss of carbon from the litter during decomposition but some fixation of gaseous nitrogen could occur. Leaching is probably responsible for the decline in potassium which occurred mainly within the first year of decomposition. Because the apparent nutrient changes are affected by loss of carbon the final concentrations were recalculated, for each sample, to the original weight of litter (concentration × fraction remaining) to show the change in absolute amount of nutrients. The absolute amount is plotted against mean weight loss rather than against time, to allow comparison of litters at similar stages of decomposition (Fig. 3). The initial increase in nitrogen in *Rubus* litter may be an artefact through loss of nitrogen-poor petioles. The net loss in nitrogen from *Rubus*, *Eriophorum*, and *Calluna* shoots is associated with their low final C:N ratios (17, 25, and 30 respectively) compared with the increase in *Calluna* stems which, at 24% weight loss, have a C:N ratio of 77. The increase in nitrogen in the stems may result from immigration of microorganisms with high nitrogen contents and could be accounted for by a microbial population of 2% of the dry weight of the stems.

There was a gradual loss of phosphorus from all litters (Fig. 3b) despite the increase in concentration (Table 6) and a very marked loss of potassium (Fig. 3c). The loss of potassium is high in the early stages of decomposition but the concentrations continue to decline within the peat (Table 6), indicating continued leaching losses. Thus, there is the expected sequence of nutrient losses, potassium > phosphorus > nitrogen, related to their degree of incorporation into organic molecules. For these elements, loss rates tend to be highest in litters with highest concentrations and most rapid decomposition and there may be transfer from nutrient-rich to nutrient-poor litters (*Calluna* stems), representing the nutrient "equilibration" discussed by Dowding (1974).

7.4 Factors Affecting Variability in Decomposition in the First Year

7.4.1 Between-Litter Variation and the Influence of Chemical Composition

Plant litters show a wide variation in rate of decomposition under the same environmental conditions, but apart from the generally recognised relationship between decomposition and certain nutrients, e.g. nitrogen, there is little information on the relative importance of different chemical fractions. To explore their influence 20 weighed samples of each of 15 plant litters were placed in a randomised block on Sike Hill. To reduce microhabitat variation the samples were all placed on litter under *Calluna*. Subsamples of the initial litter were analysed chemically, and loss in weight was measured in half the samples after one and the remainder after two years. The results are summarised in Table 7, omitting *S. acutifolium* which showed weight gains rather than losses due to growth of apparently dead litter.

In Table 7 the species are arranged in increasing order of weight loss after two years. In general first and second year losses were similar but the three litters

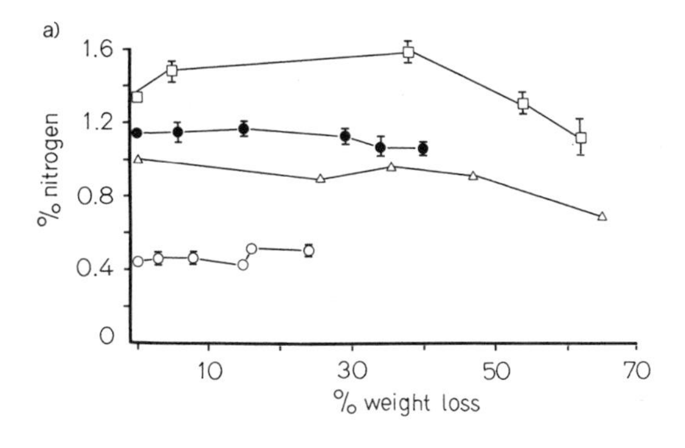

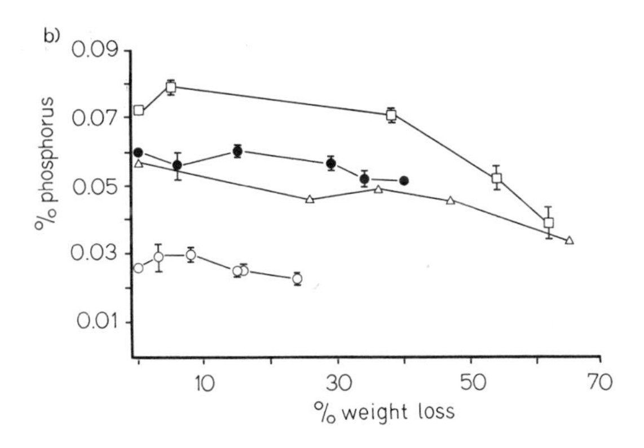

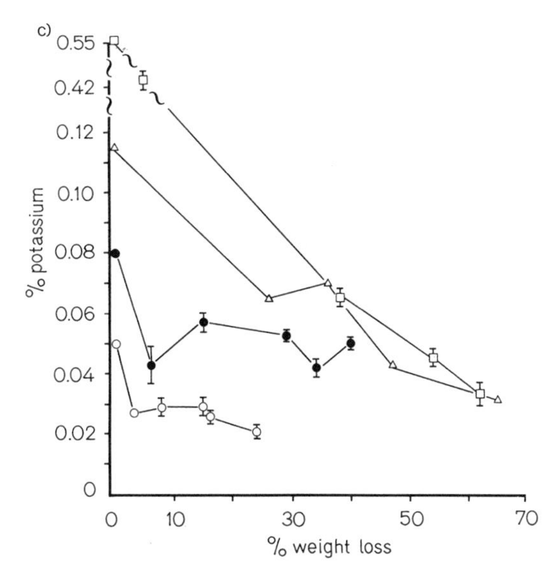

Fig. 3a–c. Amount of nutrient remaining during decomposition for *Rubus chamaemorus* leaves (□); *Eriophorum vaginatum* leaves (△); *Calluna vulgaris* shoots (●); *C. vulgaris* stems (○). Standard errors are given where sampling was replicated. (a) Nitrogen, (b) Phosphorus; (c) Potassium

Table 7. Initial chemical composition of a variety of litters placed on the bog surface (Sike Hill) and their weight loss (± se) after one and two years

	% weight loss		C	N	P	K	Ca	Mg	Sol carbo-hydrate	Sol tannin	Holo-cellulose	Lig-nin
	1st year	2nd year										
Eriophorum vaginatum (r)	0.6±0.9	1.3±1.5 (46)	50	0.50	0.056	0.210	0.11	0.081	4.5	18.0	65	34
Calluna vulgaris (bgs)	6.6±0.3	9.8±0.7 (67)	52	0.53	0.038	0.045	0.15	0.031	1.1	2.6	61	35
C. vulgaris (r)	5.0±2.0	11.4±2.1 (44)	53	0.69	0.034	0.044	0.14	0.031	2.1	3.9	51	49
C. vulgaris (ags)	7.5±0.7	13.2±1.1 (57)	51	0.56	0.038	0.034	0.11	0.020	1.3	2.6	68	31
C. vulgaris (s)	19.7±1.6	28.2±3.0 (70)	55	1.38	0.069	0.088	0.34	0.056	4.2	3.0	42	39
E. vaginatum (1)	22.0±2.2	29.0±1.7 (76)	49	0.97	0.041	0.091	0.20	0.081	3.5	4.3	61	22
E. angustifolium (1)[a]	17.6±1.5	32.5±3.5 (54)	50	0.94	0.050	0.110	0.71	0.120	5.7	nd	57	23
Scirpus cespitosus (1)	18.2±2.0	34.6±2.2 (53)	49	0.63	0.013	0.063	0.14	0.078	3.4	6.7	66	17
Pinus contorta (1)	24.5±2.0	38.1±2.3 (64)	53	0.44	0.028	0.019	0.63	0.100	5.7	9.1	54	35
Nardus stricta (1)	22.8±1.5	39.9±6.4 (57)	46	0.53	0.031	0.097	0.08	0.031	2.2	2.1	72	12
E. angustifolium (1)[b]	23.5±1.9	36.9±4.9 (64)	50	0.94	0.050	0.110	0.71	0.120	5.7	nd	57	23
Juncus effusus (1)	31.3±1.6	41.0±3.5 (76)	46	1.25	0.160	0.300	0.34	0.088	2.0	2.2	69	14
Rubus chamaemorus	36.1±2.3	50.1±3.1 (72)	50	1.31	0.066	0.094	0.85	0.530	8.2	27.0	34	6
Narthecium ossifragum (1)	44.9±2.9	49.1±3.2 (92)	48	0.69	0.006	0.280	0.67	0.140	4.3	6.2	67	22

The first year loss expressed as a % of loss in two years is shown in parentheses. r: roots, bgs: below-ground stems, ags: above-ground stems, s: shoots, 1: leaves. *E. angustifolium* was placed in both small-mesh bags[a] and large-mesh bags[b].

which had the highest loss rate in the first year had a lower rate in the second year, probably resulting from loss of readily decomposable compounds leaving the more resistant fractions. By contrast, slow decomposing litters showed little or no decline in rate in the second year. The chemical analyses are for the litter at the beginning of the experiment, therefore examination of weight loss in relation to chemical composition was concentrated on the first year losses. Weight loss (y) tended to be positively correlated with each of the elements, but only % calcium showed a significant coefficient:

$$y = 9.0 + 29.7\,\mathrm{Ca} \qquad r = 0.668 \qquad P < 0.01$$

The relationship of weight loss with % nitrogen was linear but the relationship with potassium, calcium and magnesium tended to be asymptotic (van Cleve, 1974).

Of the organic compounds % lignin showed a negative linear relationship with weight loss:

$$y = 37.7 - 0.680\,\mathrm{lignin} \qquad r = 0.649 \qquad P < 0.01\,.$$

Soluble carbohydrate, holocellulose and tannins were not significantly correlated with loss in weight. Interpretation of the data for tannin is particularly difficult because of the wide range of compounds extracted by the analytical methods and this could explain why the two litters having the highest tannin content had the highest *(Rubus)* and lowest (*Eriophorum* roots) loss rates. In most of the regressions of weight loss with inorganic and organic constituents the observed loss rates for *Juncus*, *Nardus*, and *Narthecium* leaves were higher than predicted. The reasons for the outlying position of these three litters are not obvious but they indicate that a combination of factors determines decay rate. The interactions may be chemical as in the complexing of proteins by polyphenols or direct inhibitory or stimulatory effects. The absence of a single overriding factor is not therefore unexpected. For *Eriophorum* roots the observed weight losses were repeatedly lower than predicted, but the observed rate must be regarded as suspect because, in another experiment (Sect. 7.4.5) weight losses of 8.8% were recorded for *Eriophorum* roots at the surface.

To obtain an expression of the interrelationships of chemical constituents and degree of similarity of the litters, the data (Table 7) together with moisture contents were analysed, using principal components analysis followed by cluster analysis of the significant component values. The first component appeared to represent readily soluble organic constituents, the second was a combination of cellulose and moisture reflecting the "fleshiness" of the litter. The cluster analysis of these two components (Fig. 4) was similar to that for five components and tended to group litters with similar rates of weight loss, again indicating the influence of combinations of chemical constituents. *Eriophorum* roots do not conform to the general pattern.

The interaction of the constituents is likely to be multiplicative and divisive rather than simply additive or subtractive. Thus a high content of phosphorus will

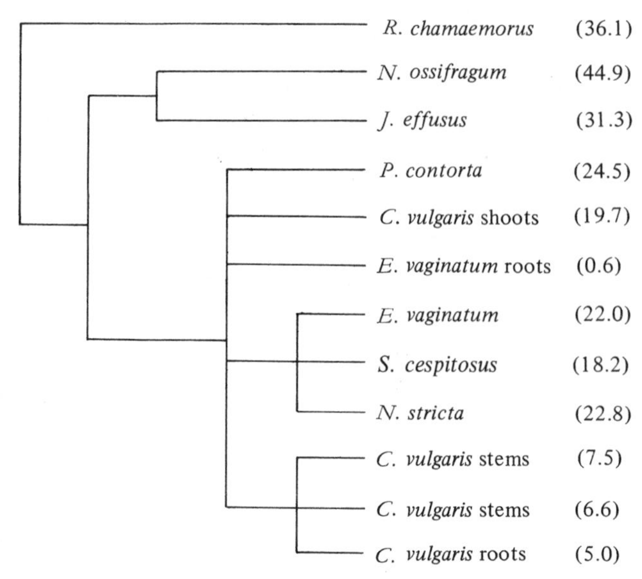

Fig. 4. Clusters of litter types based on chemical analysis data (Table 7) and derived by principal component and cluster analysis of the first two component values. Percent weight loss in the first years is given in parentheses. *E. angustifolium* litter was not included in the analysis because tannin was not determined

not result in a high decay rate when soluble carbohydrate is low, whereas moderate concentrations of both components are likely to give a high rate. Various combinations of the constituents were examined using multiple regression analysis, but the analysis was restricted partly by the small number of litter types (12–14) compared with the regressor variables (10). The equation which both explained a high proportion of the variation in first year weight loss (y) and was biologically reasonable was:

$$y = 25.3 + 30.9\,(\mathrm{P} + \mathrm{Ca}) - 0.53\,(\text{lignin} \times \text{tannin}) \qquad R = 0.977 \qquad \mathrm{P} < 0.001\,.$$

Eriophorum roots remained as an outlier, and the equation which incorporated the roots, although not the statistically best fit, included the positive multiplicative interaction of the minerals with soluble carbohydrate divided by the multiplicative interaction of lignin and tannin.

$$y = 11.2 + 149\,((\mathrm{P} + \mathrm{Ca}) \times \text{sol carbohydrate})\,(\text{lignin} \times \text{tannin})$$
$$R = 0.816 \qquad \mathrm{P} < 0.001\,.$$

The form of this relationship (Fig. 5) provides a quantitative hypothesis of the main chemical factors considered to influence the rate of decomposition, namely: nutrient elements, readily decomposable organic molecules, resistant organic molecules and inhibitors.

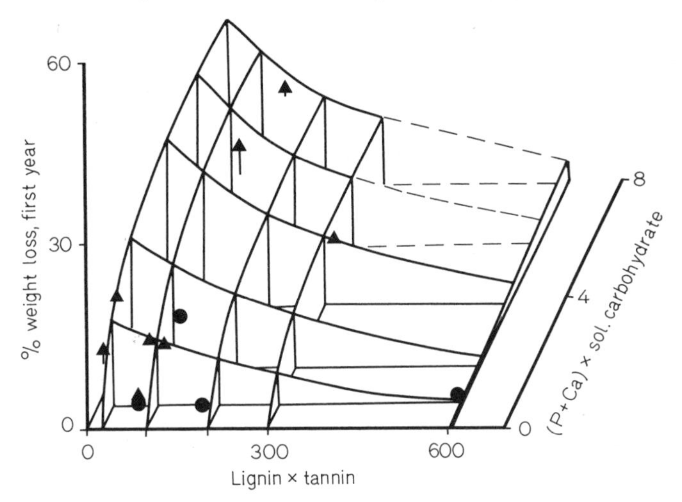

Fig. 5. Weight loss of litters in relation to initial chemical composition. Values above the regression surface shown by ▲, those below the surface by ●, the vertical line indicating distance from the surface

7.4.2 Seasonal and Between-Year Variations in Rate of Decomposition

The occurrence of respiration in litter at −5.0° C indicates that under the temperature regime at Moor House, decomposition can proceed throughout the year. Seasonal variation in decay rate has not been studied intensively but for *Rubus* leaves the high initial loss rate (0.259% d^{-1}) during September–December was followed by a low rate (0.042% d^{-1}) during December–March, then 0.113% in March–June and 0.116% in June–September. Thus summer rates are about four times those in winter and winter losses constitute about 12% of the annual weight loss. The results for *Rubus* correspond with the general pattern observed for litter of *Juncus squarrosus* (Latter and Cragg, 1967) and for *Calluna* and *Eriophorum* leaves (Standen, in press), but Clymo (1965) found, surprisingly, that *Sphagnum* had "relatively larger losses between January and April than between August and January".

The climatic conditions varied between the years of IBP, the winters of 1970–71 and 1971–72 being particularly mild. The influence of between-year variation was assessed with samples of *Rubus* leaves placed on Bog Hill in September 1967, October 1970 and September 1971. Each set was sampled after one year and the percentage weight losses of 38.1 ± 1.6, 38.6 ± 1.6, and 39.0 ± 1.8 were remarkably consistent. Although only three years were compared, and there may have been some influence of litter composition in different years, the results indicate that compared with the other sources of variation, annual climatic variation has relatively little influence on the rate of decomposition.

7.4.3 Microhabitat Variation Within Sites

The growth form of the dominant plants on the blanket bog, *Calluna*, *Eriophorum* and *Sphagnum* spp, together with the topography of the bog surface, provide

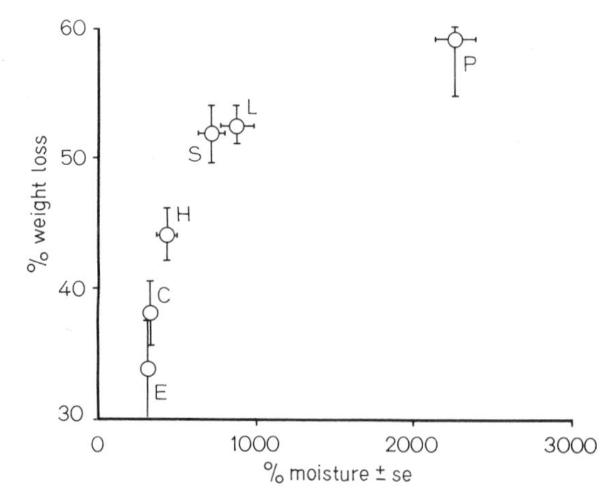

Fig. 6. Weight loss (% in first year) of leaf litter of *Rubus chamaemorus* in different microhabitats in relation to moisture content of the litter at sampling. Microhabitats are *E. vaginatum* tussock (*E*); *Calluna* (*C*); *Sphagnum* lawn (*S*) on Sike Hill; *Sphagnum* pool (*P*), lawn (*L*), hummock (*H*) on Burnt Hill

a mosaic of habitats for decomposition. On much of the bog *Sphagnum* forms a wet lawn but on the *Sphagnum* dominated bogs there is obvious variation in habitat moisture with recognisable pools, lawns and hummocks. The *Sphagnum* and *Calluna* litters have a more stable temperature regime than the litter in *Eriophorum* tussocks where maximum temperatures may reach 35° C and the sum of degree days above 0° C is about 10% higher than in other microhabitats (Chap. 1). The temperature variation probably results from differences in moisture content and exposure to radiation and wind. Thus the difference in decay rates of *Calluna* and *Eriophorum* (Table 1) results from differences in the microclimate of the habitat in which they occur as well as from substrate quality.

To determine the effect on decomposition of microclimatic differences between the six main microhabitats, ten samples of *Rubus* leaf litter were placed in each. The results showed an almost two-fold range of first year weight loss between tussocks of *Eriophorum* (33.7 ± 3.8) and *Sphagnum* pools (59.2 ± 4.4). Using moisture content of the litter as an index of the relative wetness of the microhabitats, Figure 6 shows a curvilinear relationship between weight loss (y) and moisture content (x):

$$y = -32.9 + 12.3 \log x \qquad r = 0.935 \qquad P < 0.001$$

for the 59 individual samples the regression is:

$$y = -20.9 + 10.59 \log x \qquad r = 0.663 \qquad P < 0.001 .$$

The relationship (Fig. 6) indicates that weight loss is reduced at moisture contents below about 600%, a much higher threshold than indicated by the data on respiration of *Rubus* litter (Fig. 2). The higher value is suspect because the moisture content reflects the immediate climatic conditions while weight loss is an integrated value for a period, and reflects previous climatic regimes.

The pattern of weight loss in the *Sphagnum* habitats (pools > lawns > hummocks) is not repeated by two other measures of decay rate. Loss in tensile strength of cotton strips at 0–4 cm was less in pools and lawns (13 ± 4 kg) than in hummocks (21 ± 3 kg), the tensile strength of control strips being 27.6 ± 0.6 kg (Heal et al., 1974). The discrepancy in pattern between cotton strips and *Rubus* leaves may result from leaching losses from *Rubus* litter which are probably directly related to moisture while leaching is not a component of tensile strength loss.

Clymo and Reddaway (1972) reported losses of gaseous carbon which were greater from pools (61 g m^{-2} yr^{-1}) than from hummocks (51 g m^{-2} yr^{-1}) or lawns (35 g m^{-2} yr^{-1}). Expression of the results per gram of input from *Sphagnum* showed losses in the order: hummocks > pools > lawns (0.42, 0.25 and 0.18 g g^{-1} yr^{-1} respectively). The loss of CO_2 and CH_4 from the bog surface represents losses from an undefineable depth and the quantities are influenced by the depth of the aerobic layer of the peat (Chap. 9). Different species composition of the vegetation will also cause variation in decay rate between the *Sphagnum* microhabitats (Sect. 7.3.1, and Clymo, 1965). The rate of decomposition of cotton strips declines with depth, particularly in the hummock and lawn habitats (Heal et al., 1974). When integrated over 20 cm, the loss in tensile strength is in the same sequence (hummocks > pools > lawns) as the relative decay calculated from loss of gaseous carbon.

The three measurements of decomposition in *Sphagnum* microhabitats were made on one site, Burnt Hill. The results emphasise the dangers of interpreting and generalising from a single method and the need to assess the change in decay rate with depth as well as at the surface. It is apparent, however, that a two-fold variation in decomposition can occur between microhabitats within a bog, related particularly to moisture variation.

7.4.4 Variation Between Blanket Bog Sites

The network of streams at Moor House isolates low hills covered in blanket bog with varying species composition and production, water table depth, peat conditions and depth. There is little information on variation in populations of decomposer microflora and fauna over a range of bog sites, but their activity is reflected in the rate of decomposition of organic matter. The between-site variation in the effect of environmental conditions was assessed using *Rubus* litter, Borregaard cellulose and cotton strips. On eight bog sites, at about 500 m altitude, the mean weight loss of *Rubus* litter in the first year ranged from 34 to 43% (Table 8), but analysis of variance showed that the differences between sites were not significant (df 7,71, $F = 1.2$). Moisture content of the *Rubus* litter, used as an index of the relative wetness of the sites, showed significant between-site variation (df 7,71, $F = 3.8$, $P < 0.01$), largely the result of the two Cottage Hill sites where the water table is always close to, or at, the surface. Weight losses (y) among the individual samples were weakly correlated with % moisture content (x):

$$y = 5.42 + 0.014 \log x \qquad r = 0{,}31 \qquad P < 0.01$$

Table 8. Weight loss of *Rubus chamaemorus* leaf litter and Borregaard cellulose on a number of blanket bog sites at Moor House

Site	*Rubus* leaf litter		Borregaard cellulose
	Weight loss	% Moisture	
Cottage Hill B	43.4 ± 1.9 (10)	540 ± 61 (10)	41.0 ± 6.9 (14)
Bog Hill [a]	42.5 ± 3.3 (10)	345 ± 15 (10)	—
Sike Hill B	40.6 ± 4.5 (9)	334 ± 13 (9)	—
Sike Hill A	39.2 ± 2.4 (10)	383 ± 21 (10)	28.1 ± 5.1 (12)
Bog End	38.9 ± 2.6 (10)	352 ± 28 (10)	26.6 ± 6.8 (11)
Green Burn	37.9 ± 2.1 (10)	421 ± 69 (10)	20.6 ± 4.1 (14)
Cottage Hill A	37.7 ± 1.2 (10)	558 ± 68 (10)	15.2 ± 6.0 (13)
Bog Hill [b]	34.4 ± 2.0 (10)	350 ± 42 (10)	—

Bog Hill [a] was the site used in primary production studies.
Bog Hill [b] was in a sheep exclosure (see Appendix 1, Chap. 1)

Losses are expressed as mean % loss after one year ± se. The number of samples is given in parentheses. The sites are arranged in descending order of loss rate of *Rubus* litter.

but the mean values for sites, which reflect site rather than microhabitat variation, showed no correlations. Loss of cellulose from the Borregaard strips paralleled weight loss from *Rubus* but showed higher within-site variation (Table 8).

Decomposition of cotton strips in the surface 0–4 cm showed losses in tensile strength from 27.5 ± 2.0 to 38.6 ± 1.38 kg. As with the Borregaard cellulose within-site variation was high and masked any between-site variation. The cotton strips were sampled at five depths in each site. All eight sites showed a marked decline in loss rate with depth (Heal et al., 1974), associated with waterlogged conditions, but the decay profiles of the cotton strips were not related to between-site variation in water table. Thus the three measures of decomposition, whether for surface or the profile, show that between-site variation on the bog was not sufficiently large to override within-site variation.

7.4.5 Variation Within the Peat Profile

Peat accumulation results from the resistant nature of the litter from bog plants, and from the waterlogging of the peat which produces acid and anaerobic conditions. On much of the blanket bog at Moor House peat is 0.5 to 2.0 m deep and the water table is within 20 cm of the surface (Forrest and Smith, 1975). Reducing conditions are strongest in the 10–20 cm zone in the peat and increase during the summer (Urquhart and Gore, 1973). The zone of minimum redox potential approximates to the zone within which the water table fluctuates and to the green-brown horizon. Facultative anaerobes and sulphate reducing bacteria tend to have their highest numbers in this horizon (Chap. 5).

Clymo (1965 and Chap. 9) emphasised that the rate of decomposition of organic matter declines markedly as it passes from the aerobic surface zone into the anaerobic conditions below the water table. The division into aerobic and anaerobic zones was used in the peat accumulation model of Gore and Olson (1967) who estimated mean constant fraction loss rates of 0.07 and 0.001–0.003 g g^{-1} yr^{-1} for the two zones. In the upper 20 cm of the peat there is a mixture of fresh and well decomposed organic matter, from roots and surface litter, within an environmental gradient comprising increasing moisture, decreasing oxygen and reduced temperature oscillation with increasing depth.

In the present study, the influence of the environmental profile on decomposition rate was assessed from weight loss of three types of litter inserted at different depths in the peat. *Calluna* below-ground stems taken from plants whose above-ground parts were dead and *E. vaginatum* roots in the dark yellow stage (Forrest, 1971) were chosen to represent underground litter. Leaves of *Juncus effusus* were used as an example of surface litter because easy removal of peat particles from their surface reduced experimental error in determining weight loss. Ten samples of *Calluna* and *Juncus* litter and five of *Eriophorum* roots were inserted into the peat at each of six depths (0, 5, 10, 15, 20, 25 cm) in a randomised block design over a plot of 20 × 10 m. After a year the samples were dug up and the depth and horizon recorded. The actual depth often differed from the planned depth.

Weight loss of *Juncus* leaves and *Calluna* stems decreased significantly with depth, but loss from *Eriophorum* roots increased with depth (Table 9). There is no obvious explanation for the contradiction of results although it is possible that because the *Eriophorum* roots were extracted from 10–20 cm depth they carried a microflora adapted to the conditions at the lower sample levels. The decline in rate of *Juncus* and *Calluna* corresponds with the results for cotton strips from eight bog sites (Heal et al., 1974, and Table 7) and with the losses from filter paper (Clymo, 1965). The losses from litter and cotton strips suggest a linear decrease with depth (Chap. 8) with decomposition ceasing between 15 and 41 cm. It is probable that the linear relationship reflects the small number of sample depths, and that decay continues at a very slow rate at greater depths, as indicated by the 3% weight loss per year of filter paper at 75 cm compared with 7% at 10–15 cm and 17% at 0–5 cm (Clymo, 1965). However, the results indicate that, relative to the rate at the surface, losses decline by about 3–6% cm^{-1} in the upper 20 cm of the bog. The methods used so far have not been sufficiently sensitive to detect a change in rate directly associated with the position of the water table, possibly because the depth of the table varies considerably during the year.

7.4.6 Soil Fauna in Relation to Decomposition

The soil fauna influences the rate of organic matter decomposition by direct digestion and by stimulation of microbial activity. Microbial stimulation can result from comminution of the litter to expose new surfaces and reduce particle size, from increased nutrient availability and breakdown of inhibitory compounds and from increased aeration. On the blanket bog at Moor House the fauna is characterised by the absence of earthworms, the enchytraeid *Cognettia sphagnetorum* being the dominant invertebrate, and by the concentration of the population in the surface 3 cm of the litter and peat (Chap. 4). Thus, there is little mixing and comminution of decomposing organic matter and the fauna only influences the decomposition of surface litter for a few years before it passes into lower levels.

Analysis of animals on the litter samples of *Calluna*, *Eriophorum*, and *Rubus* leaves showed that little colonisation by fauna occurred until the litter was about two years old. In *Rubus* litter, bleaching by a basidiomycete *Marasmius androsaceus* was associated with an increased weight loss (Latter, 1977). Feeding and decomposition studies concentrated on *C. sphagnetorum*, and in a laboratory preference experiment, the numbers of worms entering *Sphagnum*, *Calluna*, *Eriophorum*, and *Rubus* litters were in the ratio of 1:2:4:7. The ratio approximates to the relative rates of weight loss and respiration of these litters which are related to chemical composition. Growth rate of *C. sphagnetorum* was greater on *Calluna* and *Eriophorum* litter than on *Sphagnum* and, although the evidence is not conclusive, the enchytraeid is probably a decomposer rather than a microbial feeder (Latter and Howson, in press).

Apart from its direct effect on organic matter decomposition, *C. sphagnetorum* stimulates microbial activity. In litter bags of *Sphagnum*, *Calluna*, and *Eriophorum* leaves weight loss was about 1.35 times greater when worms were present at field densities than in litter without worms (Standen, in press). Only a small proportion of the increase is attributable to respiration of the worms. The low numbers of fauna in the bog, their restricted distribution and small influence on litter decomposition, especially of *Sphagnum*, indicates that their limited activity encourages peat accumulation.

7.5 Decomposition in Other Soils at Moor House

The studies on the blanket bog have emphasised the control of decomposition by the waterlogged conditions and the resistant character of the litter. The peaty podzols, peaty gleys and brown earths provide a range of soils with better drainage and higher pH and nutrient conditions than the bog. A few measurements on these mineral soils indicate the range of decomposition rate which can occur within the same general climatic regime as the bog. Increasing faunal activity and soil respiration, plus decreasing C:N ratio and accumulated organic matter, indicate that decomposition rate increases from the blanket bog, through the peaty podzols and gleys with *Juncus* and *Nardus* dominants, to the brown earths with *Festuca* and *Agrostis* (Cragg, 1961). This activity sequence was confirmed by the sequence of total bacterial and fungal populations and of physiological groups of bacteria (Latter et al., 1967).

Although the general sequence of activity (bare peat $<$ blanket bog $<$ podzol and gley $<$ brown earth) is established, the quantitative differences are still uncertain and there is variation in particular activities. Preliminary respiration studies suggested ratios of activity for bog:podzols and gleys:brown earths of 1:6:12 (Cragg, 1961) and 1:2.5:8 (Latter et al., 1967).

A small set of data on loss of tensile strength of cotton strips from a peaty podzol showed a slightly higher loss than on the blanket bog with a comparable decline with depth. A brown earth caused a loss in the surface layers which was 2–4 times faster than in the other two sites. However, the decline in rate with depth was slight in the brown earth resulting in an 18 times greater rate at 16–20 cm compared with the blanket bog (Table 9).

Table 9. Weight loss after one year (%) of leaf litter of *Juncus effusus*, below-ground stems of *Calluna* and roots of *Eriophorum vaginatum* at six depths on Sike Hill

Depth (cm)	*Juncus* leaves	*Calluna* stems	*Eriophorum* roots	Depth (cm)	Cotton strips	
					Sike Hill	Grassland
0	29.0 ± 1.0 (9)	6.0 ± 0.4 (7)	7.7 ± 2.0 (3)	0–4	38.6 ± 1.4 (8)	139.0 ± 23.7 (4)
1–5	31.5 ± 1.9 (9)	5.0 ± 1.0 (11)	11.8 ± 2.4 (5)	4–8	35.3 ± 2.3 (8)	73.4 ± 15.1 (4)
6–10	26.9 ± 2.6 (7)	5.5 ± 0.9 (8)	10.3 ± 2.1 (3)	8–12	12.5 ± 4.8 (8)	69.5 ± 21.7 (4)
11–15	20.6 ± 2.0 (14)	4.1 ± 0.6 (7)	13.8 ± 1.3 (7)	12–16	8.5 ± 4.0 (8)	70.1 ± 22.3 (4)
16–20	10.7 ± 1.1 (13)	3.6 ± 0.6 (12)	12.5 ± 1.7 (8)	16–20	3.7 ± 1.5 (8)	66.1 ± 16.5 (4)
21–25	8.5 ± 0.7 (7)	1.8 ± 1.4 (3)	19.2 ± 1.7 (4)	—	—	—
26–30	—	1.9 ± 0.8 (3)	—	—	—	—
F ratio	29.1[b]	3.0[a]	3.0[a]			

[a] $P<0.05$; [b] $P<0.001$.

Results are means ± se with the number of samples given in parentheses. Loss in tensile strength (kg) from cotton strips on Sike Hill, and on grassland with red-brown limestone soil corrected to annual loss, are from Heal et al. (1974).

The differences in soil conditions cause some of the variation in decomposition of cellulose between sites. The chemical and physical composition of the plant litters also differ between sites with a decreasing proportion of woody species and increasing proportion of fine-leaved grasses with relatively high nutrient concentrations on the grasslands. Few strictly comparable data are available to quantify the effect of litter quality on between-site differences in decomposition. Both *Juncus* and *Nardus* had higher loss rates (31 and 23%) than the major bog species (Table 7) and although no data are available for Moor House, the loss rates of litter from *Festuca-Agrostis* grasslands in the similar climate of Snowdonia are about 60–80% in the first year (Chap. 20). Thus the differences in decomposition between the peat and mineral sites at Moor House are a combination of both soil conditions and organic matter quality. This is reflected in the constant fraction loss rate (k) for organic matter in the system which can be estimated by:

$$k = I/Xss\,,$$

where I is the annual input of organic matter and Xss is the accumulated organic matter (Jenny et al., 1949). Although this is only an approximation because it assumes that the systems are in steady state, values for k are of the order of 0.003 for blanket bog, 0.015 for peaty podzol and 0.045 for brown earth. This approximates to a ratio of 1:5:15 for the turnover time of the decomposer subsystem in the three sites.

7.6 Discussion and Conclusions

The aim of the studies was to define the rates of decomposition of the major inputs from primary production and to identify the main factors controlling these rates. Measurement of weight loss and respiration provided reproducible results, especially for the early stages of decomposition, and although statistical errors can be attached to the rates, the extent of systematic errors is largely unknown and is rarely examined. The validity of the results is explored by Jones and Gore (Chap. 8) and Clymo (Chap. 9), where the inputs from primary production are linked to the measured rates of decomposition to calculate organic matter accumulation for comparison with observed values. The models allow particular variables to be altered to test their influence on peat accumulation, an assessment which cannot be made from field experiments. The field studies can, however, determine the range of variation which occurs in the natural environment, thus setting the limits for adjustment in computer simulations.

Clymo's (1965) factorial experiment on decomposition of *Sphagnum* showed that greatest variation occurred between depths in the peat and types of *Sphagnum* (parts and species). The series of experiments described in the present chapter, while not amenable to rigorous statistical comparison, emphasises the large variation within sites (species, depth, microhabitat) rather than between sites and years. The main source of variation is in chemical and physical composition of the plant remains with first year weight losses ranging from 5 to 8% for *Calluna* wood to about 22% for *Eriophorum* leaves and 35–45% for *Rubus* and *Narthecium* leaves. The range of variation is thus between four- and nine-fold depending on whether major or minor plant components are considered (Table 7). The second major source of variation is with depth, the same substrate showing between 2.5- and 10-fold variation between the surface and 25 cm (Table 9). Variation between microhabitats on the surface of the bog is only about two-fold (Fig. 6), similar to that between bog sites (Table 8), while the between-year variation is negligible. Although this comparison indicates the relative importance of the different factors influencing decomposition, these factors do not act in isolation. Thus the most resistant plant parts (*Calluna* roots and below-ground stems, and *Eriophorum* roots) are deposited within the peat profile where decomposition is slower than at the surface. As a result of the slow nutrient release from decomposing litters, plants have developed strategies involving low nutrient concentrations in their tissues and resorption of nutrients before death. These strategies maintain low rates of decomposition and nutrient release.

Seasonal variation is superimposed on the spatial variation in decomposition on the bog which is under the overall influence of the general climatic regime at the site. However, the range of climatic conditions occurring in Britain produces less variation than that which occurs within a site, as indicated by the comparison of *Sphagnum* decomposition in bogs at Moor House and at Thursley in southern England (Clymo, 1965). On other peat bog sites in western Ireland (Moore et al., 1975), northern Sweden (Rosswall et al., 1975a, b) and southern Canada (Reader and Stewart, 1972) the ranges of litter decomposition rates, with main variation between species and depths, are very similar to those at Moor House despite differences in research workers, methods and types of litter. The importance of litter quality and soil conditions in determining the rate of decomposition is further emphasised by the major differences between bog and grassland sites at Moor House, under the same climatic regime. The implications are that further understanding of factors controlling decomposition requires development of a chemical and physical rather than taxonomic definition of substrate quality and an increased emphasis on studies below the litter and soil surface for the analysis both of below-ground plant parts and of the later stages of decomposition.

Acknowledgements. We are very grateful to Mr. P.J.A. Howard and Mrs. D.M. Howard for providing respiration data; to Mr. D.D. French, Mr. A.J.P. Gore, Dr. H.E. Jones, Mr. D.K. Lindley, Mr. M. Rawes and many other participants in the Moor House work for their critical discussion, advice and information; to the Chemical Section at Merlewood Research Station for chemical analysis and to Mr. J. Grue, Mr. M. Moseley, Mrs. H.T. Pearce and Mr. G.R.J. Smith for assistance in litter preparation and sampling.

8. A Simulation of Production and Decay in Blanket Bog

H. E. JONES and A. J. P. GORE

The compatibility of the estimates of primary production and rates of decomposition is assessed by combining them in a mathematical model which examines the effects of variability in production and decay and calculates accumulation of peat for comparison with observed values.

8.1 Introduction

In an ecosystem study of biomass and energy flow, primary and secondary production and decomposition are the major processes to be quantified. On the blanket bog sites at Moor House, where intensive studies were carried out, the deep deposits of organic matter such as peat have developed as a result of the balance between the rates of these processes over a long period of time.

Data derived from primary production and decomposition studies (Clymo, 1965; Forrest, 1971; Forrest and Smith, 1975; Chap. 7) were used to simulate the accumulation of organic matter. Computer simulation enabled a study to be made of (1) variability in the estimates of production and decay, (2) the compatibility of independently derived data in discrete projects, and (3) the effect of altering assumptions made in the analysis of the data. The simulated accumulation was compared with observed accumulation in the field.

Because primary production measurements were not made in exclosures (Forrest, 1971), grazing was not explicitly considered, since any effect would be included in the estimates. Moreover, any effect was likely to be small because vertebrate herbivores, sheep and grouse, were few (Rawes and Welch, 1969; Chap. 10) and no effect of the main invertebrate herbivore, *Strophingia ericae*, on the production of its host plant, *Calluna vulgaris*, was detectable (Hodkinson, 1971).

8.2 Methods

8.2.1 Model Development

If X is the dry weight of organic matter per unit area of ground, then decomposition losses may be expressed as the fraction of X lost per year, and the net change in X will be the difference between the input, I, and the losses. This may be expressed as the linear differential equation:

$$\frac{dX}{dt} = I - kX \qquad (1)$$

(Jenny et al., 1949; Olson, 1963).

It was felt that, because of aeration factors, the decay rate, k, would not be constant over a depth of peat. Gore and Olson (1967) had envisaged decay as two constants, one in the aerobic peat layers, and the other in the anaerobic layers. In this model an attempt was made to introduce a more continuous rate of change of decay with depth. The peat column was represented by a series of layers, each 1 m^2 in area, and of predetermined depth. One cm was found satisfactory in that no appreciable change in simulated accumulation was caused by decreasing the layer thickness below this value. The rate of input as g m^{-2} yr^{-1} was envisaged as a uniform depth of input, added in cm yr^{-1} to the top layer. In order to convert weight to depth, assumptions and measurements were made on the density of the dead plant material (Sect. 8.6.1) and rate of change in depth was assumed proportional to rate of change in weight. The simulation model then consisted of Equation (1) simultaneously applied to weight (W) and depth (L). The introduction of an explicit depth variable was prompted by Dr. R.S. Clymo, whose model is described in detail in Chapter 9.

At the top layer of the peat column, the constant input I was made in annual steps in g m^{-2} of each litter type, and the equivalent depth of input in cm. It was assumed that the effects of seasonal variation would be insignificant over the long periods of time used. The surface decay rate, k, was modified by the depth factor at the midpoint of the layer, e.g. in a 1 cm layer, decay was modified by the depth factor at 0.5 cm. The weight and depth of the top layer continued to accumulate at this rate until the predetermined depth (1 cm in this example) was reached. The remainder of the initial input weight and depth became the input to the next layer, which began to accumulate, subject to losses at the surface rate, but now modified by the new depth factor appropriate at its midpoint, i.e. at 1.5 cm. The simulated peat accumulated until the depth at which decay was assumed to cease, then it continued to accumulate annually by the amount that remained of the original surface input, without losses through decay. A flow diagram of the model is shown in Figure 1 and a listing of the program may be obtained from the authors. The differential equations were solved numerically, for annual time steps, using the Euler "point-slope" method.

Only the three dominant blanket bog plants, *Calluna vulgaris*, *Eriophorum vaginatum* and *Sphagnum* spp, were considered, split into a total of six components, which were treated as separate input weights and depths, each subjected to their own decay rate, in the manner described above.

The importance of treating the components separately, rather than in combination, will be emphasised later, but was introduced initially because components with different surface decay rates had widely different densities, and hence input depths.

8.2.2 Derivation of Data

8.2.2.1 Sites

The intensive study site on Sike Hill (Nat Grid ref NY769331) was compared with a wetter site, Green Burn (NY774323) although the data available from Green Burn were less detailed. The same three species, viz *C. vulgaris*, *E. vaginatum*, and *Sphagnum* spp, are dominant on both sites, but *Sphagnum* is a much more important component on Green Burn, and the peat deposits are deeper.

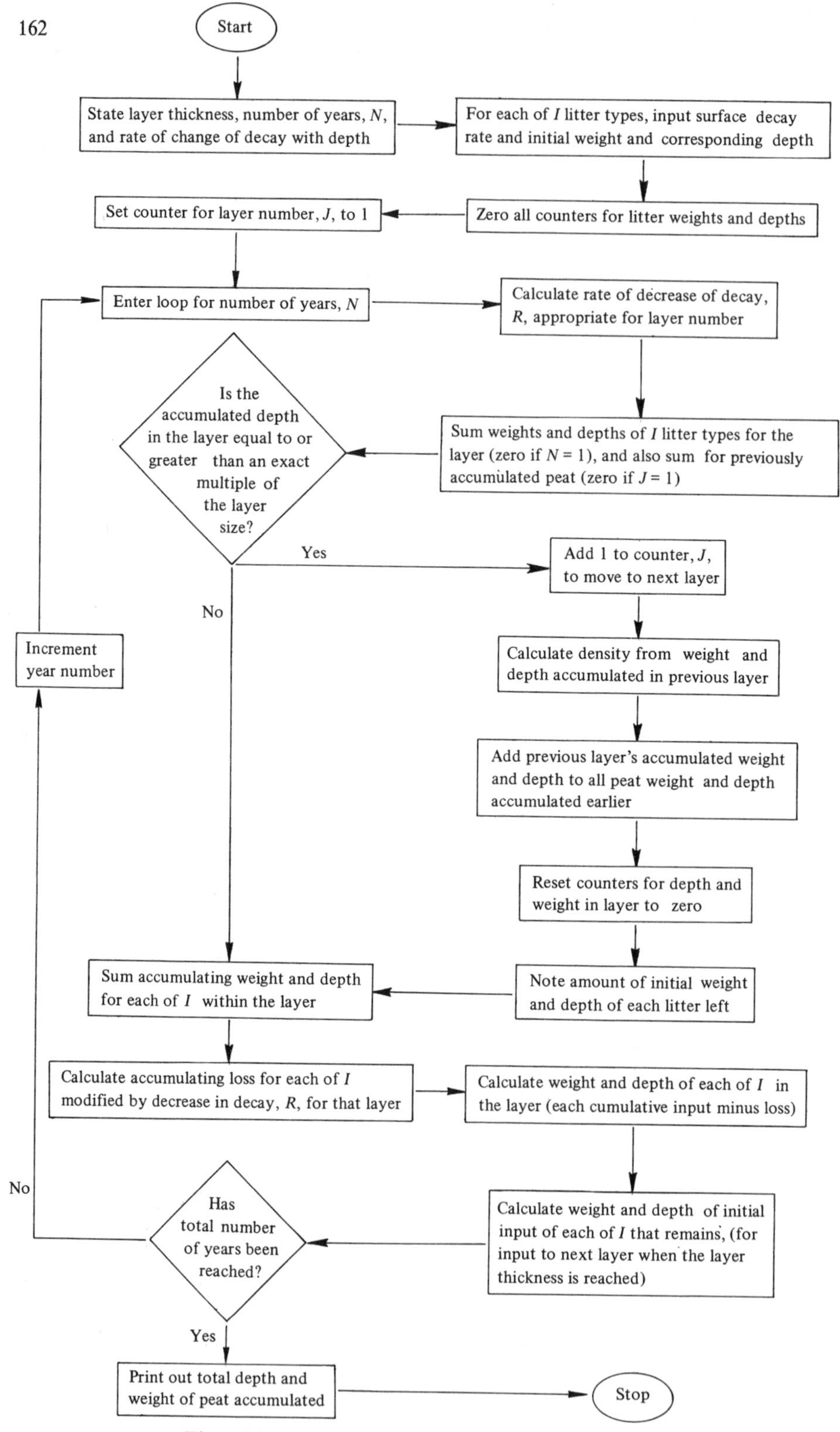

Fig. 1. Flow diagram of model for peat accumulation

Table 1. Summary of production data ($g\,m^{-2}\,yr^{-1} \pm$ se if available) for the three major blanket bog plants: *Calluna*, *Sphagnum* and *Eriophorum vaginatum*

Component	Sike Hill	Green Burn
Calluna shoots	(1) 75.3±4.9 (2) 130	(1) 50.4 (2) 71
Calluna stem	(1) 60±4.5 (2) 32.9	(1) 29.5 (2) 22.2
Eriophorum leaves	(1) 70 (2) 173	(1) 26.7 (2) 66
Sphagnum	44.8	212.9
Calluna below-ground	183	110
Eriophorum below-ground	(1) 60 {30 rhizomes, 30 roots} (2) 45 {19 rhizomes, 26 roots}	17.3 {7.3 rhizomes, 10 roots}

8.2.2.2 Primary Production Rates

A summary of the data derived and tested in the model is shown in Table 1. The six components considered were *Calluna* shoots, wood and below-ground parts, *Eriophorum* leaves and below-ground parts, and *Sphagnum*. Where two or more estimates of production of a component were available, the model was used to compare the effect of each on peat accumulation. Sike Hill estimates are from Forrest (1971) and those for Green Burn from Forrest and Smith (1975).

Calluna shoots. The current year's shoot production and the shoot contribution to litterfall were regarded as alternative estimates in the simulation, since any difference between them partly reflects interseasonal variation. The former was measured as 130 g m^{-2} yr^{-1} at Sike Hill and 71 g m^{-2} yr^{-1} at Green Burn.

Shoot contribution to litterfall was measured directly at Sike Hill as 75.3 g m^{-2} yr^{-1}, including 10.7 g flowers and capsules, assumed to decay at the same rate as leafy shoots. At Green Burn, assuming the same relative proportions of shoot and stem litter as at Sike Hill, shoot litter production 1969–1970 was estimated at 50.4 g m^{-2}, out of a total litterfall of 72.6 g m^{-2} yr^{-1}.

Calluna stems. Forrest's estimate of 60 g m^{-2} yr^{-1}, from the age structure curve for *Calluna* at Sike Hill, plus a regression of standing dead wood on age, was considered better than the 32.9 g m^{-2} yr^{-1} which he estimated from the stem contribution to litterfall. The difference between the two values would be some measure of the degree of decomposition of standing dead on the plant. Since the mean standing crop of dead wood in 1969 was 216.3 g m^{-2} and production 60 g m^{-2} yr^{-1}, some standing dead material was evidently several years old. Moreover, some aerial wood, including standing dead, would be buried by snow and wind each year, and thus not measured in stem litterfall. The Green Burn estimates, 29.5 and 22.2 g m^{-2} yr^{-1}, were similarly derived.

Calluna below-ground parts. The estimates, 183 g and 110 g m^{-2} yr^{-1} for Sike Hill and Green Burn respectively, were based on the assumption that, the bulk of the below-ground biomass consisting of buried stems, P:B ratios would be similar to those above ground. Dead root production was assumed to equal live root production in steady state conditions.

Eriophorum leaves. The estimate of 173 g m^{-2} yr^{-1} for Sike Hill (66 g m^{-2} yr^{-1} at Green Burn) was one of a series of annual transfers of dry matter between the plant parts, based largely on smoothed curves for seasonal changes in weight. The alternative estimates, 70 and 27 g m^{-2} yr^{-1} for the two sites, were derived from a seasonal dynamic model, developed to provide estimates of transfers between the *Eriophorum* plant parts, by fitting computer outputs to the pattern of field variation in all components simultaneously (Jones and Gore, in press; Chap. 2). It was felt that the lower estimates were more probable since, in the seasonal dynamic model, the higher estimates were inconsistent with the standing crops and seasonal transfers given by Forrest (1971).

Eriophorum rhizomes and roots. Forrest's estimate of 19 g m^{-2} yr^{-1} for rhizomes and 26 g m^{-2} yr^{-1} for roots at Sike Hill were considered not significantly different, based on his error terms, from the 30 g m^{-2} yr^{-1} for both components, derived from the model discussed above. Rhizomes and roots were treated as one component at Sike Hill and Green Burn (7.3 g for rhizomes and 10 g for roots) in the absence of a decay estimate for rhizomes.

Sphagnum. Production estimates at Sike Hill (44.8 g m^{-2} yr^{-1}) and Green Burn (212.9 g m^{-2} yr^{-1}) were based on the assumption that input to dead *Sphagnum* was equivalent to live production since the vegetation was not in a seral state.

8.2.2.3 Surface Decomposition Rates

The decay data derived and tested in the model (Table 2) were taken from Heal, Latter and Howson (Chap. 7) unless otherwise stated. Long-term data were obtained by extrapolation from weight losses in litter bag experiments on a nearby site (there were no significant differences between weight losses of *Rubus chamaemorus* on a series of blanket bog sites including Sike Hill and Green Burn). For *Calluna* shoots and above-ground wood and *Eriophorum* leaves, measurements were made annually, for five years, and the decay rate derived as the regression coefficient of an exponential regression ($\log_e$ final/initial weight on years). Linear regressions also provided a good fit to the data for five years, but exponential loss was thought to be more likely, because a linear form implies complete decay in a finite time. There was also some evidence that the untransformed weight losses followed a non-linear pattern, since the cx^2 term in second degree polynomial regressions of final/initial weight against years was significant in the three cases.

For *Calluna* below-ground stems, *Eriophorum* roots and *Sphagnum*, only one or two years' weight losses were obtained. To derive a decay rate from two years' data, the mean $\log_e$ values of final/initial weight were calculated and the decay rate was taken to be the slope of the line joining the mean $\log_e$ values of fraction of weight left after one and two years. Where only one year's data were available, the mean value was assumed to lie on a regression line from the origin. This involved extrapolating one year's data to *n* years, and overlooked the fact that, in the regressions for components where five years' data were available, the regression

Table 2. Summary of decay rates ($g\,g^{-1}\,yr^{-1}$) with errors where available, for the six litter components, applicable to both Sike Hill and Green Burn

Component	Decay rate	Derivation	Errors
Calluna shoots	0.1087	Regression of $\log_e$ final/initial weight against years (t) for five years $y = -0.0713 - 0.1087\,t$	sd of regression coefficient 0.0083
Calluna stem	0.0524	$y = -0.029 - 0.0524\,t$	0.0038
Eriophorum leaves	0.1853	$y = -0.1066 - 0.1853\,t$	0.0184
Sphagnum	(1) 0.0783 (2) 0.0225	One year's weight loss One year's weight loss	se = 0.0078, n = 5
Calluna below-ground	(1) 0.0349 (2) 0.0555	Two years' weight losses One year's weight loss	se = 0.0052, n = 18
Eriophorum below-ground	(1) 0.0215 (2) 0.1095	Two years' weight losses One year's weight loss	se = 0.0201, n = 8

Table 3. Summary of the rate of decrease in decay with depth

Component	Correction for depth	Derivation	Errors
Cotton strip (Sike Hill)	1 − 0.0645 × cm	linear regression of 1 − sample/control tensile strength on cm depth (d) $y = 1.068 - 0.0645\,d$	sd of regression coefficient 0.0067
Cotton strip (Green Burn)	1 − 0.0653 × cm	$y = 1.212 - 0.0653\,d$	0.0079
Calluna below-ground	1 − 0.0243 × cm	Linear regression of $\frac{\text{wt left at depth}}{\text{wt left at surface}}$ on cm depth (d) $y = 1.0065 - 0.0243\,d$	0.0058
Juncus effusus leaves	1 − 0.0368 × cm	$y = 1.1256 - 0.0368\,d$	0.0034
Juncus effusus leaves	Exponent (−0.061 × cm)	Exponential regression of $\frac{\text{wt left at depth}}{\text{wt left at surface}}$ on cm depth (d) $\log_e y = 0.2000 - 0.0610\,d$	0.0059

lines did not pass back through the origin. The decay rates so derived were therefore treated with great caution.

For *Calluna* below-ground stems, the two estimates were derived from different data: 0.0349 $g\,g^{-1}\,yr^{-1}$ from an experiment to compare weight losses over two years between litter types, 0.0555 as the mean surface value in a one year experiment on the effects of depth on decay. Forrest (1971) had included roots in total below-ground production estimates because he considered them an insignificant component. The decay rates of below-ground stems and roots were very similar. They were, therefore, not considered separately.

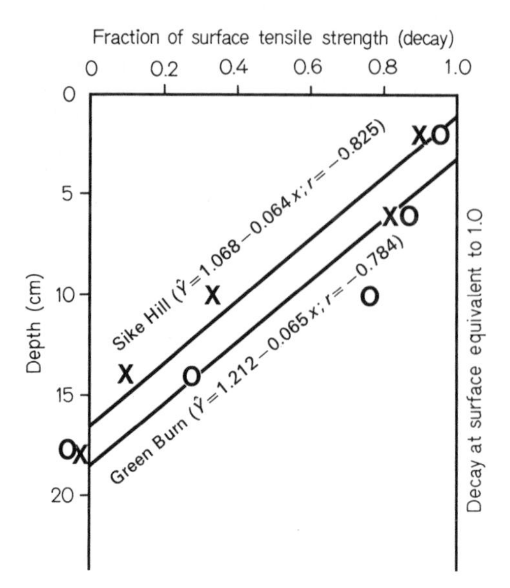

Fig. 2. Linear regressions for tensile strength (kg breaking point) of cotton strips, placed in the peat from 0 to 20 cm, against cm depth. × : Sike Hill; ○: Green Burn

For *Eriophorum* below-ground parts, rhizomes were assumed to decay at the same rate as the roots. The first estimate, 0.0215, was a very low value, obtained after two years, from the experiment to compare litter types. Some samples, which gained in weight after both one and two years in the field, were ignored. The second estimate, 0.1095, was obtained after one year as the surface value in the experiment to determine the effect of depth on decay. The large difference between the two sets of samples was not explicable, but one function of the model was to investigate the effect of such differences which tended to be much greater than the error attached to any one estimate.

For *Sphagnum*, the decay rate of 0.0225 (Table 2) was derived from one year's weight losses in the between-litter experiment. After one year in the field, some samples had lost a little weight and others had gained. After two years *Sphagnum* samples showed clear signs of growth, with up to 20% weight gains. Only samples which had lost weight were included in the calculation, but because of the clear anomaly in the second year results, the decay rate derived was considered to be very inaccurate. Reader and Stewart (1972) recorded even smaller weight losses after one year in the field (between 0.001 and 0.017 $g\,g^{-1}\,yr^{-1}$). Clymo's (1965) % weight loss values for *Sphagnum acutifolium* at Sike Hill and *S. papillosum* at Green Burn were combined and converted to $\log_e$ to provide the second estimate, 0.0783.

8.2.2.4 Rate of Change of Decay with Depth

Evidence showed that the position of the litter in the peat profile was important in determining the rate of decay (Chap. 7). Two independent sets of data indicated this relation (Table 3):

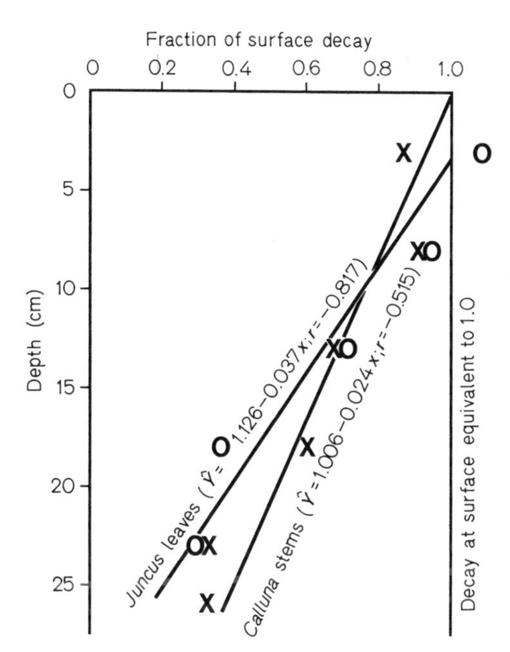

Fig. 3. Linear regressions of decay against cm depth. Values are related to a surface decay of unity. × : *Calluna* below-ground stems; ○: *Juncus effusus* leaves

1. Cotton strips had been placed in the peat profile in the top 20 cm at a number of sites. On removal, after 33 weeks, the tensile strength as kg breaking point of each 4 cm section of the strips was determined. Because tensile strength was greatest when weight loss was least, the fraction sample/control tensile strength was subtracted from unity and the regression of the fraction on depth calculated (Fig. 2). A linear regression provided a reasonable fit for the relation between tensile strength and weight loss (Latter and Howson, 1977). Therefore, the regression coefficients derived from the regression of tensile strength against depth in cm were used directly to correct the initial surface decay rates of each component as it passed down the profile (Sect. 2.1).

2. Below-ground stems of *Calluna*, and *Juncus effusus* leaves, had been placed in positions down the profile from the surface to 25 cm, and weight losses after one year determined. The results were corrected for differences in the surface decay of the two litter types, by dividing throughout by the mean surface value so that the surface decay was equivalent to unity. The regression of the adjusted values on depth was calculated (Fig. 3). The effect of depth on decay rate implied by the resulting regression coefficients was compared with those obtained from the cotton strips. The effect of an exponential decline in decay rate with depth was compared with that of a linear decline in the *Juncus effusus* leaves.

The regression coefficients for *Calluna* stems and *J. effusus* leaves indicated a smaller effect of depth on decay than in the cotton strips. The implication of linear rate of change of decay with depth, for the materials investigated, was that decay would gradually diminish to depths of 15.5, 27.2 and 41.2 cm for the cotton strip, *Juncus* and *Calluna* respectively. From data for Moor House and Thursley Common, Clymo (1965) indicated that at 75 cm depth, decay for *Sphagnum* was still 10.4% of the surface value, and for Whatman No 1 filter papers at Moor House it was 11.3%, indicating considerable decay at depths exceeding those for the data presented in Table 3.

8.2.2.5 *Respiration Data*

Weight losses from litter bags probably overestimated long-term decay (Chap. 7), because leaching and particulate losses occurred in addition to losses from microbial activity. Respiration rates, measured in a respirometer in the laboratory, for each of the five years that the litter bags of *Calluna* shoots and wood and *Eriophorum* leaves were collected, indicated a decline in respiration rate with time, although the weight loss data indicated a constant decay rate. Thus when *Eriophorum* litter had been reduced to half its initial weight, the respiration rate had declined to half its initial rate. The effect of incorporating this correction factor was investigated by later modifications to the model.

8.2.3 Validation

To provide the necessary validation data of peat accumulation over long periods, peat density profiles were obtained at Sike Hill and Green Burn by means of a Russian peat borer (Sect. 8.6.2). In order to date these profiles provisionally, further density profiles were obtained at the nearby sites, Bog Hill and Valley Bog (Johnson and Dunham, 1963), Dr. G. A. L. Johnson kindly indicating the exact positions of his sampling sites for pollen analysis. These were compared with the density profiles at Sike Hill and Green Burn. On the basis of similarities between the density profiles at Bog Hill and Sike Hill, and Valley Bog and Green Burn, and pollen dating at Bog Hill and Valley Bog, it was estimated that in the last 2500 years 56–80 cm (46–70 kg m^{-2}) peat had formed at Sike Hill, and 200–300 cm (129–204 kg m^{-2}) at Green Burn.

Samples of peat were sent to Dr. D. D. Harkness at the NERC Carbon Dating Centre. Deeper peat deposits were dated by standard C^{14} methods, and the top few cm of peat were analysed for "bomb carbon" to provide an additional check for the first few years of peat accumulation in the model (Sect. 8.6.3).

8.3 Results

8.3.1 Initial Simulation

In the initial simulation, a set of production parameters and surface decay rates was taken for each site, with the appropriate cotton strip rate of decay with depth (Table 4). The total input, with mean surface decay, weighted for the different rates of production in each component, was calculated, and the simulation run for each site for 2500 years (Table 5).

A version of the simulation to provide an analytic solution was used to check that neither annual time steps, nor the Euler method of numerical solution, caused large inaccuracies when long runs were made. The differences between the two methods of solution, relatively small compared with some of the changes resulting from variation of the parameters in the model, could be greatly reduced by making the time steps smaller than one year in the numeric solution (Table 5). This was not considered worthwhile because of the much greater computer time involved.

Table 4. Basic set of parameters used in the simulation of Sike Hill and Green Burn peat accumulation

Site	Litter component	Input weight $g\,m^{-2}\,yr^{-1}$	Input length $cm\,m^{-2}\,yr^{-1}$	Surface decay rate $g\,g^{-1}\,yr^{-1}$	Rate of change of decay with cm depth
Sike Hill	*Calluna* shoots	75	0.25	0.1087	Surface decay rate x $(1-0.0645\,x\,\text{cm})$
	Calluna stem	60	0.011	0.0524	
	Eriophorum leaves	70	0.2333	0.1853	
	Sphagnum	45	0.15	0.0783	
	Calluna below-ground	183	0.0345	0.0349	
	Eriophorum below-ground	60	0.011	0.0215	
	Overall value	493	0.689[a]	0.0719[b]	
Green Burn	*Calluna* shoots	50.4	0.168	0.1087	Surface decay rate x $(1-0.0653x\,\text{cm})$
	Calluna stem	29.5	0.0056	0.0524	
	Eriophorum leaves	26.7	0.089	0.1853	
	Sphagnum	212.9	0.709	0.0783	
	Calluna below-ground	110	0.0208	0.0349	
	Eriophorum below-ground	17.3	0.0033	0.0215	
	Overall value	446.8	0.9957[a]	0.0735[b]	

[a] From shoots, leaves and *Sphagnum* at 0.03 g cm^{-3}; wood and below-ground at 0.53 g cm^{-3}.
[b] Weighted for g input of each component.

Table 5. Results of peat accumulation at Sike Hill and Green Burn in the initial simulation run with overall values from Table 4 for 2,500 years, with a layer thickness of 1 cm, and annual time steps

Site	Estimate of accumulation from field results		Numeric solution: Annual time steps		Numeric solution: 100 steps yr^{-1}		Analytic solution	
	kg	cm	kg	cm	kg	cm	kg	cm
Sike Hill	46–70	56–80	176.8	247.4	242.4	339.1	243.0	340.1
Green Burn	129–204	200–300	422.9	942.4	479.6	1,068.9	480.3	1,070.3

Field results and those obtained with an analytic solution and a numeric solution with 100 time steps per year are shown for comparison.

Table 6. Results of peat accumulation at Sike Hill and Green Burn in the basic simulation, run with individual parameters from Table 4 for 2,500 years

Site	kg weight	cm depth
Sike Hill	17.84	9.54
Green Burn	397.93	631.27

The results in Table 5 are a considerable overestimate of peat accumulation at both sites when compared with the field estimates, although, as in the field, accumulation at Sike Hill was considerably less than at Green Burn, and the peat was denser.

8.3.2 Basic Simulation

Because woody litters have very different densities from leafy litters, and because different surface decay rates apply to all litter types (Table 4), it was felt that separating the litter types as six different inputs (Table 6) would be more realistic in the simulation than treating them in combination as was done in the initial simulation.

The very large differences between results in Tables 5 and 6 are considered to be caused by dense woody litters decaying more slowly than less dense leafy litters. The effect was very much more marked at Sike Hill (as might be expected since woody litters are a more important component there), so that although too much peat was still accumulated at Green Burn, at Sike Hill the amount was now too small.

The other consequence of the combination of dense tissues with slow decay rates, and less dense with faster rates, is that as accumulation proceeds the density of the simulated peat increases, i.e. it increases with depth. This allows an additional constraint in the simulation, since not only does the total depth and weight accumulated have to conform to the validation data, but also the pattern of increasing density in the simulation with that observed in the peat profiles.

The basic simulation, which allowed the six litter weights and depths to be treated separately, was used to study the effects of different sources of variability on the simulated peat accumulation in the following four categories:

1. Plant production rates
2. Surface decay rates
3. Rate of decrease of decay with depth
4. Litter density, affecting the weight to depth input ratios.

8.3.2.1 Variability in Plant Production Rates

The peat accumulated when plant production estimates were varied for each litter type individually was simulated, using either alternative estimates where they had been made, or the limits of error on production estimates in the basic set of parameters ($\pm$se). All changes involved manipulation of the figures presented in Table 1, and the peat accumulated, both weight and depth, was calculated as a ratio of the peat accumulated using the basic set of parameters (Table 7).

Apart from the alternative estimates for *Calluna* shoots and *Eriophorum* leaves, $\times 1.7$ and $\times 2.5$ the values in the basic set respectively, no source of variation in production rates resulted in a change $> \times 0.1$ in the simulated peat accumulation. The smaller change in production in *Calluna* shoots resulted in a larger change in the peat accumulation than that resulting from the alternative estimate in *Eriophorum* leaves, because the decay rate was lower in the *Calluna* shoots.

Table 7. Ratio, to nearest one tenth, of peat accumulation using altered production rates, to accumulation using basic set of parameters (see Table 6)

Source of variation	Production $g\ m^{-2}\ yr^{-1}$		Peat as a ratio of the basic set			
	Sike Hill	Green Burn	Sike Hill		Green Burn	
			kg	cm	kg	
(1) Alternative estimates						
Calluna shoots	130	71	8.6	5.1	1.6	1.4
Eriophorum leaves	173	66	7.1	4.0	1.1	1.2
Calluna stem	32.9	22.2	0.9	1.0	1.0	1.0
Eriophorum below-ground	45	—	0.9	1.0	—	—
(2) ± se on estimates, where available						
Calluna shoots + se	80.2	—	1.0	1.0	—	—
Calluna shoots − se	70.4	—	1.0	1.0	—	—

Table 8. Ratio, to nearest one tenth, of peat accumulation using variation in surface decay rate, to accumulation using basic set of parameters

Source of variation	Decay rate $g\ g^{-1}\ yr^{-1}$	Peat as a ratio of the basic set			
		Sike Hill		Green Burn	
		kg	cm	kg	cm
Eriophorum below-ground	0.1095	0.7	0.9	0.9	1.0
Sphagnum	0.0225	16.1	24.3	18.1	25.0
Calluna below-ground	0.0555	0.7	0.9	0.9	1.0
Eriophorum leaves + sd	0.2037	1.0	1.0	1.0	1.0
Eriophorum leaves − sd	0.1669	1.0	1.0	1.0	1.0
Calluna shoots + sd	0.1170	1.0	1.0	1.0	1.0
Calluna shoots − sd	0.1004	1.0	1.1	1.0	1.1
Calluna stem + sd	0.0562	1.0	1.0	1.0	1.0
Calluna stem − sd	0.0486	1.0	1.0	1.0	1.0

8.3.2.2 *Variability in Surface Decay Rates*

The exercise carried out to study effects of varying production rates was repeated for surface decay rates, with the alternative estimates and limits of error presented in Table 2. The ratios of these results to peat accumulated using the basic set of parameters (Table 8) indicate that the simulation is relatively insensitive to variation in decay rate, except in *Sphagnum*. The alternative estimate for this litter type, 0.0225, was particularly dubious. However, it was a three-fold reduction from the value used in the basic set of parameters, 0.0783, compared with a $> \times 4$ increase in the *Eriophorum* root rate (0.1095 instead of 0.0215), which nevertheless produced a far smaller change in the simulated peat at both sites. The differences in ratio of weight and depth were caused by the different litter densities. For example, when the *Eriophorum* root decay rate was increased, the reduction in weight of peat accumulated was greater than the reduction in depth, because it was assumed to be a dense litter.

8.3.2.3 Variability in Rate of Decrease of Decay with Depth

The simulation is very sensitive to a change in the rate of decrease of decay with depth (Table 9), using alternative estimates listed in Table 3. Many changes could be disregarded completely since simulated peat accumulation ceased well before 2500 years had elapsed. This effect was apparently much greater at Green Burn, because at Sike Hill accumulation with the basic set of parameters was already so slow that alteration of the depth factor did not cause a marked change. The exponentially and the linearly declining decay rate with depth in *Juncus* leaves caused a similar reduction in peat accumulation. Therefore, the fact that the peat continued to decay at a very low rate deep down the profile (as it would if decline in decay with depth was exponential) appeared to have less effect than differences in rates of decline in decay in the top layers.

8.3.2.4 Variability in the Density of Litters

Because very few data were available, litter densities were varied within wide limits, the resulting simulated accumulation (Table 10) indicating the sensitivity to this factor at both sites.

For example, when *Sphagnum* density was decreased, and *Calluna* wood density increased, there was less accumulation at Sike Hill and more at Green Burn, indicating the importance of wood and *Sphagnum* in contributing to peat at the two sites. The large reduction in accumulation at Green Burn with a leaf litter density increased to 0.04 and wood litter density decreased to 0.4 g cm^{-3} also indicated the greater importance of the leafy tissues at this site. When *Eriophorum* roots were separated from *Calluna* wood and set to their measured value of 0.069 g cm^{-3} (Sect. 8.6.1), the accumulation of peat at Sike Hill was 4–4.5 times that with the basic set. However, unless the change made was fairly large (e.g. the $\times 8$ reduction in density in *Eriophorum* roots) changes in litter density did not bring about large changes in the simulated peat accumulation.

8.3.3 Correction for Respiration

No combination of parameters in the basic simulation provided results that were close to the estimated field accumulation. Apart from the cotton strip decline in decay with depth at Green Burn, when accumulation was *c* twice too much, accumulation at both sites was too low, suggesting that the constant decay rates derived from the litter bag weight losses overestimated long-term decay. The density increase down the profile occurred too quickly compared with measured peat densities. There was also no further increase once the depth was reached at which decay was assumed to cease. At Sike Hill, the increase in density was too rapid, and the densities reached were too great, except when all wood and root densities were reduced to 0.2 g cm^{-3}, which was considered to be too low.

Table 9. Ratio, to nearest one tenth, of peat accumulation, using variation in decrease of decay with depth, to accumulation using the basic set of parameters

Source of variation of decay with depth	Value of coefficient		Peat as a ratio of the basic set			
	Sike Hill	Green Burn	Sike Hill		Green Burn	
			kg	cm	kg	cm
Calluna below-ground rate	0.0243	0.0243	No accumulation after 905 years		No accumulation after 1,057 years	
Juncus leaf rate	0.0368	0.0368	0.7	0.8	0.02	0.02
Cotton strip rate + sd	0.0713	0.0732	1.2	1.1	1.2	1.3
Cotton strip rate − sd	0.0578	0.0574	0.9	0.9	0.8	0.7
Calluna below-ground rate + sd	0.0301	0.0301	No accumulation after 973 years		No accumulation after 1,109 years	
Calluna below-ground rate − sd	0.0185	0.0185	No accumulation after 815 years		No accumulation after 874 years	
Juncus leaf rate + sd	0.0402	0.0402	0.7	0.8	0.03	0.03
Juncus leaf rate − sd	0.0334	0.0334	0.7	0.8	0.02	0.03
Juncus leaf exponential rate	−0.061	−0.061	0.8	0.9	0.04	0.03

Table 10. Ratio, to nearest one tenth, of peat accumulation, using variation in litter density values, to accumulation using basic set of parameters

Litter densities $g\ cm^{-3}$		Peat as a ratio of the basic set			
		Sike Hill		Green Burn	
		kg	cm	kg	cm
Shoots and leaves *Calluna* stem and below-ground *Eriophorum* roots	0.03 0.53 0.069	4.2	4.6	1.0	1.1
Shoots and leaves *Sphagnum* Wood and roots	0.0333 0.0264 0.5992	0.9	0.9	1.1	1.3
Shoots and leaves *Sphagnum* *Calluna* wood and below-ground *Eriophorum* roots	0.0333 0.0264 0.5992 0.2	1.1	1.2	1.1	1.3
Shoots and leaves *Sphagnum* Wood and roots	0.05 0.0264 0.2	2.0	2.3	1.2	1.4
Leafy tissues Wood and roots	0.04 0.4	0.9	0.8	0.5	0.3
Leafy tissues Wood and roots	0.03 0.6	1.0	1.0	1.0	1.0

Values for the basic set were $0.03\ g\ cm^{-3}$ for leafy tissues and $0.53\ g\ cm^{-3}$ for woody tissues.

Table 11. Selected results of simulated peat accumulation in 2500 years when a modification for decline in decay from respiration decrease is introduced

Site	Parameter used	Basic simulation		Adjustment for respiration losses		Field estimates	
		kg	cm	kg	cm	kg	cm
Sike Hill	Basic set	17.8	9.5	370.8	289.6	46– 70	56– 80
Green Burn		371.9	631.3	484.9	885.6	129–204	200–300
Sike Hill	With *Calluna* below-ground	No accumulation after 905 years		26.8	16.2		
Green Burn	stem rate of change of decay with depth	No accumulation after 1057 years		84.0	100.3		
Sike Hill	With *Juncus* leaf rate of change of	12.8	7.6	43.2	21.7		
Green Burn	decay with depth	11.5	15.3	261.2	441.7		
Sike Hill	With *Calluna* below-ground	No accumulation after 973 years		31.2	18.1		
Green Burn	stem rate of change of decay with depth + sd	No accumulation after 1109 years		178.3	285.6		

Results with the basic simulation are shown for comparison with those estimated from field data.

To allow for the possible overestimate of long-term decay, an additional correction was made based on the evidence of the respiration data (Sect. 8.2.2.5). When a litter component had decayed to half its initial weight, the decay was halved, and halved again when only a quarter of the initial weight remained. The simulated accumulation of peat was, in general, now considerably greater than that observed (Table 11).

The sensitivity of this version to variation in the four main parameters (production, surface decay, change of decay with depth and litter density) was tested. Apart from changes in the rate of decrease of decay with depth, to which this version, like the basic simulation, appeared to be very sensitive, parameter variation did not produce large changes in accumulation. Thus, with the smaller effects of depth on decay indicated by *Calluna* below-ground stems and *Juncus* leaves, accumulation was much reduced compared with results of the basic set in which the cotton strip correction for depth was used. The effect at Green Burn was such that, with a rate of change of decay with depth intermediate between that of the *Calluna* below-ground stems and the *Juncus* leaves (i.e. *Calluna* below-ground stem rate plus its standard deviation, 0.0301), accumulation was 285.6 cm and 178 kg which was within the limits of depth and weight estimated to have accumulated in the field in 2500 years. From the density profile (Sect. 8.6.2) it was calculated that 285 cm of Green Burn peat weighed 190 kg m^{-2}. Figure 4 shows the actual peat density profile compared with those obtained in the basic simulation and with the correction for respiration decline. In the basic simulation, with

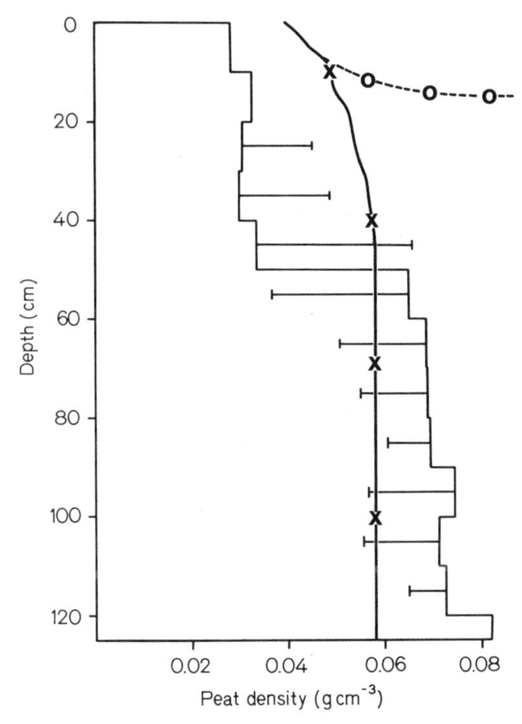

Fig. 4. Green Burn peat profile in upper layers with sd on four cores where taken. × : simulated profile when correction was made for respiration decline at half life and quarter life of litter components; ○: simulated profile with no correction for respiration decline

constant decay rates, steady state was attained in 1109 years when only 13.8 cm and 10 kg had accumulated. In the field profile there appears to be a sudden increase in density between 40 and 60 cm, whereas in the simulation corrected for respiration decline, the density increase occurred gradually until 33 cm depth when decay ceased, after which no further increase occurred.

The simulation was also run with this combination of parameters for 15 years, to check against validation data obtained from "bomb carbon": 7.65 cm and 3.82 kg peat accumulated, compared with 8–10 cm and 1.92–2.80 kg estimated from "bomb carbon" analysis, and the peat density profile. The fact that the depth accumulated in the shorter time period was almost in agreement with the validation data, but the weight was too great, underlines the fact that, in the early stages of simulation, the density of peat accumulating was rather too great.

Table 11 illustrates that, although accumulation at Green Burn was realistic with this set of parameters, the same rate of decline of decay with depth applied to the Sike Hill data caused accumulation of only about one third of the depth estimated from validation data in 2500 years. As the increase in density was too great and too rapid, the corresponding weight was a considerable overestimate. Figure 5 compares the field profile for peat density with results from both versions of the simulation.

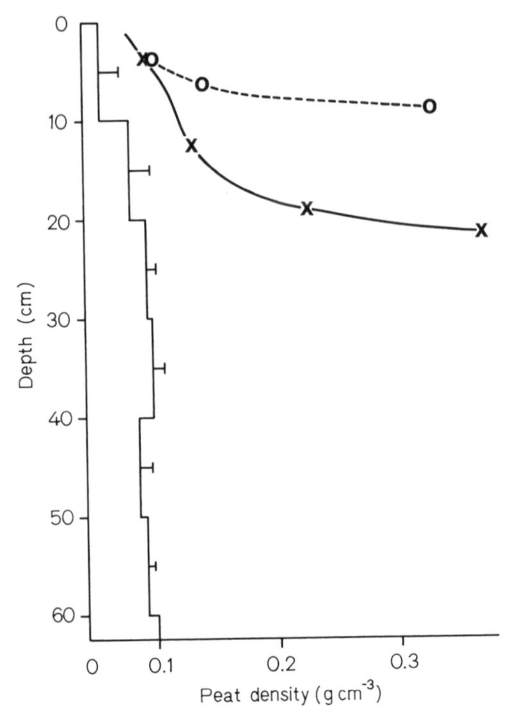

Fig. 5. Sike Hill peat profile in upper layers with sd on four cores where taken. × : simulated profile when correction was made for respiration decline at half life and quarter life of litter components; ○: simulated profile with no correction for respiration decline

8.4 Discussion

A number of assumptions which were implicit in the simulation require examination. The first was that production rates obtained now had been constant for 2500 years. Peat deposits on the Reserve, examined by Johnson and Dunham (1963), indicated that the three co-dominant species had also been present and dominant on the blanket bog areas for the sub Atlantic period (approximately the last 2500 years). However, the absolute and relative production rates of the three might well have changed in that time. For example, Johnson and Dunham (1963) considered that in the past *Eriophorum vaginatum* had been the dominant blanket bog peat former, with *Calluna* and *Sphagnum* less important, although *Calluna* is at present dominant at Sike Hill and *Sphagnum* at Green Burn.

Associated with this assumption was that of constant surface decay rates in the three species, since any climatic change sufficient to affect production would have been likely to affect decay also, so that increased production would be counteracted to some extent by increased decay rates, and vice versa.

To obtain a depth of peat corresponding to 2500 years' accumulation at Sike Hill and Green Burn, it was assumed that the ratio of peat formed in the past 2500 years to total peat depth would be constant in peats of similar types. Thus the ratio for the deep peat (160 cm) at Bog Hill, 0.4–0.6 (because the sub Boreal/sub

Atlantic transition zone was rather wide) was assumed to apply to the shallower Sike Hill peat, and the Valley Bog ratio, 0.25–0.3, to Green Burn. It is interesting to note that the ratio of peat formed before, and since, the Boreal/Atlantic transition was also 0.4–0.6 at Valley Bog, since this marked the start of peat formation at the drier Bog Hill and Sike Hill sites. From data on Welsh blanket peats (Moore, 1972), the closely similar ratios of 0.78 and 0.75 were obtained for blanket peat deposits of total depth 64 and 120 cm respectively which also supports the argument. Results in the Appendix 8.6.3 show the C^{14} dating of the deeper peat at Sike Hill and Green Burn.

In the basic simulation, with no correction for respiration rate, the rate of change of decay with depth was the most critical factor, partly because it was the most variable. This would have affected the results in two ways: first, because linear rate of decrease was assumed in most cases, it resulted in a rather artificial cut-off point for decay, so that, where peat accumulated to depths in excess of the cut-off point, it continued to accumulate by the amount of the initial input remaining at that point. The amount remaining when decay ceased was thus a more critical factor in the simulation than it would be expected to be in reality. This point was emphasised by results using the *Juncus effusus* leaf rate of declining decay, both linear and exponential, at Green Burn, when the partial correction for respiration decline was included. The depth accumulated in 2500 years with the linear rate of loss was 441 cm, much in excess of the depth at which decay ceased, and the depth accumulated with the exponential rate was 302 cm. The difference between results with the *Juncus* linear and exponential decline within the depth at which decay occurred with the linear decline, was very small (Table 9). Therefore the lower result with exponential decline was presumably due to the fact that decay would continue, if only at very low levels, for the whole depth of peat formed, and would not cease at 27.2 cm, where it ceased with a linear decline.

The second reason that decline in decay was an important factor was that, within the depth where decay occurred with linear decline, the lower the coefficient value, the less effect it had on the surface decay rates and, therefore, the less peat accumulated.

For example, a comparison of the Sike Hill cotton strip rate and the *Calluna* below-ground rate at 10 cm depth showed that, with the former, decay would only be 35.5% of the surface values whereas with the latter it would still be 75.7%.

Since the data for decline in decay were derived from cotton strips, *Juncus* leaves and *Calluna* below-ground stems, it was assumed in the simulation that one or other of these rates applied to all litter components which, in view of the difference between them, seems unlikely. Having shown the factor to be important, it would be necessary in further studies to obtain data on the rate of declining decay for each litter type included, applicable to that component only. Heal, Latter and Howson (Chap. 7) showed a slightly increasing rate of decay with depth in *Eriophorum vaginatum* roots, and Clymo (Chap. 9) showed the effect of the position of the water table to be important. Its effect in defining the aerobic and anaerobic layers would be analogous to variation in the slope of declining decay with depth. A further assumption was that the rate of declining decay with depth for fresh litter applied to material that had been decaying for many years since its point of input at the peat surface.

The sensitivity of the simulation to the lower decay rates, by adjusting for respiration losses, tended to confirm that a constant decay rate, even when adjusted for depth in the peat profile, was not realistic for litter decaying over long periods, if the primary production estimates were realistic. The simulation had thus been used to investigate the compatibility of the data which were derived in discrete projects. Without the correction for respiration, accumulation at both sites ceased long before 2500 years, using the smaller effects of depth on decay obtained with *Juncus* leaves and *Calluna* below-ground stems. Where the rate of decay is low, as it is at Moor House, it would be difficult to show that the rate might be declining with time except in a very long-term experiment. However, Minderman (1968) showed that a constant rate of decay was not sufficient to explain the curves obtained in the decomposition of forest litter.

The large change in the simulated results on correction for a decline in respiration rate emphasised the importance of obtaining more data on this aspect, and showed the suitability of simulation in examining assumptions made in data collection.

At Green Burn, a reasonable picture of peat accumulation was obtained when the lower decline of decay with depth from *Calluna* below-ground stems or *Juncus* leaves was used. The simulated density profile, although it increased nearer the surface, and more smoothly than in the field, was also an approximate estimate. The simulated peat density at Green Burn was always less than at Sike Hill, as was observed in the field. Farnham and Finney (1965) showed that *Sphagnum* peat was less dense than that from dwarf shrubs. It is possible that the sudden change in density observed in the field at Green Burn between 50 and 60 cm reflected a change from a *Calluna*- or *Eriophorum*- to a *Sphagnum*- dominated community as at present. With the same combination of parameters, the simulation did not produce a realistic description of the field situation at Sike Hill. Accumulation was too low and the density profile increased too quickly and by too much.

There has been an important omission in the development of this simulation. Although the litter types have been separated, they have all been treated as if they entered the peat system at the surface. In fact, below-ground parts enter the system at a point where they would be subjected to lower decays from the start. Forrest (1971) considered that nearly all the *Calluna* below-ground biomass was found in the top 15 cm, but *Eriophorum* roots occurred down to 30 cm, where decay would be very slow, even assuming the minimum effect of depth on decay as measured in the *Calluna* below-ground stems. The effect of inserting the roots at depth would be greater at Sike Hill where the below-ground biomass is much greater and, in terms of the results of the simulation, would cause greater peat accumulation and a more gradual increase in the density profile, since denser material would be entering the peat at depth.

8.5 Conclusion

Production and decay data of three major blanket bog species, *Calluna*, *E. vaginatum*, and *Sphagnum*, were used to simulate the accumulation of peat on two blanket bog sites of different peat depths, to study the effects of variability in production and decay and the compatibility of independently derived data from separate projects.

The peat bog system was visualised as a series of layers of equal thickness. The input to the uppermost layer corresponded to the observed annual weights and depths of plant litter deposited m^{-2} at each of the sites. The deposited depth was calculated from litter density estimates. The contents of each layer were allowed to accumulate according to the equations:

$$\frac{dW}{dt} = I_w - kW \quad \text{and} \quad \frac{dL}{dt} = I_l - kL,$$

where change in depth (L) was assumed proportional to change in weight (W), and k was the decomposition rate, which was modified by a depth factor calculated at the midpoint of each layer. When the peat accumulated to the depth of the layer, the remainder of the initial input weight and depth became the input to the next layer.

The basic assumption of the depth and weight relation meant that when a total litter input was used with an overall decay rate, the simulated profile had a constant density. However, when treated separately, the different depth and weight input and decay of each litter component resulted in a change in the density of the simulated profile, until the depth was reached where decay was assumed to cease.

Variation in litter production, litter densities, decay rates and effect of depth on these decay rates was simulated.

Constant decay rates, derived from weight losses in litter bag experiments, appeared to overestimate long-term decay, in that insufficient simulated peat was formed when constant rates were used. Respiration rates of the litters had been shown to decline with time, and decay rates were modified so that they declined in the same way. With the modified decay rates, the simulation produced a more realistic accumulation of peat.

The simulated results were compared with peat density profiles obtained independently in the field. The profiles were dated by existing pollen analyses of adjacent sites, and by the change in the $C^{14}:C^{12}$ ratio resulting from atom bomb tests.

Acknowledgements. Our thanks are due to Mr. J.N.R.Jeffers, Dr. O.W.Heal, and Miss R.A.H.Smith for reading and criticising the manuscript. St. Martin's College, Lancaster kindly lent the Russian borer for the measurement of peat density profiles, and Dr. G.A.L. Johnson provided information on the exact siting of his cores for pollen analysis. Staff at the ATLAS Computer Laboratory gave advice and assistance in editing the program to run on the 1906A computer. Miss J.M.Charnock was responsible for drawing the figures.

8.6 Appendix

8.6.1 Calculation of Depth of Input from Density of Litter Components

Assumptions on the density of the litter components were made, and used during most of the tests on the simulation. Later, measurements were made to check the validity of some of the assumptions. The assumed estimates were also varied within fairly wide limits to determine the sensitivity of the simulation to changes in litter density.

Litters were placed in one of two categories, within which densities were assumed to be the same:

1. *Calluna* shoots, *Eriophorum* leaves and *Sphagnum*. The density of these components, when freshly dead, was assumed to be 0.03 g cm^{-3}. This figure, crudely based on an estimate of specific leaf area for *Helianthus annuus* of 3 dm^2 g^{-1} (Evans, 1972), was used as follows:

$$3\ dm^2 g^{-1} = 0.333\ g\ dm^{-2}.$$

Leaves are assumed 1 mm thick.

Then 0.333 g dm^{-2} = 0.333 g dm^{-2} mm^{-1} thickness
= 0.0333 g cm^{-3} (rounded to 0.03 g cm^{-3}).

2. *Calluna* wood and below-ground parts and *Eriophorum* rhizomes and roots. Grace and Woolhouse's (1973b) estimate of 0.53 g cm^{-3} for *Calluna* wood was used in the background set.

The density of three litter components, *Calluna* standing dead wood, *Sphagnum* and *Eriophorum* roots, was measured directly to check the validity of the above assumptions. *Calluna* standing dead wood density was measured, by displacement of water by oven dried stems, at 0.5992 g cm^{-3} (se = 0.0092, $n = 30$). This value was applied to *Calluna* buried stems and *Eriophorum* rhizomes. *Sphagnum* density was measured as 0.0264 g cm^{-3} (se = 0.0012, $n = 20$) by lightly freezing large blocks of *S. papillosum*, and from these cutting small blocks of known dimensions, which were oven dried and weighed. Shoots and leaves were assumed to be slightly denser than this, 0.0333 g cm^{-3}. *Eriophorum* root density was measured on both oven dried and saturated roots, by displacement of water. The measured density of oven dried roots was 0.365 g cm^{-3} (se = 0.037, $n = 10$), and that for saturated roots 0.069 g cm^{-3} (se = 0.002, $n = 5$).

The difference between the two root sets was large. Because the oven dry density was close to the value that had been assumed initially, only the effect of a density of 0.069 g cm^{-3} was investigated. It is also likely that this is closer to the field condition since the dead roots, when they enter the decomposition cycle, must be fairly saturated.

8.6.2 Peat Density Profiles and Dating of Deeper Horizons

Peat densities were determined on cores from Sike Hill, Green Burn, Bog Hill and Valley Bog, taken with a Russian borer. Cores were cut into 10 cm lengths (20 or 40 cm in the deeper deposits) and the wet weight volume determined. The peat was dried to constant weight at 105° C and the density in g cm^{-3} (Figs. 6 and 7) and the weight as kg odw m^{-2} calculated (Tables 12 and 13).

An analysis of variance was carried out on the peat densities of the four sites at the depths common to all, i.e. 0–150 cm (Table 14).

The mean densities of the samples from each site included in the above analysis were: Sike Hill 0.0920 g cm^{-3}, Bog Hill 0.0917, Green Burn 0.0603 and Valley Bog 0.0375. From Tukey's Q test (Snedecor, 1961) the significant difference between the four sites was shown to be 0.0109, so that density profiles at Sike Hill and Bog Hill were not significantly different at

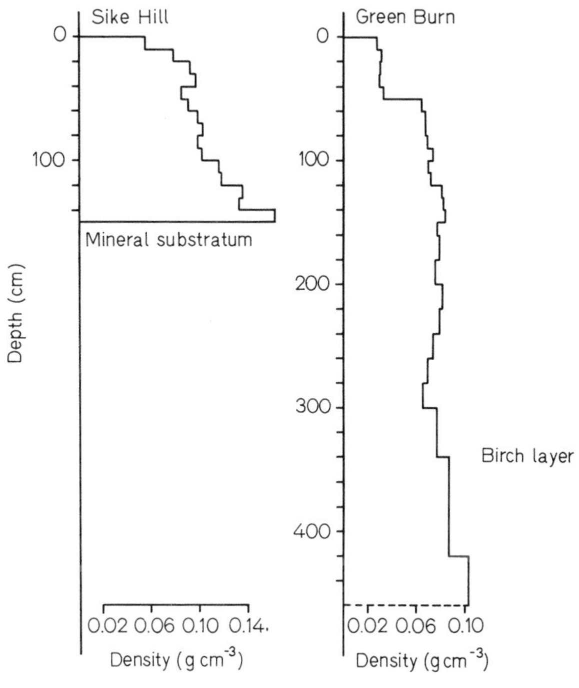

Fig. 6. Peat density profiles (g cm^{-3} per sample depth) at the two study sites

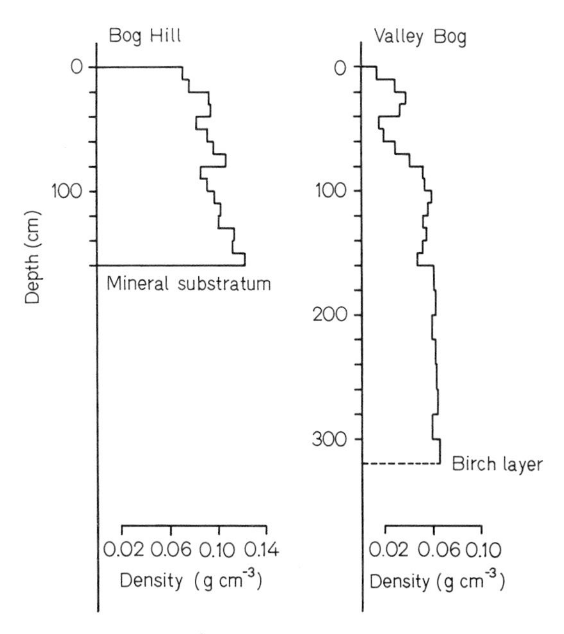

Fig. 7. Peat density profiles (g cm^{-3} per sample depth) at sites dated by pollen analysis (Johnson and Dunham, 1963)

Table 12. Weight of peat, kg m^{-2} per sample depth, to the inorganic substratum at Sike Hill, and to the birch layer at Green Burn

Depth (cm)	Sike Hill	Green Burn
0– 10	5.41 (0.87)	2.80
10– 20	7.82 (0.48)	1.62
20– 30	9.14 (0.08)	3.07 (1.46)
30– 40	9.65 (0.09)	3.02 (1.85)
40– 50	8.43 (0.11)	3.37 (3.21)
50– 60	9.04 (0.05)	6.51 (2.81)
60– 70	9.81 (0.05)	6.88 (1.77)
70– 80	10.28 (0.07)	6.88 (1.34)
80– 90	9.86 (0.09)	6.95 (0.87)
90–100	10.17 (0.10)	7.45 (1.76)
100–110	11.63 (0.11)	7.12 (1.53)
110–120	11.75 (0.07)	7.27 (0.76)
120–130	13.52	8.22 (0.78)
130–140	13.33	8.32 (1.64)
140–150	16.30	8.58 (2.22)
150–160		7.89 (0.45)
160–180		16.06 (0.52)
180–200		15.28 (0.72)
200–220		16.30 (0.46)
220–240		16.08 (0.46)
240–260		14.90 (2.56)
260–280		14.08 (2.08)
280–300		13.30 (1.30)

sd on four cores, where taken, shown in brackets.

P = 0.05, whereas at Green Burn and Valley Bog they were significantly different. Nevertheless, the Green Burn and Valley Bog profiles were judged to be sufficiently similar for the dating of Green Burn to be based on pollen analysis dating of Valley Bog. Johnson and Dunham (1963) had shown the sub Atlantic/sub Boreal transition, which occurred approximately 2500 years ago, to lie between 70 and 100 cm at Bog Hill, and between 200 and 300 cm at Valley Bog. Assuming that growth of peat at Sike Hill is proportional to that at Bog Hill (mean depth of four Sike Hill cores was 128 cm, and of three Bog Hill cores 160 cm), the transition zone at Sike Hill would lie between 56 and 80 cm. Similarly, though the density profile at Green Burn is significantly different from that at Valley Bog, the depth of peat formed in the last 2500 years there may be assumed to be between 200 and 300 cm.

8.6.3 Carbon Dating of the Peat Profile

1. The uppermost layers. Two blocks of peat were cut from within each of the two study sites. The blocks at Sike Hill were cut from areas relatively free of surface vegetation and covered by a thin layer of *Calluna* litter and bryophytes. At Green Burn areas of *S. papillosum* lawn were selected.

From the Sike Hill blocks, layers 1 cm thick were cut, corresponding to 0–1 cm, 3–4 cm, and 6–7 cm depth in the peat. From the Green Burn blocks, layers were cut 2 cm thick, corresponding to 0–2 cm, 4–6 cm, and 8–10 cm depth in the peat. As many of the adventitious roots and buried stems were removed as possible. This was relatively straightforward at Green Burn and much more difficult at Sike Hill. The samples were air dried and analysed by

Table 13. Weight of peat, kg m^{-2}, to the inorganic substratum at Bog Hill, and to the birch layer at Valley Bog

Depth (cm)	Bog Hill	Valley Bog
0– 10	7.05 (3.20)	2.71 (0.88)
10– 20	7.59 (1.75)	2.81 (0.77)
20– 30	9.24 (0.42)	3.79 (0.89)
30– 40	9.40 (0.59)	3.24 (0.68)
40– 50	8.19 (0.61)	1.49 (0.67)
50– 60	9.14 (0.30)	1.76 (0.99)
60– 70	9.58 (0.07)	2.91 (0.55)
70– 80	10.67 (0.49)	4.06 (0.80)
80– 90	8.60 (0.52)	5.23 (0.88)
90–100	9.15 (1.24)	5.29 (0.69)
100–110	9.73 (0.55)	5.94 (1.41)
110–120	10.33 (0.31)	5.61 (0.97)
120–130	10.16 (0.57)	5.24 (1.00)
130–140	11.28 (0.61)	5.50 (0.11)
140–150	11.25 (0.59)	5.30 (0.39)
150–160	12.32 (0.44)	4.81 (1.04)
160–180		12.28 (2.68)
180–200		12.66 (0.70)
200–220		12.10 (0.56)
220–240		12.54 (0.80)
240–260		12.72 (0.54)
260–280		12.84 (0.58)
280–300		12.08 (0.60)
300–320		13.40 (0.80)

sd on three cores shown in brackets.

Table 14. Analysis of variance of peat densities between surface and 150 cm at the four sites

Source	Degrees of freedom	Sum of squares	Mean square	F-ratio
Sites	3	0.1074	0.0358	73.6[a]
Error	206	0.1002	0.0005	

[a] $P<0.01$.

Table 15. C^{14} ages of peat samples from Sike Hill and Green Burn

Site	Material	Depth (cm)	C^{14} age, B P
Sike Hill	Peat	47.5– 52.5	1,362 ± 150
Sike Hill	Peat	47.5– 52.5	1,481 ± 160
Sike Hill	*Eriophorum* tussock	18 – 20	236 ± 130
Green Burn	Peat	148 –152	3,037 ± 110
Green Burn	Peat	148 –152	3,067 ± 140

Dr. D.D.Harkness at the NERC Carbon Dating Centre for "bomb barbon", i.e. the increase in C^{14}:C^{12} ratio since the advent of nuclear testing, which reached a peak in 1963 and has since declined. The dating indicated that the top 2 cm of Green Burn peat has formed in the past four years; 4–6 cm is *c* 10 years old, and 8–10 cm *c* 15 years. Sike Hill results were inconclusive, because a layer 1 cm thick would represent several years' growth where peat accumulation is slow, and contamination by living or recently dead adventitious roots would also affect the results. However, analysis of the Sike Hill surface peat showed that the deposits were very much older than at Green Burn, and thus illustrated an important difference between the two sites.

2. The deeper deposits. Cores were taken with a Hiller peat auger from a depth of 47.5–52.5 cm at Sike Hill, and 148–152 cm at Green Burn. Two cores were taken at each site at less than 0.5 m distance.

At Sike Hill, in addition to the peat samples, the remains of an *Eriophorum* tussock were extracted from 18–20 cm depth in the peat. In the laboratory, dead *Eriophorum* roots were removed from the Sike Hill peat samples, and peat that contaminated the *Eriophorum* tussock was scraped off. All samples were dried at 105° C overnight, and sent to Dr. D.D.Harkness for C^{14} analysis (Table 15).

The consequence of these directly measured ages on the assumptions made in deriving peat depths at the two sites, which corresponded to 2500 years, is examined in the Addendum.

8.6.4 An Analytic Solution to the Initial Simulation

An "analytic" solution was obtained for comparison with a solution obtained using a "numeric" or Euler point-slope method. Identical parameters were employed in the comparison. Two step sizes were used in the Euler method (1.0 and 0.01 cm). For this purpose the profile was regarded as having a number of aerobic layers of 1.0 cm each and an anaerobic layer. The cotton strip regression coefficient was used to modify the decay at the midpoints of each of the aerobic layers (Fig. 8).

8.7 Addendum

Carbon^{14} analyses of the deeper peat were not made until late in the project. The simulation runs had, therefore, been checked against the depth of peat estimated to have formed in 2500 years from pollen analysis at Bog Hill and Valley Bog (Sect. 8.6.2). The estimate was based on the assumption that the ratio of peat formed in 2500 years to total peat depth would be similar in peats of similar formation.

The direct C^{14} measured ages of peat at the two sites provided some check on the validity of these estimates. Two points were considered in the interpretation of the carbon dating:

1. Because the peat density was low, it was necessary to extract several cm to obtain sufficient weight for analysis. If the peat had accumulated very slowly, therefore, several cm would represent many years, and the C^{14} age would reflect the mean value of a wide range.

2. Below-ground biomass of vascular plants was an important component of the vegetation, particularly at Sike Hill. Insertion of roots at depths in the peat would introduce much younger material into the older deposits. Thus, at any depth, the C^{14} age determined was likely to be an underestimate. Determination of the age of an *Eriophorum* tussock was an attempt to overcome this problem.

When the second point, i.e. the effect of subsequent root growth, was considered, there was good agreement at Sike Hill between the estimated depth formed, from pollen analysis, in 2500 years (between 56 and 80 cm) and the measured age of peat between 47.5 and 52.5 cm, which was 1422 ± 110 years before present.

At Green Burn, an estimated 200–300 cm of peat had formed in 2500 years. C^{14} measurements showed that the age of peat between 148 and 152 cm was 3052 ± 90 years BP. This showed that the peat at Green Burn had accumulated more slowly than had been estimated, and emphasised that the significant difference in the density profiles of Green Burn and

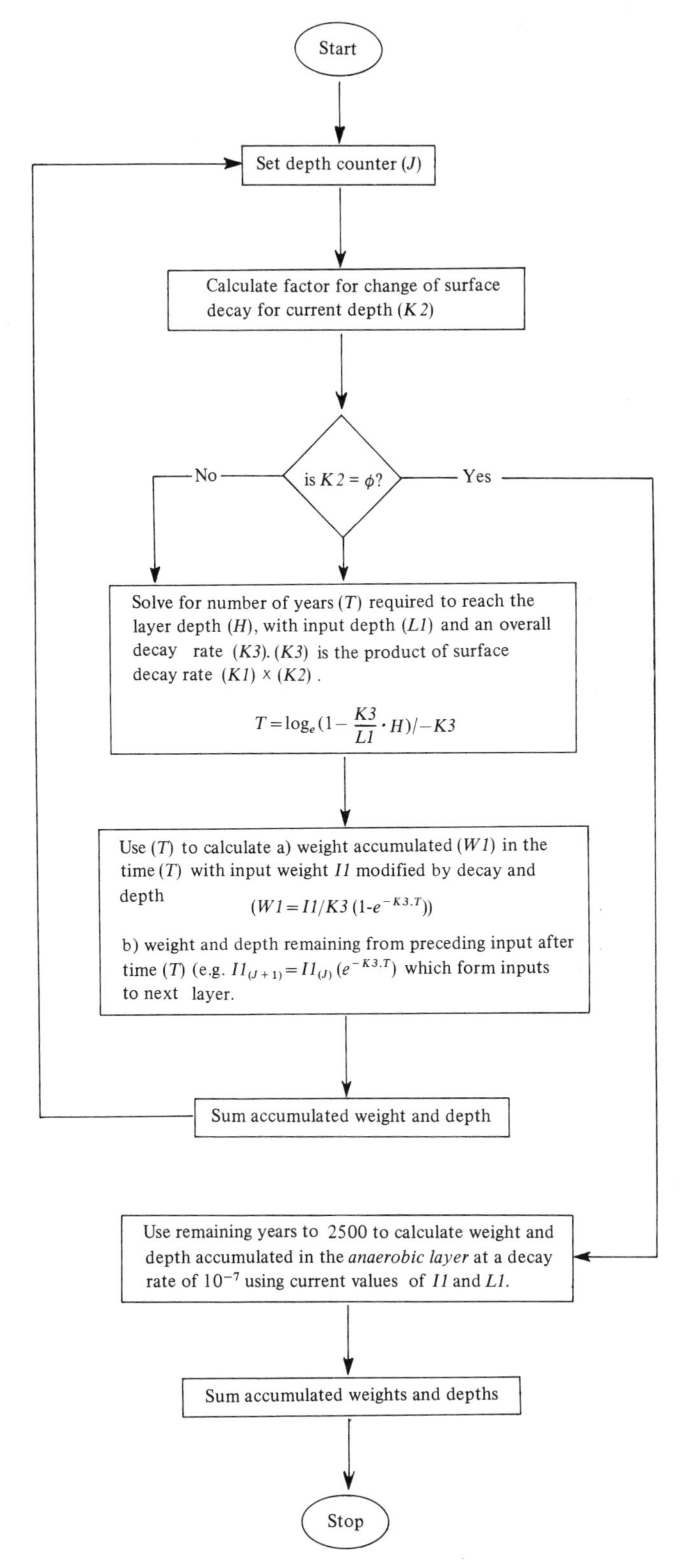

Fig. 8. Flow diagram of analytic solution used to check results of the simulation

Valley Bog should not have been ignored (Sect. 8.6.2). The difference in their density profiles and, as is now apparent, in their rate of growth, almost certainly indicates that the two deposits have been formed in different ways.

However, the fact that accumulation at Green Burn was slower than had been estimated did not significantly alter the interpretation of the simulation results, since the sensitivity to changes in the rate of decline of decay with depth was so high. Because of the major climatic changes that occurred at the sub Boreal/sub Atlantic transition, it was not considered realistic to project the simulation beyond 2500 years. Nevertheless, with mean production and decay rates at Green Burn, and the *Calluna* below-ground stem rate of decline of decay with depth, simulated peat accumulation was 100 cm and 84 kg m^{-2} in 2500 years (Table 11), compared with the observed 150 cm and 90 kg m^{-2} formed in 3052 years.

9. A Model of Peat Bog Growth

R. S. Clymo

From an analysis of the processes involved in peat growth, a mathematical model is constructed and used to predict bulk density and age profiles which are compared with observed ones. Given only bulk density and age profiles of the peat, it is possible, alternatively, to estimate the model parameters of growth, decay, compression, and consolidation.

9.1 Introduction

One recent estimate (Tibbetts, undated) is that peat covers 150×10^6 ha—between 1 and 2% of the total land surface of the earth. From estimates given by Olenin (1968) one may calculate that there are about 200×10^9 t of peat. This is the same order of magnitude as the estimated total dry matter production on Earth in one year (Fogg, 1958). In brief, there is a lot of peat, and the dynamics of peat growth would be worth studying for this reason alone.

Peat deposits are not uniform, either in structure or in rate of growth (Walker and Walker, 1961; Walker, 1970). Any attempt to account for these differences must involve some sort of conceptual model of the processes involved. At present these models are mostly qualitative. For example, Walker (1970), who is more explicit about the processes involved than most workers, writes: "... sedentary deposits are modified after their initial deposition, chemically by humification and physically by compaction. ... A rising water table leaves little time for extensive humification and ensures the rapid accumulation of little altered plant remains. Periodically or permanently dry surfaces, however, allow the dry oxidative breakdown of dead plants ..."

"In most ... mires dry oxidation is limited and humification virtually ceases as the newly deposited material sinks into the anaerobic zone. There the accumulation of more material above leads to some compaction ... compaction is soon almost totally accomplished and increases very little as more deposit accumulates above. However, compaction is the inevitable result of the removal of water from a deposit ...". Kaye and Barghoorn (1964) discuss the same problems of autocompaction for salt-marsh peat with a bulk density about seven times that of bog peat. They give a sketch graph of the way in which a given level in the peat moves with time.

One may hope that further understanding will follow the making of more precise quantitative models since the assumptions become more clear, and testable predictions can be made. Gore and Olson (1967), following Jenny et al. (1949) and Olson (1963), have made a simple model, in which dry matter was added by production, lost by decay, and moved between notional compartments. Jones and Gore (Chap. 8) have elaborated this model to allow decay rates varying with depth.

This paper describes a more complicated model, involving both dry matter and peat depth. In part 9.2.1 the processes involved in peat growth are described, functional relationships examined, and the model constructed. In part 9.2.2 data for testing the model are described, and the model is used to make predictions which are compared with measurements.

9.2 The Model

Parameters of the model

Symbol	Meaning	Dimensions	Units used in this paper
p	Net productivity	$ML^{-2}T^{-1}$	$g\ cm^{-2}\ yr^{-1}$
L	Rate of growth in length of plants at the bog surface	LT^{-1}	$cm\ yr^{-1}$
α_1	Aerobic decay constant (at depths above W)	T^{-1}	yr^{-1}
α_2	Anaerobic decay constant (at depths below W)	T^{-1}	yr^{-1}
k	Compression constant	$M^{-1}LT^2$ (1/stress)	$g\ (f)^{-1}\ cm^2$
c	Creep constant	(log time cycle)$^{-1}$	($\log_{10}$ cycle yr)$^{-1}$
W	Water table depth below surface	L	cm
p'	Rate of addition of dry matter to the anaerobic zone (below W)	$ML^{-2}T^{-1}$	$g\ cm^{-2}\ yr^{-1}$
L'	Rate of depth addition to the anaerobic zone (below W)	LT^{-1}	$cm\ yr^{-1}$

These parameters are referred to later by symbol only.

9.2.1 Making the Model

There are many types of peat-forming system, and many systems of classifying them; examples are given in Robertson (1968), Anon (undated). The general types considered here are sedentary (autochthonous)–principally oligotrophic telmatic and terrestrial systems (West, 1968). It is helpful to use the specific case of a pure carpet of a vertically growing species of *Sphagnum*. Rate of vertical increment to the surface is then easily visualised.

The rate of at least some processes involved in peat accumulation is not constant. Dry matter and length increment, for example, show diurnal and seasonal fluctuations. In this model, which is concerned with times measured in tens or hundreds of years, mean annual rates are equated with instantaneous rates: processes with a short time constant are assumed to have no effect on those with long time constants, in the same way that a moving coil galvanometer will register a drift in a DC voltage but give no apparent response to a superimposed 50 Hz alternating voltage. Forrester (1961) shows that this assumption may be untrue in systems with interacting feedback loops.

The model includes dry matter (affected by productivity and decay), depth (affected by rate of growth of plants, decay, compression and secondary consolidation) and depth of the water table. Functional relationships of these quantities are now considered.

9.2.1.1 Dry Matter

It is assumed that dry matter is added to the surface at a constant rate per unit area which may be equated with the net annual productivity p (dimensions $ML^{-2}T^{-1}$). The credibility of this assumption is reduced as time increases, since the major determinants of productivity (climate, nutrition, water supply) are all likely to change. Even if climate and nutritional conditions remain constant, it seems doubtful if the bog water table will continue to rise at the same rate as the bog surface (Granlund, 1932) and the productivity of *Sphagnum* is known to be much affected by the depth of the water table (Clymo, 1970).

It seems to be generally agreed that some decay (equated with loss of dry matter) occurs, and measures of this for *Sphagnum* are given by Clymo (1965). Oxidative breakdown results in transformation of dry matter to soluble or gaseous forms. In many peat bogs the hydraulic conductance for lateral flow in the top 20 cm is relatively great (Romanov, 1961; Chapman, 1965), though the conductance of peat at lower depths is much reduced (Boelter, 1965; Ingram et al., 1974). Both soluble and gaseous material may then be removed in the surface layers, though it may be that loss as gas is the major component at depth. Humification, i.e. the loss of plant structure, often accompanied by darkening of colour, is included within the definition of decay, insofar as it results in loss of dry matter from the system.

There is little direct evidence extending over more than a few years about the form of the functional relationship between decay rate and amount of material. Baker (1972) measured the weight of 1 cm sections of stems of *Chorisodontium aciphyllum*, which forms peat banks at least 170 cm deep on Signy Island (lat. 60° 43′S, long. 45° 38′W). He concluded that, ignoring compression if it occurred, the decay rate was a linear function of time, at least to a depth of 10 cm (corresponding to an age of about 30 years). Heal, Latter, and Howson (Chap. 7) report weight losses of plant material in nylon mesh bags at Moor House. Some of their results, and those of Baker, are shown in Figure 1a. These results could be described by:

$$x = x_o - \alpha' t \quad (1)$$

where x is the amount (per unit area) at time t, x_o is the amount (per unit area) at $t=0$, α' is a decay constant (dimensions $ML^{-2}T^{-1}$).

The difficulty comes when one tries to extrapolate these data to longer times. If Baker's linear regression is followed, his plants should have entirely decayed at 14.5 cm deep (about 45 years), but the peat bank is 170 cm deep and C-14 dating shows it to be about 1800 years old.

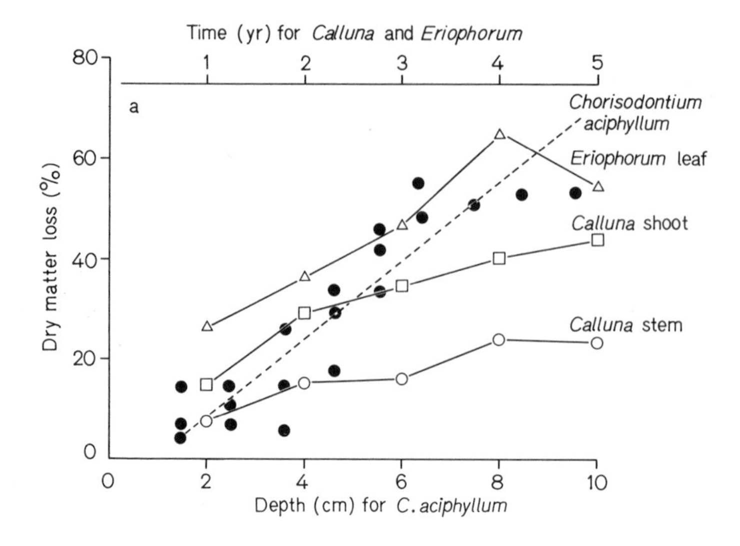

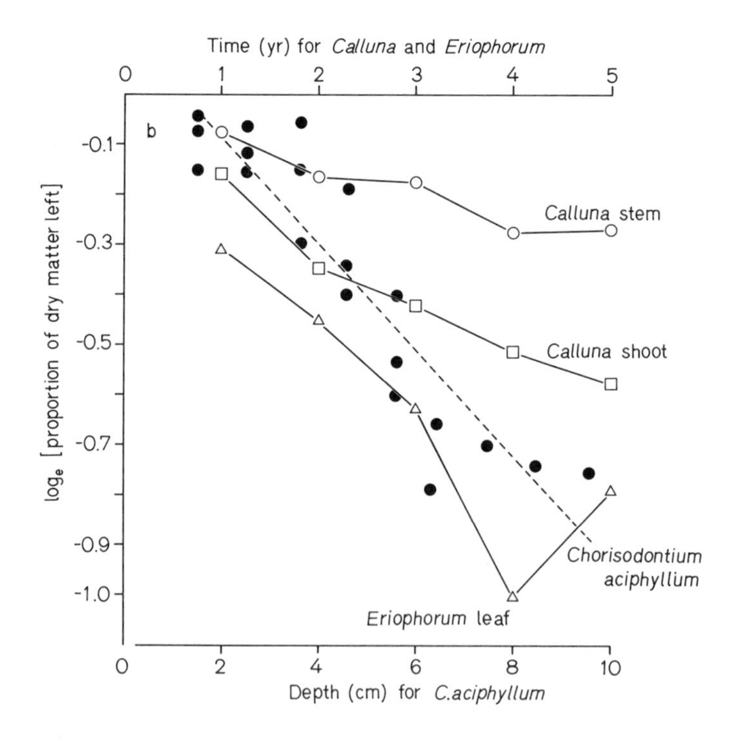

Fig. 1 a and b. Dry matter loss from four materials as a function of time. *Calluna* shoot, *Calluna* stem, and *Eriophorum* leaves: (from Chap. 7). *Chorisodontium:* ● (from Baker, 1972). *Dashed line* is the linear regression to the *Chorisodontium* points. (a) graph plotted on linear axes; (b) graph uses the same data but the dependent variable axis is logarithmic

A commoner, and more plausible, assumption is that the rate of decay is proportional to the amount of matter at the time:

$$\frac{dx}{dt} = -\alpha x \tag{2}$$

which gives

$$x = x_o e^{-\alpha t} \tag{3}$$

where α, the decay constant, has dimension T^{-1}. If this were a good description one would expect log (x/x_o) to be a linear function of t. The results of Heal, Latter, and Howson and of Baker are plotted in this way in Figure 1b. The fit (Table 1) is almost as good as in Figure 1a.

One might indeed expect the rate of decay to decline faster than this, because there is reason to believe that plant constituents are broken down at differing rates so that the material left after a few years is more resistant to decay. Minderman (1968) has shown this to be likely in a temperate pine forest. In a *Sphagnum* valley bog, *Sphagnum* peat from 60 cm brought to the surface decayed less rapidly than did surface material (Clymo, 1965), and Heal, Latter, and Howson (Chap. 7) have shown that O_2 uptake rate of *Eriophorum vaginatum* leaf litter recovered from the field (with attendant microorganisms) declines by a factor of about two to three over five years. *Calluna* shoot and stem litter did not show this decline, and the results in Figure 1 do not suggest such an effect either. For the model, therefore, the functional relation [Eq.(2)] has been used.

Apart from the effects already mentioned there is direct evidence (Clymo, 1965) that the rate of decay of comparable age *Sphagnum* is lower at depth in the peat than at the surface. Heal, Latter, and Howson (Chap. 7) found that *Calluna* stems and *Juncus effusus* leaves gave similar results (though *E. vaginatum* roots showed a puzzling small increase in decay rate at greater depth). The change in tensile strength of unbleached calico strips also suggests a reduction in decay rate with depth. The major decrease in decay rate was correlated (Clymo, 1965) with the change from aerobic to anaerobic conditions at about the water table. The

Table 1. Fit of decay data to two models

Material	Number of esti- mates	Correlation coefficient		Variance: residual/total	
		Linear model (1)	Negative exponential (2)	Linear model (1)	Negative exponential (2)
Chorisodontium aciphyllum (whole plants)	20	0.987	−0.917	0.026	0.160
Calluna vulgaris (stems)	5	0.944	−0.911	0.108	0.156
Calluna vulgaris (leaves)	5	0.937	−0.929	0.122	0.138
Eriophorum vaginatum (leaves)	5	0.882	−0.863	0.223	0.256

calico data can be interpreted as approximating such a steplike change. Type and activity of microorganisms change from aerobic peat to anaerobic (Chap. 5), though there is dispute about the viability and activity of microorganisms at depths of 1 m or more in peat (Waksman and Stevens, 1929; Waksman and Purvis, 1932; Burgeff, 1961).

Jones and Gore (Chap. 8) prefer to fit a linear regression with depth to the tensile strength data (Chap. 7), and such a description is certainly preferable for the data on decay of *Calluna* stems and *J. effusus* leaves. They also show that their model is sensitive to the slope of this regression.

As a compromise it has been assumed here that one decay parameter, α_1, is applicable throughout the aerobic zone, and a second parameter, α_2, applies in the anaerobic zone below the water table at depth W. The functional relation [Eq.(2)] is used in both. In this model, sensitivity to small changes in W is equivalent to Jones and Gore's change of decay/depth regression coefficient.

9.2.1.2 Depth

As in the case of dry matter and with the same reservations, it is assumed that the positive rate of growth in depth, L, attributable to upward growth of *Sphagnum*, is constant. For a sward of *Eriophorum* or *Calluna* the concept of upward growth of the plants is less obvious, and L would be a notional parameter not easily equated with any linear measure of leaf or shoot growth.

Removal of material during decay may be expected to reduce depth, just as the removal of a brick near the bottom of a pile will reduce the height of the pile. This simple view is the one adopted here, for lack of direct evidence to the contrary. Where visible structure has been lost this assumption is plausible. Where structure is still present, it seems more likely that decay may not directly reduce the depth, just as the removal of isolated bricks from a wall does not at first reduce the height of the wall (though eventually there is a complete and sudden collapse). Decay may however have a secondary effect by altering the compressive and creep properties. Again, evidence is lacking.

From the assumption that decay affects dry matter and depth in the same way, it follows that bulk density, defined as $\Delta x/\Delta l$ (where l is the depth at time t corresponding to the dry matter per unit area, x) would be constant with depth, although the age profile would depend on α, p and L.

Nevertheless, the bulk density of peat from 50 cm below the surface is commonly 0.1 g cm^{-3}, about 10 times that at the surface. Peat *can* be compacted further: the "Finbloc" of commerce has a bulk density of about 1.25, about 100 times the original value. Jones and Gore (Chap. 8) point out that selective decay of less dense plant materials can of itself lead to increased bulk density. The argument could be extended to different fractions of one plant. Such an effect could not account, however, for the observed increases in bulk density by a factor of ten, particularly for a peat formed almost entirely from *Sphagnum*. It seems then that a model of peat bog growth should attempt to include compaction, however imperfectly.

The general features of peat consolidation have been investigated by engineers, and are discussed by Barden and Berry (undated) and by Berry and Poskitt

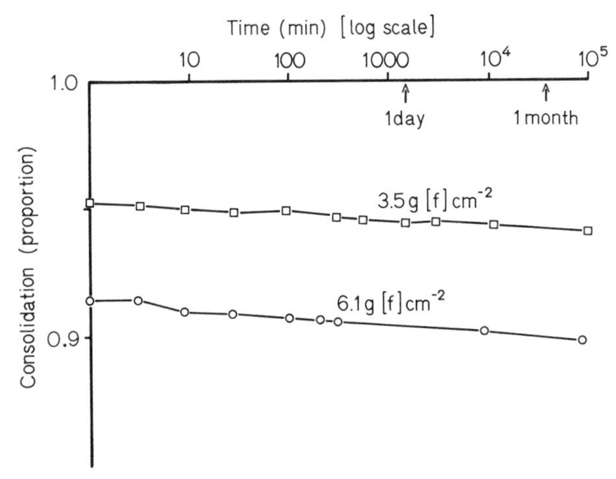

Fig. 2. Consolidation of *S. magellanicum* peat cylinders 21 cm diameter and 5 cm thick under different stresses applied for eight weeks

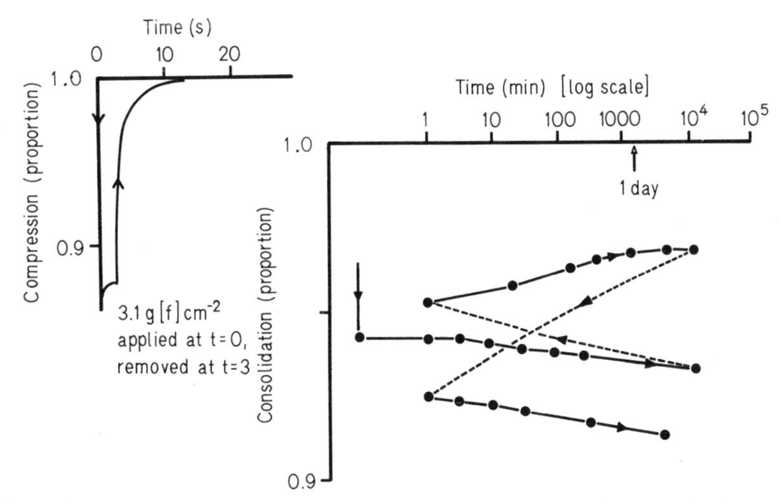

Fig. 3. Consolidation and recovery under repeated application and removal of stress, of cylinders of peat similar to those of Figure 2. *Dashed lines* show rapid (< 10 s) responses when stress was first applied or removed. *Inset:* short term compression and recovery

(1972). They refer to much of the other work in this field. The stresses of interest to engineers are usually an order of magnitude greater than those developed naturally. However, the same general features are shown at stresses of interest to ecologists. Examples of one dimensional strain vs compressive stress curves for peat are shown in Figure 2. They are for discs of *Sphagnum magellanicum* peat, 5 cm thick by 21 cm diameter. The discs were cut from a core, and span the depth 6–11 cm. (The apparatus and method used are described in Sect. 9.5.2).

There were two phases of consolidation, the first with a time constant of a few seconds at the most, and the second which was long continued. The strain during the first phase was nearly reversible for at least 22 cycles if the stress was removed

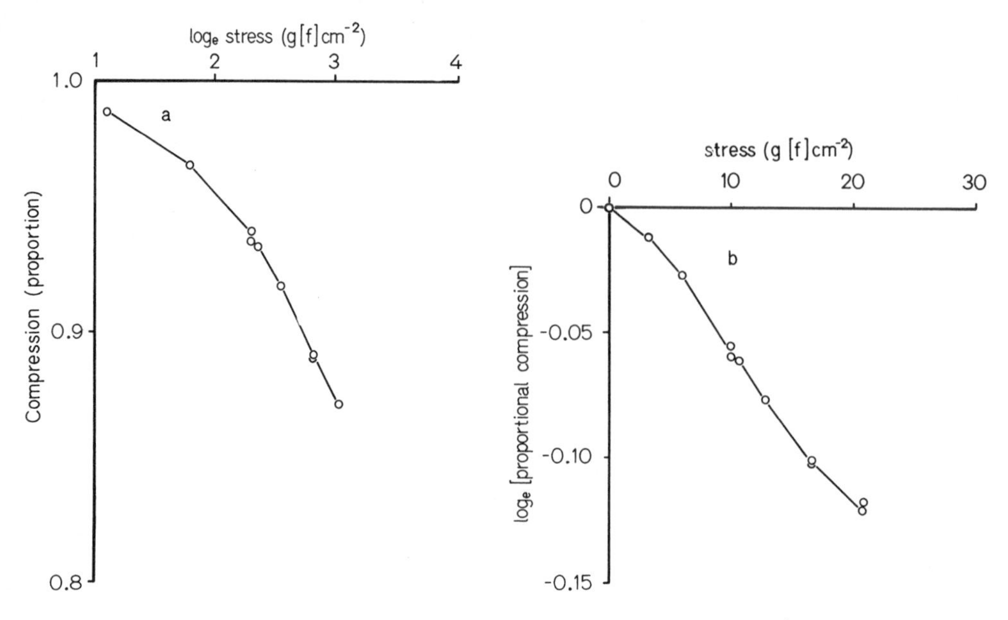

Fig. 4 a and b. Compression under stress of short duration of peat cylinders similar to those in Figure 2. The two graphs show the same data plotted using different axes

rapidly (Fig. 3 inset), and might be called elastic. Barden and Berry postulate that during this primary phase any excess pressure produced in the macrocapillary pores will be dissipated at a rate depending on the thickness of the peat slice. The relation of strain to stress for brief applications is shown in Figure 4. For these large strains (equivalent to stresses produced by about 3 m depth of peat) a linear relation is perhaps not to be expected. Barden and Berry (undated) assume a linear relation between "void ratio" in the peat and log stress. Figure 4a shows data which should give a straight line on this assumption. The fit is not particularly close, and there are difficulties in implementing this relationship in a model where the applied stress starts at zero. For descriptive purposes a negative exponential (Fig. 4b) is also a fairly close approximation at least within the range of interest:

$$\frac{dZ}{d\varepsilon} = -kZ \tag{4}$$

where ε is the stress, Z the slice thickness, and k the compression constant (dimensions $M^{-1}LT^2$).

During the second phase (secondary consolidation by creep of the peat skeleton) the strain is approximately proportional to log time.

For "granular amorphous" peat the rate of creep is little affected by the applied stress (Fig. 5). This rather surprising feature is shown in several sets of experimental measurements (Hanrahan, 1954; Barden and Berry, undated; Berry and Poskitt, 1972). The results in Figure 5 indicate that even the fresh *Sphagnum* peat behaved as if it were "amorphous". This behaviour is described by:

$$Z/Z_o = 1 - c_a \log(t + m) \tag{5a}$$

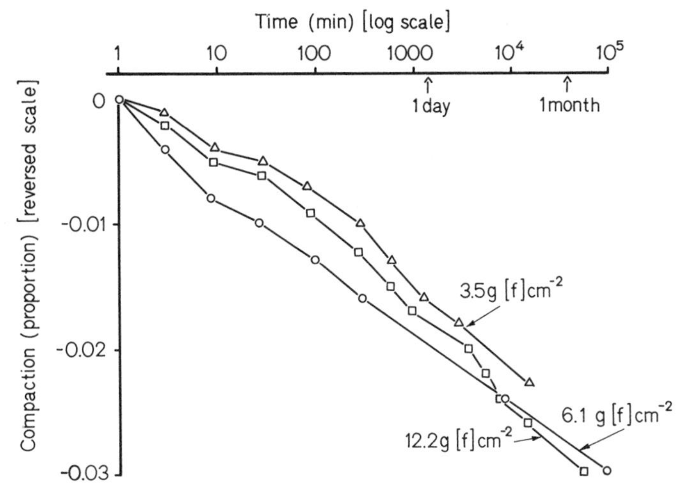

Fig. 5. Secondary consolidation (creep) of cylinders of peat similar to those of Figure 2 under different stress. The lines are arranged to coincide at $t = 1$ min, and the proportional compression is expressed as the proportional decrease from the thickness at 1 min

where Z is the slice thickness at time t, Z_o the initial thickness, c_a is the creep constant for amorphous peat, dimensions (log time cycle)$^{-1}$, and m is an arbitrary number, small compared to the range of t, defining the origin of the log time axis.

The mechanism of creep in these peats is postulated to be the relative movement of separate elements of the peat skeleton, the rate being affected by the thin layer of highly viscous water around each particle.

For "fibrous peat", although the linear strain/log time relationship still holds, the slope depends on the applied stress too. From Barden and Berry's Figure 9a:

$$Z/Z_o = 1 - \varepsilon c_f \log(t+m) \tag{5b}$$

where c_f is the creep constant for fibrous peat. The rate of creep in this case is thought to be controlled by the slow squeezing out of water from microcapillary spaces inside the peat skeleton.

The existence of two relationships makes model construction awkward, or undesirably arbitrary, since any one peat mass may start behaving according to Equation (5b) but gradually change to Equation (5a). For tests, both Equation (5a) and Equation (5b) were used on one core, and the results compared.

It seems to be generally agreed that the rate of the first phase depends on sample dimensions—presumably because the dissipation of pore pressure occurs as a result of flow of water out of the sample, and the longer the path the greater the resistance. Hanrahan (1954) finds the same "scaling" effect for the second phase too, but Berry and Poskitt (1972) were unable to confirm this. In the first phase, even with a 3.5 m slice of peat the response time is less than two months. Since the conditions in a growing peat column are changing so slowly, and since there is doubt about whether the second phase rate is affected by scale at all, no allowance is made in the model for such effects in either phase.

Kaye and Barghoorn (1964) used the relationship between creep and log time, and noted that the rate of upward growth of deep peat may, as a result of creep throughout the mass, become very sensitive to small changes in surface conditions. Even without environmental changes however, it is implicit in Equation (5) that at some value of t the slice of peat vanishes (and at longer times has negative thickness). This is obviously impossible and eventually the relationships of Equation (5) must become poor descriptions. The "vanishing time" is therefore, a statistic of this model which is worth checking. A corollary is that although the model predicts a steady state mass equal to p/α, the bulk density should continue to increase.

If the first phase consolidation were truly elastic, one would expect that on removing the top layers of peat, the lower layers would expand noticeably. This happens only to a very limited extent, possibly because other changes in the peat make the initial elastic change permanent as is shown in Figure 3. It is still possible, however, for peat from some depth to undergo further initially elastic compression if a stress is applied.

The experimental measurements for creep cover a year at most and, as with decay, one is forced to extrapolate grossly. The model also assumes that the compression and creep parameters (k and c) remain constant, which seems unlikely. The decay parameter, α, does now affect the bulk density profile, but it does so indirectly through effects on compressive stress and creep time.

Assembly and implementation of this model are described in Section 9.5.1.

9.2.2 Checking the Model

9.2.2.1 Introduction

The model needs values for the seven parameters (p, L, α_1, α_2, k, c, W) and specified depths below the surface, and predicts values both for age at each specified depth and for bulk density of the layer between depths.

There are two ways of checking the model. Much the simpler is to measure the seven parameters and bulk density profile, and if possible date the profile, then compare the predicted and observed age and bulk density profiles. The model can also be stimulated by varying the parameters and observing its responses. This method of checking will be referred to as the "direct method".

The second, more complicated, method is to provide a bulk density profile and if possible an age profile, and attempt to minimise some function of the difference between observed and predicted profile(s) by adjusting the parameters. This will be called the "indirect method". The results are estimates of all seven parameters and, if no age profile is used, a predicted age profile. These may be compared with such measurements as happen to be available—one does not need all seven as in the first method. Unfortunately the second method costs about 10^4 times as much in computer time/space, and does not produce a unique solution. Some of the problems are discussed by Plinston (1972). For two parameters, one may visualise the problem as equivalent to locating the lowest point on an undulating surface, where altitude represents "badness of fit", and parameter values are given by co-ordinates in a N–S and in an E–W direction.

One attempts to find the "least bad fit". All methods for doing this may settle into any hollow in the surface, which need not be the lowest one. Whether they do so or not depends on starting point and step length. There may be a long flat-bottomed valley, rather than a nearly circular hollow. If this valley is oriented N–S or E–W, the implication is that one of the parameters can vary a great deal without much effect on the "badness of fit", i.e. that the model could do without it. If the valley runs NE–SW (or on the other diagonal) a change in one parameter can be compensated by a change in the second; the parameters are correlated. Again the model gains little from having two separate rather than one composite parameter. The more parameters, the greater the likelihood of correlations of this kind.

Both direct and indirect types of check have been tried.

9.2.2.2 *Methods*

The methods are described in Section 9.5.2.

9.2.2.3 *Results and Discussion*

Detailed Investigations on a Single Site. A detailed study was made, by both direct and indirect methods, of cores from a specific site (MH 1) 2 m by 1 m on Burnt Hill at Moor House (nat. grid. ref. NY 754328). The site is an almost pure *S. magellanicum* lawn with a few small plants of *Calluna vulgaris* and *Eriophorum angustifolium*. Cores were taken from places with no emergent vascular plants.

Peat Characteristics. The seven parameter values measured are shown in Table 2. Values for p and L were measured directly. The aerobic decay rate, α_1, was measured on four similar lawn sites, within 200 m, as integrated carbon dioxide flux using

$$\alpha_1 = -\log(1 - x_f/x_r) \tag{6}$$

where x_f is the integrated flux of carbon, expressed as CH_2O equivalent, and x_r is the total dry matter "at risk"; in this case dry matter in the aerobic zone. If

$$x_f \ll x_r \quad \text{then}$$

$$\alpha_1 \simeq 1 - x_f/x_r\,. \tag{7}$$

The first estimate of α_1 in Table 2 uses this method. The second estimate was obtained from direct measurement of dry weight lost from material in nylon mesh bags at a similar site about 150 m distant, using Equation (3). The third estimate, 0.062 from Al, Ti, and Mg concentrations, is described later.

The estimate of α_2 was derived from integrated methane fluxes during the same time as the carbon dioxide fluxes. A peat depth of 3 m with average bulk density of 0.1 g cm^{-3} was used to estimate x_r for use in Equation (7). The samples for this "long core" were collected in cylinders 7.5 cm diameter by 10 cm long from a peat face exposed in an erosion gulley 250 m from the main core site.

Table 2. Measured parameter values for *Sphagnum magellanicum* lawn

Productivity (p)	$0.016 \text{ g cm}^{-2} \text{ yr}^{-1}$ (0.007, $n=12$)
Plant length growth rate (L)	1.5 cm yr^{-1} (0.3, $n=12$)
Aerobic decay (α_1)	0.032 ($n=4$) 0.050 (0.021, $n=40$) 0.062 yr^{-1} ($n=1$)
Anaerobic decay (α_2)	10^{-5} yr^{-1} ($n=4$)
Primary compression (k)	$0.005 \text{ g(f)}^{-1} \text{ cm}^2$ ($n=3$)
Secondary creep (c)	0.018 ($\log_{10}$ time cycle)$^{-1}$ ($n=3$)
Water table (W)	8 cm ($n=3$)

Figures in parentheses are standard error of estimate and number of observations.

Some check on this estimate of α_2 is possible. If the dry matter part of the model is correct, then from Equation (A2) in appendix, 9.5.1:

$$\alpha_2 < p'/x_b \tag{8}$$

where x_b is the dry matter per unit area accumulated to date, and p' is the dry matter per unit area passing into the anaerobic zone. Since $x_b < x_\infty$, taking $p' = p$ sets an upper limit on α_2. In this case the maximum value is 0.7×10^{-3}, an order of magnitude greater than the measured value.

It is also of interest that Svensson (1973) has found similar fluxes of CO_2 and CH_4 from a subarctic mire near Abisko (Sweden).

Estimates of compression *(k)* and creep *(c)* parameters were obtained from the slopes of lines similar to those of Figures 4 and 5. The water table depth *(W)* was measured (in the holes left when cores were removed) at three times in the following year.

Where possible, standard errors are shown in Table 2. From the nature of the methods, however, it seems likely that systematic error is greater than random error in the estimates of α_2, k, c, and W. In consequence significance tests could be misleading, and are not reported.

Bulk density, β-activity, lead concentration and non-destructive relative Cs-137 estimates were made on one core. Much more reliable absolute Cs-137 estimates were made on a second core about 30 cm from the first. Al, Ti, Fe, Mg, Cu, Cd, Zn, and Pb concentrations were measured on a third core.

The metal concentration and bulk density profiles are shown in Figure 6 and the Cs-137 and β-activity profiles in Figure 15 (Sect. 9.5.3). The two lead profiles were measured by different analysts on different cores. The correlation between them is 0.83. The concentrations of Mg and Fe in the top 5 cm are 860 ppm and 2420 ppm, similar to the values (740 ppm and 2160 ppm) recorded by Gore and Allen (1956) from Moor House blanket bog with *S. rubellum*.

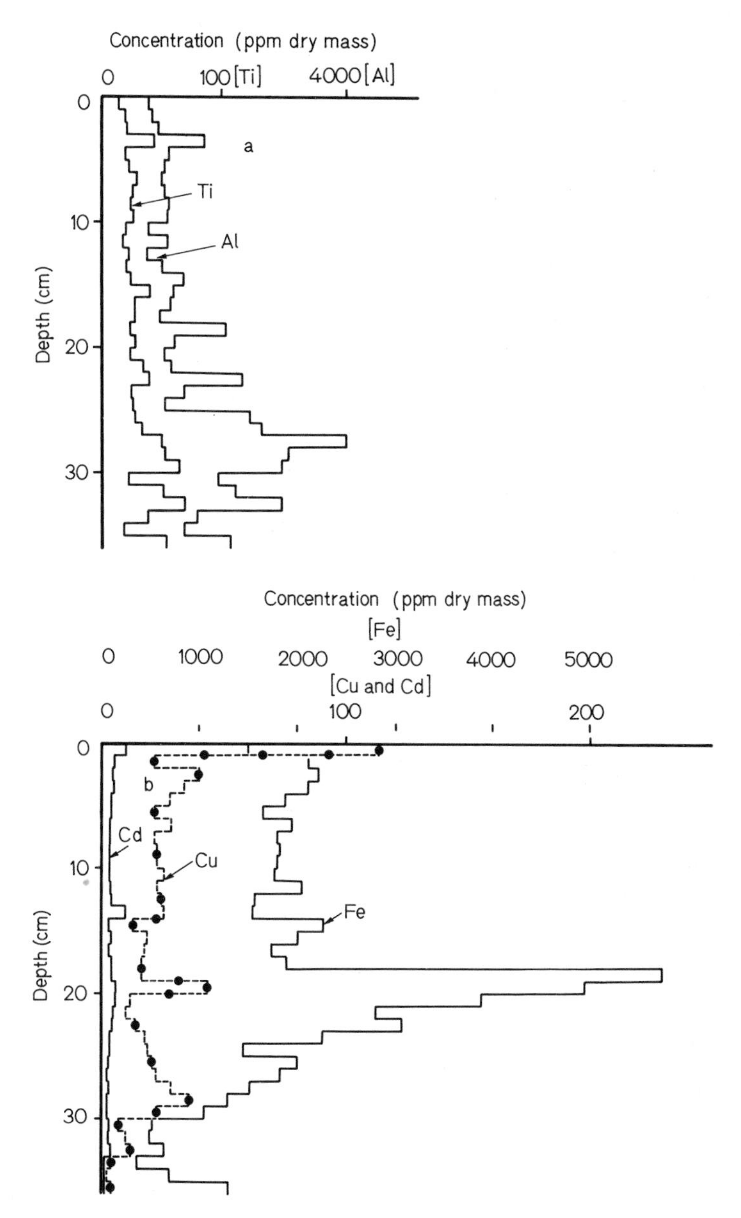

Fig. 6 a–d. Bulk density and concentrations of Ti, Al, Cd, Cu, Fe, Mg, Pb, and Zn in a 21 cm diameter core from a *S. magellanicum* lawn on Burnt Hill, Moor House National Nature Reserve

Bulk density increases fairly steadily with depth from 0.01 g cm^{-3} at the surface to 0.06 g cm^{-3} at 28 cm. Below this it increases sharply to 0.2 g cm^{-3}—a value as high as any recorded for organic peat. Above 28 cm *Sphagnum* leaves could be seen, but below 28 cm they could not be found even with a microscope.

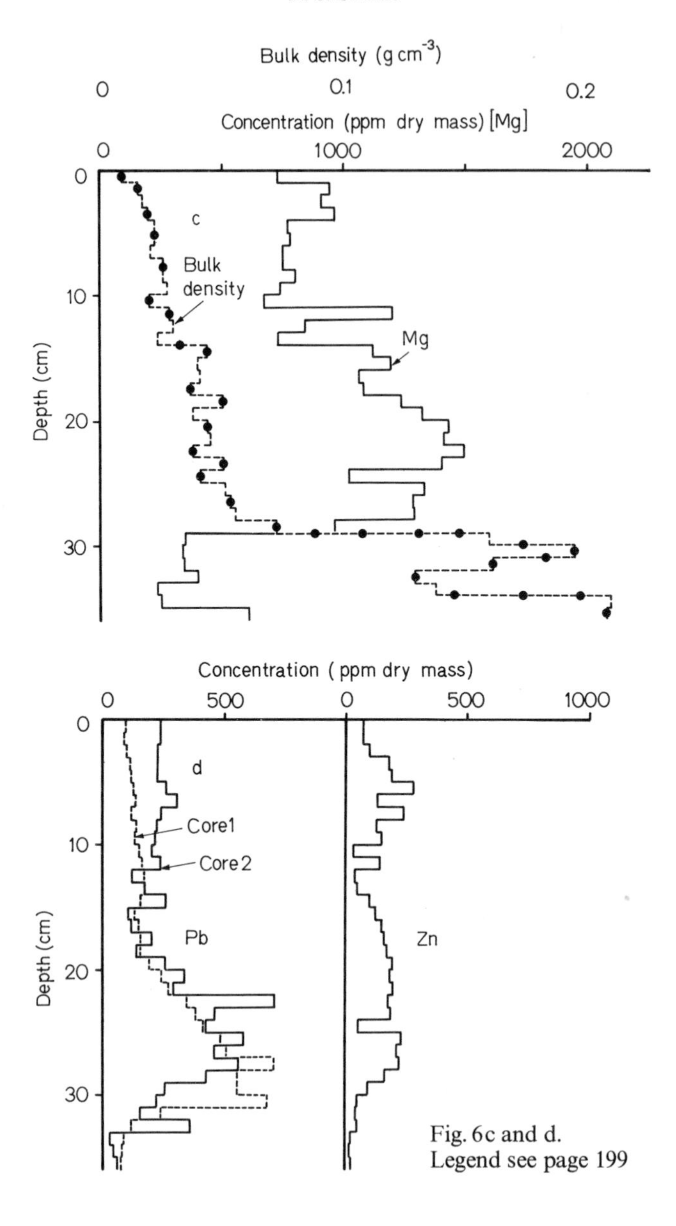

Fig. 6c and d.
Legend see page 199

There was a marked visible change in colour (from light brown to dark brown) and in texture (from rough to smooth) at the same level. A similar discontinuity is seen in the concentrations of Al, Ti, Mg, Pb (and perhaps Zn).

If conditions had been constant during the formation of the whole core it would be surprising to find the sharp change in bulk density and loss of structure at 28 cm, though it is possible that this is a critical point at which structure collapses. The associated decrease of concentration of so many elements of such diverse origins makes such an explanation unlikely. More probably there was at this time a marked change in environment, possibly connected with pool drainage. The model assumes that the parameter values do not change so it seems best to apply the model to the top 28 cm only of this core.

Age Profile. A continuous age profile may be established from the cumulative total amounts of Al, Ti, and Mg calibrated by the date of a single horizon established from a peak in Cs-137 concentration corresponding to the peak in fallout in 1963. Details are given in Section 9.5.3.

The Al, Ti, and Mg concentrations may also be used to check the aerobic decay rate α_1. If the deposition rate has been constant then from Equation (3):

$$\alpha_1 \simeq \log(C_o/C_t)/t \tag{9}$$

where C_o is the concentration of metal at the surface, and C_t the concentration at a depth of age t, the concentrations being mass of metal per unit mass of dry peat. Using the 0–2 cm and 11–13 cm depth, with $t=9$ yr, the third estimate in Table 2, $\alpha_1=0.062$, is obtained.

Direct Method Comparison of Observed and Predicted Profiles. The bulk density and age profiles predicted by the model, using the direct method, may now be compared with the observed bulk density, the age at 12 cm, and less reliably with the Al/Ti/Mg age profile. Figure 7 shows observed values and ones predicted using the "amorphous peat" [Eq. (5a)] model. The $P=0.05$ error bounds on the observed profiles are those for the least squares polynomial with the same number of parameters as the model.

The predicted bulk density profile is in fair agreement with the observed one to about 14 cm depth, but below that level the predicted profile is systematically below the observed one. The predicted age is systematically too great near the surface and too little lower down.

The age profile comparison is important because for this particular observed bulk density profile the model contains a high degree of redundancy. A linear regression (with intercept) accounts for 0.85 of the variance, whilst the sixth degree polynomial accounts for only 0.05 more. The age profile provides a comparison independent of the bulk density.

The effect of stimuli (changing the parameter values) on the "amorphous peat" model is shown in Table 3 and effects on the "fibrous peat" model, using Equation (5b) in Table 4. The ratio of predicted/observed bulk density is shown at four depths. These were selected so that the observed bulk density was close to the trend line at that depth. The shallowest (4 cm) shows whether the predictions are seriously in error near the start. The 12 cm depth is one for which the age is best established. If the bog did start growth anew on a bare peat surface at 28 cm it would be reasonable to suppose that productivity, at least, varied in the early stage of re-establishment. The 22 cm depth is some way above such a point. The 28 cm level was chosen for reasons already described.

It is apparent from Tables 3 and 4 that changing the decay or creep parameters has relatively little effect on the bulk density predictions. The creep parameter has little effect on predicted age either, but the aerobic decay parameter α_1 has a marked effect on age. Varying the compression parameter k or water table W has a bigger effect, but the model is most sensitive to variations in productivity p, and the rate of growth in length L.

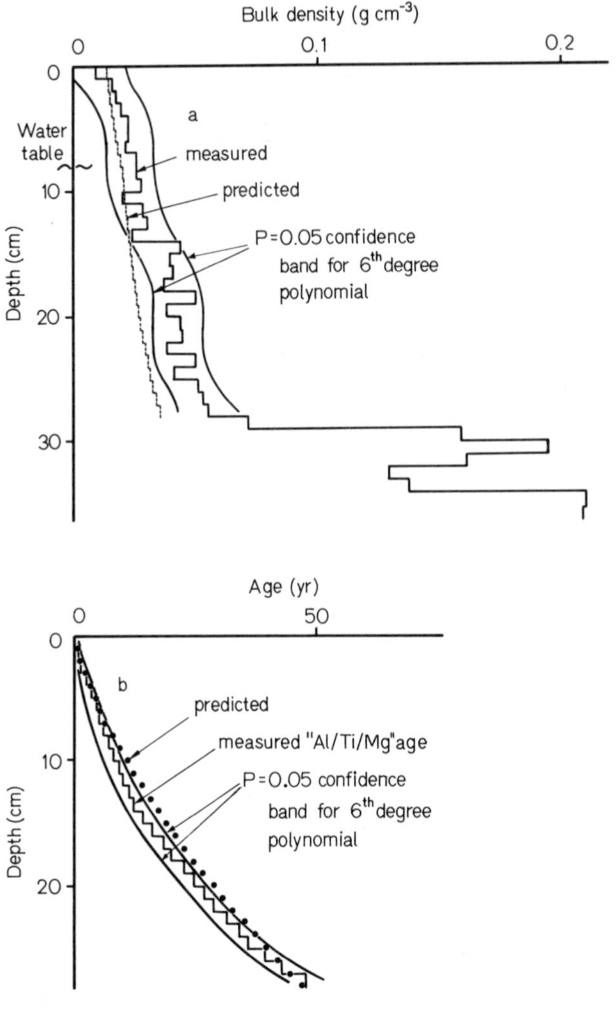

Fig. 7 a and b. Measured and predicted bulk density and age profiles for the same core as Figure 6 using measured parameter values and the "direct" method—see text. The $P = 0.05$ confidence bands for a 6th degree polynomial fitted to the observed values is also shown. For the "Al/Ti/Mg age" the polynomial was fitted to the concentration for each slice and the band shown was obtained from the cumulative error variance

Very approximately one may say that the ratio p/L determines the general position of the predicted profile, whilst the slope and curvature are determined by k and to a lesser extent c and indirectly, through effects on stress and creep time, by α. Water level has an indirect effect by altering the time for aerobic decay and by affecting the mass of peat exerting its full weight in air.

None of these tests produces a good fit to both the observed bulk density profile and to the age profile, although it seems that the age profile might be fitted by varying α_1 with relatively little effect on the bulk density profile.

Table 3. Effect of varying basic parameter set of Table 2 in "amorphous peat" model

	Bulk density: predicted/observed				Age (yr)	
Depth (cm)	4	12	22	28	12	22
Observed bulk density (g cm^{-3})	(0.020)	(0.029)	(0.045)	(0.055)		
set=	1.0	1.0	1.0	1.0	9	30
Basic set	0.78	0.75	0.64	0.64	13.9	32.5
0.5 p	0.36	0.29	0.21	0.18	11.5	24.0
1.5 p	1.28	1.59	2.10	4.56	18.0	53.1
0.5 L	0.53	3.77	np	np	174.0	np
1.5 L	0.49	0.42	0.31	0.28	7.4	15.5
0.5 α_1	0.78	0.75	0.64	0.64	12.0	26.9
1.5 α_1	0.78	0.75	0.64	0.64	16.9	41.3
0.5 α_2	0.78	0.75	0.64	0.64	nr	32.5
1.5 α_2	0.78	0.75	0.64	0.64	nr	32.5
0.5 k	0.72	0.58	0.41	0.36	11.5	24.0
1.5 k	0.85	1.06	1.40	3.04	18.0	53.1
0.5 c	0.78	0.74	0.62	0.62	13.8	32.0
1.5 c	0.78	0.76	0.65	0.66	14.1	33.0
W= 4 cm	0.78	0.65	0.53	0.41	11.2	23.6
W=12 cm	0.78	0.87	0.79	0.87	15.9	50.2
W=16 cm	0.78	0.87	1.11	1.35	nr	134.5

np: no prediction; nr: not relevant.

Table 4. Effect of varying basic parameters of Table 2 in "fibrous peat" model

	Bulk density: predicted/observed				Age (yr)	
Depth (cm)	4	12	22	28	12	22
Observed bulk density (g cm^{-3})	(0.020)	(0.029)	(0.045)	(0.055)		
set=	1.0	1.0	1.0	1.0	9	30
Basic set	0.87	1.19	2.07	np	19.6	65.6
0.5 p	0.38	0.33	0.26	0.24	12.8	28.3
1.5 p	1.53	8.29	np	np	59.1	np
0.5 L	2.49	np	np	np	np	np
1.5 L	0.53	0.52	0.46	0.49	8.6	20.0
0.5 α_1	0.87	1.19	2.07	np	16.1	51.3
1.5 α_1	0.87	1.19	2.08	np	25.6	91.8
0.5 α_2	0.87	1.19	2.07	np	nr	65.5
1.5 α_2	0.87	1.19	2.07	np	nr	65.6
0.5 k	0.76	0.67	0.52	0.49	12.8	28.3
1.5 k	1.02	5.50	np	np	58.9	np
0.5 c	0.87	1.19	2.06	np	19.6	65.4
1.5 c	0.87	1.19	2.09	np	19.6	65.7
W= 4 cm	0.87	0.84	0.97	1.51	13.3	32.7
W=12 cm	0.87	0.84	20.2	np	30.7	700

np: no prediction; nr: not relevant.

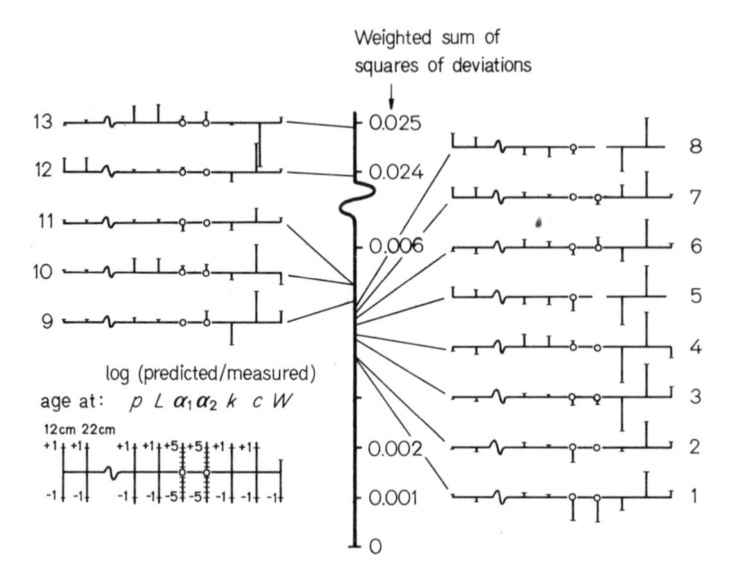

Fig. 8. Agreement between observed and predicted values of parameters, and (independently) of age at two depths using the "indirect" method (see text) on the same core as in Figure 6. Thirteen solutions of varied success are shown. *Vertical bars* show predicted/observed values on a logarithmic scale. The scale of the bars for the decay parameters α_1 and α_2 is one-fifth that of the others. The shorter the bars, the better the agreement. In solutions 5 and 8 the predicted water table, W, was at the bottom of the core, so the parameter α_2 has no value

Indirect Method Comparison. It is now worth examining the results of the indirect tests, in which only the bulk density profile was supplied, and best estimates of the parameters were made.

Because this is an expensive procedure, it was not considered worth-while testing both "amorphous" and "fibrous" peat models. The "amorphous peat" model was chosen because it was the more stable, produced results with the direct tests which were closer to the observed ones, because in any extension to other peats and depths the "amorphous" peat predominates, and because the rate of creep of the fresh *Sphagnum* peat of this core (Fig. 2b) seems in practice to be almost independent of applied stress in the range of interest.

The first indirect tests were made with all seven parameters (p, L, α_1, α_2, k, c, W) free to vary. A lower limit of 0.0 (or 0.1 cm for W) was set, to avoid wasting time with unreal (negative) values. An upper limit of 200 (or, in the case of W, the depth of the core) was set for the same reason. Start points varied by a factor of 100 on either side of the measured parameter values. These constraints are very light ones. Thirteen runs were made, and each located a different minimum (Fig. 8). For each solution the ratio of "best fit"/measured value for each parameter is shown, and the same ratio for age at 12 cm and 22 cm depth (using the "Al/Ti/Mg age" already described). A log scale is used, that for the decay parameters being five times smaller than for the others. The shorter the bars, the better the agreement between measured and "best estimates". It is worth emphasising that

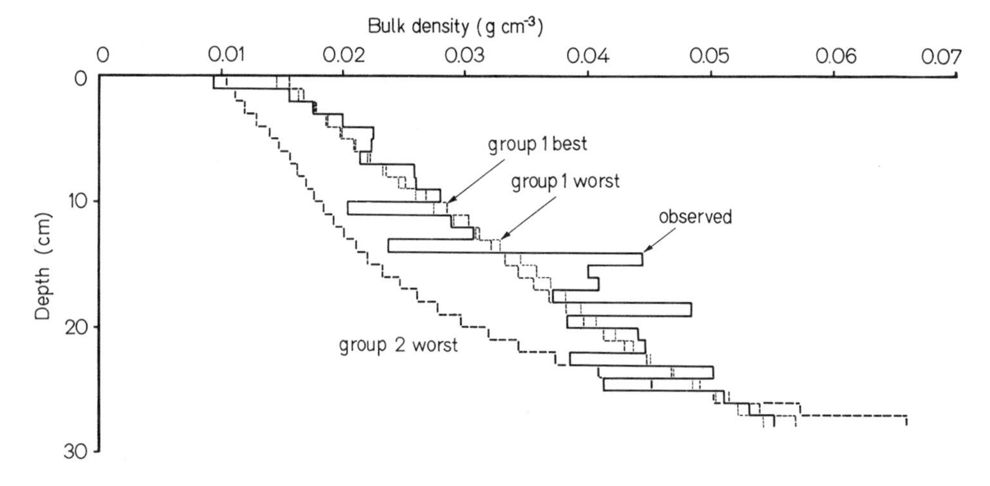

Fig. 9. Observed bulk density and that predicted for solutions 1 ("group 1 best"), 11 ("group 1 worst"), and 13 ("group 2 worst") of Figure 8

the "best estimates" are based solely on the bulk density profiles and, with the light constraints, differences from the measured values (except for *W*) of tens of orders of magnitude might occur.

The solutions are in two groups. The sum of weighted squares of deviations for the lower 11 are within a small range; the largest is 1.38 times the smallest. The other two are six times the smallest. The bulk density profiles for the least and greatest in the first group, and for the worst of all, are shown in Figure 9. It seems unrealistic to claim that any solution within the first group is demonstrably preferable to any other, but the second group are a distinctly worse fit to the bulk density profile. One may also conclude that in view of the general variability of the observed profile, and the necessarily smooth nature of any predicted profile, it is unlikely that there is any solution which is a much better fit than any yet found: that the hypersurface does not have any much lower points on it.

The hypersurfaces around the minima—Figure 10a shows an example for the two parameters which were always most precisely estimated—is some way from parabolic, so the correlation matrices must be interpreted with caution. There is much intercorrelation: in six solutions five values (out of the 21 possible) are greater than 0.6 or less than -0.6. The correlation illustrated in Figure 10a between productivity and length growth rate is 0.98. Since a linear least squares regression provides a good fit this is not surprising. Although other solutions show similar numbers of high correlations, there is no obvious pattern: it is not always the same pairs of parameters which show a high correlation.

Taken together these results suggest that the surface is equivalent in two dimensions to an elongated trough with an uneven floor of shallow depressions with no clear pattern. The trough seems to be broader for some parameters (α_1, α_2) than for others, since the range of solutions is much greater for them, and the precision of the estimates of α_1 and α_2 in any one solution is low.

Fig. 10a and b. Two-dimensional surfaces of minimisation function in cases showing the highest correlation between parameters. The parameter units are relative to a value of 1.0 at the minimum in all cases. A contour is sketched close to the minimum. In (a) (solution 6 of Fig. 8), where only the deviations from bulk density were minimised, the contour is much elongated and off centre. In (b) where both bulk density and age deviations were minimised, the contour is less elongated and less off centre. Ideally the contour would be circular and centred on the minimum

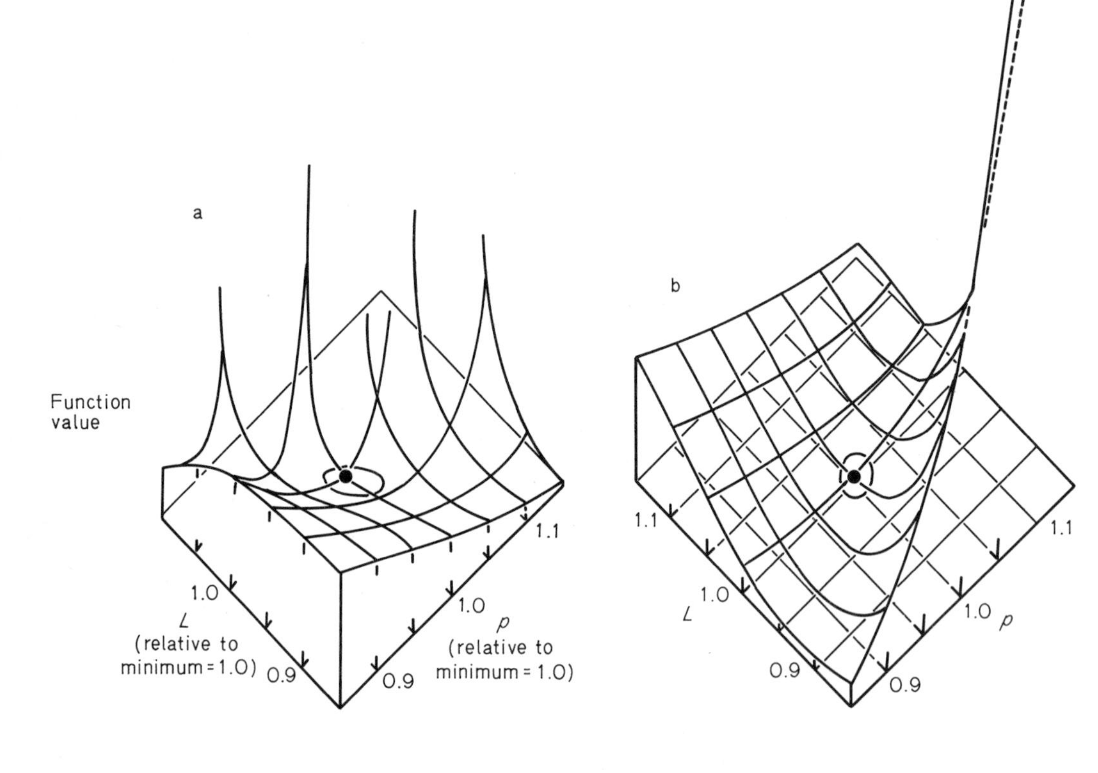

In view of the enormous possible range of the parameters, however, the convergence of *p*, *L*, *k*, *c* and to a lesser extent *W* on a relatively narrow range may perhaps indicate that the model is moderately realistic.

That the general region of the hollow coincides with the measured position (in many cases) is additional and independent support. The decay parameters however are not centred on a well-defined region. The major anomaly is in the estimates of the creep parameter, which are consistently an order of magnitude greater than the measured values. This may indicate that the rate of creep in partly humified peat is markedly different from that measured in unhumified peat.

With 21 correlations, and only 28 observations, it seemed desirable to reduce the number of parameters. Water table level seemed the obvious one to choose. In the real system it is as tightly constrained as any; there is a tendency for solutions

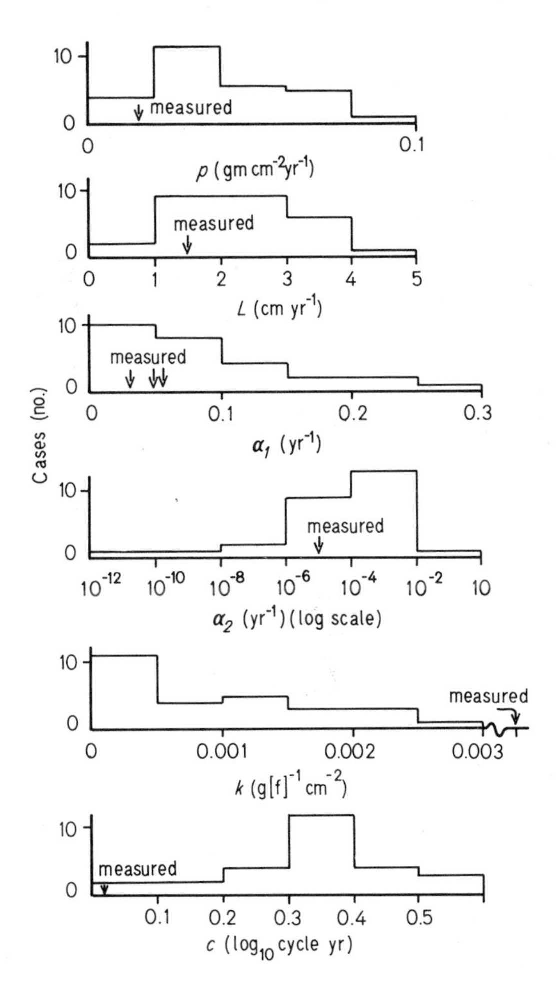

Fig. 11. Histogram showing distribution of 28 sets of parameter estimates, using the "indirect" method (see text) and the same core as shown in Figure 6, but fixing water table, *W*, at 8 cm. *Arrow* = measured values

to be found in the impossible situations of water table above the surface or below the bottom of the peat, which then inactivates one or other of the decay parameters; and it is a parameter whose value can be relatively easily measured.

A second set of tests was therefore made on the same data, but with the water table fixed at 8 cm; see Figure 11. The weighted sum of squares of deviations ranged from 0.0042 to 0.0053 (compared with 0.0038 to 0.0052 for the first group of solutions). The results are in general similar to the seven parameter tests, except that α_1 now shows a much reduced range, as might be expected. The degree of parameter correlation was also reduced; only two were outside the range -0.6 to $+0.6$. The agreement with measured parameter values was not very good, however, when the effectiveness of a linear regression fit is recalled.

Table 5. Best estimates of parameters minimising deviations from both age and bulk density

Parameter	Solution			Measured value
	I	II	III	
p	0.029	0.032	0.032	0.016
L	1.53	1.78	1.80	1.5
α_1	0.020	0.038	0.039	0.032, 0.050, 0.062
α_2	0.4×10^{-7}	0.0065	0.0054	10^{-5}
k	0.0015	0.0022	0.0022	0.005
c	0.033	0.0001	0.0023	0.018
W	18.7	12.0	12.1	8
p'	0.018	0.021	0.021	
L'	0.44	0.64	0.64	

The extent of intercorrelation of parameters may also be reduced if deviations from both bulk density and the complete Al/Ti/Mg age profile are minimised simultaneously.

A third set of tests was, therefore, made. The random error in the Al/Ti/Mg measurements (judged from the residual sum of squares of deviations from a sixth degree polynomial) was about the same (0.10 of the total) as it was in the bulk density measurements. There was no need therefore to give different overall weights to bulk density and age, though the individual measurements of age were given weights in the same way as were the bulk density measurements. The quantity to be minimised was the sum of the geometric mean of the squares of the weighted deviations at each level. The hypersurfaces of the minima (for example Fig. 10b) were much closer to parabolic than in the first tests. Intercorrelation of parameters was much reduced too; in the best case the most significant correlation (shown in Fig. 10b) was only -0.42 and in no others reached the 0.25 level. The spread of the weighted sum of squares of deviations was much greater (for the same starting points as in the first group in Fig. 8), from 0.71 to 5.13. The three best results, all nearly equally good, are shown in Table 5. Again, bearing in mind that the parameters might vary over tens of orders of magnitude, the agreement of estimated and measured parameters is fair, except for α_2 and c which are in any case rather imprecisely estimated.

To provide an equally good polynomial fit to both bulk density and age profiles also needs five parameters (two for bulk density and three for age).

In conclusion the model predicts profiles or parameters which are in moderate agreement with the measured ones. Whether the differences arise from deficiencies in either or both the model and measurements is unclear.

Investigations on Other Sites. The tests described have used data from a single site. The results might be due to chance properties of this data set, so it seemed important to investigate other sets too.

Three groups of cores have been examined. First, three cores taken from Moor House using the same methods as already described, and for which Cs-137 pro-

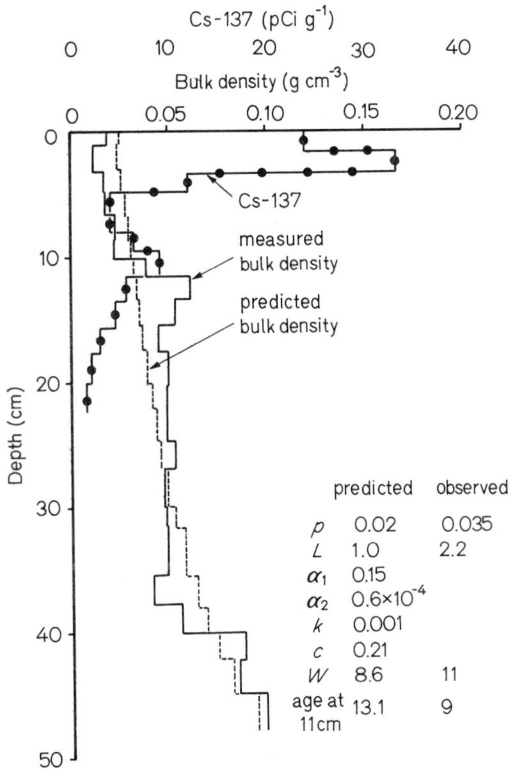

Fig. 12. Observed and predicted bulk density, and Cs-137 concentration for a core (MH2) from a *S. papillosum* lawn at Moor House National Nature Reserve. The "indirect" method (see text) was used, and this is the best of three solutions. The predicted, and three measured, parameter values are shown. The observed age is obtained by taking the Cs-137 peak as representing 1963

files were made. Second, cores made to much greater depths using a Russian pattern (West, 1968) borer (Chap. 8). For one of these cores two C-14 ages are available. Third, cores from other areas; the only one reported here is from Abisko (N Sweden), and a C-14 date is available for it too.

Short Cores from Moor House. MH2 came from a *Sphagnum papillosum* lawn on Burnt Hill, about 200 m from the cores already examined in detail (grid ref NY753329). MH5 came from a *S. papillosum* lawn in a small bog about 250 m from Green Burn (grid ref NY774322). It should be emphasised that the core site is not Forrest's site, described in Smith (1973). MH6 came from a *S. magellanicum* hummock, about 3 m diameter, 5 m from MH5.

The model was fitted using the indirect method (minimising a weighted sum of squares of deviations of the observed bulk density profile from that predicted). The results from the best of three runs for each core are shown in Figures 12, 13, and 14. Measurements of p, L, W and the Cs-137 1963 peak were available. The agreement with predicted values is again moderately good, with the same reservations as for the first core.

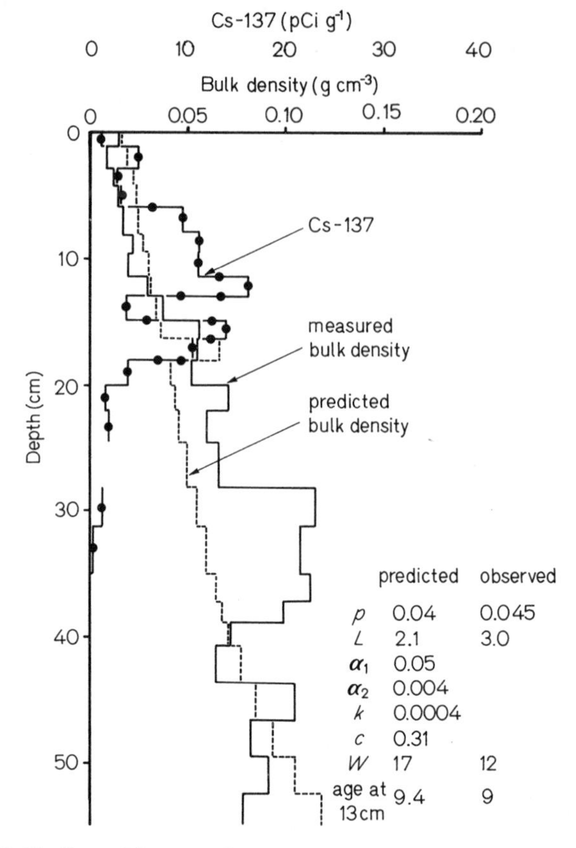

Fig. 13. Similar to Figure 12, for a core (MH5) from *S. papillosum* lawn

Long Cores from Moor House. The "steady state" assumptions underlying the model reported here cannot be expected to hold for the ages and depths of the long cores from Moor House blanket bog peat (Chap. 8). Nor is it to be expected that the functional relationships can be extrapolated far. Nevertheless, it seemed of interest to see how far the model could be stretched. The indirect method of fitting was therefore applied to Jones and Gore's Valley Bog data (Chap. 8, Table 8.13). The best of three results is shown in Table 6. The estimates of k and W seem to be far from the likely real values, and the predicted ages are low by a factor of 3. For these colossal extrapolations however it is perhaps surprising that there is even order of magnitude agreement.

Core from Abisko. Finally the model was fitted to bulk density data, collected by Prof. Mats Sonesson, from Stordalen, Abisko, N Sweden. The same method as before was used. The best of four results is shown in Table 7. This core is interesting because the bulk densities are much greater than others used so far, and the peat was formed in very different climatic conditions. The measured parameter values are in better agreement with the estimates than were those from long cores at Moor House.

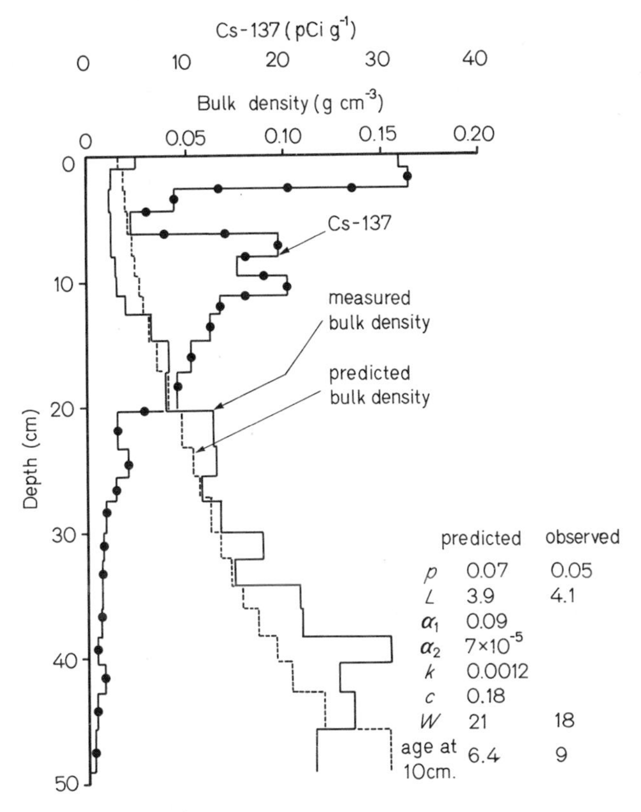

Fig. 14. Similar to Figure 12, for a core (MH6) from a *S. magellanicum* hummock

9.3 General Discussion

This model, although crude, produced predictions which were in moderately good agreement with measurements. The sensitivity tests (Tables 3 and 4) indicate that the parameters which most affect peat bulk density are productivity and length growth rate. Next in importance are variations in water table and in the compression parameter. Least important are the creep and decay parameters. The importance of water table variation can be compared directly with the finding by Jones and Gore (Chap. 8) that in their model the slope of the regression of decay rate on depth has a large effect on the bulk density profile. It is not so much the rate of aerobic decay but the length of time for which it operates which matters. This is not a new conclusion: the model reflects old beliefs. This point is emphasised when one examines the apparent average rate of peat accumulation (Table 8). There are two types of estimate of peat growth included. The C-14 ones are simply average rates. The first three sites in Table 8 were chosen from many available (Walker, 1970, gives a longer list) for relevance to the material in this paper, or in the case of Tregaron because a wide range of rates is shown at one site. The other two estimates are of the annual increment to the anaerobic zone.

Table 6. Best estimates of parameters for Valley Bog, Moor House

Parameter	Estimate	Measured
p	0.034	
L	1.4	
α_1	0.028	
α_2	1.5×10^{-5}	
k	0.00013	
c	0.202	
W	27	
Age at 160 cm (yr)	604	2,200[a] ± 50
Age at 310 cm (yr)	1,500	4,700[a] ± 57
p'	0.011	
L'	0.97	

[a] Turner et al. (1972).

Table 7. Best estimates of parameters for Abisko, N Sweden

Parameter	Estimate	Measured
p	0.033	
L	0.32	0.4
α_1	0.004	about 0.03
α_2	10^{-5}	
k	0.004	
c	0.23	
W	32	30
Age at 36 cm	637	1,110
Age at 72 cm	3,770	5,240
p'	0.0048	
L'	0.019	

Table 8. Some estimates of peat growth rate

Place	Method	Growth rate (cm yr^{-1})	Growth rate (g cm^{-2} yr^{-1})	Source
Abisko (Sweden)	C-14	0.01		Sonesson (unpubl.)
Moor House Valley Bog (UK)	C-14	0.057 to 0.098		Turner et al. (1972)
Tregaron (Wales)	C-14	0.01 to 0.3		Turner (1964)
Glenamoy (Ireland)	p, α_1 and W model	0.04[a]	0.0032[a]	Moore (1972)
Moor House MHI data (UK)	7 parameter model	0.4[a]	0.018[a]	This paper

[a] These are the predicted input to the anaerobic zone.

Since these make no allowance for later decay and compaction, one would expect them to be greater than the C-14 averages. The estimates from the model used on the core MH1 are, however, well above most of the C-14 estimates. Those for Glenamoy are close to the C-14 estimates. The measured aerobic decay rate is slightly greater at Glenamoy though productivity is similar at both sites. The Glenamoy model, however, allows aerobic decay to operate to a depth of 25 cm. This points to one of many weaknesses in this *Sphagnum* peat model: it has been assumed that the water table is at a fixed position, and that this is coincident with the change from aerobic to anaerobic decay. In fact of course the transition is probably some way below the water table, and the water table fluctuates during the year (Clymo, 1965; Goode, 1970; Forrest and Smith, 1975) being lower in summer. Decay is faster during the summer in field conditions (Clymo, 1965) and at higher temperatures and higher water content in laboratory experiments (Rosswall and Berg, 1973). In consequence the mean water level is probably well above the notional aerobic/anaerobic transition.

The indirect method of model testing, which seems attractive at first sight, proves less helpful than hoped because the shape of the bulk density profile is relatively simple. Using the age profile as well gives a marked improvement and it would be valuable in any development to introduce a third variable if possible. This is desirable because it is likely that all realistic peat growth models based on steady state hypotheses will produce predictions which are monotonic functions of depth. The present model does not produce monotonic predictions—when the creep term begins to exceed the others the bulk density becomes positive infinite then negative—but this is simply due to extrapolating beyond the limits of physical reality. It is indeed surprising that predictions on the long and older cores are as close to observation as they are.

Walker (1970), having examined the apparent average rate of peat accumulation of bog peats at various depths below the surface, concluded tentatively that "there is no correlation between sample depth and apparent accumulation rate confirming that, beyond the earliest stages of deposition, progressive compaction does not normally occur". Kaye and Barghoorn (1964) expect autocompaction in *Sphagnum* peat to be negligible, and refer to a C-14 dated core indicating a steady rate of accumulation of about 1 mm yr^{-1}. Lee and Tallis (1973) find that the rate of peat accumulation at a site in the south Pennine hills has been roughly constant for three hundred years to the present time.

By contrast the age profile for the *Sphagnum* peats (reported here) indicates that compaction *has* occurred in the top 0.5 m at least, accompanied by a five-fold increase in bulk density. The time scale of this change is relatively so short however that it is still consistent with the conclusions of Walker and of Kaye and Barghoorn. The peat examined by Lee and Tallis has a present day cover of *Eriophorum*, and the model parameters for such a system are probably very different from those reported here.

The great importance of events in the top 50 cm of these ecosystems as determinants of peat accumulation rates is indicated.

There are some unresolved difficulties however. A rate of accumulation constant over thousands of years is rather surprising when one remembers the concomitant changes in climate. Furthermore, if the hypotheses of Granlund (1932) and Wickman (1951) concerning the maximum height for bog growth are correct, one might expect a generally decreasing rate of peat accumulation with time. In the model presented here it is possible for the age to vary independently of the bulk density and the sensitivity of age to variation in parameter values is not always the same as that of bulk density. In particular the age is more sensitive to variation in the aerobic decay rate, α_1, than is the bulk density. The consequence is that a change in peat accumulation rate might leave no trace in the bulk density profile. The reciprocal effect—a change of bulk density such as may be found at a recurrence horizon—with less marked change in accumulation rate is not predicted.

The bulk density profiles, to depths of 3 m and more, reported by Jones and Gore (Chap. 8) do seem to show a gradual increase of bulk density with depth. If the model were applicable to such cases one would expect a curvilinear relationship between age and depth.

Compaction, selective decay of less dense materials, and progressive changes in climate, plant cover and bog surface microenvironment may all contribute to the age and bulk density profiles. The model described here has allowed only the first of these. It is hardly surprising therefore that the whole range of phenomena concerned with peat growth cannot yet be reconciled.

9.4 Conclusion

A peat bog gains dry matter by growth of plants at the surface, and loses it by decay. Rate of growth in depth is the net result of gain by plant growth, and loss by decay, primary compression and secondary consolidation (creep). The evidence for functional relations is presented, and a seven-parameter model constructed.

The data used for checking the model are parameter measurements, bulk density profiles, and an age profile, based on measurements of Cs-137 and on Al, Ti, and Mg concentrations.

The model is used in two ways:

1. Given measurements of the parameters, age and bulk density profiles are predicted.

2. Given a bulk density profile, best estimates are made of the seven parameters and the age profile. Alternatively both bulk density and age profiles are supplied, and the seven parameter values estimated.

Productivity, the rate of growth in length of plants, aerobic decay rate and water table position seem to be the most important parameters in this model.

The importance of the surface layers in determining peat accumulation rates is emphasised.

Acknowledgements. I record my thanks to Mr. R.S. Cambray and Mr. M.C. French for making the estimates of Cs-137 and lead; to Mr. E.J.F. Reddaway for the estimates of carbon dioxide and methane flux and for some of the growth estimates; to Mrs. P. Ratnesar for expert and unfailingly careful assistance with most of the other experimental work; to Dr. P. Osmon for guidance with the radioactive counting; to Prof. Mats Sonesson for permission to use an unpublished peat density profile; to Prof. J. Essam, Dr. D. Knowles and Dr. C. Place for help with mathematical problems; to Dr. P. Cawse, Dr. K.E. Clymo, and Mrs. B. Thake for commenting on the typescript, and to other colleagues for occasional help in many ways.

I am grateful to the Natural Environment Research Council for financial support for part of this work.

9.5 Appendix

9.5.1 Model Construction

The justification for functional relations used is given in Section 9.2.1. Here they are assumed.

Dry Matter

$$\frac{dx}{dt}=p-\alpha x \tag{A1}$$

where p = rate of addition of dry matter per unit area
x = total accumulated dry matter per unit area
α = decay parameter
t = time since growth began.

The solution of Equation (A1) is

$$x=\frac{p}{\alpha}(1-e^{-\alpha t}). \tag{A2}$$

It is implicit in Equation (A2) that as t increases, x approaches p/α which may be called the steady state mass.

Depth. Consider a finite slice of peat which is now, at n time units, of thickness Z; and which was formed j time units from the start. When first formed it was of thickness Z_1. Then if T represents time in discrete units from 1 to n,

$$Z_j=Z_1 f(T), \quad f(1)=1. \tag{A3}$$

This equation describes compaction with time for a single slice. The peat bog is assumed to have a new slice added each time unit, so the depth of peat, l, is given by:

$$l=Z_1 \sum_{j=1}^{n} f(T_n-T_j). \tag{A4}$$

Now consider the process as a continuous one. Analogous to Equation (A 4) we may write:

$$l=L\int_0^t f(t-t')\,dt' \tag{A 5}$$

where Z_1, the depth added in unit time, is replaced by L, the instantaneous rate of addition, and T_n and T_j are replaced by t and t'.

Then:

$$\begin{aligned}\frac{dl}{dt}&=L+L\int_0^t f'(t-t')dt'\\ &=L-L[f(t-t')]_0^t\\ &=Lf(t).\end{aligned} \tag{A 6}$$

This simple but unobvious result allows one to write the rate of growth of the whole peat column given simply a knowledge of the behaviour of one part, provided that this is a function of time alone. This is obviously so for creep [Equ. (5)].

For compression, the stress in Equation (4) is due to the weight of peat (per unit area) above the slice. Where the peat is under water, the stress ε is:

$$\varepsilon=nxg(1-\rho_m/\rho_s) \tag{A 7}$$

where g is the weight per unit mass ("acceleration due to gravity") at this point on the Earth's surface, ρ_m is the density of the medium and ρ_s is the density of the peat dry matter; 1.6 g cm^{-3} for *Sphagnum* (Clymo, 1970). The factor n (explained below) is 1. The quantity x, a function of time, may be found from Equation (A 2).

Where the peat is in air above the water table, the effective weight is due not only to the dry matter, but also to associated water in capillary spaces. Field measurements show that the water content of *Sphagnum* varies from about 10 times the dry mass at the surface to about 40 times just above the water table. An empirical function could be used, but implementing the model is then more awkward and much more time consuming, so for simplicity the approximation that the water content is on average 20 times the dry mass has been used. For computations using Equation (A 7) above the water table ρ_m is $\simeq 0.0$, and $n=20$.

In some circumstances the stress of snow and ice might be very important. Certainly *Calluna vulgaris* at Moor House can be crushed in this way. West (1968) gives the density of old snow as about 0.2 g cm^{-3} so a 100 cm deep drift might produce a stress of 20 g (*f*) cm^{-2}, which is roughly equivalent to the same depth of wet *Sphagnum* in air. If the surface peat layer is frozen the stress acts at an unknown depth in the peat. The stress is temporary however, and for want of knowledge of its effects has not been included in the model. Compression can thus be written as a function of time.

For decay, the analogue of (A 3) for slice thickness as a function of time may be written.

Assuming that all four processes operate independently, and making use of Equation (A 6) gives:

$$\frac{dl}{dt} = L \exp\left\{-\alpha t - kng(1-\rho_m/\rho_s)\frac{p}{\alpha}(1-e^{-\alpha t})\right\}(1-c\log[m+t]) \qquad \text{(A 8)}$$

for "amorphous peat". For "fibrous peat" the term log $[m+t]$ is multiplied by the stress. No explicit solution for Equation (A8) is known, but it may be solved numerically.

Aerobic and anaerobic layers are incorporated by redefining a local t at zero as the water table is passed at depth W. At this time the mass of aerobic peat has reached a steady value, and so has the contribution to compressive stress from this source. A new value of p, which now represents the input to the anaerobic zone, is calculated from Equation (A 1) whilst ρ_m becomes 1.0 g cm^{-3}. The parameter α, which was α_1 relating to aerobic conditions, is replaced by α_2 relating to anaerobic conditions. The calculations are continued, adding an extra term in Equation (A 8) for αt at the water table, and another for the stress due to material in the aerobic zone. Time for the creep term refers to the original time.

Implementing the model: A FORTRAN subroutine takes values of the parameters p, L, α_1, α_2, k, c, and W and a series of depths defining slices of peat parallel to the surface. Between the two values of l defining each slice and the value of t for the upper surface, Equation (A 8) is integrated to give t for the lower surface. A simple Gaussian quadrature method could be used, but in practice to allow possible addition of terms involving l, a fourth-order Runge-Kutta technique (Merson, 1957) is used, involving a fifth evaluation of Equation (A 8) from which some estimate of error is derived. The Runge-Kutta method was chosen because the integration needs restarting for each depth, must finish at a predetermined time or depth, and need not be of accuracy better than about 1 in 10^4. These requirements are not so well satisfied by predictor-corrector, nor by polynomial extrapolation methods. In critical cases a check is made by repeating the calculation with reducing step sizes. The step size is normally adjusted to minimise computing time. From this value for t, and using Equation (A2) x is calculated. The bulk density of the peat in the slice may then be calculated.

This process is repeated for each slice. The main output from the subroutine is a series of times for slice boundaries, and bulk densities of slices. The use made of these is described in Section 9.2.2.

9.5.2 Methods

9.5.2.1 Minimisation Technique

The CERN program MINUITS (library D 506, D 516) was employed, using the simplex option (Nelder and Mead, 1965) to locate minima, and Davidon's (1968) method to refine the estimate and to estimate parameter correlations. Davidon's method assumes that the surface is quadratic about the minimum. In a few cases values about the minimum were calculated to investigate this point.

The quantity minimised was usually a weighted sum of squares of differences between observed and calculated bulk density. The weighting was made proportional to layer thickness, and inversely proportional to the variance about the straight line of best fit of the nearest five observed densities and depths. This weighting was used to equalise the deviations as far as possible in data sets with different characteristics.

9.5.2.2 Bulk Density Profiles

Sites were selected where *Sphagnum* formed nearly pure carpets. Cylindrical cores 21 cm diameter down to 75 cm were taken using a cylindrical cutter with 3 cm long teeth with sharpened edges. The cores were removed either by forcing a spade below the cylinder and hauling out at an angle of 45°, or using a network of nylon cords drawn across the bottom of the core. The first method is simpler, and was used if the peat was coherent or fibrous. The second method worked better on semi-fluid humified peats.

Cores were transferred to plastic tubes with wooden bases to minimise loss of water (and consequent compaction) during transport.

It was possible to check the length of the cores at each stage. Any cores showing a change of more than 1 cm were rejected; these were mostly from very wet sites with *Sphagnum cuspidatum*.

The cores were frozen solid, then cut into slices with a sharp carpenter's saw. Distances were all measured from the core base, so errors in slice thickness should be neither cumulative nor proportional to thickness.

The slices were allowed to thaw, separated into small pieces, dried for 24 h at 104° C, cooled in a desiccator, and weighed. This time was sufficient to reduce weight changes to less than 0.1% per day for the most humified samples.

9.5.2.3 Radioactivity Measurements

It was hoped that the age of one horizon in each core might be established by a peak in the activity of radioactive nuclides attributable to bomb tests. The most suitable nuclides, because of their relatively large amounts and long half lives, are Cs-137 and Sr-90 (Harley et al., 1965). These same reasons (which make them useful for dating) make them potentially hazardous to health, so a lot is known about their distribution in time and space (e.g. Bartlett, 1971; Cambray et al., 1971; Peirson, 1971). "Fixation" of Cs-137 in *Sphagnum* has already been reported by Bovard and Grauby (1967), though their reference to the "roots" of *Sphagnum* suggests they may have misinterpreted their results.

Slices were prepared for counting by dry ashing at 450° C with severely restricted air supply to prevent self combustion raising the temperature. It was established that loss of Cs or Sr chloride or carbonate was not detectable (less than 0.1%) in these conditions. Recovery of added Cs Cl and Sr CO_3 was nearly complete (0.99 ± 0.028; 0.97 ± 0.035; eight samples).

On most cores, absolute amounts of Cs-137 were estimated using a Ge/Li detector at liquid nitrogen temperature. These estimations were made by Mr. R.S. Cambray at the Health Physics Division of the AERE, Harwell.

On one core the spectrum of β-activity was recorded from dry ashed samples using a $^1/_4'' \times 2^1/_2''$ plastic scintillator. The energy range corresponding to a 5% window centred on the broad peak of a standard Sr-90 source of the same geometry was assumed to be mainly due to Sr-90. The counts due to natural K-40 were always less than 5% of the total in this range. On the same core, and before dry ashing the peat, relative Cs-137 amounts were estimated with a NaI detector.

9.5.2.4 *Lead Measurements*

It is known that the amount of lead in the aerial environment has increased (Jaworowski, 1967; Rühling and Tyler, 1968; Lee and Tallis, 1973; Persson, Holm, and Lidén, pers. comm.) and it seemed possible that the time of increase of lead concentration might be used for dating peat. Samples were wet-ashed with nitric acid, and lead concentration estimated by atomic absorption flame spectrophotometry. These analyses were made by Mr. M.C. French at the Monks Wood Laboratory of what was at that time the Nature Conservancy.

9.5.2.5 *Other Metals*

Interpretation of the lead data proved difficult, so additional analyses of Al, Ti, Fe, Mg, Cu, Cd, Zn, and Pb were made on a parallel core using an atomic absorption flame spectrophotometer (not available to me when the first set of Pb analyses was made). The samples were wet ashed with nitric acid. A residue of insoluble organic matter was dry ashed at 450° C and the ash combined with the nitric acid- soluble fraction. Internal standards were used, and interference in the Mg determination was suppressed with 2500 ppm $Sr(NO_3)_2$. Overall recovery (including the ashing procedure) of soluble salts of those elements analysed ranged from 0.95 to 1.02.

9.5.2.6 *Annual Growth Rates*

In one case direct estimates of growth in length and dry matter of *Sphagnum* were made using plants cut to known length and replaced in position (Clymo, 1970). In other cases innate markers (branch crowding and length) or growth against an external wire marker (Clymo, 1970) were used.

9.5.2.7 *Compression and Secondary Consolidation Parameters*

The parameter estimates were obtained from the slopes of graphs similar to those of Figure 4 and Figure 5. It should be noted that the model incorporates the functional relationships of those graphs but not the specific values of slopes.

The apparatus was a hollow rigid cylinder, 21 cm internal diameter, with perforated sides and base. In this was put a 5 cm thick slice of peat of the same diameter. A second hollow cylinder with perforated end was a loose sliding fit inside the first. The weight of the second cylinder was counterbalanced, and compressive stress applied to the peat by adding weights to the sliding cylinder. Strain was measured ± 0.01 mm against a reference mark on the central wire support of the sliding cylinder, using a travelling microscope.

The peat was saturated to field capacity before measurements began, and a polythene cover minimised evaporation.

For measurements of the short term (10 sec) time course of compression, a compressible variable resistor was put beneath the counter weight, and the output from a resistance bridge connected to the resistor was recorded. Movements up to 0.5 cm were readily recorded with trivial disturbance to the system.

9.5.2.8 *Decay Measurements*

Decay rates in the aerobic zone have been measured using *Sphagnum* in nylon mesh bags (Clymo, 1965), and for other species of peat bog plants by Heal et al. (Chap. 7). These methods are not very sensitive, and interpretation of results may be uncertain; Heal finds that the rate of loss of dry matter continues apparently unchecked whilst the respiration rate, admittedly in artificial conditions, declines. For the anaerobic zone annual loss rates may be less than 1%, and estimates from bagged material are much too inaccurate and imprecise. That anaerobic breakdown does continue is suggested, though certainly not shown, by the presence of quite large amounts of methane gas at some depth in peat, perhaps especially where bulk density is low. Any bog pool will, if stirred, yield gas bubbles, which a mass spectrometer analysis shows to contain mainly methane, nitrogen and some argon. The production of methane in waterlogged organic habitats has been known for a long time (Dalton, 1802).

Apart from methane, the only carbon-containing gas leaving the bog surface in relatively large amounts is carbon dioxide; (there may also be traces of higher paraffins or olefines). Loss of dry matter in solution may also occur in lateral runoff.

The simplest hypothesis, used here, is that carbon dioxide and solution losses come from the aerobic zone, and methane from the anaerobic zone.

The methane might, of course, have been formed at one particular stage of bog growth and simply have been trapped since, but methane is slightly soluble in water, so could have diffused upwards. The Henry's law constant for methane is 20×10^6 at 6° C (Washburn, 1926). For pressure corresponding to a depth of 5 m, gas bubbles containing 50% methane would be in equilibrium with about 1 cm^3 dm^{-3} in solution. Methane is given off by the bog surface. How much is oxidised by methane-oxidising bacteria is not known. Such bacteria occur in some lakes at least (Cappenberg, 1972) but Collins, D'Sylva, and Latter (Chap. 5) were unable to isolate any from Moor House peat.

Fluxes of methane and carbon dioxide were measured at twelve sites on three microhabitat types at Moor House, at roughly monthly intervals in April–October, and less often in winter, using the method described by Clymo and Reddaway (1972). This involves collecting the gas evolved from a defined area, bare of vegetation, over 24 h, and measuring the concentration of methane and carbon dioxide.

The dry matter is assumed to have the general formula of CH_2O.

Losses of dry matter in solution were estimated for the same area and time (Clymo and Reddaway, 1972).

9.5.3 Age Profiles

The age of a single horizon in each core may be obtained from the Cs-137 and β-activity profiles (Fig. 15). Both show a surface peak, a second smaller peak at about 11 to 13 cm, and declining values at greater depths. The surface peaks may perhaps be due to Cs and Sr which are effectively recycled within the live capitula and branches of *Sphagnum*, and are thus carried up with the surface. The smaller peaks are interpreted as indicating peat laid down in 1963—the peak time of fallout (Bartlett, 1971; Cambray et al., 1971). It is conceivable that the peaks relate to the Windscale nuclear reactor accident in 1957, but Windscale is about 70 km away, and Gorham (1958a) concluded that in the Lake District, which is closer to Windscale and nearly on a direct line between Windscale and Moor House, there was no clear evidence of the accident in the radionuclides accumulated in *Sphagnum*.

The apparent 2 cm difference in peak position for β-activity and absolute Cs-137 concentration is probably due to difficulty in defining the surface more accurately. The estimates were made on different cores. The cruder relative Cs-137 estimate (Fig. 15) on the core used for β-activity showed a peak at 12 cm, coinciding with that of the β-activity. It is also possible of course that the peat growth rate was slightly different in the two cores.

The concentrations are expressed per unit dry matter. If unit thickness of peat slice is used, then the lower peak is at about 16 cm. The difference is due to bulk density increasing with depth, so that a sample of given volume from lower down in the profile contains a larger amount of dry matter. Ideally the measure of radioactivity should be expressed per unit time, and the peak would then be nearer the surface than that based on unit dry matter. Assuming exponential decay of dry matter, the peak would be about 1 cm above that shown in Figure 15.

Transport of the radionuclides (other than by recycling within plants) may have occurred, just as it does during elution from an ion exchange resin column. Unfortunately there appears to be no direct experimental evidence bearing on this point. Some upward transport may

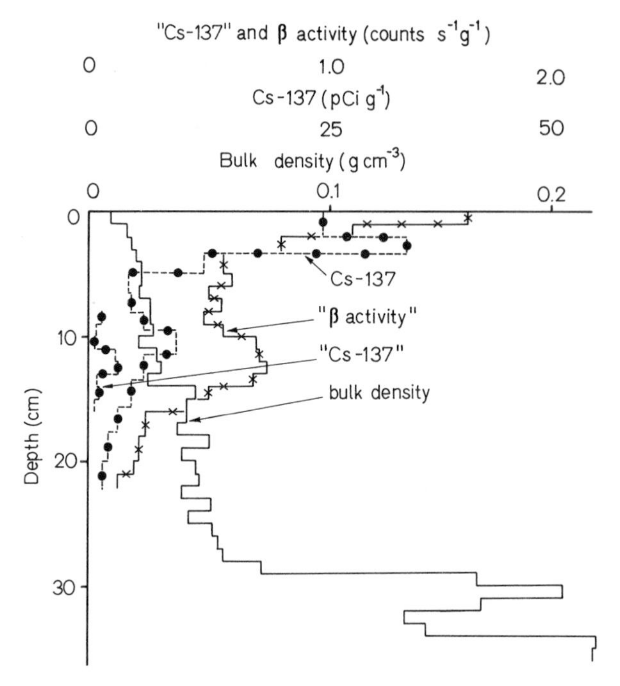

Fig. 15. Bulk density, relative "β-activity" and relative "Cs-137" activity in a core parallel to that of Figure 6. Also shown are absolute Cs-137 concentrations in a parallel core

happen on the occasions when evaporation exceeds precipitation, but in general the movement, if it occurs, is probably downwards. If the Cs-137 and Sr-90 are present as cations one would expect the monovalent Cs^+ to move more rapidly than the divalent Sr^{2+}, since the peat has a large cation exchange capacity (Clymo, 1963). If they are present as uncharged molecules this would not apply. Mattsson (1972) presents some evidence to suggest that at least part of the Cs-137 in Swedish lichen carpets is present in particulate form. The complete Cs-137 profile on a *S. magellanicum* core (Fig. 14) shows a narrow "tail" down to 50 cm, but there is no indication that the main Cs and Sr peaks in the same core are at different levels, so it is assumed that peak movement, if it has occurred, is negligible. The 12 cm level is therefore thought to have been formed in 1963.

Establishing a complete age profile involves finding some substance which has been added to the surface at a constant rate, and which has subsequently either stayed in situ or moved in a known way. Carbon-14 is not suitable for detailed continuous dating within the last hundred years, though it can be used in the same sort of way as Cs-137 for locating the bomb-test peak, as Jones and Gore (Chap. 8) show. Some of the metal elements may be suitable, but one might expect that others would not be. To sort the metals in the Moor House data into groups showing similar behaviour, the correlation coefficient matrix for the concentrations shown in Figure 6 was calculated. These coefficients were then arranged in a dendrogram using an agglomerative technique and centroid strategy (Williams et al., 1966). The dendrogram is shown in Figure 16. Al and Ti are clearly separated from the rest. Peirson et al. (1973) have evidence that Al deposited at Wraymires in the English Lake District is derived from soil, and Mattsson and Koutler-Andersson (1954) considered that both Ti and Al in a peat deposit originated from soil dust. Mg, which probably derives mainly from sea spray (Gorham, 1958b), appears similar in behaviour to Cu, Cd, and Fe. This is because Mg has a correlation of 0.65 with Fe, which in turn has a correlation of 0.28 with Cu and Cd. The correlation of Mg with Cu is only 0.01 however; Mg is linked to Cu and Cd through mutual partial similarity to Fe, but to different parts of the Fe profile.

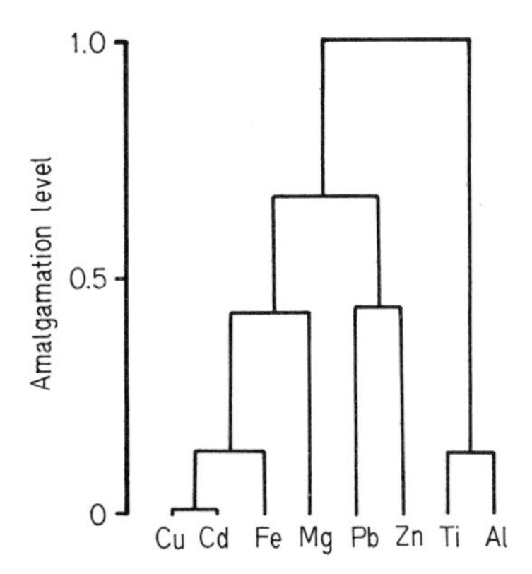

Fig. 16. Dendrogram showing relationships between the metals of Figure 6

The metals Cu, Cd, Pb, and Zn are ones whose deposition rate has increased as a result of industrial activity (Jaworowski, 1967; Goodman and Roberts, 1971; Rühling and Tyler, 1971; Lee and Tallis, 1973) even though the increase may be local and may now be declining in some cases. Peirson et al. (1973) conclude that Cu, Pb, and Zn deposited at Wraymires come mainly from artificial or industrial sources. In Table 9 the surface concentration and accumulation rate at Moor House of some of the metals are compared with those in *S. magellanicum* growing in Sweden and with total (soluble and insoluble) deposition rates at Wraymires. The Moor House figures are from the total in the top 12 cm assumed accumulated since 1963. Only Cu and Cd show notably higher concentration in the top (live) cm. The Swedish figures were calculated from 2.5 years' growth. The rates are similar, though the accumulation of Cu, Cd, and Pb is rather greater at Moor House whilst Mg and Fe are rather less; (the Zn profile is erratic for reasons not known, and may be unreliable.) The accumulation rate of Pb (63 mg

Table 9. Surface concentration (ppm) and accumulation rate (mg m^{-2} yr^{-1}) of metals on *Sphagnum* at Moor House and in S Sweden[a] and deposition rate (mg m^{-2} yr^{-1}) at Wraymires in the English Lake District[b]

Element	Al	Ti	Mg	Fe	Cu	Cd	Pb	Zn
Concentration								
Moor House	856	19	860 (740)	1,960 (2,160)	5.8[c]	6.5[c]	235	80
S Sweden	—	—	980	1,060	10.2	0.99	68[d]	90
Accumulation rate								
Moor House	292	7	215	490	8	1.1	63	42
S Sweden	—	—	281[d]	600[d]	3	0.7	45	60
Deposition rate								
Wraymires	230	—	—	280	about 30	—	55	120

[a] Rühling and Tyler (1971).
[b] Peirson et al. (1973).
[c] Concentration in the top cm about twice that in the next two cm.
[d] Calculated from other data given by Rühling and Tyler (1971).
For Moor House "surface" is the 0–3 cm layer; for S Sweden it is the surface 3 segments of *Sphagnum*.
Figures in parentheses are for total Mg and Fe in peat from *Sphagnum*-dominated unburnt blanket bog at Moor House (Gore and Allen, 1956).

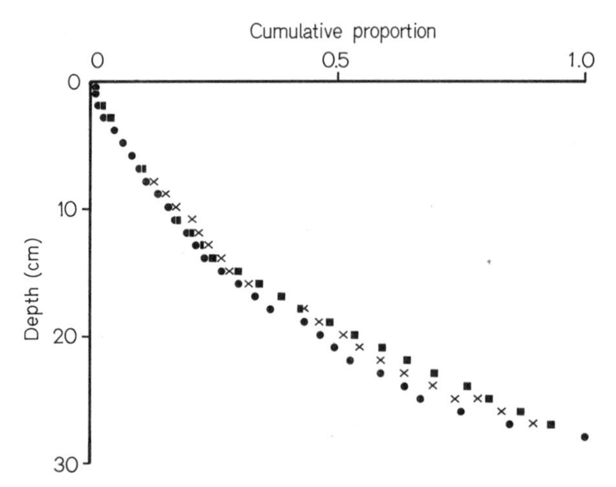

Fig. 17. Cumulative profiles for Al, Ti, and Mg, adjusted to coincide at 28 cm depth (where only the Al point is shown)

$m^{-2}\ yr^{-1}$) is similar to the deposition rate (55 mg $m^{-2}\ yr^{-1}$) at Wraymires. Chamberlain (unpubl.) calculates the average rate of emission of Pb from automobile exhausts in Britain, excluding Devon, Cornwall, W Wales and NW Scotland, to be about 20 mg $m^{-2}\ yr^{-1}$, and from smoke concentrations and particulate fallout rates in country districts he calculates Pb fallout to be on average 50 mg $m^{-2}\ yr^{-1}$. Peirson et al. (1973) conclude that Pb deposited at Wraymires is probably of both local and more distant origin. This may be the explanation for the profile of Pb concentration, which is conspicuously different from that found by Lee and Tallis (1973) at sites about 140 km away in the southern Pennines. At Moor House there are conspicuous local sources in the largely bare spoil heaps left by lead miners. The most recently worked mine, about 2 km distant, was abandoned about 70 years ago. A smelter only 0.5 km away was last used about 150 years ago (M. Rawes, pers. comm.). It must be added, however, that the climate at Moor House is not conducive to the spread of dust.

The assumption of constant rate of deposition at least to 28 cm depth seems (from Fig. 6) to be untenable for all but Al, Ti, Mg, and Pb, since the other metals show either erratic fluctuations or an increase in concentration near the surface. Although Pb shows the same general increase in concentration down to 28 cm as do Al, Ti, and Mg, the uncertainty about the relative importance of local and regional components and the likelihood that deposition rates have changed recently make it suspect.

The extent of vertical movement subsequent to deposition is uncertain. The Cs-137 and Sr-90 peak coincidence, the high cation exchange capacity, the similarity of profiles of cations of differing valence, and the abrupt changes in concentration below 28 cm, all argue against the existence of a great deal of vertical movement.

On the other hand, based on measurements of Mg concentration in rain (Clymo, unpubl.), about two thirds of the deposited Mg is not retained in the peat. Further, the total Cs-137 in the *S. magellanicum* peat profile of Figure 17 is 10.2 p Ci cm^{-2}. At Milford Haven, to the same date, 10.5 p Ci cm^{-2} were deposited (Cambray et al., 1971). Deposition is closely related to rainfall however, and whilst the mean annual rainfall at Milford Haven was 100 cm (16 yr average) that at Moor House is about 200 cm. The half life of Cs-137 is 30 yr, which accounts for only a small part (perhaps 10%) of the difference. Presumably the rest is lost in runoff. The concentration in the peat—up to about 3×10^4 p Ci g^{-1}—is unusually high, though the normalised specific activity (=activity kg^{-1} of crop/activity deposited $m^{-2}\ d^{-1}$) of 120, is not exceptional (Chamberlain, 1970).

The cumulative totals of Al, Ti, and Mg, adjusted to coincide at 28 cm, are shown in Figure 17. All three curves are similar. Assuming constant accumulation rate, limited vertical movement, and a known age for one point (12 cm), these curves may be used to give a continuous age profile.

Two features of the metal profiles (Fig. 6) remain unexplained. First is the peak in Fe at 20 cm, and second the decrease in concentration per unit dry mass of peat and for Mg, Fe, Pb, and Zn (though not for Al and Ti) in concentration per unit volume of peat at depths below 28 cm. Tyler (1972) reports a similar finding for Ca, Fe, Pb, Zn, and Cd (and less obviously for Mg and Mn). He suggests tentatively that it may be connected with oxidation reduction conditions and hydrology. This is plausible for Fe and Mn, and for Cu, Cd, Zn, and Pb which form rather insoluble sulphides but as Tyler recognises, can hardly account for a peak in Ca or Mg concentration. In the Moor House profile where a distinct change in stratigraphy is apparent, it appears preferable to postulate an earlier (below 28 cm) phase of higher productivity and/or lower decay rate. The 20 cm iron peak, uncorrelated with any stratigraphic change, may however be related to water table.

Work in progress shows that there is much variation in the profiles of metal concentration in peat; a great deal remains to be discovered about the factors controlling vertical movement of metals in peat-forming systems.

Note added in proof. Four years later it is apparent that the simple concept of static elements and concentration increase as a result of decay of the organic matrix is wholly wrong for many elements in the anaerobic zone, although it may not be too incorrect for the depths to which it has been used here, and especially for Mg.

10. The Blanket Bog as Part of a Pennine Moorland

M. Rawes and O. W. Heal

In the previous chapters the blanket bog has been considered as a discrete, self-contained ecosystem. The bog is now considered in combination with other vegetation—soil communities, to examine the contribution of each community to the diversity of the moorland, to determine the transfers between the communities and between the land and the streams, and to indicate the relationship between man and the moorland.

10.1 Introduction

The IBP studies, and their predecessors, concentrated on the analysis of discrete ecosystems recognisable by their distinct vegetation and soil type. The synthesis of this information in the preceding chapters and associated papers (Heal et al., 1975; Heal and Perkins, 1976) has shown that on the blanket bog <1% of the annual primary production is consumed by herbivores, *c* 5% is assimilated by decomposer fauna and, assuming that 10% passes below the water table, *c* 85% is decomposed by the peat microflora. In contrast, the *Agrostis-Festuca* grasslands show a greater herbivore consumption (<40%) and soil fauna assimilation (20–40%) apparently with a smaller amount of organic matter decomposed by the microflora annually, although at a much higher rate, than on the bog. The distinction between these two ecosystems results from the higher nutrient concentration and availability in the grassland through the mineral soil which in the bog is divorced from the vegetation. The nutrient limitation on the bog has not markedly reduced primary production but, through low nutrient quality, has decreased the rate of organic matter circulation through herbivores and decomposers.

The production, energy and nutrient studies have emphasised the integrity of the recognisable ecosystems, i.e. plant–soil communities. In reality the blanket bog is part of a mosaic of terrestrial and aquatic habitats. The organisms in the habitats are influenced by the composition and arrangement of the habitats, by the movement of water and materials between the habitats and by man in the past and present. The mosaic thus constitutes a larger unit of ecological organisation—the watershed or catchment. This is "the minimum ecosystem unit which must be considered when it comes to man's interests" (Odum, 1971). The interactions within the mosaic are shown at other IBP sites: at Point Barrow, Alaska, where the small scale mosaic of polygons provides gradients rather than discrete systems; at Devon Island, Canada, a number of discrete plant–soil systems comprise a biologically and geographically distinct lowland; and in Hardangervidda, Norway, the plant–soil systems combine in a plateau where water and reindeer link the systems (Bunnell et al., 1975; Bliss, 1975; Østbye, 1975).

To put the IBP studies into the context of the local landscape we will discuss:
1. the present ecological pattern of the moor
2. the interrelationships of its habitats, and
3. some changes in the landscape with time.

The account cannot adequately quantify these aspects of moorland ecology, despite drawing upon twenty years of research of a wide variety. There have been few projects on the Reserve in which the relationships between plants and animals and between habitats have been studied. However, we can examine qualitatively the broader ecological boundaries and assess the effect man has had on what is often considered to be a near-natural landscape.

Streams and pools are an integral part of the landscape. Although they have not been included in the IBP research, we have drawn on information to show some of the relationships between the terrestrial and aquatic habitats and to compare the production characteristics of these contrasting ecosystems.

10.2 The Present Ecological Pattern

10.2.1 The Components of the Moorland Landscape

The moorland on which most of the IBP projects were sited shares with the larger area of the Reserve, and of the northern Pennines, similar physical characteristics of soil and climate, and has been subjected to similar influences of management, principally those of sheep grazing, grouse shooting and lead mining.

A recent vegetation survey (unpubl.) by Moor House staff covered a 65,000 ha area of the northern Pennines, above 330 m, within which the Reserve lies. This survey, together with data from the vegetation map of the Reserve (Eddy et al., 1969) shows (Table 1) that there is proportionally more blanket bog on the east side of the Reserve and within the altitude range of 500 to 700 m where the IBP sites were located.

Areas of uninterrupted bog are usually less than 10 ha, and are separated by streams, erosion channels, grasslands, flushes and man-induced features such as mining artefacts and footpaths. This is illustrated in an 87 ha area, including some of the IBP sites, of gentle relief with the recognised vegetation types showing little seral development (Fig. 1). With few expections—one is described by Crisp et al. (1964)—large-scale erosion seems to have ended, but the type of water-induced erosion described by Bower (1959) continues with dissection of the peat, sheet erosion and the development of peat faces on the edge of the peat mass. Mining, also responsible for much erosion, has long since ceased, but sheep grazing continues as it has for centuries, and probably exerts more influence on the vegetation and fauna than any other biological factor.

The mosaic of terrestrial habitats provides a small range of vegetation structure and of temperature. As a result of the presence of the small areas of grasslands and mineral soils, a relatively wide range of moisture, nutrient and pH conditions occurs in the soil (Chap. 1).

The mean precipitation is 1900 mm and with many streams and pools present, water is a major feature of the landscape. The streams of the Tees catchment

Table 1. The extent of different vegetation types within a Pennine area, expressed as percentage of area

	Whole area (1)	Moor House NNR (2)	E side of MH NNR (2)	IBP sample area
Area (ha)	65,600	3,842	2,549	87
Blanket bog				
Calluneto-Eriophoretum[a]	26.6	31.2	47.1	—
Trichophoretum[a]	—	0.4	0.6	—
Erosion complexes[a]	4.0	8.4	11.3	—
Recolonisation complexes[a]	—	6.5	8.9	—
Eriophoretum	6.8	10.9	11.3	—
Total	37.4	57.4	79.2	85
Flushes				
Sphagneto-Juncetum effusi[a]	—	1.9	2.2	—
Carex rostrata facies[a]	—	0.1	0.1	—
Sphagneto-Caricetum	—	0.9	0.9	—
Calcareous flushes	—	0.4	0.4	—
Flushed gleys[a]	—	0.8	0.5	—
Total (including water)	2.1	4.1	4.1	7
Grasslands				
Juncetum squarrosi[a]	1.6	9.8	4.2	2
Nardetum[a]	11.5	16.0	6.3	3
Agrosto-Festucetum[a]	0.9	3.9	1.8	
Festucetum	—	4.8	3.8	3
Shingle and scree	0.8	2.4	0.1	
Mines and tracks[a]	0.1	0.9	0.7	
Pteridietum	0.5	0.9	0.0	—
Miscellaneous acid grassland	43.1	—	—	—
Total	58.5	38.7	16.9	8
Woodland	2.0	+	+	—

(1) Unpublished survey.
(2) After Eddy et al. (1969).
[a] Vegetation types found in IBP sample areas.
– Details not available.
\+ <1%.

generally rise from peat drainage or springs. They have a gentle gradient, rarely exceeding 1 in 15, and are usually less than 3 m wide and 1 m deep. The substrata are usually of limestone, sandstone or shale, with boulders and shingle, and banks commonly of exposed peat. On the blanket bog, with the water table often at or near the surface (Chap. 1) temporary and permanent pools are frequent, usually less than 10 m^2 in area and containing *Sphagnum cuspidatum* or *Sphagnum subsecundum*, though some have a bottom of bare peat.

The chemical and physical properties of the water are closely related to those of the adjacent terrestrial communities. The concentrations of elements in the bog

Surveyed by B. Marsh & L. Teasdale, April 1973.

Fig. 1. Vegetation map of an 87 ha area at Moor House, including three of the IBP sites

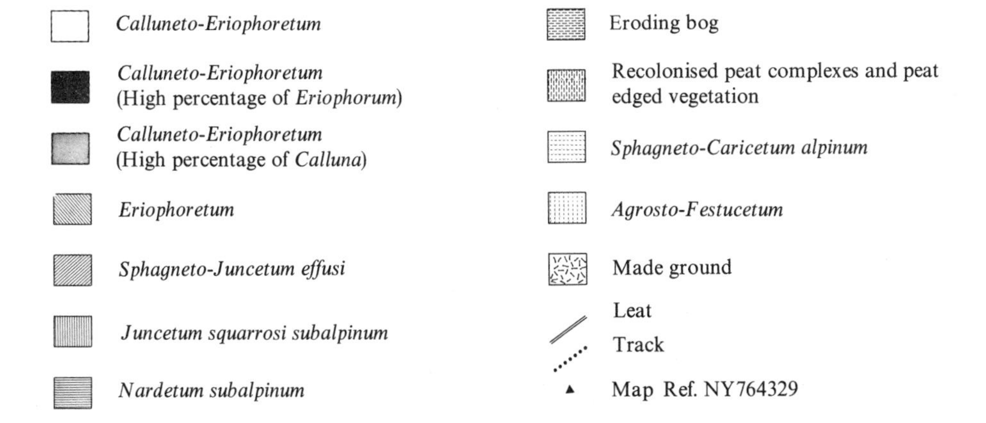

Table 2. Chemical analysis of water from various sources at Moor House and from a lowland stream in southern England (4)

	1 Bog pools	2 Trout Beck	3 Rough Sike	4 East Stoke Mill Stream	5 Peat leachate	6 Grassland leachate	7 Mine water	8 Rain
K	0.1 –1.0	0.3 – 0.7	0.2 –1.0	1.1 – 4.2	0.9	0.05 –3.6	6.3– 6.5	0.01–0.8
Ca	0.3 –1.4	3.4 –19.2	0.9 –8.5	50.0 –96.0	4.0	0.9 –4.4	74.0–80.0	0.10–1.0
Na	1.6 –5.3	1.6 – 3.3	2.0 –3.4	11.0 –15.0	—	0.5 –6.3	3.7– 5.7	0.1 –4.0
Mg	0.2 –1.1	0.5 – 0.6	—	1.2 – 3.6	0.9	0.15 –0.3	5.9–10.0	0.04–0.50
P	—	—	0.01–0.04	0.04– 0.32	0.05	0.02 –0.15	—	0.02–0.21
N	0.9 –2.9	—	0.08–0.25	1.14– 3.46	—	—	—	0.15–1.40
NO_3–N	0.02–0.05	0.05	—	—	—	0.005–0.062	—	—

Data ($mg\,l^{-1}$) extracted from: 1, 2, 7: Gorham (1956); 3: Crisp (1966); 4: Crisp and Gledhill (1970); 5: Allen (1964); 6: Park et al. (1962); 8: Gore (1968).

Table 3. The contribution of four habitats to the floral and faunal diversity of the moorland, indicated by the % of species unique to each habitat or common to all habitats

	Blanket bog	Flush	Marginal grassland	Rich grassland	All sites	Number of species
Higher plants	4	24	0	23	4	152
Mosses	6	28	2	28	7	112
Liverworts	6	19	2	19	4	48
Lichens	39	0	0	24	0	33
All plants	8	22	1	24	5	345
Insects	8	5	7	13	16	522
Soil fauna	9	1	10	10	20	252

Habitat types as in Figure 2. Data from Banage (1963), Block (1965), Coulson (1959), Nelson (1965), Hale (1966a), Eddy et al. (1969) and Springett (1970).

pools and in the streams are very similar, apart from calcium which has a high concentration in the streams. The concentrations of elements in water leached through peat and through mineral soil are similar, indicative that, despite the higher concentration of calcium in the grassland soils compared with bog peat (Chap. 1, Table 1.2) the high calcium levels in the streams result from solution of rocks in the stream bed and banks.

Exceptionally high concentrations of elements in water associated with mines appear to be very localised and do not influence the bog pool and stream water. The concentration of elements varies considerably with the rate of discharge of the streams (Crisp, 1966) but even at the highest concentrations, the levels are lower than those occurring in lowland streams (Table 2).

The weekly mean temperature of the stream water is usually 1–2° C higher than that on the adjacent bog surface but is about 3–4° C lower than that in a similar lowland stream at the same latitude (Crisp and LeCren, 1970).

10.2.2 Diversity and Productivity of the Terrestrial Habitats

On the Reserve, including terrestrial and aquatic habitats, about 300 species of higher plants, 350 species of mosses, liverworts and lichens (Eddy et al., 1969) and about 1250 species of invertebrates have been recorded (reference to fauna publications in Chap. 4).

To what extent do the different parts of the landscape contribute to the diversity and productivity of its flora and fauna? Using the numbers of species as an index of habitat diversity, the bog is shown to be relatively poor in plant species, especially higher plants, while calcareous flushes and limestone grassland sites are particularly rich (Fig. 2). In contrast, data for invertebrate numbers, from a more limited range of sites, show no major difference in species abundance between the sites (Fig. 3).

The percentage of species unique to one habitat or common to all habitats indicates that the small areas of flushes and rich grasslands contribute most, and the marginal grasslands least, to the floral variety of the area (Table 3). Lichens are mainly unique to the bog and marginal grasslands while only about 3% of species are common to all the habitats.

In contrast, a similar analysis of invertebrate data (Table 3) shows that the habitats contribute approximately equally to the diversity of the area, about 20% of the species being common to all the habitats. This is not obviously related to the mobility of the fauna because the less mobile soil fauna (Collembola, mites, tipulid larvae, enchytraeids and nematodes) show a pattern similar to that of the more mobile surface (spiders and carabid beetles) and aerial fauna.

While there is no obvious correlation between floral and faunal diversity it is possible that increasing height of the plant canopy and decreasing soil fertility over the series limestone grassland to marginal grassland to bog, are compensating mechanisms, the numbers of aerial and surface fauna being influenced by canopy structure and soil fauna by soil fertility. Such trends are not shown in the data presented, but Cherrett (1964) showed that "coarse" vegetation, resulting from the withdrawal of grazing pressure by sheep, supported larger numbers of individuals and species of spiders than did grazed vegetation.

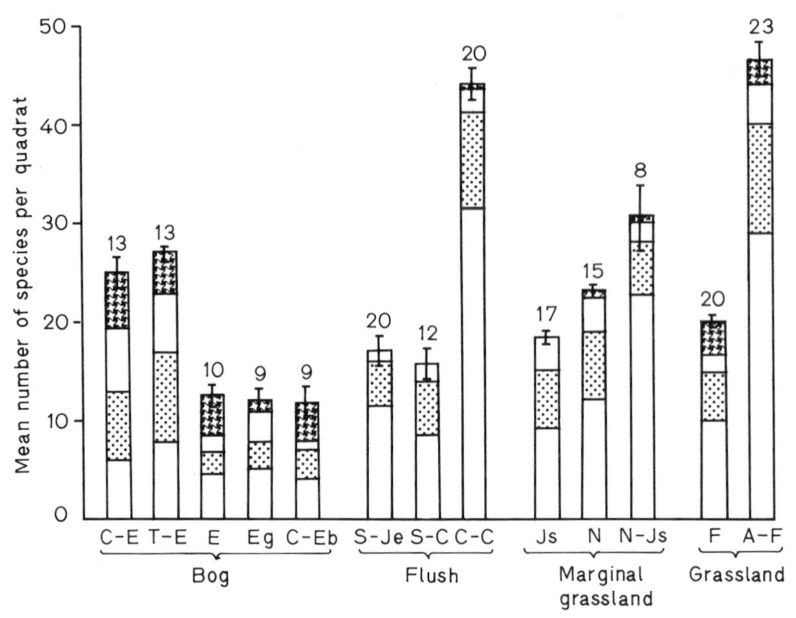

Fig. 2. Botanical diversity indicated by the numbers of species of higher plants, mosses, liverworts and lichens in various plant communities. The mean number of species for each community was calculated from quadrat data in Eddy et al. (1969); the standard error of the mean of the total number of species, and the number of replicates, are shown at the top of each histogram. *C–E*: *Calluneto-Eriophoretum typicum*; *T-E*: *Trichophoretum-Eriophoretum typicum*; *E*: *Eriophoretum* high level facies; *Eg*: *Eriophoretum* grazed facies; *C-Eb*: *Calluneto-Eriophoretum* burnt facies; *S-Je*: *Sphagneto-Juncetum effusi typicum*; *S-C*: *Sphagneto-Caricetum alpinum*; *C-C*: *Cratoneuron-Carex nodum*; *Js*: species-poor *Juncetum squarrosi sub-alpinum*; *N*: species-poor *Nardetum sub-alpinum*; *N-Js*: species-rich *Nardeto-Juncetum squarrosi*; *F*: *Festucetum*; *A-F*: *Agrosto-Festucetum*

Microflora data are limited but the list of fungal species isolated from bog, *Juncus squarrosus* and limestone grassland sites (Latter et al., 1967) show no major differences between habitats in their contribution to the species diversity of the area.

Although the small areas of grasslands contribute greatly to the species diversity of the moorland, their contribution to production is insignificant in quantity. On the east side of the Reserve less than 1% of the above-ground annual primary production is produced by the *Agrosto-Festuceta* and 94% by the blanket bog. However, the "quality" of the primary production of these grasslands is important.

Soil nutrient concentrations are reflected in the vegetation, being higher on the *Agrostis-Festuca* grasslands than on the bog, with the *Juncus* and *Nardus* intermediate (Chap. 1). This relationship is shown using the C:N ratio of the soil and the ash concentration in the vegetation (Fig. 4). The fibre content of the vegetation, an index of palatability to herbivores, declines as the soil C:N ratio narrows. The increasing ash and decreasing fibre content, expressing increasing vegetation "quality", are positively correlated with faunal and microbial activity and with assimilation by herbivores and decomposer invertebrates (Fig. 5).

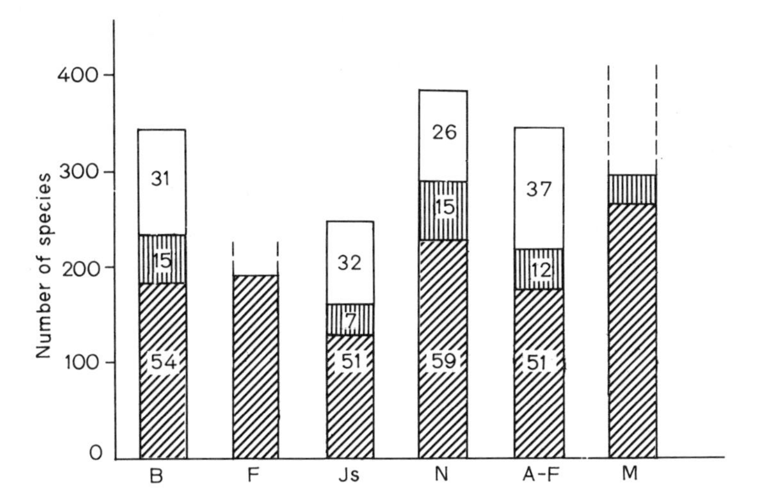

Fig. 3. Faunal diversity, indicated by the number of species of aerial ▨, surface ▥, and soil □ invertebrates in major vegetation types. Within each vegetation type, the % of species in each category is shown except in flush and meadow, where this is impossible because total numbers of soil invertebrates are not known. Data from Coulson (1959), Banage (1963), Cherrett (1964), Block (1965), Hale (1966a), Nelson (1971) and Houston (1973). *B*: blanket bog; *F*: flush; *Js*: *Juncetum squarrossi*; *N*: *Nardetum*; *A-F*: *Agrosto-Festucetum; M*: meadow

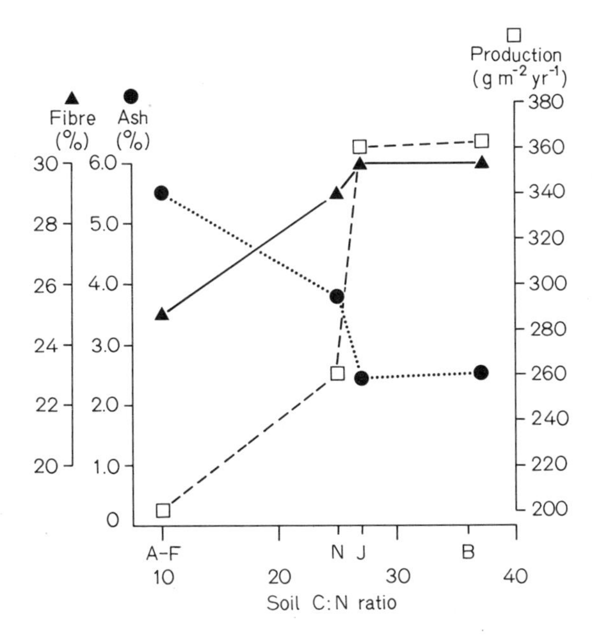

Fig. 4. The relationship between the C:N ratio of the soil in four sites and the above-ground annual primary production (□) and the % of ash (●) and fibre (▲) in the vegetation. *A-F*: *Agrosto-Festucetum*; *N*: *Nardetum*; *J*: *Juncetum*; *B*: blanket bog

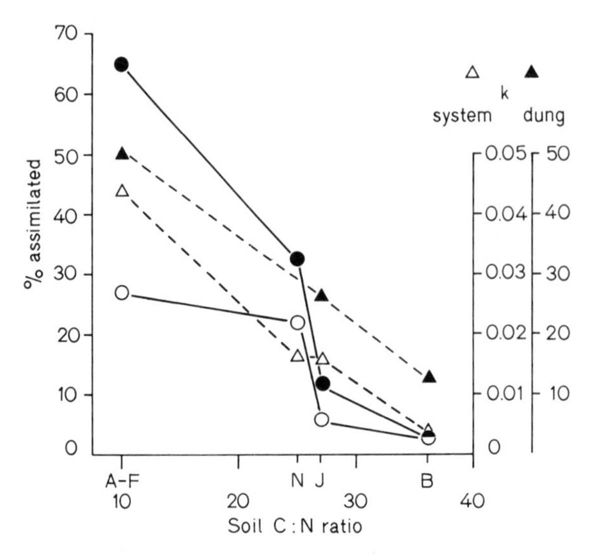

Fig. 5. The relationship between the C:N ratio of the soil in four sites and the % of above ground primary production assimilated by herbivores (●), the % of total primary production assimilated by decomposer fauna (○) and the fractional loss rate, *k*, of sheep dung (▲) and of the system (△). Data from Coulson and Whittaker (Chap. 4) and White (1960a). Vegetation types as in Figure 4

A number of microbial population and activity measurements are high in the *Agrostis-Festuca* grasslands, intermediate in the marginal grasslands and low in the bog (Latter et al., 1967). These are reflected in the rate of turnover, crudely expressed as *k*, the annual fractional loss rate of particular substrates such as sheep dung, and of the total organic matter in the systems (Fig. 5).

The causal relationships between these soil–vegetation–fauna–microflora characteristics are complex and cyclic. However, although the grasslands contribute little to the dry matter production of the bog-dominated moorland, they provide variety of species composition, quality of primary production, soil conditions and rates of biological processes.

10.2.3 Production in Aquatic Habitats

The interaction between terrestrial and aquatic habitats is discussed later (Sect. 10.3) but first the production characteristics of the ponds and streams are examined for comparison with those of terrestrial habitats. The data from aquatic habitats are very sparse but indicate four main features. First, the number of species of plants in the pools and streams is low, probably less than ten species of higher plants and about 30 mosses and liverworts being present (Eddy et al., 1969). Second, over 100 species of aquatic invertebrates have been recorded on the Reserve and the number present is probably nearer 200, i.e. 10–15% of the moorland invertebrates are dependent on habitats which comprise less than 5% of the area. Third, primary production is probably maximal in pools with 100% cover of *Sphagnum cuspidatum*, reaching about 500 g m^{-2} yr^{-1} with little contribution from higher plants (Clymo, 1970; Chap. 2). Primary production in streams has not been studied but, as cover is limited to small areas of mosses, it is likely to be lower than that in pools. Fourth, the annual production of trout, a carnivore, is similar to that of carnivores in the terrestrial habitats. The data on trout are more complete than those of other stream components and were generously provided by Dr. D.T. Crisp of the Freshwater Biological Association, Windermere. Additional data are from Crisp (1963, 1966) and Crisp et al. (1974).

A limited study of trout populations in Great Dodgen Pot Sike (Fig. 1) was carried out in a larger study of the Cow Green Reservoir and its catchment. Data on trout caught by electro-fishing show a mean August density, over a five-year period, of 0.25 fish m^{-2} with a biomass 1.1 g dw m^{-2}, assuming 80% moisture content. Based on growth and mortality curves derived from data from this and similar streams, annual production is estimated at 0.3–0.7 g dw m^{-2}. The total food requirement of this population was estimated, using metabolic relationships from Winberg (1960), as 8.1–8.6 g dw m^{-2} (0.3–0.7 g m^{-2} for growth and 6.2 g m^{-2} for metabolism).

In another site in the same stream production (0.1–0.2 g dw m^{-2} yr^{-1} from a biomass of 0.7 g m^{-2}) and food consumption (4.1–4.2 g m^{-2}) are low. This is due partly to the low population density (0.08 fish m^{-2}) and partly to the population being composed of older fish whose production and metabolic rates are low compared to smaller fish. This population is uncharacteristic of streams in the Cow Green catchment.

Probably 10–30%, i.e. 1.6 g out of an annual consumption of 8.4 g m^{-2} yr^{-1}, of the food of the trout is of terrestrial origin, the remaining 6.8 g m^{-2} yr^{-1} consisting of aquatic fauna. This is a minimal estimate of the production of stream invertebrates, since trout are, at best, capable of catching only 20% of the prey produced; 34.0 g m^{-2} is, therefore, a more realistic value for invertebrate production. The number of stream bottom fauna has been very approximately estimated at 1000–6000 m^{-2}, with a biomass of 0.5–1.0 g dw m^{-2}.

The trout production in Great Dodgen Pot Sike is at the low end of the range (3.0–12.1 g lw or 0.8–3.0 g dw m^{-2} yr^{-1}) for a variety of stream types (Le Cren, 1969). Hynes (1961) estimated invertebrate production at 3.0 g lw m^{-2} yr^{-1} in a Welsh mountain stream but this may represent only about 20% of the true value (Hynes and Coleman, 1968), indicating a production of 15 g lw or 3.8 g dw m^{-2} which is below the range of estimates for the Moor House stream.

10.3 Interrelationships Between Habitats

Although each major habitat has a distinct soil–plant–animal complex, and has been treated as a discrete ecosystem (Heal et al., 1975; Heal and Perkins, 1976) there is interaction between them. The fauna uses more than one ecosystem for its ecological requirements and transfer occurs, largely by water, of nutrients and organic matter between the ecosystems.

Few animals are confined to one habitat of the moor. The Red grouse, for instance, nesting and usually found on the blanket bog, visits streamsides for grit and water, and may feed on swards like the *Juncetum squarrosi* when the blanket

Table 4. Estimated maximum transfer by sheep of dry matter and nutrients (mg m^{-2} yr^{-1}) between blanket bog and grasslands

	Bog to grassland[a]	Grassland to bog[b]	Net gain or loss by bog
Dry matter	1,040	580	−460
N	49	15	− 34
P	4	1	− 3
K	19	13	− 6
Ca	3	8	+ 5

[a] Estimated from intake sheep $^{-1}$ d^{-1} × number of sheep m^{-2} × number of days on bog × nutrient concentration of *Eriophorum vaginatum*.
[b] Estimated from dung volume m^{-2} on bog × dung density × proportion of intake which is defaecated × nutrient concentration of *Festuca-Agrostis* grassland.

bog is fully occupied by territorial birds. Sheep largely neglect the bog as a food source at the present stocking density, numbering about 0.2 ha^{-1} compared with about 2.0 ha^{-1} on the *Juncus*, *Nardus*, and *Festuca* swards and 7.0 ha^{-1} on the *Agrostis-Festuca* grasslands. Numbers are directly related to food quality but are also influenced by other aspects of their ecology, for example shelter and social behaviour.

Territory size for sheep, seldom less than 2.5 ha, varies with vegetation and landscape. At Moor House territories tend to be linear, the sheep following streams, gullies and paths in a fairly set diurnal pattern of movement covering about 3 km. A group usually consists of only a ewe and a lamb, rarely a flock. Thus distribution is widespread unless bad weather forces them to seek common shelter, usually on coarse vegetation types, compared with their preferred resting sites on the relatively dry grasslands.

Although the density of sheep on the bog is low, the large area of bog results in about 20% of the total sheep population being on the bog at any moment as they move across the bog from one preferred grazing to another or shelter on the bog slopes (Rawes and Welch, 1969). Thus grazing of the bog vegetation, unmeasurable by standard cropping techniques, may nevertheless make a significant contribution to the total food consumption by the population and reduce competition for the more nutritious herbage. This is indicated by a 30% higher weight increment on the east side of the Reserve (0.5 sheep ha^{-1}) compared with the grass covered west side (3.1 sheep ha^{-1}). Grazing of the *Agrosto-Festuceta* tends to be at a set density related to the territorial behaviour of the sheep. Any change in stocking density would affect the grazing intensity first on the bog, then on *Juncetum* and *Nardetum*, and finally on the *Agrosto-Festucetum* (Rawes and Welch, 1969).

The movement of animals and the differential grazing pressures, digestibility and nutrient contents of the swards, cause a transfer of material from one habitat to another.

Table 5. Estimated inputs to and losses from a catchment, and net transfers between blanket bog and grasslands within a catchment

	mg m^{-2} yr^{-1}			
	N	P	K	Ca
Input precipitation[a]	+ 816	+22–+27	+ 196	+ 824
nitrogen fixation[b]	+ 110–+3,790			
Net transfer to bog from grasslands[c]:				
Sheep[d]	− 34	− 3	− 6	+ 5
Grouse	+ 4	+ 1	+ 3	+ 1
Meadow pipits	+ 6	+ 1	+ 1	+ 0.1
Spiders	+ 47	+ 4	+ 4	+ 1
Output				
Sheep	− 5	− 1	− 0.5	− 2
Insects 1[d]	− 6	− 0.5	− 0.5	− 0.04
Insects 2[d]	− 60	− 5	− 5	− 0.4
Burning[e]	− 249	− 1	− 12	− 6
Stream water[d]	− 294	−40	− 896	−5,375
Peat erosion[d]	−1,463	−45	− 206	− 483
Total output	−2,077	−93	−1,120	−5,866
Balance	−1,151–+2,529	−66–−71	− 924	−5,042

+ and − represent net gain or loss by the bog. [a] from Gore (1968), [b] from Martin and Holding (Chap. 6), [c] see text, [d] from Crisp (1966), [e] from Allen (1964). Estimate 1) for insects is for the total habitat while 2) assumes most is derived from the grasslands of the stream margins, which comprise 10% of the catchment.

On the basis of estimates of intake, and dung and urine deposition on different swards, Rawes and Welch (1969) suggested a net loss of Ca of 0.1–0.4 g m^{-2} yr^{-1} from *Agrosto-Festucetum* and *J. squarrosi* and a transfer of 0.1–1.0 g m^{-2} yr^{-1} N and K from *J. squarrosi* to *Agrosto-Festucetum*. Assuming that all food ingested on the blanket bog was transferred to other vegetation types and that all the dung and urine deposited on the bog originated from the nutrient rich *Agrosto-Festucetum*, then the difference in amounts represents the maximum transfer between bog and grassland. The calculations (Table 4) indicate that any net income to, or loss from, the bog is very small and is probably not significant in view of the errors involved in the estimate.

Grouse probably obtain 95% of the food supply of 0.8–8.0 g m^{-2} yr^{-1} from the blanket bog but capsules of *Polytrichum commune* and seed heads of *Juncus squarrosus* are taken from *Juncetum squarrosi* which usually borders areas of blanket bog. Using estimates of nutrient concentrations in *Polytrichum* capsules, first approximations of nutrients transferred from *Juncetum* to blanket bog are of the same order of magnitude as the net transfer by sheep (Table 5).

Coulson and Whittaker (Chap. 4) emphasised the association of predators with a wide habitat range compared with the relatively restricted range in herbivores. All the predators studied at Moor House take prey from a number of terrestrial and aquatic habitats—the Meadow pipit *(Anthus pratense)*, frog *(Rana temporaria)*, trout *(Salmo trutta)*, bullhead *(Cottus gobio)* and Orb-web spiders *(Meta merianae* and *Araneus cornutus)* (Coulson, 1959; Crisp, 1963; Cherrett, 1964; Houston, 1973).

The importance of the mosaic of habitats to the survival of a species, emphasised by Cragg (1961), is also shown by two invertebrates. In 1955 *Tipula subnodicornis* larvae showed high mortality through drought in many of the peat sites but 50% of the 4th instar larvae survived in the wet flushes (Chap. 4, Table 11). Similarly exhaustion of the available food supply may cause the rush moth *Coleophora alticolella* to become locally extinct on its primary host *J. squarrosus* but recolonisation takes place from low density populations surviving on a subsidiary host, *J. effusus* (Reay, 1964). In both cases high survival in sub-optimal habitats at times of stress shows that habitat diversity is necessary for exceptional as well as normal circumstances.

Nutrient transfer between habitats through predation is very difficult to quantify. However, prey consumption by Meadow pipits and spiders is estimated at 0.07 and 0.50 g m^{-2} yr^{-1} respectively based on estimated annual respiration and production (Chap. 4) and an assimilation efficiency, including rejecta, of 0.6. Maximum estimates of transfer to the bog would assume that all prey is derived from grassland and stream habitats, and that no subsequent transfer occurs e.g. by emigration of the predator. Despite the high nutrient concentrations in prey, the estimates are low and of the same order as the estimated transfer by herbivores (Table 5).

The amount of other nutrient transfers through animal dispersal is unknown, apart from estimates of invertebrate transfer from terrestrial habitats to streams and of removal of sheep from the catchment (Table 5). Crisp (1966) and Nelson (1965) estimated that in an 83 ha catchment about 50 kg insects yr^{-1} were transferred from terrestrial habitats to the stream surface, equivalent to a loss of about 6 mg m^{-2} yr^{-1} from the area as a whole. As most of the material was, however, probably derived from the grassland comprising the stream margins, i.e. about 10% of the catchment, the loss would be 60 mg m^{-2} yr^{-1}. The range of estimated loss (6–60 mg m^{-2} yr^{-1}) is very similar to that for transfers by herbivores and predators. The net loss of nutrients from the catchment by sale of sheep and wool is *c* 0.5–5.0 mg m^{-2} yr^{-1} (Crisp, 1966), although estimates by Rawes and Welch (1969) for the whole Reserve are an order of magnitude higher.

The transfers of nutrients between ecosystems by fauna are small compared with transport by water. The high rainfall and high water table on the blanket bog cause surface run-off which may pass over and through the alluvial grasslands as well as along drainage channels into the streams. Estimates of nutrient movement by water between terrestrial habitats are not available but the inputs in precipitation (Gore, 1968), losses in burning (Allen, 1964; Allen et al., 1969; Evans and Allen, 1971) and output via streams (Crisp, 1966) provide a balance sheet for a Moor House catchment (Table 5). The incorporation of tentative estimates for nitrogen fixation, burning and transfers between the terrestrial habitats extends the results presented by Crisp (1966) but does little to alter his general conclusions.

The large nitrogen losses in peat, mostly from the small area of eroding bog, are probably unimportant in the nutrient balance and functioning of the develop-

Table 6. Average summer numbers and biomass of sheep on different vegetation types and estimated annual production of green food (data from Rawes and Welch, 1969)

	Sheep		Green food production (kg dw ha^{-1})
	Numbers (ha^{-1})	Biomass (kg dw ha^{-1})	
Calluneto-Eriophoretum	0.13	2.6	2,500
Trichophoretum	0.14	2.8	1,800
Erosion and recolonisation complexes	0.21	4.2	—
Juncetum squarrosi	1.30	26.0	3,000
Nardetum	2.00	40.0	1,500
Sphagneto-Juncetum effusi	2.20	44.0	1,500
Mines, tracks	5.00	100.0	1,000
Flushed gleys	6.00	120.0	2,000
Agrosto-Festucetum	7.40	148.0	2,000

Biomass estimates assume 50 kg live weight per sheep, and 60% water content [water content of a ewe is 50–70% over the annual cycle of production (D. Sykes, pers. comm.)].

ing and mature blanket bog. The losses in burning, calculated on an annual basis assuming a 15 year burning rotation, are probably maximal because some nitrogen in smoke is redeposited within about 100 m downwind of the burn. Considerable variation in the pattern of loss and redeposition can be expected through variation in the temperature of the burn and in the wind speed and direction (Evans and Allen, 1971). Fixation of atmospheric nitrogen by microorganisms has been measured for a number of sites and horizons (Chap. 6). The range of tentative annual estimates for five bog sites (Table 5) indicates that fixation could balance losses from streams.

Large amounts of nutrients are removed in stream water and peat, Crisp (1966) estimating a loss of *c* 93,000 kg of peat from an 83 ha catchment, but some of the suspended peat is redeposited on the alluvial soils of the streamside. Rawes (unpubl.) observed an increase of 115 mm in soil depth over a period of nine years on a streamside alluvium, but because these areas constitute only a small proportion of a catchment (Table 1 and Fig. 1) probably less than 1% of the peat is redeposited.

The large output of potassium and calcium in stream water results from solution from the rock and mineral soil within and adjacent to the streams (Crisp, 1966). With the large capital of these elements in the mineral soils, the ecology of the grasslands is unlikely to be affected by the losses, but losses from blanket bog are important because of the small capital, especially of potassium, and the large percentage (38%) which is utilised by the organisms (Heal and Perkins, 1976). The nutrient budgets (Table 5) lack information on between-year and between-site variation but show the order of magnitude of the transfers. To detect a net loss or gain in nutrients by the catchment requires a major research effort to achieve the necessary accuracy in a large number of variables.

It is apparent that nutrient transfers between habitats, e.g. blanket bog and grasslands, are small compared with transfers within a habitat, or into and out of the whole catchment. Two of the main attributes of an ecosystem (Odum, 1971) are that it is an interacting complex of components—plants, animals, microorganisms and the abiotic environment—forming a relatively discrete system, and that most of the primary production is produced and recycled within the system through herbivory, decomposition and mineralisation.

The criteria of interaction and circulation, applied to Moor House, indicate two levels of ecosystem. First, the plant–soil–animal system, i.e. the blanket bog or alluvial or limestone grassland distinguished by the structure of the plant community and soil, and characterised by a well-developed internal circulation of matter. Second, the catchment whose recognisable boundaries of structure and functioning allow its analysis as a discrete system, and within which material transfers occur, especially between the terrestrial and aquatic components.

10.4 Sheep, Grouse, Man, and Time

Since man is currently involved in the mosaic of terrestrial habitats mainly through sheep farming and grouse shooting, we will examine the productivity of these animals, the extent of their dependence and influence upon the composition of the moorland, and the changes induced by management.

10.4.1 Sheep and Grouse

Animal productivity is closely linked to the nutritional quality rather than quantity of the primary production. At Moor House the high fibre content and low digestibility of much of the vegetation inevitably result in low rates of utilisation. Hunter (1962) showed a clear relationship between the fibre content of vegetation and sheep grazing intensity, whilst Arnold (1964) demonstrated that small changes in concentration of chemical constituents (P and K) can influence selection of herbage by sheep. Grouse, 90% of whose food is *Calluna*, select for P in autumn, winter and spring and Ca and soluble carbohydrates in winter and spring, selection being greatest when the mineral concentration is lowest (Moss, 1972).

The numbers and biomass of sheep on the blanket bog are low (Table 6), there being more than 50 times as many sheep on the *Agrosto-Festucetum* although green food production is only 60% of that on the bog.

A 50 kg (live weight) hill ewe eats about 500 kg dw yr^{-1}, with an average digestibility of 65% (Eadie, 1970). In trials at Moor House sheep fed on *Calluna* shoots and *Eriophorum* leaves, together and separately, maintained body weight but the apparent digestibility was no more than 45 and 59% respectively and animal health deteriorated over the ten day feeding periods (Rawes and Williams, 1973). In field trials the standing crop of blanket bog vegetation was not altered significantly when grazing was prevented, despite selective grazing of flowers and leaves of *Eriophorum vaginatum* and *Rubus chamaemorus* which have relatively high mineral contents at times in the year when they are grazed. However, Taylor and Marks (1971) showed that even very low grazing pressures significantly reduce aerial production and fruit yield of *R. chamaemorus*, and the ecological

Table 7. The numbers of adult grouse in spring and late summer, and the adult to young bird ratio

	Spring		Late summer		Adult : young
	Number ha^{-1}	Number of counts	Number ha^{-1}	Number of counts	
1965	0.24[a]	6[b]	0.55[d]	9	1:1.3
1966	0.33 ± 0.33	9[b]	0.99[d]	9	1:2.0
1967	0.17 ± 0.26	9[b]	0.36 ± 0.23	15	1:1.1
1971	1.14 ± 0.41	—[c]	4.90[d]	33	1:3.3
1972	1.85 ± 1.08	—[c]	5.55[d]	42	1:2.0

[a] Only one area surveyed, 3–5 areas surveyed in other years.
[b] By flushing birds with dogs.
[c] By recording nests.
[d] Derived from sample counts of adult: young bird ratio giving a maximum estimate.

effects of as few as 0.13 sheep ha^{-1} may be considerable despite the apparently negligible effect on the total dry weight production.

Sheep graze much of the above-ground vegetation produced on the better grasslands. Between 47 and 110 g m^{-2} of the above-ground vegetation of the *Agrosto-Festucetum* is consumed, a mean of 60% of the annual production, compared with 35% on *J. squarrosi*, with an annual production of 340 g m^{-2}. The frequency of sheep on the *Juncetum* was low; a case of intensive grazing of relatively large production over short periods (Rawes and Welch, 1969). These are high figures for utilisation compared with Eadie's (1967) estimate that year-round stocking of a grassy hill by sheep resulted in only 30% consumption of the 200 g dw m^{-2} production. However, grazing on the moorland is restricted to April–October and is heavily concentrated on the small preferred grasslands. On the whole of the east side of the Reserve, less than 5% (0.25×10^6 kg) of the above-ground primary production (5×10^6 kg) is consumed by 1000 sheep, each animal consuming 1.37 kg of herbage daily (Rawes and Welch, 1969).

Whilst accurate estimates of grouse production are prevented by limited population data at Moor House (Taylor and Rawes, 1974), Table 7 illustrates the expected range. Breeding birds were estimated by counting birds or nests in 10–30 ha sampling areas, but total summer counts are difficult on this ground and the more accurate measure of the adult to young ratio is emphasised. The estimate of bird numbers unit $area^{-1}$ in late summer is therefore maximal, assuming survival of all breeding birds. Numbers were exceptionally high in 1971 and 1972 and those in 1965 and 1966 probably approach the more usual situation.

Density of birds also varies over the moor, for example in 1972 breeding density ranged from 1.48 to 2.35 birds ha^{-1}. In 1971 and 1972 average clutch size for the Reserve was 8.2 ± 0.33 and 8.9 ± 0.31 respectively, with between-site variation in 1972 of 7.1 ± 0.38 to 9.5 ± 0.33. Hatching success was 93.3 and 94.1% in the two years, and chick mortality on the nest 1.99 and 7.5% with a total chick mortality after hatching possibly up to 20%. A day old chick averages 18 g lw, increasing in the first 23 days by about 86 g. Growth rate then appears to increase and most young birds reach adult weight (600 g) in August. Body weight is maintained until the end of the year after which there is a loss so that the breeding birds in April weigh about 500 g.

From the population and growth data production, estimated from Allen graphs, was 0.21 kg lw ha^{-1} in 1967 (biomass 0.20 kg ha^{-1} on 31 August) and 2.7 kg ha^{-1} in 1971 (biomass 2.14 kg). This compares with the live production of sheep on the Reserve at 22.8 kg ha^{-1} (Rawes and Welch, 1969).

Because the numbers of grouse, unlike sheep, vary between years, their impact on the blanket bog differs markedly, but for both animals there is always an apparent large annual surplus of food. Even when numbers are high, as in 1972, grouse consume only 8.4 g or 6% of the annual production of *c* 150 g m^{-2} of *Calluna* green shoots (Forrest, 1971; Forrest and Smith, 1975).

10.4.2 The Effects of Management

The importance of past human activity in the development of hill land habitats is often overlooked because the vegetation appears natural. However, the major land uses in the northern Pennines—lead mining, settlement, sheep grazing and grouse shooting—have each, in isolation and in combination, markedly affected the landscape.

The influence of lead mining, at its height in the 18th and 19th centuries, is varied. Tracks and pathways were made, some across the bog, and buildings, now mostly derelict, were erected. Vast heaps of limestone, sandstone and shale were brought to the surface, and their colonisation by plants has sometimes been slow because of the high concentrations of heavy metals. Smelting has left small areas, covered by toxic waste (soil samples from one enclosure showed a Zn level of 4.3%), uncolonised by vegetation for over 100 years. Small reservoirs formed by diverting and damming streams and by draining the bog through a network of channels (leats in Fig. 1), were subsequently breached to wash away surface material and expose mineral bearing rock, or to transport rock into the valley bottom. Mine shafts and levels, acting as drains, further dried the bog, although drainage channels became blocked within a few years. In time the old tracks and mine workings became colonised, adding to the diversity of the moor. Species such as *Minuartia verna*, *Thlaspi alpestre* and *Gentianella amarella* are found, the first two specific to these habitats, which also have a very rich rock lichen flora. Tracks across the bog, often made by teams of ore-carrying ponies, have become vegetated by a sward dominated by *J. squarrosus*, which is perpetuated by sheep grazing and treading.

Settlement, usually on drier habitats, has involved attempts at management, including small enclosures. At the Moor House Field Station, on a limestone outcrop, a one hectare area was enclosed with a stone wall and has been cropped annually for at least 100 years. As a result of this, and of feeding cattle, ponies and sheep in the hectare, a meadow flora has developed containing some plant species not otherwise found on the Reserve.

Sheep have grazed the northern Pennines for 1000 years and the pattern of grazing has changed little in the past 100 years although numbers have steadily increased and shepherding (control) has become less frequent (Rawes and Welch,

Table 8. The effect of sheep grazing on botanical composition of blanket bog expressed as percentage cover (data from Welch and Rawes, 1966)

	Year-round grazing	Summer grazing
Species stimulated by grazing		
Carex nigra	0.5	—
Deschampsia flexuosa	5.0	—
Eriophorum vaginatum	45.3	22.5
Festuca ovina	4.1	—
Juncus squarrosus	5.9	—
Luzula campestris	0.1	—
Species reduced by grazing		
Calluna vulgaris	10.5	59.5
Erica tetralix	—	1.7
Species indifferent to the treatment: *Drosera rotundifolia, Empetrum nigrum, Eriophorum angustifolium, Listera cordata, Narthecium ossifragum, Rubus chamaemorus, Scirpus cespitosus, Vaccinium myrtillus, V. oxycoccus*		
Total number of species	48	42

1969). The age structure of the flocks has altered, and only ewes, lambs and gimmer hoggs (the previous year's female lambs) are now turned on to the fell. The absence of wethers (castrated males) which utilised poorer vegetation, such as *Nardetum*, and exposed habitats, may have resulted in an increase in *Nardus* (Roberts, 1959).

Long-continued sheep grazing has resulted in an increase in the amount of grassland at the expense of the bog, and prevented the introduction or survival of plant species, such as trees and shrubs, that cannot withstand cropping. Grazing on grassland encourages the growth of gramineous and rosette species, promotes tillering and shoot production, but reduces flowering. Large numbers of herbaceous species may occur under sheep grazing, but there is growing evidence of an increase in unpalatable species, such as *Nardus stricta*, *J. squarrosus*, and *Cirsium* spp, which flourish as a result of reduced competition for light (Rawes, unpubl.).

Prevention of sheep grazing by exclosure of grasslands results in increased primary production, Rawes and Welch (1969) recording a 15–20% increase in growth of an *Agrosto-Festucetum* after one year and a doubling of annual production after 15 years. Although the standing crop of the sward was 50% higher after seven years exclosure (Welch and Rawes, 1964) the number of plant species fell from 93 to 67. Those plants which, under grazing, had failed to reach their potential in growth, e.g. *Deschampsia cespitosa*, *Deschampsia flexuosa*, and *Agrostis* spp, flourished, whilst light demanding species like *J. squarrosus*, *Cerastium holosteoides*, *Minuartia verna*, *Thymus drucei*, *Trifolium repens*, and *Viola* spp declined. Thus the absence of grazing caused a change in the structure of the sward, with an increase in the size of individuals, more flowering and a build up of slowly decaying litter. With other gramineous swards, such as *J. squarrosi* and *Narde-*

tum, similar changes occurred but, because fewer species are involved, no marked change in composition apart from the considerable decrease of *J. squarrosus*.

Grazing pressure on Moor House blanket bog is so low that exclosure has little noticeable effect, but comparison (Table 8) with a bog that had been grazed continuously by 0.5 sheep ha^{-1} for many years (Welch and Rawes, 1966) shows marked differences in composition. Increased grazing reduces *Calluna* and increases *E. vaginatum* cover and the total number of species, *J. squarrosus* and some grasses appearing. The standing crop of *Calluna* was 670 g m^{-2} under the summer grazing regime of Moor House compared with 80 g m^{-2} under the year-round long-term regime. Similar changes in the bog vegetation have been induced in a controlled grazing experiment over five years, with a mean stocking density of four sheep ha^{-1} (Rawes and Williams, 1973). These observations suggest that an increase in the present grazing pressure on the bog would produce a sward which is more suitable for sheep grazing, and an increased concentration of nutrients.

The direct relationship between carrying capacity and nutrient concentration of the sward suggests the use of fertilisers as a means of sward improvement. Fertilising has dramatic effects on the botanical composition, production, and utilisation of some gramineous swards (Rawes, 1963; Jones, 1967) but, unless accompanied by surface treatment—drainage and cropping—has little effect on the acid *Nardetum* and blanket bog vegetations. Agronomic treatments other than stock control are probably uneconomic in areas like Moor House.

Miller and Watson (Chap. 13) have shown grouse production to be directly related to nutrient concentration in *Calluna* shoots, and unrelated to dry matter production per unit area of moor. The recommended management technique to achieve the main habitat requirements for grouse, viz *Calluna* of good nutritional quality and sufficient cover and vantage points for nesting and territorial behaviour, is to burn numerous small strips regularly so that ample young heather, up to seven years in age, is available. However, Rawes and Williams (1973) suggest that this may not be the best treatment for wet blanket bog and their view is reinforced by the findings of Grace (unpubl.) and Forrest (1971) who have shown that the average age of *Calluna* in unburnt blanket bog is seven to eight years, with a range from one to 34 years. Rawes and Williams (1973) point out the deleterious effects of burning on the bog surface especially where even minimal sheep grazing follows the burn. In trials, better *Calluna* shoot production followed light grazing by sheep alone (Fig. 6) and possibly the moor as a whole would be improved for grouse if sheep numbers on the bog were increased rather than burning continued. Burning might have an advantageous short-term fertilising effect in the addition of plant ash, but if burning proved a necessary adjunct to improving grouse productivity it is unlikely to be compatible with sheep grazing, even of minimal pressure.

Fridrikson (1972), in his examination of the ecological deterioration in the vegetation and soils of Iceland since settlement, suggests a direct relationship between the herbivore and human populations. The basic sustenance per man per year can be provided by six sheep and half a cow (6 ewes = 1 cow), equivalent, in terms of hay, to primary production of 6300 kg yr^{-1}. Under the present system of sheep farming the animal production from 2500 ha of Moor House could provide sustenance for about 50 people.

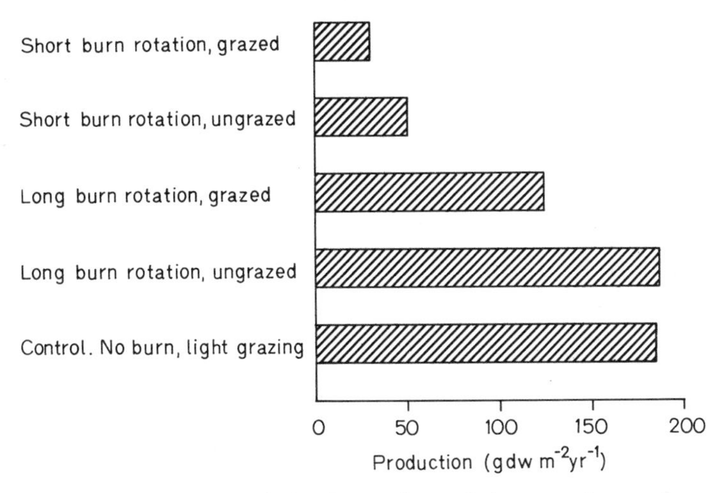

Fig. 6. The effect of experimental variation in burning and sheep grazing on the production of shoots of *Calluna*. Details of experiments in Rawes and Williams (1973)

This could be a standard against which other uses could be measured. In the past the options have been for combined use for mining, sheep farming and grouse shooting, but there are now greater pressures on hill land for recreation other than shooting, and the importance of water and nature conservation is more widely recognised, as is the suitability of the low fell land for afforestation. However it is used, low rates of exploitation with small input are both economically and rationally sensible, the low levels of secondary production and the fragility of the system safeguarding a nutritionally and species-poor resource.

Acknowledgements. We are very grateful to Dr. D.T.Crisp for providing advice and information on the freshwater fauna.

II. Supporting Studies – Dwarf Shrub Communities

11. The Productivity of a Calluna Heathland in Southern England

S. B. Chapman and N. R. Webb

In contrast with the main site studies in upland Britain, the primary production, litter production and accumulation of dwarf shrub heaths in lowland England is described; burning of the heath provides an aged series of sites which are used to determine rates of production and accumulation and to analyse the contribution of oribatid mites.

11.1 Introduction

An analysis of processes such as production and decomposition is important in attempting to understand the function of an ecosystem. Depending upon the characteristics of a particular ecosystem certain aspects of production studies can assume even greater importance. Lowland heathland is a type of vegetation occurring upon soils where minerals capable of weathering and releasing plant nutrients are present in only small quantities. In a case such as this, where losses of nutrients can be considerable due to leaching and burning, it becomes particularly important to consider the overall nutrient budget of the ecosystem and to quantify the inputs and losses of nutrients to and from the system.

A general nutrient budget for heathland in south eastern Dorset has been described by Chapman (1967). Nutrient or energy budgets can only be calculated when suitable production, decomposition and accumulation data are available. While the published nutrient budget was based upon earlier and less complete information than is now available the results reviewed in this present paper do not materially alter the conclusions that were made but allow more detailed considerations of the heathland ecosystem and more valid comparisons with dwarf shrub communities elsewhere in the British Isles (Barclay-Estrup, 1970; Forrest, 1971; Chaps. 2, 12, and 13) and in other parts of the world (Groves and Specht, 1965; Egunjobi, 1971; Burrows, 1972).

Some of the results in this paper are given only in general terms and details are in Chapman et al. (1975a, b). In some cases the results represent work that is still in progress and far from complete; these will be published later.

The term heathland is taken to mean an area where *Calluna vulgaris*, or some other member of the Ericaceae, is an important constituent of the vegetation. It is difficult to say precisely what are the differences between the terms heathland and moorland but, while they both refer to dwarf shrub communities, the term heathland is generally used to describe a lowland type of vegetation growing upon a mineral soil, while the term moorland describes a similar type of vegetation

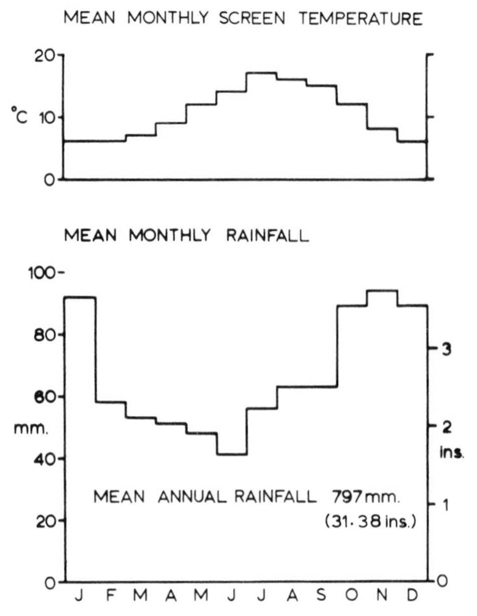

Fig. 1. Mean monthly screen temperatures and rainfall for the south east Dorset heathland area

growing upon an organic soil in upland Britain. Lowland heathland is used as a general term for a series of plant communities that range from dry heathland to valley bog.

Dry heathland is found in lowland Britain where free draining acid soils occur and where some factor such as grazing, burning or extreme exposure prevents the establishment and growth of scrub and trees. The principal areas of heathland in lowland Britain are found on the Tertiary deposits of the Hampshire and London Basins, on a number of the Cretaceous deposits of south eastern England, the Breckland, the east Suffolk Sandlings, the east Devon Commons, the Lizard and Land's End peninsulas. Other areas such as Dartmoor, Exmoor and the Shropshire Hills provide areas of heathland that are transitional between lowland and upland in their character.

Although the number of plant species associated with dry heathland is small, several of the important constituents of the vegetation show restricted distributions within the British Isles. Such species include *Ulex minor*, *Ulex gallii*, and *Agrostis setacea*. As a result of these distribution patterns a range of types of dry heathland vegetation exists across the country.

The heathlands described in this paper occur on soils derived from deposits of Bagshot Sand within the Poole basin of south eastern Dorset, and lie between zero and about 60 m above sea level. The soil profiles under the areas of dry heath in this area are mostly well-developed humus-iron podzols. The mean monthly temperatures and precipitation are summarised in Figure 1.

The vegetation of these heathlands has been described by Chapman (1967) in a paper outlining the overall nutrient budgets of the ecosystem, but it can be summarised as being of a type dominated by *Calluna vulgaris* in association with *Erica cinerea*, *Ulex minor*, *Agrostis setacea*, and *Molinia caerulea*. The vegetation and part of the litter layer are burnt periodically and the above-ground vegetation is regenerated mainly from existing rootstocks, even in relatively old and degenerate stands of heather.

These heathlands, which are not now grazed by vertebrates to any significant extent, support a rich invertebrate fauna consisting of a mixture of continental and oceanic species whose distributions overlap in the area. Comparisons of this fauna with that of the IBP moorland site at Moor House have been made in Chapter 4. Many species of Odonata, Orthoptera, Lepidoptera, and Hymenoptera occur on these heathlands and vertebrate species such as the Dartford warbler (*Sylvia undata*), the Sand lizard (*Lacerta agilis*) and Smooth snake (*Coronella austriaca*) are now almost entirely restricted to this habitat in southern Britain. Extensive ecological surveys of these heathlands were carried out from 1931 onward by Capt. C. Diver (reviewed in Merrett, 1971) and more recent studies on the ecology of ants, spiders and dragonflies have been carried out by Brian (1964), Merrett (1967, 1969) and Moore (1964) respectively. Moore (1962) has given a general account of the ecology and conservation problems of the heathland in this area.

11.2 Primary Production

Net primary production (P_n) can be estimated in two ways (Newbould, 1967):

1. Estimates can be obtained by adding the increment in the standing crop (ΔS) to litter production (L) and to any losses by grazing (G)

$$P_n = \Delta S + L + G .$$

2. An alternative method is to add together a series of direct estimates of production by different components of the vegetation

$$P_n = P_{\text{green}} + P_{\text{wood}} + P_{\text{flowers}} + P_{\text{root}} \quad \text{etc.}$$

The growth and production by *Calluna* in the heathland vegetation is summarised in Figure 2, where primary production is represented by the sum of the compartments on the upper line of the diagram.

The pattern of production throughout a vegetation succession can be studied from a series of experimental plots or from a series of existing stands of vegetation of known age. The results described in this paper have been obtained from a combination of these two methods. Experimental plots have been used to obtain information from areas up to six years of age, and existing areas of heathland have been aged, to provide data for older stands of heather. Details of these sites have been given by Chapman (1967).

The standing crop of the above-ground vegetation and the accumulated litter and standing dead was measured by a direct harvesting technique with at least ten random quadrats (0.5 × 0.5 m) sampled for each determination. Each site was

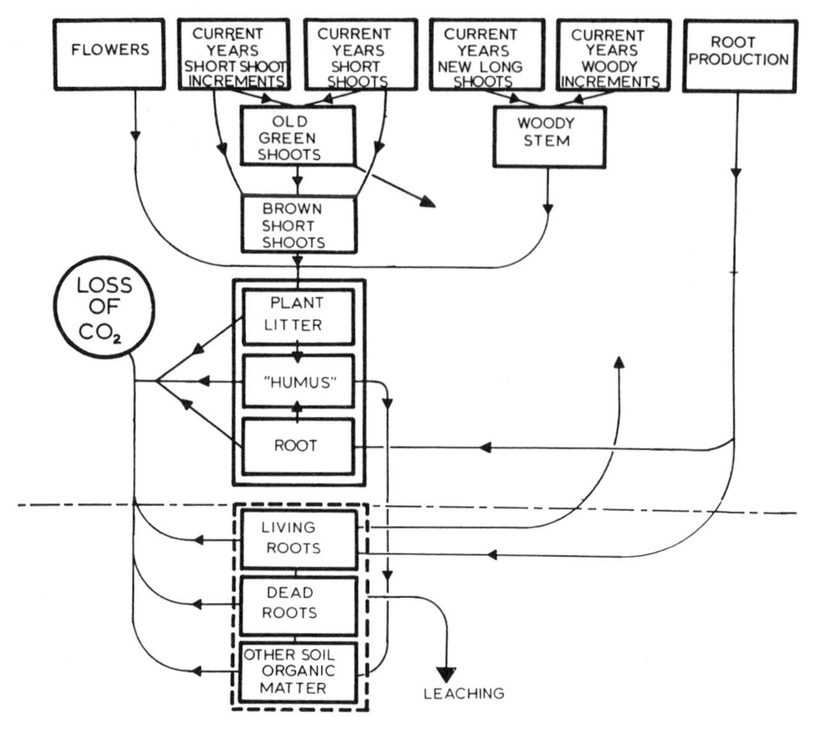

Fig. 2. Diagram showing the relationships between production of different components of *Calluna vulgaris*, the transfer of organic matter and the accumulation of litter and soil organic matter

sampled in alternate years at the end of the growing season over a period of eight years. The total above-ground standing crop from different-aged stands of heathland is shown in Figure 3, the curve for *Calluna* being similar (Chapman et al., 1975a), with a maximum of 2000 g m^{-2} represented by the equation

$$log_e Y = 7.6179 - 2.2685\,(0.8670^{x-3})\,.$$

The fitted growth curve in Figure 3 is a Gompertz curve having the general form,

$$\log_e Y = C + Ab^x$$

where Y = weight of the standing crop in g m^{-2}
x = age of the vegetation in years
A, b, and C are constants.

The use of these fitted growth curves does not imply any particular biological significance in the values of the derived constants, but their use summarises and describes the data in a manner convenient for subsequent examination and discussion.

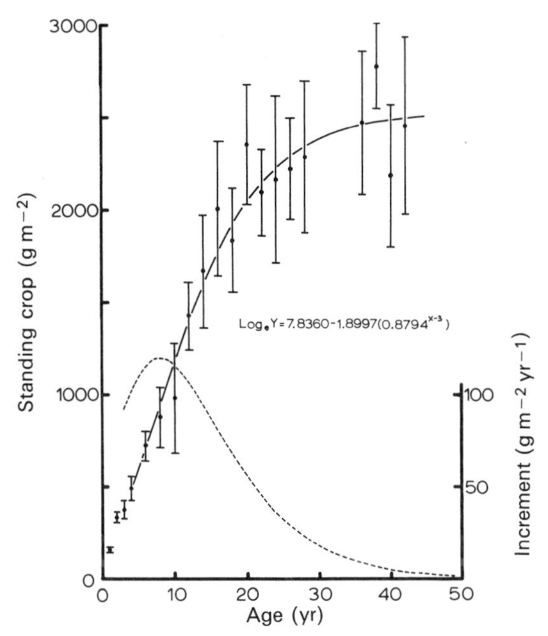

Fig. 3. Standing crop (——) and rate of increment (---) of the total heathland vegetation in relation to the age of the vegetation since the last burn (with 95% confidence limits)

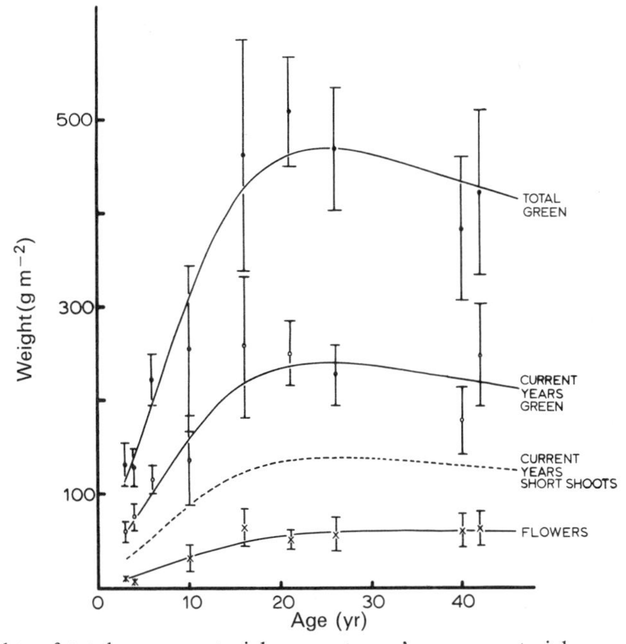

Fig. 4. Weights of total green material, current year's green material, current year's short shoots and flowers of *Calluna* in relation to the age of the heathland since the last burn (with 95% confidence limits). The individual points for current year's short shoots have been omitted for clarity

The growth curves have been fitted to the data from stands of an age of three years onwards as there is a marked departure from the Gompertz growth curve in the first three years that follow a fire. This effect can be partly explained by the pattern of shoot and litter production (Chapman et al., 1975a). The litter that has accumulated on the soil surface is not all burnt during a heathland fire, but most of that which remains is removed by wind action over the next few years (Sect. 11.3). For these reasons it is convenient to fit the growth and accumulation curves from the end of the "post burn" period. The rate of increment of the standing crop of total vegetation is plotted in Figure 3 and is derived from the differential of the fitted growth curve.

The weights of the standing crop of a series of components of *Calluna* obtained from field quadrat data are shown in Figure 4. The curves have been derived from a combination of the Gompertz growth curve for the standing crop of *Calluna*, and power regressions relating the proportion of a particular component of *Calluna* to the age of the vegetation

$$Y = aX^b$$

where Y = proportion of some component of *Calluna*
X = age of vegetation in years
a and b are constants.

A recurrent problem in the calculation of net primary production is the estimation of the weight of the wood increment. In this work the growth curve fitted to the standing crop of woody material on heathland of different ages was

$$\log_e Y = 7.1331 - 4.4598\ (0.8634^x)$$

and the rate of increment was estimated from the differential, showing a maximum of nearly 70 g m^{-2} yr^{-1} at ten years (Chapman et al., 1975a). This method of estimating wood production will produce underestimates if significant amounts of wood are lost to the litter layer. Observations suggest that in the present case such losses are negligible, at least for the younger stands of heather. It may be that significant losses of woody material to the litter occur in older areas of heathland and that the estimates of wood production from stands of 25–30 years of age onwards are too low.

To estimate net primary production by the first of the two methods described, litter production was measured by weighing the material collected in litter traps. These traps consisted of plastic dishes 16.5 cm in diameter, 5 cm deep and with drainage holes in the base (Chapman, 1967). The results from 20 such traps at each of the five different sampling sites over a four-year period showed a rise from the nine and ten year old sites (40 and 70 g m^{-2} yr^{-1}) to about 180 g at 15–18 years and between 200 and 250 g at 20–40 years (Chapman et al., 1975a). The results obtained from the nine- and ten-year old sites are underestimates, due to unavoidable over-exposure of the traps. This was indicated by the lower numbers of seed capsules collected in the traps compared with the numbers of flowers that were present per unit area of the heather canopy.

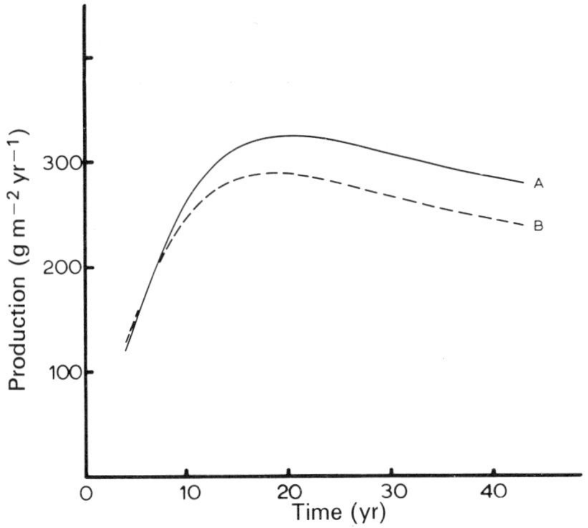

Fig. 5. Estimates of net aerial production by *Calluna* in relation to the age of the heathland since the last burn. *A*: Current year's growth + wood increment + flowers. *B*: Litter + standing crop increment

An independent estimate of litter production by *Calluna* has been obtained by adding flower production to the difference between the total weight of green material present in one year (Σ_{gx-1}) and the amount remaining as old, short shoots at the end of the following year (OSS_x). Before shoot material falls from the plant as litter it loses weight by reabsorption. The magnitude of this loss of weight was estimated to be 20% by comparing the weights of standard lengths of green and brown shoot, assuming that little or no change in length accompanies the change in colour. This result is very similar to those reported by Bray and Gorham (1964) and by Burrows (1972) from different plant species. To correct for this 20% loss of weight in the indirect estimation of litter production the differences between the weights of total green and old short shoots have been multiplied by a factor of 0.8. The estimates of litter production obtained by this method agree well with the results obtained from the litter traps and have been used in the calculation of net primary production.

The estimates of net aerial production of heathland obtained by the two different methods described earlier show the same pattern of change with increasing age, but over most of the age range estimate B is in the order of 12–13% lower than estimate A (Fig. 5). The estimates of litter production used to calculate B assumed no contribution from woody material and may, therefore, be low, especially in older stands, but it has also been shown that the estimates of wood increment used to calculate A may be low because losses to litter have been neglected. If significant, either of these factors would lead to an underestimation of net production. Even more speculative is a consideration of the effects of invertebrate grazing upon these two estimates of production. It has been assumed that such grazing is absent or at the most present only at a very low level; if this is not in fact the case, then it can be argued that grazing would produce a more serious error in an estimate of type B because grazing will have had its full effect whereas in A some parts of the production have been measured before the plant material has been subjected to a full season of grazing.

Plant species other than *Calluna* have been shown to contribute amounts to the net aerial primary production that decrease from about 40 g m^{-2} yr^{-1} at an age of three years to about 15 g m^{-2} yr^{-1} after 40 years. Between 10 and 20 g of dry matter are contributed to the litter layer by plant species other than *Calluna* each year. The total net above-ground production by the heathland vegetation is in the order of 320 g m^{-2} yr^{-1} for stands of age 15–40 years and somewhat lower in younger stands.

At present it is not possible to provide more than an approximation of root production based upon data obtained from a number of different approaches to the problem. The dune-heath system at Studland in Dorset (Diver, 1933; Wilson, 1960) provides a series of dune ridges covered with heathland communities of different ages. The heathland ridges at Studland are approximately 50, 150 and 250 years of age. The roots have been extracted from these soils and when their weights are combined with the additional root biomass data given by Chapman (1970) for older heathland sites, they can be examined against a monomolecular accumulation curve as used by Jenny et al. (1949) and by Olson (1963).

$$Y = A(1 - e^{-kt})$$

where Y = biomass of roots present at time (t)
$A = I/k$ = biomass of roots present at steady state
I = net annual root production
t = age of site in years
k = fractional loss rate.

When the weights of root present in the soils of this series of sites are plotted graphically against age the points are contained between a pair of monomolecular accumulation curves having inputs of 100 and 25 g m^{-2} yr^{-1} and steady state values of 5000 and 8500 g m^{-2} respectively. The use of accumulation curves in this way assumes constant rates of input and decay, and Minderman (1968) has discussed the shortcomings and arguments against such an assumption of a constant decay rate (k). If a simple calculation is made upon synthetic data of the type used by Minderman it can be shown that the assumption of a constant decay rate may well lead to underestimation of the magnitude of the true input to the system.

The evolution of CO_2 from the soil, excluding the litter layer, has been measured by means of conductimetric soil respirometers (Chapman, 1971) using 0·3N KOH as the electrolyte. The individual determinations of soil respiration (R) were made over periods of 24 h and related to mean daily soil temperature (T).

$$\log_e R = 0.084\,T - 2.075 \;(r = 0.82,\; P < 0.001)$$

$$Q_{10} = 2.32$$

This relationship has been used in conjunction with recordings of mean daily soil temperature to calculate the annual evolution of CO_2 from the soil as being about 3630 g CO_2 m^{-2} yr^{-1}, equivalent to 900 g carbon m^{-2} yr^{-1}.

Determinations of soil respiration have been carried out on the series of dune-heath soils and ploughed firebreaks of different ages. These soils provide a series of different combinations of weight of root and soil organic matter. A multiple regression of soil respiration upon the different components of soil organic matter and roots suggests that live roots produce 72% of the carbon dioxide evolved from these heathland soils. This would mean that from about 200 g carbon $m^{-2} yr^{-1}$, or the equivalent of 400 g $m^{-2} yr^{-1}$ of organic matter, are being lost from the root zone of the soil by decomposition. If the root zones of older heathland areas are in a steady state (Chapman, 1970) then the inputs of organic matter must be balanced by the losses. At present there is insufficient information regarding the transfer of organic matter from the litter layer into the root zone, or about the rate of movement of organic matter down the soil profile to complete a budget for the organic matter in the root zone, but the indications are of an input of organic matter to the root zone in the order of 400 g $m^{-2} yr^{-1}$.

Preliminary measurements of the rate of respiration of excised *Calluna* roots are of the same order as those reported by Crapo and Coleman (1972) and by Newton (1923). If these rates of respiration are combined with the standing crop of roots they more than account for the quantities of carbon dioxide evolved from the soil.

From these indirect and as yet rather incomplete sources of data it is suggested that the net production of root material on these heathlands is of the order of 400 g m^{-2} yr^{-1}. If accepted, this estimate suggests that the total net primary production of the heathland is in the order of 700 g m^{-2} yr^{-1}.

11.3 Litter Loss and Litter Accumulation

The weight of litter accumulated on the soil surface was measured when quadrats were harvested to estimate the above-ground standing crop. An appreciable amount of litter remains after a fire (about 500 g m^{-2} at 1 year) but it becomes dry and friable through exposure and is blown from the site by wind. Minimum weights of litter accumulation (about 200 g m^{-2}) are found four or five years after a fire, followed by an increase to approximately 3000 g m^{-2} after about 25 years, when accumulation appears to approach a steady state (Chapman et al., 1975b).

Litter accumulation is the result of the balance of litter production and loss. Samples of the entire litter layer were contained in mesh bags to determine the percentage of the litter layer that was lost each year in three different-aged stands of vegetation (Chapman, 1967).

To combine these rather limited data with those of litter production, it is necessary to make some assumptions about the relationship between the age of the heathland and the rate of litter loss. Two alternative assumptions have been made from the data in Table 1: firstly that the annual rate of litter loss remains constant at 10%, and secondly that the rate of loss increases linearly from 3% to 15% over the period from three to 40 years. When these two rate-loss relationships are combined with estimates of litter production the derived curves correspond closely with field data up to about 20 years (1500 g m^{-2}) but in later years

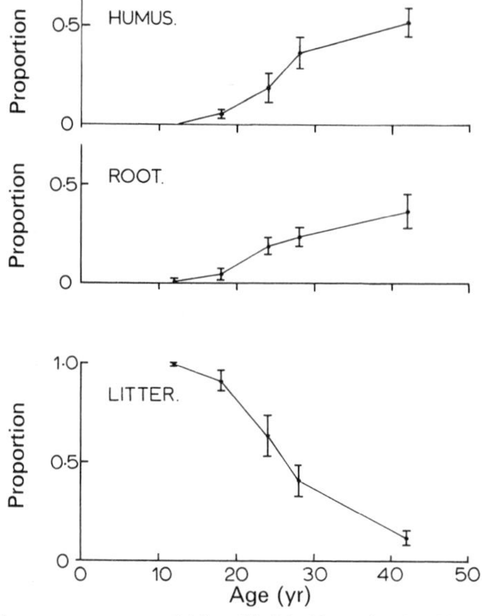

Fig. 6. Proportions of humus, roots and identifiable litter in the litter layer in relation to the age of the heathland since the last burn (with 95% confidence limits)

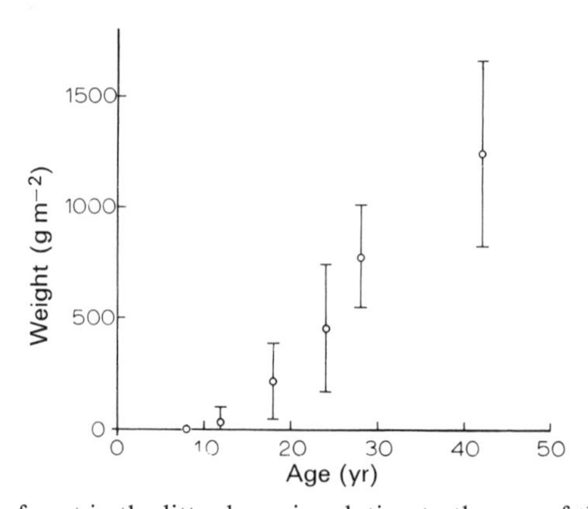

Fig. 7. The weights of root in the litter layer in relation to the age of the vegetation since the last burn (with 95% confidence limits)

Table 1. Estimates of the percentage of the litter layer lost each year from heathlands of three different ages (from Chapman, 1967)

Age (yr)	9	19	33
% litter layer lost yr^{-1} (with 95% confidence limits)	5.4 ±0.30	9.8 ±0.84	12.8 ± 2.19

the predicted weight of litter is much lower than observed values (Chapman et al., 1975b). However, the litter layer from heathlands older than about ten years of age contains significant amounts of root material, and by the time the vegetation is 40 years old the litter layer may contain up to 40% root material (Fig. 6).

This weight of root in the litter can explain the discrepancy between the observed and predicted levels of litter accumulation. The weights of root present in the litter are shown in Figure 7 and the amounts are about the same as the discrepancies between the predicted and actual levels of litter accumulation. Good agreement exists between the curve, based upon an increasing rate of litter loss, and the weight of root present, but as the roots have contributed to the "humus" component of the litter layer it may be that even higher rates of litter loss are required to explain the field data completely.

Additional litter bag experiments have been carried out over a longer period but invasion of the litter contained in the bags by roots has made it difficult to interpret the results.

11.4 Contribution of Oribatid Mites to Decomposition

Oribatid mites (Acari: Cryptostigmata) are a common feature of the fauna of most soil types, and are associated with decomposing plant material. They are thought to act as secondary decomposers, mainly feeding on microorganisms, although it has been suggested (Luxton, 1972) that some species digest plant tissue and thus may act as primary decomposers. The numbers of oribatids have been estimated in heathlands of different ages in Dorset and these data have been combined with laboratory measurements of respiratory metabolism to quantify the role of these mites in the turnover of material in the heathland ecosystem. The annual production of these populations was also calculated.

Up to 35 species of oribatid mites have been recorded from heathland soils in southern England (Webb, 1972) but only half of them occur regularly. The number of species and the size of their populations increases with the time since the heathland was last burnt.

During a heathland fire it is unlikely that many of the animals living in the soil and litter are killed; the insulating properties of the litter prevent lethal temperatures occurring at any appreciable depth in the profile. However, populations of soil animals decline rapidly following a fire because without plant cover the litter becomes much drier, and some of the upper litter layers are removed by wind.

Table 2. Some characteristics of the vegetation and fauna for different aged heathland sites in Dorset

Age (yr)	0.5	1	4	8	15	27
Max height of veg (cm)	7	10	13	15	42	53
Depth of litter (mm)	0	1	1	10	25	47
Total number of arthropod spp recorded	14	15	22	27	37	36
Total number of oribatid spp recorded	6	10	10	11	16	15

The weight of accumulated litter and its depth at the individual sampling sites (Table 2) show that during the first eight years hardly any litter accumulates; during this time the number of species of soil arthropods increases from 14 to 27 and that of oribatids from 6 to 11 (Table 2). Similar changes occur in the population density of oribatids which increases from 4360 to 15,760 m^{-2}. After eight years the canopy closes and the numbers of species and individuals begin to level off, a pattern which closely follows the growth of the vegetation and the accumulation of litter.

Population respiration has been calculated from the wide range of respiratory data now available for oribatid mites (Berthet, 1964; Zinkler, 1966; Webb, 1968, 1969; Webb and Elmes, 1972). Where data were not available for particular species, the rate of oxygen consumption has been estimated from a regression of weight on respiration derived from the published results. The relationship obtained was:

$$\log_e R = 0.710 \log_e W - 0.606$$

where R = oxygen consumption in $\mu l \times 10^{-3}\ h^{-1}$ and W = live weight in μg. The rates of oxygen consumption of nine of the fifteen most important species of oribatids are given in Table 3.

The mean annual population 25 cm^{-2} at each site is given in Table 4 together with the maximum and minimum numbers recorded. These latter figures are presented because some of the species, especially during the early stages of the succession, occurred in such small numbers that their distribution in the sample material was not normal. In these cases it is unsatisfactory to apply the standard tests of confidence and these extreme values can be taken to indicate the degree of variation associated with the estimates of population density.

Population respiration (Table 5) has been calculated by the combination of respiration rates from Table 3 with the estimates of population density from Table 4. Temperature corrections have been made by the application of Krogh's Curve (Krogh, 1914). The result for each species is expressed as ml of oxygen consumed $m^{-2}\ yr^{-1}$.

The mean annual production at each site (Table 5) has been calculated from the relationship of production on population respiration given by McNeill and

Table 3. The weight and oxygen consumption of soil Oribatei in heathland soils in Dorset

	Fresh weight per individual (μg)	Oxygen consumption per individual ($\mu l\ O_2 \times 10^{-3}\ hr^{-1}$ at 15° C)	
Steganacarus magnus (Nic.)	236	17.0	Webb and Elmes (1972)
Rhysotritia ardua (C. L. K.)	57	6.2	Berthet (1964)
Nothrus silvestris Nic.	52	12.2	Webb (1969)
Carabodes minusculus Berl.	27	5.7	Calculated
Tectocephus velatus (Mich.)	4.2	1.0	Berthet (1964)
Oppia spp	2.4	1.0	Berthet (1964)
Oribella castanea (Herm.)	1.6	0.8	Calculated
Chamobates schützi (Ouds.)	16	3.9	Calculated
Oribatula tibialis (Nic.)	10	2.4	Calculated

Table 4. The population density of the nine species of soil Oribatei from different aged heathland in Dorset for which population respiration has been calculated

Age (yr)	0			1			4			8			15			27		
	1	2	3	1	2	3	1	2	3	1	2	3	1	2	3	1	2	3
S. magnus		0			0			0			0		0	1.3	5	0	1.3	4
R. ardua	0	8.2	30	0	1.7	14	0	0.9	5		0		0	0.9	5	0	2.4	6
N. silvestris		0		0	0.1	1	0	0.3	4	0	4.7	40	0	13.5	43	8	22.3	63
C. minusculus	0	1.5	6	0	10.9	39	1	37.7	124	0	28.7	95	0	6.0	261	0	6.1	29
T. velatus	0	0.1	1	0	0.2	1	0	1.1	4	0	7.0	22	0	3.7	7	0	3.5	9
Oppia spp	0	0.1	1		0		0	1.2	9	0	1.9	6	0	9.2	38	6	14.1	24
O. castanea		0		0	0.2	2	0	0.3	2	0	1.0	5	0	0.4	2		0	
C. schützi	0	0.1	1	0	0.1	1	0	1.0	5	0	1.9	8	0	1.4	4	0	2.9	6
O. tibialis		0		0	0.1	1	0	0.5	3	0	0.1	1	0	0.5	2	0	0.3	2
Total 25 cm^{-2}		10.0			13.3			43.0			45.3			36.9			52.9	

Columns 1, 2, and 3 are the minimum, mean and maximum numbers 25 cm^{-2} respectively.

Table 5. The annual respiration, R (ml O_2 m^{-2}) and production, P (kJ m^{-2}) of soil Oribatei from different aged heathlands in Dorset

Age (yr)	0		1		4		8		15		27	
	R	P	R	P	R	P	R	P	R	P	R	P
S. magnus	0	0	0	0	0	0	0	0	52.8	1.09	45.7	0.96
R. ardua	112.9	2.05	21.8	0.54	11.5	0.29	1.3	0.04	15.4	0.38	30.8	0.71
N. silvestris	0	0	2.5	0.08	7.8	0.21	118.6	2.13	174.1	2.93	587.9	7.99
C. minusculus	17.6	0.46	127.7	2.26	445.0	6.36	265.9	4.18	305.8	4.64	71.5	1.42
T. velatus	0.2	0.04	0.4	0.04	2.3	0.08	14.3	0.33	7.7	0.21	7.2	0.21
Oppia spp	0.2	0.04	0	0	2.5	0.08	4.0	0.13	19.2	0.46	29.1	0.67
O. castanea	0	0	0.3	0.04	0.5	0.04	1.6	0.04	0.6	0.04	0	0
C. schützi	0.8	0.04	0.8	0.04	5.7	0.17	15.4	0.38	11.3	0.29	23.5	0.54
O. tibialis	4.2	0.13	1.4	0.04	7.0	0.21	1.4	0.04	7.0	0.21	4.2	0.13
Totals	135.9	2.66	154.9	3.00	482.3	7.44	422.4	7.28	593.9	10.25	799.9	12.63

Annual production calculated from McNeill and Lawton (1970).

Table 6. The estimated consumption of plant litter by soil Oribatei on different aged heathlands in Dorset

Age of heathland (yr)	0	1	4	8	15	27
Litter consumed (mg dw m^{-2} yr^{-1})	202	187	567	500	811	969

Lawton (1970) and the results are included to facilitate comparison with other localities. Population respiration increases from about 150 ml O_2 m^{-2} yr^{-1} in an area freshly burnt, to about 800 ml in a mature stand of heather. Assuming a respiratory quotient of 0.82 and that 1 mg of plant litter requires 0.85 ml of oxygen for its oxidation (Petrusewicz and Macfadyen, 1970), then the populations of oribatids on heathland respire the equivalent of less than 1% of the quantity of plant litter falling on the soil surface in one year (Table 6). These population respiration estimates for oribatids are, of course, minimal and would be increased two- or three-fold if the immature stages were included (Webb, 1970) and further increases would be obtained if other groups of soil animals, especially the Collembola, were included. Nevertheless, populations of heathland soil animals are small by comparison with those of other habitats and since there are no large decomposers it is concluded that the soil fauna contributes little to decomposition processes directly. The most important agents of decomposition are probably the microflora, the activity of which may be stimulated by the soil fauna.

11.5 The Phases of Heathland Development in Relation to Production and Nutrient Economy

The phases of development described by Watt (1955) in his studies of competition between *Calluna* and *Pteridium aquilinum* have been applied to the development of other *Calluna*-dominated plant communites by a number of workers. The characteristic features of the pioneer, building, mature and degenerate phases are usually described in physiognomic terms (Gimingham, 1960; Barclay-Estrup and Gimingham, 1969, and summarised by Gimingham, 1972). In some studies of heathland development these phases have been discussed with reference to changes in primary production, organic matter accumulation and nutrient turnover (Chapman, 1967; Barclay-Estrup, 1970).

It is important when discussing any data within the context of such seral or potentially cyclical situations that the scale of the investigation be appreciated. In the present and earlier work (Chapman, 1967) the heathland was examined on a relatively small scale basis, thus the results refer to mean values for large areas (ha). In the work of Barclay-Estrup (1970) the results were from a relatively large scale study and must be discussed in relation to cyclical changes taking place within smaller areas (m^2). These distinctions are important if confusion is to be avoided when results from apparently similar studies are compared.

The relationships between the production and accumulation data from the Dorset heathlands, and the phases of heathland development, are summarised in Figure 8. The pioneer phase, as generally defined, is not well represented upon

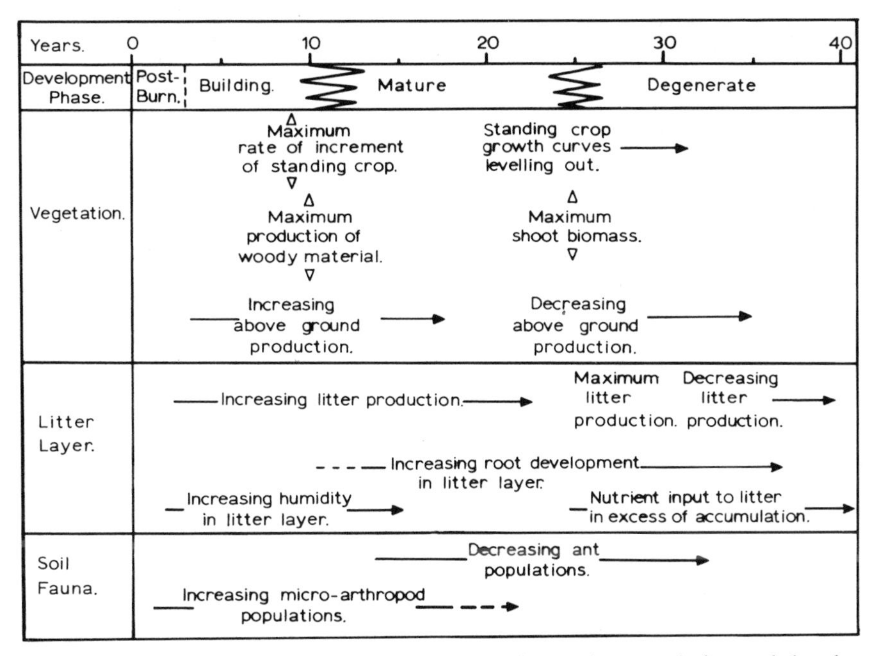

Fig. 8. A summary of the relationship between production and accumulation and the phases of development of Dorset heathlands

southern heathlands, where regeneration of the vegetation following a fire is mainly from existing rootstocks (Chapman, 1967). It has been shown in the discussion of the application of growth curves to the development of the aboveground vegetation that a "post burn" phase exists where there is an initial rapid increase in the standing crop, with a loss of any litter material that remains after a fire. It is suggested that the term "post burn" is more suitable than "pioneer" for the initial phase of lowland heathland development in situations where vegetative regeneration replaces regrowth of the vegetation by seed germination.

It was shown by Chapman (1967) that in older stands of heathland the rates of accumulation of nutrients were lower than the rates of input of the same nutrients to the litter layer. The same effect can be seen more clearly from the data now available. The rates of accumulation in, and the rate of input to, the litter in relation to the age of the heathland are shown in Figure 9; the curve for the rate of accumulation of phosphorus has been derived from a smooth curve drawn by eye through estimates of the amounts of phosphorus contained in the litter layer. The dashed curve represents estimates of the input to the litter layer obtained from indirect estimates of litter production, and the phosphorus content of brown shoots and flowers of *Calluna*. The crosses in Figure 9 represent independent estimates of the annual phosphorus input from analyses of the material collected in litter traps.

It can be seen that when heathland reaches an age of just over 20 years the rate of accumulation of phosphorus decreases and phosphorus is lost from the litter layer; the same pattern can be demonstrated for other nutrients. It is significant that over the same period of time there is also considerable root development in

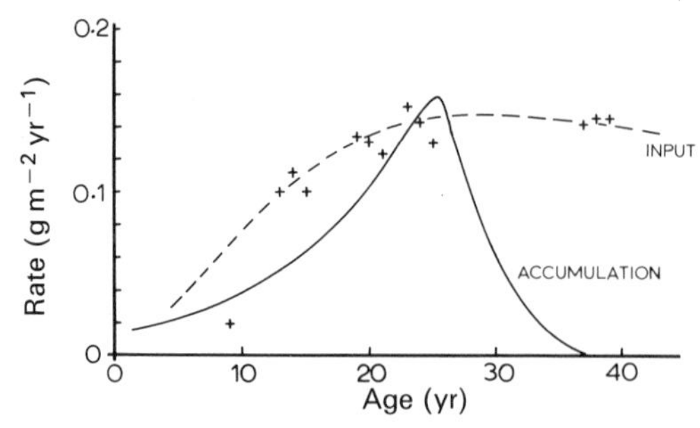

Fig. 9. Rates of accumulation (———) and inputs (--- and + +) of phosphorus to the litter layer in relation to the age of the heathland

the litter layer (Figs. 6 and 7). It could be argued that it is in the litter layer that conditions for nutrient uptake might be most favourable and that losses of nutrients from the ecosystem would be minimised. The relative activity of the root system in the soil and litter layer of mature and degenerate heathland is therefore of considerable interest and importance in the maintenance of the nutrient status of heathland upon poor soils.

It has also been shown (Chapman, 1967) that the nutrients contained in rainfall are capable of replacing losses of nutrients due to burning with the exception of phosphorus and nitrogen. The estimates of nutrients contained in rainfall in that paper were based upon filtered samples as it was considered that particulate matter contained in the samples could have originated within the limits of the ecosystem and therefore did not necessarily represent a true input. Comparison with the results obtained by other workers (Gore, 1968; Allen et al., 1968) for nutrient input by rainfall suggest that these levels for phosphorus are very low. Work by Crisp (1970) on nutrient budgets of a water cress bed in the same general area of Dorset also shows higher annual inputs of phosphorus (about 0.3 kg ha^{-1} yr^{-1}) in rainfall. If these higher levels are accepted as representing true inputs then rainfall is capable of replacing the losses of phosphorus due to burning at intervals of about ten years.

The situation regarding the nitrogen budget is also far from being completely understood. Rainfall can by no means make up the losses of nitrogen over the period of a heathland burning cycle. However, in the case of the Dorset heathlands *Ulex minor* is an important constituent of the vegetation, especially in the post-burn and building phases of development, and nodules on the roots of *U. minor* have been shown to be capable of fixing nitrogen (Thake and Chapman, unpubl.). It is likely that fixation is important in balancing the nitrogen budget for these heathlands. Dwarf gorse (*U. minor* or *U. gallii*) is absent or present in only small amounts on heathlands in some parts of lowland Britain and it would be of interest to investigate the nitrogen economy of a number of such sites in greater detail and to compare them with sites where *Ulex* spp are present in varying degrees of abundance.

Acknowledgements. Thanks are due to a number of people for help with the collection of the field data over a period of several years, and to the Chemical Service at Merlewood for assistance with the chemical analyses.

12. Production in Montane Dwarf Shrub Communities

C. F. SUMMERS

In the north east of Scotland measurements on a series of dwarf shrub communities provide an analysis of the variation in primary production over a range of environmental conditions, supported by field and laboratory experiments to determine the influence of soil fertility and climate.

12.1 Introduction

Much of the recent interest in the productivity of British dwarf shrub heaths has centred on vegetation at altitudes below 500 m, particularly on even-aged stands of heather, *Calluna vulgaris*. At altitudes above 500 m in north east Scotland heather moors give way to a different vegetation type. These areas, which are not subject to rotational burning, contain uneven-aged mixtures of dwarf shrubs including *Loiseleuria procumbens*, *Vaccinium myrtillus*, *Vaccinium vitis-idaea*, *Vaccinium uliginosum*, *Empetrum hermaphroditum*, and *Calluna*. These species, each of which can attain local dominance, vary greatly in cover and abundance, forming a mosaic of vegetation which is characteristically short and poorly stratified. The description of this type of vegetation in the Cairngorm mountains, given in a series of papers by Watt and Jones (1948), Metcalfe (1950), Burges (1951), and Ingram (1958), remains the authorative work on the subject. These authors recog-

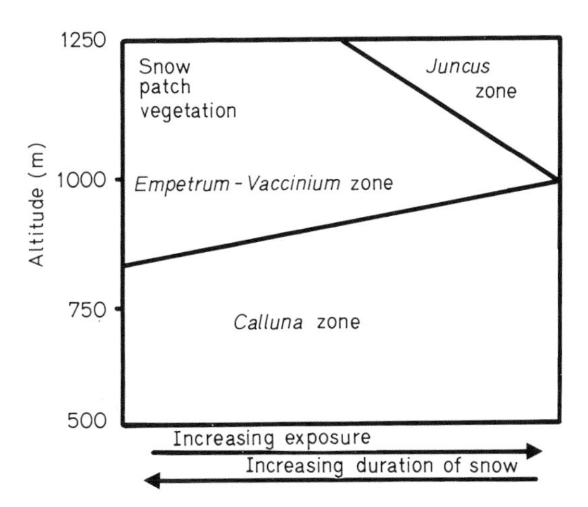

Fig. 1. Effect of altitude, exposure and snow cover on the distribution of dwarf shrub heaths (modified from Watt and Jones, 1948)

Table 1. Summarized site descriptions

Locality	Site number	Vegetation type[b]	Zone[a]	% cover	Altitude (m)
Beinn a'Bhuird	10	Lichen rich *Calluna* heath	C	100	930
	34	Lichen rich *Empetrum-Vaccinium* heath	EV	100	1,000
	44	*Nardus* snow patch	EV	80–90	1,050
	53	Eroded *J. trifidus–D. flexuosa* hummocks	J	10–20	1,100
	61	Lichen rich *Empetrum-Vaccinium* heath	EV	90–100	1,000
	62	Lichen rich *Calluna* mixed heath	C/EV	100	915
	63	Bryophyte rich *Calluna* mixed heath	C	100	915
	64	Exposed *Callunetum* on eroded terraces	C	40–50	840
	65	Mixed dwarf shrub heath on scree	C	85–95	820
	69	Bryophyte rich *Calluna* mixed heath	C	100	900
Cairnwell	66	Exposed *Calluna-Empetrum* eroded terraces	C	45–55	870
	67	Moderately lichen rich *Calluna* mixed heath	C	95–100	870
	68	Bryophyte rich *Calluna* mixed heath	C	100	850
Cairngorm	70	Bryophyte rich *V. myrtillus–V. uliginosum-Empetrum*	EV	100	1,000
	71	*Calluna* wet heath	C	80–90	915
	72	*Calluna* wet heath	C	75–80	850
Kerloch	73	6 year old stand of *Calluna*	—	90–100	150

[a] Each vegetation type is related to the scheme illustrated in Figure 1 by the notation:
C = *Calluna* zone
EV = *Empetrum-Vaccinium* zone or
J = *Juncus* zone.
[b] For details of cover abundance of individual species see Table 2.

nised three zones determined by altitude and exposure, *viz* the *Calluna* zone, the *Empetrum-Vaccinium* zone and the *Juncus* zone (Fig. 1). Principal component analysis and group analysis, used to analyse data from a phytosociological survey at Beinn a'Bhuird in the eastern Cairngorms, gave results which agreed closely with the Watt and Jones' scheme for Cairngorm (Summers, 1972).

In the present study three localities at high altitude and one at low altitude were used. These were Cairngorm (National grid reference NJ 996049-NJ 999049) and Beinn a'Bhuird (NJ 072978-NJ 087995) both in the Cairngorm mountains, Cairnwell (NO 135774) in Glen Shee and Kerloch, a lowland heather moor in Kincardineshire (NO 700905). Within each of the montane localities sites were chosen from the *Empetrum-Vaccinium* and *Calluna* zones to represent as wide a range of vegetation types as possible (Table 1). *Calluna* was present in some but not all sites of the *Empetrum-Vaccinium* zone, while *Empetrum* and *Vaccinium* in the *Calluna* zone varied from absent to locally dominant. Sites from both zones varied in bryophyte and lichen richness. The montane soils ranged from skeletal types to fully developed podsols, though profile development was frequently interrupted by burial or erosion. The lowland site at Kerloch supported an even aged stand of heather.

Table 2. Estimates of annual and daily above-ground production

Site	Growing season (d)	Production ($g\ m^{-2}\ yr^{-1}$: $g\ m^{-2}$ growing season d^{-1}) 1968	1969	1970	
10	126			74:0.58	
34	117			212:1.81	
44	108			66:0.61	
53	101			45:0.45	
61	117	233:1.99	158:1.35	216:1.85	
62	129	200:1.55	146:1.13	122:0.95	
63	129	432:3.35	257:1.99	210:1.63	
64	137	110:0.80	90:0.65	53:0.39	A
65	141	239:1.70			
66	132	91:0.69	100:0.76	34:0.26	
67	134	254:1.89	270:2.01	147:1.09	
68	137	300:2.19	383:2.79	156:1.14	
69	132			165:1.25	
70	117			215:1.84	
71	129			243:1.88	
72	137			259:1.89	
73	204			201:0.98	
				B	

Details of analyses of variance (annual production data).
A: Significant difference between years $P<0.01$
Significant difference between sites $P<0.01$
Significant interaction between year and site $P<0.01$
se of difference between the means = 20.
B: Significant difference between sites $P<0.01$
se of difference between the means = 34.
For key to sites see Table 1.

12.2 Measurement of Production

Annual shoot production and above-ground standing crop were measured in autumn for each site by clipping and sorting eight $^1/_8$ m^2 quadrats. No measurements were made on wood increment, roots or litter. The method of harvesting used (the direct method of Gimingham and Miller, 1968) depends on recognition of the current year's growth of the plants. This was possible for dwarf shrubs, giving true values for net production, excluding wood increment, while for herbs, annual production was assumed to equal the standing crop. Empirically determined conversion factors were applied to the standing crop values of bryophytes (Traczyk, 1967) and lichens (Scotter, 1963; Thomson, 1967) to estimate production. Cages were used to exclude grazing animals and the difference between harvests from plots inside and outside the cages provided an estimate of the

Table 3. Autumn above-ground standing crop

Site	Standing crop (g cm^{-2})			
	1968	1969	1970	
10			403	
34			574	
44			71	
53			73	
61	1,031	864	872	
62	1,047	775	604	
63	1,591	827	646	
64	377	357	233	A
65	906			
66	549	356	261	
67	1,218	998	770	
68	1,052	1,005	633	
69			750	
70			698	
71			867	
72			988	
73			666	
			B	

Details of analyses of variance.
A: Significant difference between years $P<0.01$
Significant difference between sites $P<0.01$
No significant interaction
se of difference between the means = 121
B: Significant difference between sites $P<0.01$
se of difference between the means = 55.
For key to sites see Table 1.

amount of vegetation removed by herbivores, *viz* ptarmigan, *Lagopus mutus* (Montin), Red deer, *Cervus elaphus* L., and Mountain hares, *Lepus timidus scoticus* (Hilzheimer).

The results show a wide range of values for above-ground production and standing crop (Tables 2 and 3). Bryophyte-rich *Calluna* mixed heaths (sites 63, 68, 69) tended to be the most productive and have the highest standing crop, while eroded *Calluneta* (64, 66) had the lowest values for both parameters. *Calluna, Empetrum hermaphroditum*, and *Vaccinium myrtillus* were the most productive of the different species of vascular plants, though the combined contribution of herbaceous species often exceeded 25% of the total production. The relative importance of bryophytes and lichens varied considerably, presumably reflecting the differences in drainage between sites, but, considering non-vascular plants as a unit, their contribution to total production often exceeded 30%. To express production on the basis of g m^{-2} d^{-1} of growing season, the duration of the growing season was estimated for each site from local Meteorological Office data by taking 5.6° C as the minimum mean daily temperature for growth and allowing a decrease of 1° C for each 165 m increase in altitude (Gloyne, 1958). The estimated growing season varied from 101 days at 1100 m to 204 days at 150 m. Daily production was estimated to be in the range 0.26 to 3.35 g m^{-2} (Table 2).

Though the production data listed in Table 4 do not represent a thorough survey of published material they do allow comparison between the tundra of the Cairngorm mountains, other British heathlands and other tundras. Taking the lowest and highest values for dwarf shrub production in the Cairngorm mountains, the eroded *Callunetum* of Cairnwell is less productive than any dwarf shrub community listed. After allowing for its shorter growing season, a bryophyte rich, *Calluna* dominated mixed heath at Beinn a'Bhuird is substantially more productive than its lowland British counterparts. Production of the Cairngorm sites is within the range of production from other tundra regions but the highest values do not reach the maximum daily rates observed in USA.

Comparison of montane standing crops with those of heaths from lower altitudes (Table 5) is complicated because the gradual increase of biomass with age, recorded in managed moors, does not occur in the mixed-aged heaths of the Cairngorm mountains. The standing crop also fluctuates considerably from year to year (Table 3). However the values recorded were generally smaller than all but the youngest *Calluneta* from lowland sites.

12.3 Factors Influencing Production

It is clear that not all the observed differences in productivity between sites in any year can be attributed to differences in floristic composition. In attempting to explain the wide range of production from floristically similar sites environmental factors which vary locally were investigated.

12.3.1 Grazing

Grazing pressure at all sites was so light that it was not detected by the method used (sect. 12.2). An estimate of the quantity of vegetation consumed by herbivores in the dwarf shrub zones at Beinn a'Bhuird was obtained by calculations based on the size of the herbivore population, the daily energy requirements of the animals, the calorific value of their diets and the quantity of food plants in the area (Summers, 1972). It was estimated that less than 1% of the above-ground primary production would be removed annually. Grazing is, therefore, considered to be an unimportant influence on primary production in these areas.

12.3.2 Exposure

An index of site exposure was obtained by growing *Festuca rubra* at each site from a fixed weight of seed in a seed mixture compost. The pots were placed inside exclosures on the sites in June 1970 and were harvested in August of the same year. The reciprocals of the dry weights were used as indices of exposure. Above-ground production of dwarf shrubs was negatively correlated with exposure at montane sites (Fig. 2). However, dwarf shrub production at some montane sites

Table 4. Comparative above-ground production data from heathland and tundra ecosystems

Locality	Vegetation type	Altitude (m)	Cover (%)	Growing season (d)	Production	
					($g\,m^{-2}\,yr^{-1}$)	($g\,m^{-2}\,d^{-1}$)
Arctic Russia[a]	Shrubby, mossy, tundra	—	35–60	50–70	119–185	1.0–3.0
Rocky Mts. USA[b]	*Carex-Geum* turf	—	15	60	6.5	0.1
Medicine Bow Mts. USA[b]	*Deschampsia* wet meadow	—	—	70	348	4.97
Mt. Washington USA[b]	*Vaccinium* heath	approx 1,700	95	71	283	4.00
Cairngorm Mts. UK	Bryophyte rich, *Calluna*-dominated mixed heath (Site 63)	915	100	129	300[g]	2.32[g]
	Eroded *Calluna-Empetrum* terraces (Site 66)	870	45–55	132	75[g]	0.57[g]
Moor House UK[c]	*Calluna-Eriophorum* bog	550	—	184	290	1.58
Kerloch UK[d]	*Callunetum*	150	95	204	240–270	1.17–1.32
Elsick Heath UK[e]	Bryophyte rich, *Calluna*-dominated mixed heath	107	—	200[h]	195–470	2.35
Dorset UK[f]	*Calluna* dominated dry heathland 15–40 years old	50	100	250[h]	320	1.28

References: [a] Bliss (1962); [b] Bliss (1966); [c] Forrest (1971); [d] Miller and Watson (Chap. 13); [e] Barclay-Estrup (1970); [f] Chapman and Webb (Chap. 11); [g] means for 1968–1970; [h] minimum estimates.

Table 5. Comparative above-ground standing crop data from British heathlands

Locality	Vegetation type	Altitude (m)	Standing crop ($g\ m^{-2}$)	References
Dorset	*Calluna* dominated dry heath 6 yr old	50	723	Chapman and Webb (Chap. 11)
	12 yr old		1,428	
	24 yr old		2,166	
	42 yr old		2,458	
Elsick Heath (Kincardineshire)	Pioneer stage — Bryophyte rich, *Calluna*-dominated mixed heath	107	889	Barcley-Estrup (1970)
	Building stage		1,720	
	Mature stage		2,305	
	Degenerate stage		1,561	
Kerloch (Kincardineshire)	Pioneer stage (2–7 yr)	150	420	Miller and Watson (Chap. 13)
	Building stage (8–19 yr)		1,180	
	Mature stage (18–29 yr)		2,000	
	Degenerate stage (37–42 yr)		2,200	
	25 yr old *Callunetum*		1,840	Kyall (1966)
Dartmoor	*Calluna*-dominated heathland 2–18 yr old	425	326–2,218	Chapman (1967)
Moor House	Mixed age *Calluna-Eriophorum* bog	550	1,300	Forrest (1971)
Cairngorm Mts.	Eroded *Callunetum* 14–24 yr old (site 64)	840	322[a]	
	Bryophyte rich, *Calluna*-dominated mixed heath 13–30 yr old (site 63)	915	1,021[a]	

[a] Mean for 1968–1970.

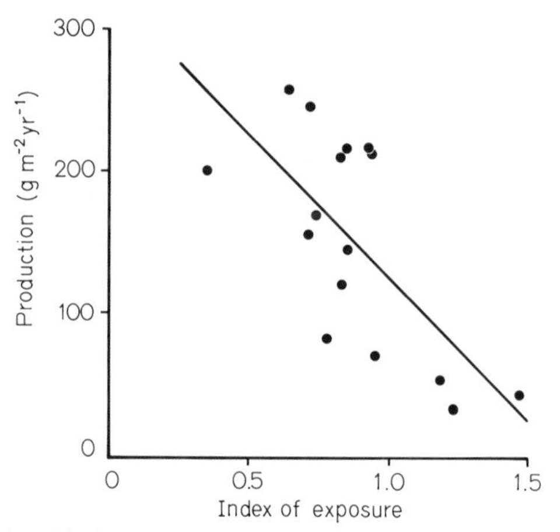

Fig. 2. Relationship between site exposure and production. Regression equation $y = 323.8 - 197.4\,x$, $r = -0.67$, $P < 0.01$

exceeded that at Kerloch, which was indicated by *Festuca rubra* to be the least exposed site. This suggests that the alpine vegetation is adapted to growing in exposed situations.

12.3.3 Soil Fertility

There was no indication from soil analysis of correlation between above-ground production and the chemical composition of the soil. However, by agricultural standards, all the soils were deficient in nitrogen, phosphorus, potassium, and calcium. To determine the influence of soil fertility on production, a fertiliser experiment was carried out in a fenced plot in the *Empetrum-Vaccinium* zone of Beinn a'Bhuird. A randomised block design was used in which seven fertiliser treatments were applied to 25 m^2 plots in seven blocks. The treatments were:

1. control without fertiliser,
2. nitrogen as ammonium nitrate at 14 g N m^{-2},
3. phosphorus as triple superphosphate at 10 g P m^{-2},
4. calcium as ground limestone at 70 g Ca m^{-2},
5. nitrogen and phosphorus, levels as above,
6. nitrogen, phosphorus and calcium, levels as above,
7. nitrogen, phosphorus, calcium, and potassium, the last applied as muriate of potash at 17 g K m^{-2}; other levels as above.

Except in the case of nitrate, these amounts were applied in autumn 1968 and 1969. Half the nitrate was applied each autumn and the remainder each following spring. Cover and performance parameters were measured in autumn 1969 and 1970.

The results showed that the size of individual plants increased in multiple fertiliser treatments and that this response was greater in *Carex bigelowii*, *Deschampsia flexuosa* and *Juncus trifidus* than in *V. myrtillus* and *E. hermaphroditum* (Table 6). In addition, multiple fertiliser treatments also altered the species composition, bringing about an increase in the importance of monocotyledonous

Table 6. Fertilizer experiment: comparison of means of performance data

Variable	Means of fertiliser treatments averaged between years							Means of 1969–70 parameters for all fertiliser treatments	
Carex bigelowii leaf length (cm)	g 8.8	e 8.4	f 8.3	c 6.9	b 6.7	a 5.7	d 5.5	1969 7.3	1970 7.0
Deschampsia flexuosa leaf length (cm)	e 9.3	g 8.6	f 7.6	c 4.9	d 4.6	b 4.4	a 3.8	1969 6.6	1970 5.8
Deschampsia flexuosa influorescence height (cm)	e 39.2	g 38.8	f 36.7	d 27.3	c 26.8	b 25.1	a 22.1	1969 32.9	1970 28.8
Deschampsia flexuosa influorescences tussock^{-1}	e 33.1	g 31.1	f 18.2	d 2.3	c 1.9	b 1.8	a 1.2	1969 17.7	1970 7.9
Deschampsia flexuosa tussock diameter (cm)	e 17.5	f 16.6	g 16.0	d 8.5	c 7.1	b 7.1	a 6.9	1969 11.4	1970 11.4
Juncus trifidus shoot height (cm)	f 22.8	e 20.8	g 19.9	b 16.8	d 14.4	c 14.1	a 12.6	1969 17.8	1970 17.0
Empetrum hermaphroditum shoot length (cm)	f 3.7	c 3.4	d 2.9	e 2.9	g 2.9	a 2.7	b 2.6	1969 2.8	1970 3.2
Vaccinium myrtillus shoot length (cm)	e 4.7	g 4.6	f 4.5	d 4.4	c 4.2	a 4.1	b 4.0	1969 4.6	1970 4.2
Vaccinium myrtillus leaf area (mm^2)	f 114	g 111	e 97	b 93	c 86	a 80	d 79	1969 93	1970 95

a: control; b: N; c: P; d: Ca; e: N+P; f: N+P+Ca; g: N+P+Ca+K.

Values joined by solid line are not significantly different.
Values joined by dashed line are significantly different at 5% level of probability.
Values not joined by lines are significantly different at 1% level of probability.

Significant differences between means indicated by analysis of variance and located by Tukey Q test (Snedecor, 1961).

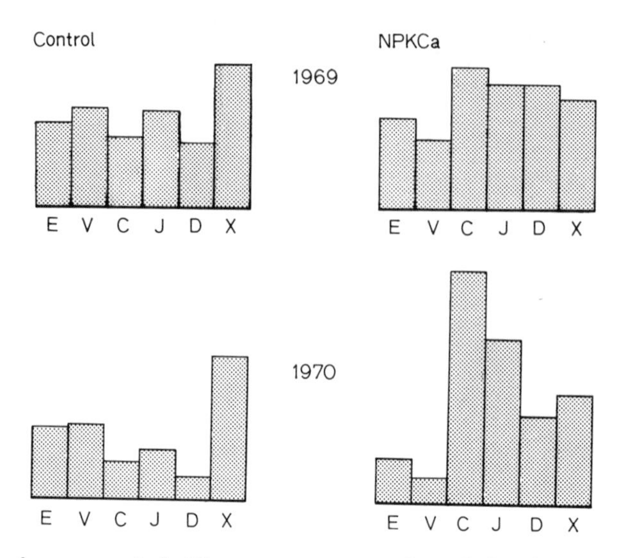

Fig. 3. Effect of compound fertiliser treatment on the relative importance of species (based on Domin estimates of cover-abundance transformed to linearity to allow averaging, see Bannister, 1966). *E*: *Empetrum hermaphroditum*; *V*: *Vaccinium myrtillus*; *C*: *Carex bigelowii*; *J*: *Juncus trifidus*; *D*: *Deschampsia flexuosa*; *X*: bare ground

species at the expense of dwarf shrubs (Fig. 3). Thus, although differences in production between sites were not correlated with the chemical composition of the soil, the experiment indicates that production is limited by the nutrient status of the soil, and by implication nutrients are equally limiting at all sites.

12.3.4 Growing Season

With one exception, aerial production followed the same general trends of year-to-year variation for all the sites at each locality (Fig. 4). Since relative exposure and soil fertility are unlikely to vary considerably from year to year and the light grazing regime does not appear to be an important influence on production, it is probable that these trends were climatically determined, perhaps as a result of fluctuations in the length of the growing season. Bliss (1966), who found very little year-to-year fluctuation in productivity, and Tranquillini (1964) have shown, at Mount Washington (USA) and the European Alps respectively, that arctic-alpine plants do not need all of the available growing season to accumulate adequate storage products. This suggests that fluctuations in the length of the growing season are unimportant. However, Mount Washington and the European Alps have a continental climate, whereas that of the Cairngorm mountains is oceanic. Comparison of the mean monthly temperature curves of Mount Washington and Cairngorm (Fig. 5) shows that although the growing season at Cairngorm is longer, that at Mount Washington is warmer and consequently is likely to give a higher daily growth rate. It is probable that in oceanic localities plants require a longer growing season to accumulate sufficient dry matter for storage, and, in

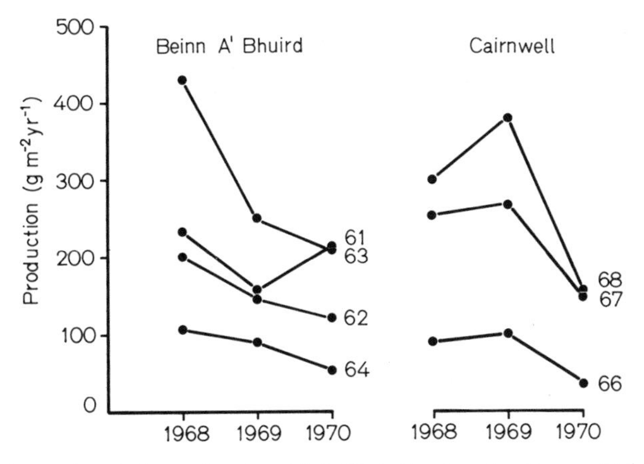

Fig. 4. Trends in production 1968–1970. For key to sites see Table 1

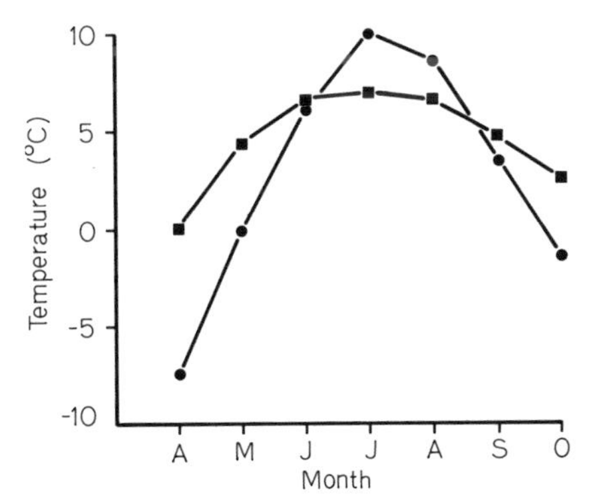

Fig. 5. Mean monthly temperature curves (April–October) for Cairngorm (■) and Mount Washington (●) (from Met. Office tables of temperature, relative humidity and precipitation for the world)

these localities, fluctuations in the length of the growing season could have a significant effect on production. Although this may explain year-to-year fluctuations in production, the fact remains that, even in poor years, some of the montane sites are as productive as lowland areas, suggesting that the tundra plants are adapted to production under cool summer conditions.

12.4 Adaptation to Environment

The hypothesis that montane plants are adapted to their environment was tested by growing *Calluna* seedlings from different altitudes in a range of controlled environments for sixteen weeks. *Calluna*, which was chosen as an example of a common dwarf shrub with a wide altitudinal range, displays a continuous range of growth form from erect to prostrate. All the seedlings from the lowland site gave rise to plants with the erect habit but those from the upland site gave rise

Table 7. Growth-chamber experiment. Each value is the mean biomass (g) of six replicates

Sites	Soil	Temperature (°C)		
		4–7	10–13	15–18
Kerloch, Kincardineshire (150 m OD)	Fertilised	2.8	7.8	9.8
	Unfertilised	3.3	3.9	4.1
Carn Aosda, Aberdeenshire (650 m OD)	Fertilised	3.6	10.8	9.5
	Unfertilised	3.4	5.4	4.5

In the fertilised soil, N, P, K, and Ca were applied at the same levels as in the fertiliser experiment (Sect. 12.3.3).

Table 8. Summary of results from light response curves at 20° C

Plant source and growth form	Dark respiration ($mg\,dm^{-2}\,h^{-1}$)	Maximum rate of photo-synthesis ($mg\,dm^{-2}\,h^{-1}$)	Quantum efficiency (mol CO_2 $Einstein^{-1}$)	Light compensation point ($W\,m^{-2}$)
Lowland erect habit	1.4[a]	11.1	0.023	7.8
Montane erect habit	1.9	12.7[a]	0.028	8.7
Montane prostrate habit	2.0	10.1	0.022	11.8[b]

Each value is the mean of six replicates.
Significant differences calculated from analysis of variance are shown by
[a] $=P<0.1$, [b]$=P<0.01$.

to both erect and prostrate forms. The experiment was designed as a $2\times2\times3$ factorial trial with 6 replicates. The factors were: site of origin (150 m and 650 m OD), soil fertility (with and without compound fertiliser) and temperature (4–7° C, 10–13° C, 15–18° C). The results of an analysis of variance show highly significant effects ($P<0.01$) of temperature and soil fertility on production (Table 7). Although the effect of site of origin was significant at a lower level of probability ($P<0.2$), plants from high altitude showed a consistently higher production at low and medium temperatures compared with plants from the low altitude site. There was also highly significant interaction ($P<0.01$) between temperature and soil fertility.

If some *Calluna* plants from high altitude are more efficient than their low altitude counterparts in terms of dry matter production, it seems likely that there are differences in their rates of photosynthesis. In order to compare photosynthesis of *Calluna* from different altitudes net assimilation was measured over ranges of light intensity and temperature in erect and prostrate montane plants and in erect lowland plants.

The measurements were made using the infra red gas analyser (IRGA) and assimilation chamber equipment built by P. G. Jarvis and D. K. L. McKerron at the University of Aberdeen and described by Ludlow and Jarvis (1971). Photosynthesis in relation to light shows several differences between plants from different sites (Table 8). Montane plants with an erect growth form had the highest maxi-

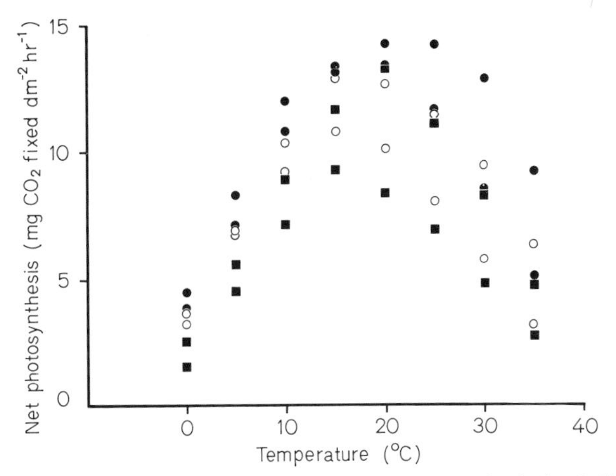

Fig. 6. Light saturated temperature response of net photosynthesis in *Calluna*. (●: montane erect habit: ○: montane prostrate habit; ■: lowland erect habit)

mum rate of net photosynthesis; erect and prostrate plants from high altitude had higher respiration rates than lowland plants; montane plants with the prostrate habit had the highest light compensation point. The effect of temperature on photosynthesis (Fig. 6) suggests that plants from high altitude, particularly those with an erect growth form, can photosynthesise over a wider temperature range than lowland plants. The high maximum rate of photosynthesis, low light compensation point and wide temperature range for photosynthesis of erect plants from the montane site, are features which are presumably important in bringing about the high productivity at sites with this growth form (see Table 8).

The higher respiration rates observed in plants of both growth forms from high altitude, in comparison with the rates of plants from low altitude, could explain the consistently lower standing crop present on montane heaths compared with lowland vegetation of similar age. Alternatively, mechanical damage by wind action could result in the loss of large quantities of dry matter as litter. Grant and Hunter (1968) point out that wind pruning of montane Callunetum maintains physiological vigour and this presumably contributes to higher productivity. However, the IRGA results imply that adaptations of the prostrate plants to the montane environment are principally morphological. In this respect, Grant and Hunter (1962) have noted that the growth habit of *Calluna* forms an altitudinal ecocline, with an increasing proportion of individuals having a genotypically determined prostrate habit as altitude increases. This raises the question of whether the adaptations described above are genetically determined or are plastic responses to the environment.

12.5 Conclusion

Above-ground production and standing crop from montane dwarf shrub heath in northeast Scotland vary considerably between sites and from one year to another. The most productive sites are at least as productive, in terms of g

m^{-2} d^{-1} of growing season, as comparable lowland vegetation. Production is inversely related to exposure and results suggest that all sites are limited by soil infertility. Annual fluctuations in production are possibly related to annual variation in the length of the growing season. *Calluna vulgaris* is adapted physiologically as well as morphologically to the alpine environment.

Acknowledgements. I am indebted to colleagues in the University of Aberdeen and the Nature Conservancy (Banchory) for advice, assistance, and discussion. I am especially grateful to Professor C.H.Gimingham and Dr. G.R.Miller for their encouragement throughout. The work was supported by grants from the University of Aberdeen and the Natural Environment Research Council.

13. Heather Productivity and Its Relevance to the Regulation of Red Grouse Populations

G. R. Miller and A. Watson

Red grouse graze heather selectively and eat only a negligible proportion of the primary production on their territories. Thus, unless the palatability and nutritive value of the plant material are known, simple measurements of production are of little value in explaining grouse population processes. Detailed studies in the Scottish highlands augment the limited studies at Moor House, where it is also a main herbivore.

13.1 Introduction

In Scotland, large numbers of Red grouse *(Lagopus lagopus scoticus)* are cropped from tracts of heathland dominated by heather *(Calluna vulgaris)*. The grouse are territorial birds that depend heavily on heather for both food and cover (Jenkins et al., 1963). Their habitat is managed by periodic burning of the heather in small patches, and there is evidence that both this and the experimental application of nitrogenous fertiliser can increase the density of grouse populations (Miller et al., 1966, 1970; Picozzi, 1968). However, the exact mechanism of the link between heather and grouse population size is not fully understood and it was necessary to study how the birds utilise their main food plant as part of a wider investigation of this problem.

This paper presents data on (1) the size of grouse territories and the amount and type of heather found there, (2) the annual production of dry matter by heather in relation to the development of the stand after burning, and (3) the consumption of dry matter by grouse. By combining these data it was possible to estimate the degree of utilisation of heather by grouse, and assess whether changes in the population density of the birds might be caused by variations in heather productivity.

13.2 Study Area

The study area lay at an altitude of 150 m on part of Kerloch moor, Kincardineshire and was typical of the treeless, burnt and grazed grouse moors of north east Scotland. The soils were imperfectly drained peaty podzols, often with a well-defined iron pan. Over 80% of the area was occupied by heather-dominant vegetation that had been burnt experimentally in three successive years to produce a mosaic of stands at different stages of development (Miller et al., 1970). Along with heather there usually grew a scattering of *Erica cinerea*, *Erica tetralix*, *Trichophorum cespitosum* and various moorland grasses. *Juncus-Sphagnum* mire occurred in a few poorly drained patches and in these places heather was sparse or absent.

Although the area had previously been used as hill grazing for sheep and cattle, both were excluded during the period of study, 1964–69. Apart from grouse, only a few rabbits (*Oryctolagus cuniculus*) and up to ten Mountain hares (*Lepus timidus*) km^{-2} grazed the heather freely.

13.3 Size and Composition of Grouse Territories

13.3.1 Methods

Boundaries between different vegetation types and between patches of heather at different stages of development were traced from vertical aerial photographs of the study area. The map was enlarged to a scale of roughly 1:2500 and duplicated for use in the field. There was enough detail to locate individual grouse to within a few metres of their true position.

Territories were plotted in the winter of 1964–65 by regularly watching for interactions between neighbouring cock grouse. We used the second of the methods described by Jenkins et al., (1963, p. 324) for mapping territorial boundaries. On the vegetation map we marked where adjoining territory owners were seen showing the "walking-in-line" display described by Watson and Jenkins (1964). Cocks met and paraded daily in fine weather and repeated observation enabled the lines dividing territories to be plotted on the field map. By the end of the winter, 13 territories were mapped and their size measured by planimeter, to the nearest 0.1 ha.

On the study area four types of heather were recognised, corresponding to the "pioneer", "building", "mature" and "degenerate" phases of growth described by Watt (1955). The average ground cover of heather in each phase was calculated from visual estimates, made to the nearest 10%, within 10–20 quadrats of 0.5 m^2 placed randomly in every patch of heather. For each territory, the proportions of ground occupied by the different growth phases, and by vegetation with no heather, were also estimated.

13.3.2 Results and Discussion

The sample of 13 territories ranged in size from 0.8 to 4.6 ha, with a mean of 2.09 ha (Table 1). Individual cocks can defend areas varying between 0.2 and 13.2 ha (Watson and Miller, 1971) but on most grouse moors the average territory size is usually 2–5 ha, depending on the size of the breeding stock. Thus, there was a relatively large stock of grouse with small territories on the Kerloch study area in 1964–65.

Table 1 shows that pioneer heather (aged 1–5 yr) occupied some 26% of an average territory. Older heather in the building (aged about 6–15 yr), mature (16–

Table 1. Area occupied by different growth phases of *Calluna* on Red grouse territories in 1964–65

Territory size	Growth phase of *Calluna*				Other vegetation	Total
	Pioneer	Building	Mature	Degenerate		
Mean	0.54±0.18	0.98±0.47	0.28±0.21	0.02±0.03	0.27±0.17	2.09±0.64
Smallest	0.2	0.3	0.3	0.0	0.0	0.8
Largest	0.8	2.7	0.7	0.0	0.4	4.6
% *Calluna*	25	60	85	85	0	45

Results are given as ha (mean ±95% confidence limits), and as % of ground covered by *Calluna*,in 13 territories.

25 yr) and degenerate (25+ yr) phases occupied 47, 13, and 1% respectively. About 13% of the study area had no heather.

Because the experimental burning had been done in the three years previous to 1964–65, recovery of the heather on the burnt patches was incomplete and pioneer growth covered only some 25% of the ground. On average, about 60% of the ground was covered in patches of building heather, whereas cover was almost complete in mature and degenerate patches.

13.4 Production of Dry Matter by Heather

13.4.1 Methods

The main aim was to measure the annual production of heather as potential food for grouse—long and short shoots and flowers. As wood and roots are not eaten, their annual increments were not measured.

Six uniform stands of more or less pure heather were fenced to exclude all vertebrate herbivores except birds and were studied for six years, 1964–1969. The stands were selected to represent a series of increasing age, with a gap of about six years between each. In 1964 they were aged 2, 8, 14, 18, 24, and 37 yr and so, by the end of the six-year study, there were data on the productivity of heather throughout all its four phases of growth.

Methods of measuring the dry matter production of dwarf shrubs have been discussed by Gimingham and Miller (1968). Their so-called "direct" method was appropriate since the six stands were even-aged and estimates of woody increments were not needed. Every year, the production of green shoots and flowers was measured from a single harvest in October after growth had ceased.

In each stand all vegetation rooted within nine random 50×50 cm plots was cut to ground level. Other species were discarded and the heather was weighed fresh to ±10 g. A subsample of this heather was oven-dried at 80° C and weighed to estimate the total oven-dry weight from each plot. A second subsample of 3–5 whole shoots cut at ground level was separated into (1) current year's flowers and flower buds, (2) current year's shoots, including increments on previous year's short shoots, and (3) the remainder, comprising all green shoots aged more than

Table 2. Aerial biomass and annual production of shoots and flowers in different growth phases of *Calluna* in 1964–69

	Growth phase of *Calluna*			
	Pioneer	Building	Mature	Degenerate
Age (yr)	2–7	8–19	18–29	37–42
Cover (%)	55	85	95	95
Height (cm)	17	33	41	40
Aerial biomass (t ha^{-1})	4.2 ± 0.8	11.8 ± 0.9	20.0 ± 1.5	22.0 ± 1.8
Annual production				
(kg ha^{-1}): shoots	1,392 ± 94	1,766 ± 64	2,152 ± 94	1,934 ± 115
flowers	286 ± 40	326 ± 28	546 ± 52	454 ± 47
total	1,678 ± 121	2,092 ± 74	2,698 ± 117	2,388 ± 132

(mean ±95% confidence limits)

one year, dead material, and wood. Each component was dried at 80° C, weighed, and the weight expressed as a proportion of the total dry weight of the separated subsample. These proportions were used to estimate the production of shoots and flowers. A more detailed account of this method and of its accuracy will be given elsewhere.

13.4.2 Results and Discussion

Both shoot and flower production varied widely from one year to another. In stands with a complete cover of heather, the annual production of shoots ranged from 1680 kg ha^{-1} in 1966 to 2440 kg ha^{-1} in 1964, and of flowers, from 270 kg ha^{-1} in 1965 to 880 kg ha^{-1} in 1966. These data will be critically appraised in another paper and it will be shown that the variations in shoot production were closely related to the rainfall and air temperature in the different growing seasons.

In Table 2, data for the six years have been pooled for each growth phase. In the early stages after burning, the annual yield increased as the heather increased in its ground cover, height and biomass. By about ten years after burning, heather cover exceeded 80%; thereafter, annual production did not change greatly although biomass continued to increase and had doubled by 40 yr. There is some evidence of a peak during the mature phase but this is by no means clear cut. In summary, when heather cover was more or less complete during the mature and degenerate phases, annual production averaged 2595 kg ha^{-1}; flowers accounted for about 20% of this total.

Barclay-Estrup (1970) found that production of heather shoots and flowers on an unburnt moor reached a maximum in the building phase and then declined appreciably in the mature and degenerate phases. However, there was a parallel decline in heather cover (Barclay-Estrup and Gimingham, 1969) and, if this is taken into account, there is no evidence of a decrease in productivity before the degenerate phase is reached.

The annual yields reported by Barclay-Estrup (1970) near the coast of north east Scotland are much greater than those found elsewhere. Thus, his building and mature heather yielded 3600–4400 kg ha^{-1} compared with the 1300–2600 kg

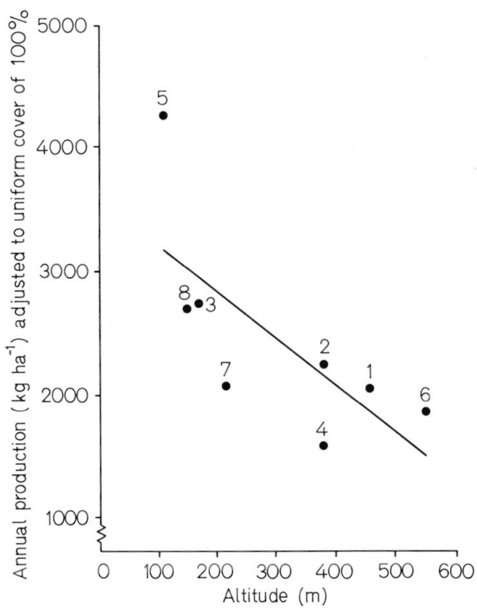

Fig. 1. Annual production of shoots and flowers by *Calluna vulgaris* in relation to altitude. Yields have been adjusted to 100% uniform cover of *Calluna*. Sources of data: *1–4*: Moss (1969); *5*: Barclay-Estrup (1970); *6*: Forrest (1971); *7*: Grant (1971); *8*: present study
Regression equation: $y = 3565 - 3.69x$
$(S_b = 1.45, P < 0.05)$

ha^{-1} found by us at Kerloch, by Moss (1969) and Grant (1971) at other sites in north east Scotland, and by Forrest (1971) in northern England. Much of this variation can be attributed to differences amongst sites in altitude and in heather cover. When the various production estimates are adjusted to represent yields from heather that completely covers the ground, there is a clear negative correlation with altitude (Fig. 1). No direct comparison is possible with data from southern English heathland (Chapman, 1967) because these include increments of wood, which we did not measure.

13.5 Consumption of Dry Matter by Grouse

13.5.1 Methods

All work was done on a uniform patch of nine-year-old building heather in February–March 1964. We measured standing crops of green heather shoots before and after periods of grazing by caged grouse. Differences between the two sets of measurements provided estimates of the consumption of heather by the birds.

Firstly, we cropped 20 random 0.1 m^2 quadrats to ground level. By subsampling the heather as described previously (13.4.1), we found that the standing crop of green shoots (i.e. long shoots of the current year and all green short shoots) was

2980 ± 220 kg ha^{-1}. There was no growth at that time of year and grazing was negligible, so this was regarded as a basic datum that would not change appreciably during the three weeks of the study.

Three tame cock grouse were taken from captive stock and caged individually on the heather. They were all in good condition with a live weight of 650–750 g. The floor area of each cage was 4.5 m^2 so that each bird had access to about 1.34 kg of green heather shoots. Water and grit were supplied ad lib. but nothing else. The cages were left in position for three periods of 4–8 days at a time. At the end of each grazing trial, the cages were moved, the birds were weighed, and five 0.1 m^2 quadrats were harvested to estimate the amount of green heather remaining.

13.5.2 Results and Discussion

All the birds lost live weight during the trials and, indeed, one died after losing 200 g. Nonetheless, the estimates of the daily consumption of heather in Table 3 are relatively consistent between birds and between trials. The overall mean is 0.10 ± 0.01 kg of dry matter per day.

Using magnesium as a tracer, Moss and Parkinson (1972) calculated that the intake of dry matter by caged grouse feeding on fresh heather varied between 0.06 and 0.08 kg d^{-1}. Moreover, Savory (1974), by watching grouse feeding in the wild, found an average daily intake of about 0.06 kg. Thus, despite the crudeness of the method, the large losses of body weight by the grouse, and even the death of one of them, it is unlikely that our estimate of the consumption of heather by grouse is grossly in error; if anything, it is an overestimate.

Moss and Parkinson (1972) also found daily losses in live weight of up to 12 g. They suggested that this was due to the captive birds being imperfectly adapted to a diet of fresh heather (they are normally given an artificial diet of pellets with only a small supplement of heather). This seems the most probable explanation of the weight losses recorded in our feeding trials. Clearly the change from an artificial to a natural diet should be imposed gradually.

Table 3. Estimates of daily consumption of *Calluna* shoots and losses in live weight by three captive Red grouse during field feeding trials in February–March 1964

Trial	Duration (d)	Consumption of *Calluna* shoots (kg d^{-1}) by bird no:		
		2,181	2,200	2,219
1	4	0.07	0.11	0.16
2	8	0.09	0.11	0.08
3	6	0.12	died	0.07
Overall		0.10	0.11	0.10
Loss in live weight (g d^{-1})		8	15	8

Table 4. Estimated annual production of shoots and flowers by *Calluna* on territories of Red grouse at high population density

Territory size	Area (ha)		Annual production (kg) from:				Total
			Pioneer	Building	Mature	Degenerate	
Average	2.09	shoots	750	1,730	600	40	3,120
		flowers	150	320	150	10	630
		total	900	2,050	750	50	3,750
Smallest	0.8	shoots	280	530	650	0	1,460
		flowers	60	100	160	0	320
		total	340	630	810	0	1,780
Largest	4.6	shoots	1,110	4,770	1,510	0	7,390
		flowers	230	880	380	0	1,490
		total	1,340	5,650	1,890	0	8,880

13.6 Utilisation of Heather by Grouse

By combining the data in Tables 1 and 2, the average annual production of heather shoots and flowers in territories of Red grouse at high population density can be calculated. Allowing for the amount of ground actually covered by heather, production on the average territory would be about 3750 kg of dry matter (Table 4). The largest territory had five times the total production of the smallest territory studied.

Each territory contained a cock and a hen. If one assumes (from Table 3) that their combined daily intake was 0.2 kg at all seasons, the pair would consume about 75 kg of dry matter in one year. Chicks usually hatch in late May and disperse in August–September (Jenkins et al., 1963). Many of the dispersing birds become transients and may continue to visit the area to feed until January (Watson, 1967). Thus young birds may depend on the resources of the study area for about 30 weeks at most.

Grouse actually bred poorly at Kerloch in 1965 (Jenkins et al., 1967) and only 11 young birds were reared from the 13 territories. However, in exceptionally good years, an average of up to six young can be reared by each pair of adults. If they were to eat heather at the same rate as adults, six young grouse would ingest about 125 kg in 30 weeks.

Thus the combined consumption by adults and young might amount to about 200 kg, i.e. *c* 5% of the estimated annual production of dry matter on the average territory of 2.09 ha. On the largest territory studied, utilisation would have been $<2\%$ of production, and even on the smallest 0.8 ha territory only 11%. Even allowing for gross errors, it is clear that the grouse could have exploited only a small fraction of the heather's annual production of shoots.

The above calculations assume that (1) territories were relatively small, (2) all cocks were mated, (3) many young were reared and all fed on the area for 30 weeks, (4) the birds ate nothing else but heather, and (5) the young ate heather at the same rate as adult grouse. In fact it is most unlikely that all these suppositions would hold true at any one time. For instance, average territory size is seldom as small as 2 ha and can be 5 ha or more. Moreover, cocks are commonly unmated, breeding frequently fails (as it did on the study area), some of the young do not become transients, and it is known that grouse eat from a variety of food plants even to the extent that heather comprises 50% or less of their summer diet (Jenkins et al., 1963; Eastman and Jenkins, 1970). Therefore, the estimate of 5% for the average utilisation of heather by grouse is probably excessive for the Kerloch study area. It might be more than this at higher altitudes where heather productivity is less than at Kerloch moor (Fig. 1). However, on most heather moors in north east Scotland, utilisation of the primary production by grouse must be no more than 1–2%. A similar conclusion was reached by Savory (1974), who used a different method based on watching wild birds feeding.

13.7 Population Regulation of Grouse

Field observations and experiments (Jenkins et al., 1963; Watson and Jenkins, 1968) have shown that the density of breeding stocks of Red grouse is determined by the average size of the territories taken by the cock birds each autumn. Our observations indicate that these territories are so large in relation to the amount of heather eaten by birds that only a minute fraction of the annual yield of heather shoots and flowers is utilised. It seems unlikely that grouse should take territories containing such large amounts of heather to buffer themselves against heavy grazing by other herbivores. Even allowing for grazing by sheep, Red deer *(Cervus elaphus)* and Mountain hares, total utilisation probably exceeds 10% only rarely and locally (Miller and Watson, 1974). Indeed, grouse moors must be burnt periodically to dispose of the excess primary production that has accumulated as wood, dead material and litter.

One possible reason for grouse having relatively large territories is that they are selective feeders and do not have equal preferences for all types of heather. They much prefer feeding on pioneer heather (Miller and Watson, 1974), presumably because it is more accessible to them and/or because the shoots have a greater content of nutrients than older heather (Moss et al., 1972). Nevertheless, a total annual consumption of 200 kg by a family of grouse at Kerloch would still have amounted to only about 20% of the annual production from pioneer heather on a 2 ha territory (Table 4). However, not all pioneer heather may be equally acceptable to grouse. There is evidence (Moss et al., 1972) that the birds' greatest preference is for heather 2–3 years after it has been burned. Moreover, Moss (1972) has shown that they may be able to select for individual plants and plant parts within an apparently uniform stand, and that hen grouse are particularly selective in spring, just before they lay their eggs. It is thus an oversimplification to regard all heather shoots of the current year as equally palatable grouse food.

Spring is known to be a critical time as much of the subsequent variation in chick survival (the main process affecting breeding success) can be accounted for by differences in egg quality which in turn are related to the state of the parents' food supply (Watson and Moss, 1972). Even with such large territories, it is common for many hens to fail rear any young or to raise only small broods. If grouse were to take smaller territories and eat more of the heather's annual production, the hens would have less to select from in spring and their breeding success might become poorer. This simple model may explain how the present territory–food relationship might have evolved, producing an apparent, though probably illusory, excess of food.

It is axiomatic that an animal population cannot perform beyond the limits of its food supply. Yet in an ecosystem where only a small proportion of the primary production is consumed, simple measurements of production are of little value in understanding the processes by which herbivore populations are regulated. This does not mean that food is unimportant in the case of Red grouse. On the contrary, there is much evidence (e.g. Miller et al., 1970) that it is the critical factor determining their numbers and breeding success. However, it is the availability of sufficient plant material of adequate nutritive value that is crucial rather than simply the total amount produced each year.

Because grouse are very selective feeders and have largely unknown requirements, their nutrition is poorly understood and we cannot yet state what is minimal for their maintenance and reproduction. Once this can be done, it may then be appropriate to consider the precisely defined food resource in terms of its productivity and availability.

Acknowledgements. We are grateful to Dr. R.M.F.S.Sadleir and Mrs. Ann Miles for their help with this work.

III. Snowdonia Grasslands

Y Wyddfa (Snowdon) National Nature Reserve showing sheep exclosure cages on *Agrosto-Festucetum*. (Photo D. F. Perkins)

14. Snowdonia Grassland: Introduction, Vegetation and Climate

D. F. PERKINS

Grassland ecosystem research was concentrated on a montane area in Snowdonia, North Wales, with particular emphasis on primary production and herbivores. The characteristics of the site, its vegetation, climate and management are described.

14.1 Introduction

The IBP PT project (UK/PT/3) was a productivity study of montane grasslands in Snowdonia. These grasslands consist of a complex of communities ranging from acidic to near neutral types. Over short distances there occurs a wide range of geological, pedological and climatological conditions which are reflected in the botanical composition, structure and productivity of those communities. All the grasslands are grazed to a greater or lesser degree by sheep. The comprehensive IBP study, of which the following chapters were part, was initiated under the direction of R. Elfyn Hughes and arose out of his research studies (Hughes, 1949, 1958a, 1973) concerned with the relationship of sheep populations and environment in Snowdonia. The work which developed at Bangor was closely aligned to the conservation of the scientific interest of the mountain ecosystem and, until 1973, formed part of the Nature Conservancy's research programme. The IBP project was funded by the Nature Conservancy, then a component body of the Natural Environment Research Council, in association with the Royal Society.

The study concentrated on an *Agrostis-Festuca* grassland at Llyn Llydaw situated in Cwm Dyli, one of the five major glaciated cwms of Snowdon and within the Y Wyddfa (Snowdon) National Nature Reserve, Wales (lat 53° 05′N, long 4° 02′W) at 488 m. It is surrounded to the west by Y Wyddfa (1085 m), the highest mountain south of the Scottish Highlands, to the north west by Crib y Ddysgl (1065 m) and Crib Goch (921 m) and to the south by Y Lliwedd (898 m). Llyn Llydaw is the principal lake within the cwm and lies adjacent to the site to the south. Snowdon summit is the severely denuded remnant of a plateau formed possibly during the Cretaceous period and elevated by stages during the Tertiary period (Davis, 1909). The geology is complex (Williams, 1927); most of the rocks within the Reserve are igneous and of quite varied character and chemical composition. Rhyolite flows are the most extensive, but there are considerable areas of calcareous volcanic ashes. It is on and around exposure of these ashes that the site is situated and in the vicinity of the location of arctic-alpine species occurring near their southernmost limit in Britain. There are extensive outcrops of rock, particularly at higher altitude, and soils developed from these parent materials are to a large extent free of glacial drift (Ball et al., 1969) and form the substratum for the many different grassland communities.

The vegetation of Snowdon is highly modified from its original state as a result of centuries of management by man. Below about 600 m the natural vegetation climax is the oak *(Quercus)* woodland and above the tree line dwarf shrub heath was probably dominant (Seddon, 1962). During Roman and Mediaeval times the forests of Wales were severely cleared (Moore and Chater, 1969) and subsequently were replaced by scrub and eventually grasslands following the greater use of the mountains for grazing by domesticated stock. Sheep were numerically of first importance as grazers of mountain pasture in northwest Snowdonia although both cattle and goats were also present (Hughes et al., 1973). An ancient hill camp at Dinas Emrys, occupied intermittently in the Bronze Age, the Romano-British Iron Age and the Dark Ages (Savory, 1956, 1961), provides some of the earliest evidence of human activity in the area. Pollen analysis of peat deposits revealed during archaeological investigations indicated that the natural forest cover of the slopes of Snowdon had been cleared locally by Roman times and that hazel scrub *(Corylus avellana)* had taken its place (Seddon, 1962). The Aberconwy Charter (*c* 1198), by which certain lands including parts of Cwm Dyli were made over to the Cistercian Abbey of Aberconwy by Llewelyn ab Iorwerth (Llewelyn the Great) provides the first documented evidence of land occupation on Snowdon (Gresham, 1939).

A traditional form of animal husbandry involving the use of the *hafod* or summer dwelling took place at Cwm Dyli as late as 1861 (Roberts, 1937). Cattle were moved up in early March, fed in the stall in the morning and evening on hay cut the previous year, and allowed to graze pasture during the day. After June the cattle were taken down and the pastures given over to sheep, hay was cut to *c* 427 m, cheese and butter were made and sold in the autumn markets. Intensive sheep farming dates from about 1780 to 1820 but grazing throughout the year was replaced at the beginning of the present century by a period of mainly summer grazing (April to October) by ewes and lambs. This practice has continued to the present day and has led to a reduction in grazing of such plant species as *Nardus stricta*, *Molinia caerulea*, and *Juncus squarrosus* that had previously taken place during the winter after the more palatable species had been selected during the summer. Under the present system of management *Nardus* is virtually ungrazed throughout the year and this has allowed the plant to spread onto more productive pastures, its tiller phenology and nutrient economy (Perkins, 1968) being enhanced by the absence of grazing.

The chapters that follow present a synthesis of the productivity studies undertaken at the main site during the period 1966 to 1973. The study was an integrated investigation of production and the inter-relationships of soil, vegetation and herbivores at a single site. Production of organic matter is dependent upon solar energy, moisture and nutrients, supplies of which are important factors in ecosystem processes and exert a measure of regulation upon production which is the source of fixed energy to the whole system. IBP productivity studies have laid stress upon the determination of primary and secondary production in relation to trophic levels, the unifying theme being energy flow. Additional emphasis has been placed in this study on the transfer of five major nutrients through the grazing and litter pathways and spatial variation in production parameters caused by abiotic and biotic factors.

14.2 Vegetation of the Main Site

The vegetation of the area around Snowdon is a mosaic of community types comprising grassland, mire and heath. The predominant grasslands range between two extreme types: mesotrophic—dominated by *Agrostis tenuis* and *Festuca ovina* associated with a large number of minor species; and acidic—dominated by *N. stricta* with many fewer associated species. Floristic and structural composition are related to gradients in soil moisture, soil nutrient supply and sheep grazing intensity. Many intermediate types of grassland occur, differing in the proportion of the dominant species. The main site is a moderately herb rich *Agrostis-Festuca* type developed on a Brown Earth soil (Chap. 15). The surrounding vegetation (Fig. 1) is *Agrostis-Festuca-Nardus* and *Nardus-Festuca-Agrostis* on peaty podzolic soils with interspersed wet peaty soils dominated by *Juncus effusus* or *Eriophorum angustifolium*.

The main experimental area (0.74 ha) was delineated into 32 plots each 232 m^2 (Fig. 2). Considerable local heterogeneity resulted in variation in species composition and quantity (recorded on a grid point basis) between, but also within, plots. Five groups of plots were distinguished from the spatial relationship (Fig. 3) derived from reciprocal averaging ordination (Hill, 1973), axis 1 reflecting a gradient of increasing soil organic matter content and decreasing nutrient supply from A to E. Mean % cover of species in the plot groups, arranged according to axis 1 scores (Table 1), indicates a central group of species occurring more or less constantly throughout the plots but exhibiting differing cover values. Thus, *Agrostis tenuis* and *Anthoxanthum odoratum* decrease from A to E whereas *Festuca ovina* increases. Similarly the moss *Rhytidiadelphus squarrosus* decreases from A to E whilst *Hypnum cupressiforme* increases. The more oligotrophic plots D and E are differentiated by the presence of small amounts of *Sieglingia decumbens*, *Juncus squarrosus*, and *N. stricta*. *Carex pilulifera* is present on all plots but increases from A to E. In contrast, the more eutrophic plot groups A and B are characterised by higher proportions of e.g. *Holcus lanatus* and *Trifolium repens* and the presence of *Cardamine pratensis*. *Juncus effusus* is also present as small discrete patches, an indication of water flushing downwards from rocks above the plots during wet weather. In dry weather plots D and E retain more moisture, a reflection of greater soil organic matter content.

14.3 Climate

The climate of Snowdonia is essentially maritime, being subject to the effects of Atlantic air masses resulting in moderate temperatures and much rainfall throughout the year. With high amounts of cloud cover and frequent hill fog the

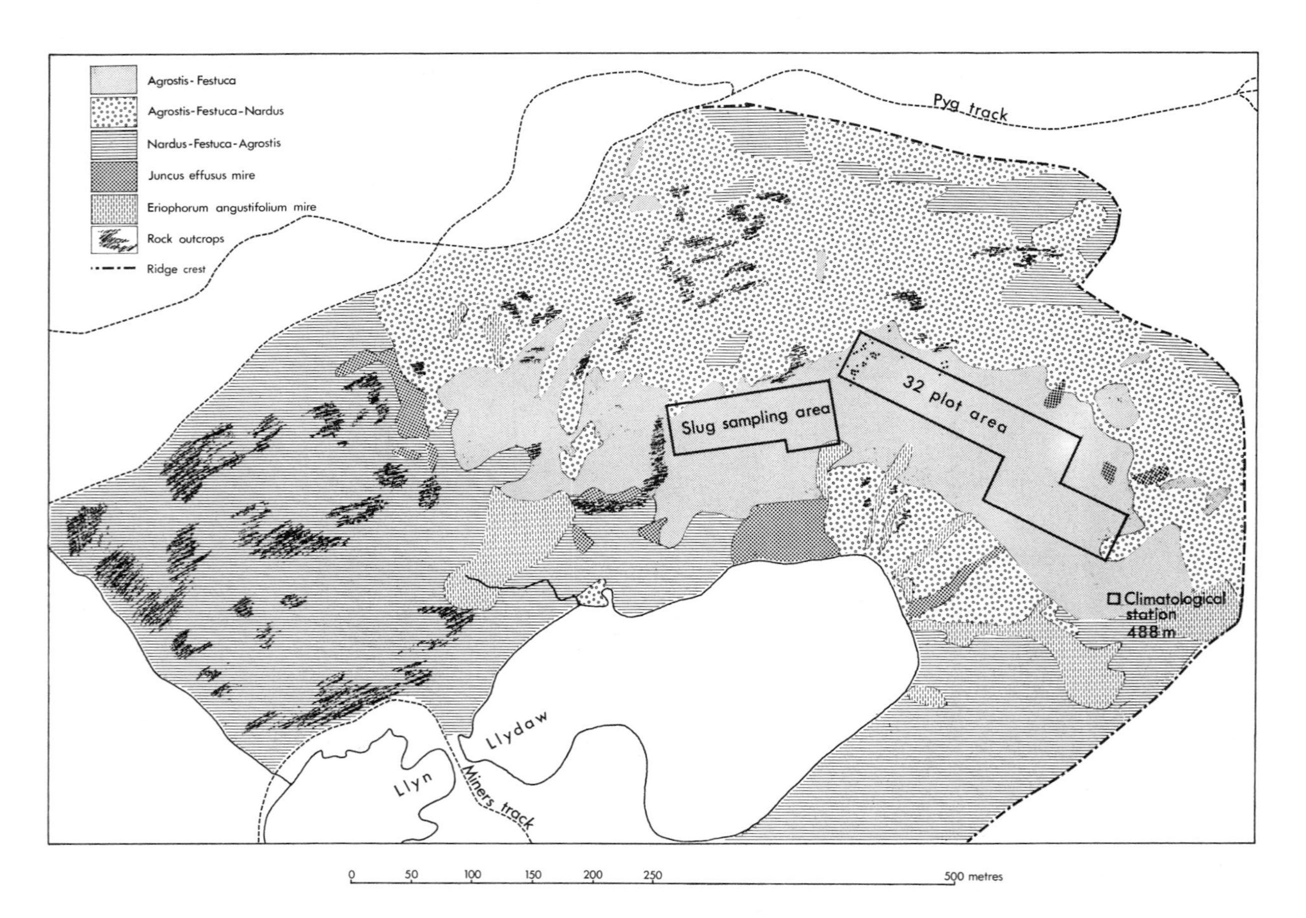

Fig. 1. Location and map of the vegetation of the Llyn Llydaw IBP main site (Y Wyddfa—Snowdon National Nature Reserve)

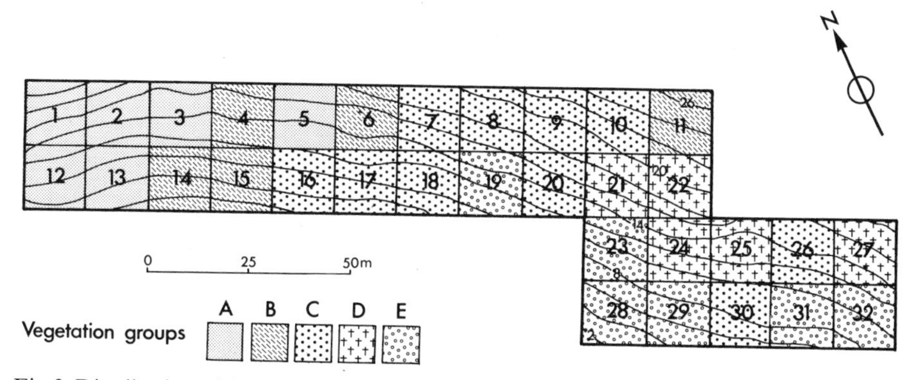

Fig. 2. Distribution of the 32 plots on *Agrostis-Festuca* grassland. Plots are grouped into types derived by ordination and contours indicated at 2 m intervals

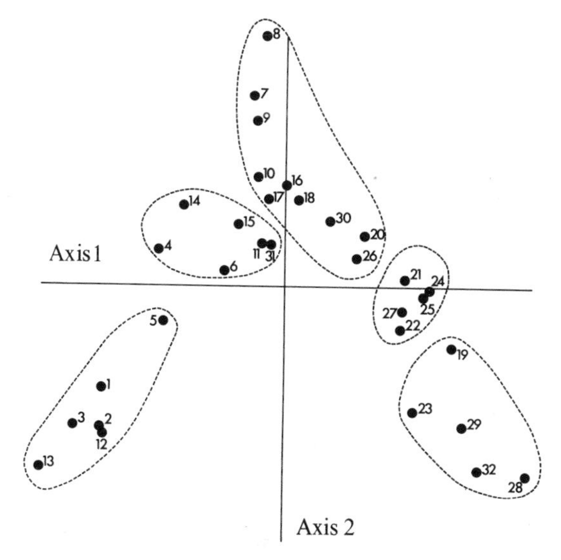

Fig. 3. Ordination diagram showing the spatial distribution of plots derived by reciprocal averaging (using plant species % cover data)

duration of sunshine is reduced compared with adjacent coastal areas and the topography of the area causes marked local differences in rainfall. Atmospheric humidity is high and cumulative temperature is low which together with high exposure to south westerly winds results in a climate more rigorous than might be expected from the moderate (*c* 1000 m) altitude. These conditions are reflected in western British mountains by the absence of trees above *c* 600 m, and the restriction of useful cultivation to a zone $< c$ 300 m (Pearsall, 1950).

Solar radiation recorded at the main site climatological station at 488 m averaged 308 kJ cm^{-2} (Table 2), maximum values occurring in June (1.83 kJ cm^{-2} d^{-1}). Although the mountains are subject to rapidly changing weather, temperature variation is small. Mean air temperature (1966–1971) was 7.2° C

Table 1. % cover value of species occurring in the plot groups (vegetation types A to E)

Species	Ordination score		Plot groups					Site mean
	Axis 1	Axis 2	A	B	C	D	E	
Dicranella heteromalla	95	−35	—	—	—	0.8	0.8	0.3
Carex pulicaris	95	−62	—	—	—	—	0.8	0.2
Sieglingia decumbens	86	−47	—	—	—	0.8	2.0	0.6
Juncus squarrosus	85	−36	—	—	—	0.4	1.4	0.4
Nardus stricta	83	−41	—	—	—	1.2	6.2	1.5
Bellis perennis	72	− 4	—	—	—	0.6	—	0.1
Euphrasia curta	64	−12	—	0.3	0.4	3.4	1.6	1.1
Viola riviniana	63	5	—	—	0.2	1.8	0.4	0.5
Plantago lanceolata	61	−13	—	1.2	3.2	6.8	8.4	3.9
Thymus drucei	59	− 7	0.3	0.3	5.3	12.0	11.2	5.8
Prunella vulgaris	47	4	—	—	0.2	—	0.4	0.1
Carex pilulifera	46	−10	0.3	4.0	4.6	5.8	12.0	5.3
Polygala serpyllifolia	40	20	—	—	0.2	—	—	0.0
Hylocomium splendens	22	27	—	—	1.2	0.8	—	0.4
Hypnum cupressiforme	18	0	30.8	34.8	53.4	57.6	63.6	48.0
Festuca ovina	16	0	36.2	36.2	52.7	59.8	54.2	47.8
Galium saxatile	12	2	15.8	17.5	24.6	23.6	22.0	20.7
Lophocolea bidentata	9	11	7.3	12.3	15.6	12.4	7.8	11.1
Mnium punctatum	9	11	—	1.3	0.4	1.2	—	0.6
Polytrichum commune	8	11	16.2	12.8	28.5	22.2	15.0	18.9
Achillea millefolium	5	19	1.5	2.8	4.4	2.6	2.4	2.7
Cirsium palustre	3	12	7.2	10.2	14.4	8.6	6.6	9.4
Agrostis tenuis	0	1	70.0	70.3	67.3	61.6	53.8	64.6
Deschampsia caespitosa	0	6	1.8	1.8	1.6	3.6	0.6	1.9
Polytrichum piliferum	0	41	—	—	0.6	—	—	0.1
Anthoxanthum odoratum	− 2	− 2	17.7	13.7	13.3	13.4	11.8	13.9
Cynosurus cristatus	− 2	−16	2.8	2.0	0.6	1.8	1.8	1.8
Luzula campestris	− 2	0	10.8	11.8	9.0	10.0	7.0	9.7
Mnium undulatum	− 8	4	14.3	17.2	13.3	8.2	9.8	12.6
Agrostis canina	− 9	18	0.3	1.0	0.5	—	0.8	0.5
Pleurozium schreberi	− 9	39	2.3	0.8	8.4	3.2	—	2.9
Rhytidiadelphus squarrosus	− 9	− 9	76.2	29.7	40.7	43.0	37.4	45.4
Dicranum scoparium	−10	100	—	—	0.4	—	—	0.1
Rhacomitrium lanuginosum	−11	−18	0.7	—	0.6	—	0.8	0.4
Cirsium arvense	−12	3	1.3	2.3	1.2	0.4	0.6	1.2
Ranunculus repens	−14	12	8.0	12.0	8.9	4.8	3.4	7.4
Sagina procumbens	−15	2	0.3	1.0	0.2	0.6	—	0.4
Holcus lanatus	−25	− 3	10.7	8.2	5.1	3.0	3.2	6.0
Thuidium tamariscinum	−26	−13	5.7	—	2.5	0.8	1.6	2.1
Cerastium holosteoides	−27	22	0.7	1.0	1.2	—	—	0.6
Poa annua	−33	− 3	6.7	7.3	2.6	—	2.2	3.9
Trifolium repens	−34	− 2	16.2	10.5	7.4	3.8	2.6	8.1
Hieracium pilosella	−40	7	0.3	—	0.2	—	—	0.1
Cardamine pratensis	−54	17	0.5	0.7	0.2	—	—	0.3
Festuca rubra	−63	−14	0.3	—	—	—	—	0.1
Rumex acetosella	−72	−26	6.2	1.8	0.6	—	0.4	1.8
Pseudoscleropodium purum	−83	−32	13.3	2.5	0.8	—	—	3.3
Poa pratensis	−95	−41	0.3	—	—	—	—	0.1
Juncus effusus	−100	−51	17.2	1.2	—	—	—	3.7

Table 2. Mean monthly climatological data for the Llyn Llydaw IBP site

	J	F	M	A	M	J	J	A	S	O	N	D	Year
Solar radiation (kJ cm^{-2}) 1969–1970	6.6	11.6	18.5	31.8	44.2	54.8	44.3	39.6	23.6	18.8	7.7	6.6	308.2
Air temperature (°C) 1966–1971													
mean maximum	4.5	2.8	4.3	7.0	10.7	14.6	15.2	15.3	14.0	11.2	6.0	5.2	9.2
mean minimum	1.6	−0.5	0.5	2.3	5.7	9.3	10.3	10.7	9.7	7.5	2.6	1.5	5.1
$\frac{1}{2}$ (max + min)	3.0	1.2	2.4	4.6	8.2	12.0	12.7	13.0	11.9	9.3	4.3	3.4	7.2
highest maximum	13.4	9.9	18.8	17.3	21.0	24.0	25.8	23.0	22.4	22.5	12.8	17.2	25.8
lowest minimum	−7.5	−8.2	−4.8	−5.3	−4.1	3.2	5.9	6.7	4.9	0.3	−3.8	−6.7	−8.2
Lowest grass minimum (°C)	−11.5	−10.0	−13.0	−8.0	−5.5	−3.0	1.0	−1.0	−2.8	−3.6	−6.7	−15.0	−15.0
Soil temperature (°C) 1968–1971													
15 cm	3.5	2.7	2.8	4.6	8.3	11.9	12.8	13.5	12.3	10.6	6.8	4.5	7.9
30 cm	3.9	3.2	2.6	4.8	7.8	11.2	12.3	12.7	12.1	10.7	7.2	5.2	7.8
60 cm	4.1	3.6	2.8	4.6	7.1	8.8	11.5	12.4	11.9	11.6	7.8	5.5	7.6
Rainfall (mm) 1945–1965[a]	409	292	262	241	193	252	300	343	399	338	376	417	3,820
1966–1971[b]	272	297	214	225	179	212	112	257	229	318	329	289	2,932
Wind velocity, average (m s^{-1}) 1967–1971	4.46	4.86	4.53	3.18	3.62	3.13	3.45	3.66	3.36	4.50	5.12	4.58	4.04

[a] Mean of three gauges in vicinity of the main site (Grid refs: SH 623550, SH 638551, SH 654540).
[b] Gauge maintained at the main site climatological station (Grid ref: SH 636553).

(mean maximum 9.2° C, mean minimum 5.1° C). Summers are cool with the highest mean occurring in August (13.0° C). The highest maximum temperature was 25.8° C (July) and the lowest minimum −8.2° C (February). Low grass minimum values were recorded throughout the year, frost occurring in each month except July. Mean soil temperatures were 7.9° C (15 cm), 7.8° C (30 cm) and 7.6° C (60 cm) with maximum values occurring in August. Upland soil temperatures are much influenced by the temperature of the air (and rainfall) reflecting annual air temperature variation. Commencement of plant growth in the spring varied considerably between years (Chap. 16) but the season corresponded well to a period based on mean air and soil temperature $>5.6°$ C and was about 200–220 days from May to October.

The area around Snowdon summit is one of the wettest locations in the British Isles, being one of the first high land masses encountered by moisture-laden prevailing south westerly winds. Annual rainfall recorded by three gauges within Cwm Dyli nearest to the IBP site averaged 3820 mm between 1945 and 1965. Monthly distribution indicates that the wettest period is in the autumn and winter, whilst the driest months are from March to July. The gauge at the site in a more exposed location showed a similar distribution pattern but recorded an average of 2932 mm between 1966 and 1971. Estimated soil moisture deficits (Meteorological Office, 1962–1971) were small for the summer months (0–13 mm) and, except only rarely in the most free draining areas, moisture was adequate throughout the year. Number of wet days (>1 mm d^{-1}) averages 200 yr^{-1} and relative humidity 82% throughout the growing season. High humidity values are characteristic of western upland areas where $>90\%$ can be expected for 70% of the year (Oliver, 1964). Snow lies for between 16 and 100 d yr^{-1} in the area depending on altitude, but although late snow does sometimes remain in gullies it is insufficient for the development of late snowbed vegetation. Moderate average wind velocities occur throughout the year (mean 4.04 m s^{-1}) ranging from 3.13 m s^{-1} in June to 5.12 m s^{-1} in November.

Acknowledgements. These studies were undertaken and funded as part of the Nature Conservancy's (Natural Environment Research Council) research programme in association with the Royal Society, and the latter stages continued with the support of the Institute of Terrestrial Ecology (Natural Environment Research Council). The work was part of the Snowdonia IBP investigations initiated and directed by Dr. R. Elfyn Hughes and was made possible by the cooperation of Mr. D. Jones and Mr. G. Jones, Gwastadannas. I am grateful to Mrs. V. Jones and Mrs. P. Neep for recording the vegetation data; Mr. I. Ellis Williams and Miss S. Brasher for information incorporated in Figure 1; and Mr. D. Meikle, Mr. S. Rees, and Mr. D. Edwards for undertaking climatological observations. Instrumentation throughout the IBP project was prepared and maintained by Mr. G. H. Owen.

15. Physiography, Geology and Soils of the Grassland Site at Llyn Llydaw

D. F. Ball

Variations in soil conditions have a considerable influence on the pattern of vegetation and grazing in Snowdonia. This paper provides a short description of the soil characteristics and the factors influencing their development.

15.1 Physiography and Geology

The IBP study area (Fig. 1, Chap. 14) rises from *c* 425 m at the lake shore to *c* 510 m at its upper margin, above which are steeper dissected and gullied cliffs of weathering rock. There are breaks of slope in the lower half of the area and marked microrelief produced by small gulley dissection and by active and healed slip features. The main site slopes at 25 to 32° between altitudes of 460 and 490 m.

Geologically, the Cwm Dyli sector of Snowdon consists of volcanic rocks of Ordovician age, and subordinate intrusions of dolerite (Williams, 1927). The study area is almost entirely underlain by pumice-tuffs of the Bedded Pyroclastic Series. The pumice-tuffs are rapidly weathering fissile volcanic ashes, iron and magnesium rich and usually slightly to moderately calcareous. These rocks, in their ease of weathering, combined with relatively moderately high base status, contrast sharply with the other igneous rocks, slates and grits of the Snowdon area which either weather slowly, or effectively do not weather at all. On the shore of Llyn Llydaw outcrops of rhyolite and rhyolitic ash of the Lower Rhyolitic Series occur as ice-smoothed and striated surfaces, but contribute nothing to the parent material of the soils on the slope.

A range of different superficial deposits were mapped by Williams (1927) within a single "drift" unit, which is shown as occupying the study area. Ball et al. (1969) obtained more detail on the distribution of scree, till and peat. Hummocky morainic till occurs around part of the Llyn Llydaw shoreline but the study area is free of boulder clay or any contribution from former boulder clay cover. The slope is mantled with unconsolidated loamy material derived from weathering of the pumice-tuff. The absence of drift may be attributable to erosion of the relatively soft pumice-tuff by late-glacial ice into near vertical cliffs. The cliffs could not retain drift and post-glacial subaerial erosion created a colluvial slope in place of the original cliff. Alternatively, or additionally, any drift present on this south-facing slope was removed by late-glacial solifluction or water erosion quite rapidly after ice retreat. Whatever the causes, the important point is that the soils of the study area are entirely derived from more or less in situ weathering of a single rock type, without contamination by material of other geological origin.

Table 1. Total chemical composition of pumice-tuff rocks and clays at Llyn Llydaw and adjacent areas (wt %)

Material and source	SiO_2	Al_2O_3	Fe_2O_3	FeO	MgO	CaO	Na_2O	K_2O	TiO_2	MnO	H_2O^+	H_2O^-
A. Pumice-tuff, Snowdon summit (*c* SH 610543) (Williams, 1927)	44.7	15.8	1.8	12.8	14.5	0.4	0.1	tr	1.0	0.5	7.8	0.3
B. Pumice-tuff, Llydaw site (*c* SH 635552) (Partial analysis by X-ray fluorescence, Ball et al., 1969)	46	nd	Total as Fe_2O_3 13.0		nd	7.9	nd	1.8	2.0	0.14	6.5	0.1
C. Alluvium derived from pumice-tuff, bed of Llyn Llydaw below study area (*c* SH 355550) (analysis as B)	48	nd	13.3		nd	0.3	nd	0.5	3.0	0.29	6.5	2.1
D. Clay fraction from weathered pumice-tuff, near Pen-y-Gwryd (*c* SH 658532) (analysis by J.H. Scoon in Ball, 1966)	32.7	17.3	3.9	13.4	19.1	0.4	0.5	0.8	0.63	0.22	10.6	0.5

nd: not determined; tr: trace.

Chemical analyses of pumice-tuff rocks confirm their general high magnesium-iron character (Table 1). The Snowdon summit example (A) contained no calcium carbonate ($CaCO_3$) as analysed, but the calcium oxide (CaO) found in the partial analysis of the Llydaw fresh rock (B) was virtually all present as $CaCO_3$ in vesicles. The alluvium (C) deposited on the lake floor by downslope movement of soil derived from rocks comparable to those analysed (B) has lost the carbonate through weathering. The clay fraction of the weathered pumice-tuff (D) is generally representative of clays from these rocks and derived soils in Snowdonia, although a range of chlorite compositions is found (Ball, 1966).

Mineralogically the fissile pumice-tuffs are dominated by micaceous minerals of the chlorite group, which have remained essentially unaltered in the virtually mono-mineralic clay and silt fractions of soils derived from these rocks (Ball, 1966; Hayes, 1970). This suggests the relative immaturity of these soils because in other parent materials in Wales chlorite in mixed clay mineral assemblages weathers substantially (Adams et al., 1971).

15.2 General Soil Profile Description

The soil on the colluvial loam at the study site is classifiable as a Brown Earth (= Brown Earth of low base status or Typical Brown Earth) in British classifications, and probably as a Typic Cryochrept[1] in the USDA (1960, 1967) system. These soils, recorded by Hughes (1949, 1958a), were defined in a regional survey (Ball, 1963) as the *Crafnant* series.

A type profile description and analyses (Ball et al., 1969) are given below and in Table 2.

Profile: IBPS3 (S. 6)
Location: Northeastern shore of Llyn Llydaw, Cwm Dyli (SH 639550)
Site: 25° uniform colluvial slope below pumice-tuff cliffs at *c* 485 m OD
Parent material: Colluvium derived from pumice-tuff rocks
Drainage: Free profile drainage on normal draining site

Profile: 0–15 cm Very dark greyish brown (10 YR 3/2) loam to silty loam with frequent small and occasional large stones of pumice-tuff; strong large crumb structure; good porosity; friable; moderate organic matter content; abundant roots; moist; few earthworms and earthworm channels; merging boundary to

15–30 cm Dark greyish brown (10 YR 4/2) loam to silty loam as above; moderate cloddy structure breaking to medium crumb; good porosity; slightly compact; moderate organic matter content; frequent roots; moist, few earthworms and earthworm channels; merging boundary to

30–40 cm Dark yellowish brown (10 YR 4/4-4/6), otherwise as horizon above, except for fewer roots and low organic matter content; merging boundary to

40 cm As above but mottled with faintly greyer brown of fresher weathering tuff; earthworms and channels to 60 cm; horizon continuing to 90 cm over weathered pumice-tuff.

The type profile data give a general indication of soil morphological and chemical characteristics in the study area. Locally there are patches of shallower

[1] The soil temperatures at the adjacent climatological site, at 60 cm depth the were: annual mean 7.6° C; mean June–August 10.9° C; mean December–February 4.4° C. Alternate names for this soil in the USDA system depend on the classification assigned to the 15–30 cm horizon, which has transitional characteristics for the defining criteria.

Table 2. Chemical analyses of type profile (IBP 53) of *Crafnant* series soil

		Horizon depth (cm)			
		0–15	15–30	30–40	40–60
pH (field moist soil)		5.2	5.3	5.4	5.1
Loss-on-ignition (wt %)		12	7	4	5
Organic carbon (wt %)		5.5	3.2	1.8	2.3
Total nitrogen (wt %)		0.59	0.37	0.20	0.21
C/N ratio		9.3	8.7	9.0	10.9
Exchangeable cations (me 100 g^{-1})	Ca	1.1	0.6	0.4	0.4
	Mg	1.0	0.3	0.2	0.1
	K	0.38	0.06	0.02	0.01
	Na	0.38	0.35	0.28	0.28
	Mn	0.22	0.04	0.01	0.01
Exchangeable P_2O_5 (mg 100 g^{-1})		1.0	0.1	<0.06	<0.06

Analytical methods: pH—1 : 2.5 soil : water ratio; loss-on-ignition—as wt % loss after ignition at 375° C, 16 h; carbon—calculated from loss-on-ignition (as in Ball, 1964); nitrogen—wt % total nitrogen; exchangeable cations—as me 100 g^{-1} determined in neutral N ammonium acetate extractant; and extractable P_2O_5—as mg 100 g^{-1} in 0.5N acetic acid.

soils, especially in recent or old landslip areas. In a few sectors slight mottling in the upper horizons indicates some drainage impedance or significant flushing at spring or deep drainage gulley sites. The former profiles are of Brown Ranker type [cf *Cwm Glas* series (Ball, 1963)] and the latter approach an unnamed Gleyed Brown Earth (Brown Earth with gleying or Stagnogleyic Brown Earth) variant of the *Crafnant* series. These subordinate soils do not make a substantial contribution to the intensively studied sector of the slope. The study area is, for this montain region, a good example of a moderately extensive area of morphologically uniform freely drained soil.

The survival of Brown Earth soils in an area with such heavy rainfall and hence leaching cannot be attributed simply to the effect of a parent material of relatively high base status. Where pumice-tuff has weathered on a nearby level stable site a podzolic profile has developed (Ball et al., 1969). The existence of a Brown Earth profile with a crumb-structured mull humus surface horizon requires the survival of burrowing earthworms which depend on moderately high pH and available calcium levels. Such levels are sustained on the study area through the relative instability of the soils on slopes. Earthworms and moles bring material from lower horizons to the surface. Subsurface rock weathering provides fresh mineral material at the base of the soil profile. Run-off water from the moderately lime rich rocks above the site, and water which re-emerges at the ground surface after percolating through lower horizons of the soil, flush the upper soil horizons. Mass movement of material occurs by accretion of material fallen and washed from the weathering outcrops above, from soil bared by minor slips and gullies which affect the slope, and by rain wash of material brought to the surface by faunal activity.

The continuing supply and recycling of dissolved and solid materials do not cause destruction of the slope soil cover or of soil horizons, but they sustain surface pH and nutrient levels which permit survival of the fauna necessary to maintain Brown Earth soil profiles. On a stable site in Snowdonia, the "climax" soil on the fissile chloritic ashes among the pumice-tuffs is a podzolic soil. On a completely unstable site shallow, immature Brown Ranker soils are found in a patchy occurrence on sheltered ledges among rock outcrops. On the quasi-stable colluvial slopes the low base status Brown Earth of the study area is the typical soil occurring on these rocks. In fresh exposures in road sections breakdown of the greenish grey unaltered massive rock produces a brown loamy matrix within a few years.

15.3 Soil Chemical Characteristics

15.3.1 Total Element Composition

The total chemical composition of a soil indicates the ultimate mineral resources of elements potentially available for release by weathering.

The major element soil analyses (Table 3) and those of the rocks and clays (Table 1) confirm the uniformity of soil parent material across the site that was indicated by field evidence. A higher Na content in the soils compared to the rocks is the only major distinction but local variation in Na and K contents is a feature of the pumice-tuff rocks, and the Na levels are consistent throughout the sampled soils. Na supply in rain water is probably also a significant factor. Although there are no substantial differences between the total chemical composition at different depths (Table 3) there is considerable variation between sites in total K and Ca. The total nutrient reserves in the non-ignited soil are estimated as: CaO 0.16% (=0.11% Ca); MgO 7.5% (=4.5% Mg); Na_2O 2.8% (=2.1% Na); K_2O 0.65% (=0.54% K); P_2O_5 0.28% (=0.12% P).

Trace element concentrations (Table 4) are not exceptional in relation to the general ranges found in soils (Swaine, 1955). Rb, Sr, Pb, Ni are relatively low but within the general range; Cu, Y, Zr, Nb are close to average values; Zn, Cr, Co are relatively high but within the general range.

15.3.2 pH, Organic Matter and Exchangeable Cations

The chief source of variation in soil chemistry at 0–14 cm depth in the main site is spatial. Much of this variation occurs within plots, at the cm scale (Ball and Williams, 1968). Seasonal variations, if present, were much smaller than spatial variability and not detected. Detailed soil sampling of 11 of the 32 plots was carried out for an investigation of sampling programmes (Ball and Williams, 1971). These data, combined with analyses from the remainder of the plots, provide the most comprehensive soil chemistry data for the main site (Table 5). In addition to changes over short distances there are broader trends between plots on a scale seen also for biotic factors (Chap. 16). In plots with higher soil organic matter and exchangeable cations (particularly K) there was slower decomposition of surface plant litter and lower net primary production. In the more productive plots, where sheep grazing was more intense, soil P_2O_5 values were higher, related

Table 3. Total chemical composition of soils from Llyn Llydaw study area (wt % of ignited samples). (Analyses by Dr. G. Hornung, Department of Earth Sciences, University of Leeds)

Soil		Organic matter (as loss-on-ignition)	SiO_2	Al_2O_3	Fe_2O_3 (Total)	MgO	CaO	Na_2O	K_2O	TiO_2	MnO	H_2O^+	P_2O_5
Profile sampled 1973 near site A 1 of Ball and Williams (1968)	0–15 cm	7.7	47.11	15.40	14.88	8.53	0.16	2.8	0.70	3.53	0.31	6.25	0.29
	15–30 cm	4.7	48.84	15.29	14.81	8.35	0.15	2.8	0.75	3.75	0.29	5.21	0.26
	30–45 cm	4.8	46.59	15.59	14.86	8.65	0.13	2.8	0.67	3.42	0.30	5.70	0.25
	45–60 cm	4.9	48.50	15.96	14.74	8.71	0.13	3.0	0.72	3.16	0.31	5.43	0.28
Profile sampled 1973 near site C 1 of Ball and Williams (1968) (in IBP plot)	0–15 cm	9.9	51.11	15.30	14.01	7.45	0.26	3.4	0.58	3.21	0.28	4.82	0.26
	15–30 cm	8.6	48.80	15.99	14.87	7.74	0.23	3.0	0.55	3.00	0.29	5.55	0.26
	30–40 cm	6.8	48.43	16.38	15.32	8.22	0.16	2.8	0.65	2.83	0.33	5.56	0.32
	40–60 cm	7.4	47.15	16.59	15.71	8.51	0.17	3.1	0.44	2.91	0.30	5.62	0.30
Surface soil random sample 1 (0–15 cm) (in IBP plot)	0–15 cm	10.0	50.16	15.97	13.55	8.39	0.11	3.1	1.07	3.03	0.24	5.06	0.27
Surface soil random sample 2 (0–15 cm) (in IBP plot)	0–15 cm	10.1	49.44	15.66	13.39	8.60	0.19	2.9	0.79	2.93	0.24	5.45	0.28
Mean of all soil samples above		7.49	48.63	15.81	14.61	8.31	0.17	3.0	0.69	3.18	0.29	5.47	0.28
Mean of the four surface soil horizons above	0–15 cm	9.42	49.45	15.58	13.95	8.21	0.18	3.1	0.78	3.17	0.27	5.40	0.28
Mean of the six deeper soil horizons above	15–60 cm	6.20	48.09	15.96	15.05	8.36	0.16	2.9	0.63	3.17	0.30	5.51	0.28

Table 4. Total trace element content of soils from Llyn Llydaw study area (ppm). (Analyses by Dr. G. Hornung, Department of Earth Sciences, University of Leeds)

Soil		Rb	Cu	Sr	Zn	Y	Zr	Pb	Nb	Cr	Co	Ni
Profile sampled 1973	0–15 cm	13	29	43	137	35	227	50	15	191	53	39
near site A 1 of Ball	15–30 cm	16	23	46	134	35	218	23	17	193	51	43
and Williams (1968)	30–45 cm	14	22	42	137	39	222	15	14	208	51	44
	45–60 cm	13	23	44	133	36	216	17	12	217	48	49
Profile sampled 1973	0–15 cm	21	16	63	103	41	208	nd	16	183	45	45
near site C 1 of Ball	15–30 cm	18	7	58	101	26	216	nd	14	201	48	57
and Williams (1968)	30–40 cm	22	25	53	109	37	218	nd	13	201	55	52
	40–60 cm	15	17	55	109	25	208	nd	13	231	57	54
Surface soil random sample 1 (0–15 cm) (in IBP plot)	0–15 cm	23	31	67	97	26	181	43	13	203	50	56
Surface soil random sample 2 (0–15 cm) (in IBP plot)	0–15 cm	20	22	60	103	25	200	47	16	186	46	60

nd: not determined.

Table 5. Soil chemical analyses for the 32 plot study area (0–15 cm soil depth)

	pH (air-dry soil)	Loss-on-ignition (wt %)	C (calculated wt %)	Total N (wt %)	Exchangeable cations (me 100 g^{-1})					Extractable P_2O_5 (mg 100 g^{-1})
					Ca	Mg	K	Na	Mn	
Mean	4.73	12.11	5.14	0.51	1.98	0.84	0.47	0.49	0.24	2.95
Standard error	0.03	0.37	0.17	0.03	0.07	0.02	0.02	0.01	0.03	0.30
Coefficient of variation (%)	3	17	18	20	19	13	21	6	58	58

Analytical methods as in Table 2.

to a greater deposition of sheep faeces containing appreciable quantities of phosphorus (Chap. 19). Whilst the surface soil chemistry is influenced by the addition of plant litter and sheep faeces, sub surface horizons showed little variation. Only subsequent monitoring could determine whether the indicated trends proceed at rates that will lead to substantial changes in the course of soil development and its associated biotic cycle.

16. Primary Production, Mineral Nutrients and Litter Decomposition in the Grassland Ecosystem

D. F. PERKINS, V. JONES, R. O. MILLAR and P. NEEP

Measurements of seasonal changes in the standing crop of live vegetation are adjusted for transfers to herbivores, standing dead litter, and for litter decomposition, to estimate primary production of the grassland. The dynamics of nutrient concentrations are related to phenology of the grasses, and spatial variation in productivity to species composition, soil chemistry and intensity of sheep grazing.

16.1 Introduction

The vegetation of the ecosystem studied at the main IBP grassland site at Llyn Llydaw is a semi-natural, moderately herb rich, *Agrostis-Festuca* grassland (Chap. 14). It represents one of the most productive grasslands (in terms of supporting sheep) in the mountains of Snowdonia at high rainfall and medium altitude range (Chap. 18). This chapter describes the main features of the distribution of dry matter, energy and some major mineral nutrients and nitrogen in the primary producers of the grassland ecosystem. The assessment of net primary production and the rate at which dry matter and its contained nutrients become available to herbivores has been a principal objective.

Closely associated with production of living material is the appearance or production of dead material through death of leaves and other parts of plants. The rate of disappearance of dead material from the soil surface was studied to assess the activity of decomposers in the plant litter. The nutrient concentration in living and dead plant material per unit area of land surface allowed an assessment not only of the nutritional quality of the vegetation for herbivores but also of some of the parameters necessary to construct nutrient transfer models of the ecosystem. Similarly, the distribution and fate of energy in the primary producers provided parameters for a model of the energy flow through the ecosystem (Chap. 20).

Considerable sampling problems were caused by the heterogeneity of the site, reflected in the spatial pattern of species distribution, but dynamic inter-relationships could be related to the underlying causes of the variability. Thus biomass and net primary production of the main vegetation components (grasses, sedges, herbs and mosses) were related to soil chemical composition, plant species composition and effects of grazing.

16.2 Plant Biomass and Net Primary Production

16.2.1 Methods

Assessment of plant biomass, net primary production and the quantity of vegetation consumed by large herbivores (sheep) involved harvesting material in the field by methods outlined by Milner and Hughes (1968) and similar to those used at Moor House (Chap. 2). Changes in annual amounts of biomass and production related primarily to variation in climate. Whilst these fluctuations must be taken into account in an overall assessment, it is assumed (as at Moor House) that the system approaches a "steady state" i.e. the site is assumed to be in equilibrium with local environmental factors, the state of biotic or plagioclimax described by Tansley (1939).

16.2.1.1 Biomass (Standing Crop)

Biomass, which here is taken to include all living and dead plant material together with litter on the soil surface, was determined by harvesting material (the standing crop) by either of two methods:

1. Hand clipping 0.5 m × 0.5 m quadrats above the plant bases or crowns within 1–2 cm of soil surface. At Llyn Llydaw, this coincided with the surface of a dense layer of bryophytes, and

2. Sampling of turf cores (6.0 cm diameter, 8.5 cm deep) and separating the aerial samples into component parts in the laboratory. Grasses were separated and the number of tillers counted, but only the most abundant species, *Agrostis tenuis*, *Festuca ovina* and *Anthoxanthum odoratum*, are reported individually. Other plants have been grouped into sedges and rushes, herbs and mosses. Each component was further separated into living (green leaves and non-green bases) and dead (brown parts), referred to as "standing dead". Litter was harvested from the surface of the soil core but not separated into species. Roots were sub sampled from the primary core using a 3.2 cm diameter corer, washed free of soil (no dispersant), the roots being retained on a 2.0 mm mesh sieve. Soil remaining in the primary core was retained for determination of moisture content and chemical constituents. Samples of vegetation were dried at 70° C in a forced draught oven; biomass and production estimates are, therefore, reported on a dry weight basis which includes mineral ash.

16.2.1.2 Net Primary Production

All organisms in an ecosystem are dependent upon the energy and nutrients fixed in photosynthetic plants—the primary production. The amount of above ground net primary production (P) of grazed grasslands in excess of that respired during plant metabolism and translocated to roots, during a specified period of time (t) can be assessed from Equation (1) reported by Milner and Hughes (1968):

$$P = \Delta B + L + G \quad (1)$$

where ΔB = change in biomass; L = plant losses to standing dead and litter by death and shedding of leaves and stems; G = plant losses by grazing. The method can be used satisfactorily for biomass measured either by harvesting clipped quadrats or turf cores. Sheep grazing was prevented during the experiments by using portable animal exclosures 1.5 m × 1.5 m and 1.0 m high, of 5 cm mesh wire netting. Production was, therefore, estimated from:

$$P=(B_2-B_1)+L+(B_3-B_2) \quad \text{or} \quad P=(B_3-B_1)+L \tag{2}$$

where B_1 is the biomass at the beginning of the sampling period (usually *c* 28 days, the minimum time in which a change of biomass could be measured with reasonable accuracy by weighing); B_2 is the biomass outside and B_3 the biomass inside the exclosure at the end of the period.

The amount of herbage consumed by sheep (C) was estimated (Chap. 19) from:

$$C=B_3-B_2\,. \tag{3}$$

The value of C is unaffected by L, assuming that death and shedding of plant parts are comparable in short term ungrazed and grazed plots. Consumption by other consumers is small compared with that by sheep and could be ignored [therefore $G=C$ in Eq. (1)] but consumption by slugs (Chap. 17) has been added to the consumption by sheep to estimate an overall production value for the site.

Distinction must be made in the assessment of net primary production by harvest methods in the absence of grazing (in exclosures) between:

1. the production yield, viz the weight of living and/or dead material harvested at the end of a specified period of growth; and

2. the adjusted production, viz the weight of living material harvested at the end of a specified period of growth plus the dead material (L) formed in the same time, corrected for loss in weight upon senescence.

Method (1) results in an underestimate of production in which only living material is harvested and L in Equation (2) is not or is only partially estimated. If attached dead is harvested together with living material, the underestimate is confined to the reduction in weight from living to dead material and the loss from standing dead to litter and disappearance of dead material during the sampling period. Method (2) requires the measurement of dead production (L) and is possible from turf core data, L being added to the living weight increment if both are significant values ($P<0.1$). The method is similar to that proposed by Weigert and Evans (1964). A correction is necessary for losses of standing dead to litter and for the very rapid rates of disappearance of the dead biomass during the sampling period. If a significant increase in both living and dead material occurs the loss of dead biomass (d) in time (t) is estimated as:

$$d=\frac{B_1+B_3}{2}\times k\times t \tag{4}$$

where B_1 = the biomass of dead material at the beginning of the sampling period; B_3 = the biomass inside the exclosure at the end of the period; k = instantaneous rate of disappearance of dead material determined in situ by the modified core method. A correction factor (mean for vascular plants = $\times 1.52$), obtained by comparing mineral ash contents of living, dead and litter samples, is used to reconstitute the living weight of dead material. The inaccuracies in such a calculation, which tends to underestimate the living weight of dead material, are small compared with those associated with the estimate which makes no allowance for L. The above method attempts to limit error due to the summing procedures by considering dominant species and groups of the minor species and accepting only significant changes of living, dead and litter biomass (Weigert and McGinnis, 1975: Singh et al., 1975).

A third way (method 3) of estimating net primary production is similar to that used by Weigert and McGinnis (1975). It requires accurate measurement of the biomass of standing dead and litter and its instantaneous rate of disappearance and thus provides a check on the above method (2) calculations. Thus L is estimated from:

$$L = \bar{B} \times k \times t \quad (5)$$

where $\bar{B}$ = mean biomass of standing dead and litter; k = instantaneous rate of disappearance of dead material determined in situ by experiment; and t = time (365 days). The quantity of dead material disappearing each year under steady state conditions should equal dead production resulting in zero change in the biomass. Primary production is then estimated from:

$$P = L + G\,. \quad (6)$$

The measurements of root biomass in the surface 0–1 cm and 1–8.5 cm horizon of soil are used to assess root production, and turnover rates (T) calculated as suggested by Dahlman and Kucera (1965) and Dahlman (1968) from:

$$T = \frac{\text{maximum biomass} - \text{minimum biomass}}{\text{mean biomass}} \quad (7)$$

for a 12 month period. Root production is estimated from the product of the maximum biomass, usually at midsummer at Llyn Llydaw, and the turnover rate (T). The results obtained from these estimates are tentative until more precise measurements have been undertaken.

Samples for all harvest determinations were taken in each of the 32 plots (Chap. 14) and the sample location recorded so that results could be related to soil and animal data. Site values are, therefore, based on a mean of 32 observations at each sampling time.

16.2.2 Distribution and Dynamics of the Biomass of Living Components

Agrostis-Festuca grasslands in Snowdonia generally have low standing crops as a result of intense defoliation by sheep. Comparison of clipping and turf core methods of estimating biomass in 1968 indicated that, as well as being difficult and inaccurate due to the small quantities of vegetation, clipping left 100–

200 g m^{-2} of material on the soil surface—about 45% bryophyte and 55% vascular plants, *c* 50% of each component being dead. The turf core samples were therefore considered to give the more satisfactory estimate of total biomass. Similar core sampling methods were used by de Vries (1937) and Hutchinson (1967) to measure biomass in pasture of low availability to sheep.

Biomass data for 1969 and January 1970 for the site at Llyn Llydaw (Table 1) show *A. tenuis* to be the main constituent, with a mean biomass of 76 g m^{-2}, 35% of the total living dry matter. There was much bryophyte material (28%), mainly mosses, with a mean biomass of 61 g m^{-2}. Grasses accounted for about 54% of the living components of the sward (75% if bryophytes are excluded): *F. ovina* 13% and *A. odoratum* 4% of the total biomass. Herbs, and sedges and rushes, whilst representing a large number of species, constituted only 12% and 6% of the biomass respectively.

The biomass of each component varied throughout the year. Although the sward is grazed during the summer the biomass throughout the period investigated, 1965–1970, tended to increase as the season progressed with a peak of 290 g m^{-2} in September. This is reflected in peaks for *A. tenuis* and *A. odoratum* in August and *F. ovina* in September. The increase may reflect an excess of plant production over removal by grazing or result from new tillers and leaves which are too small to be grazed. Numbers of tillers of all species (Table 2) were lowest during the winter months, increasing in spring to a maximum in late summer. Tiller weight, particularly for *A. tenuis*, is low in early summer when the numbers are increasing, and at a maximum in August and September. Grazing is very selective, not only between different vegetation types but also between individual plants and parts of plants (Arnold, 1964; Hughes et al., 1964). At Llyn Llydaw the increase in biomass of *F. ovina* from April to a September peak is greater than the increase in *A. tenuis* because of preferential consumption of *A. tenuis* (Chap. 19). The pattern in the vegetation is thought to result from selective sheep grazing and to a lesser extent abiotic environmental factors.

The mean biomass of roots from 0–30 cm depth was 578 g m^{-2} in 1969 and 627 g m^{-2} in 1970. Biomass was at a minimum during the winter and spring, rising to a maximum in summer (Table 1, 1969 data). Troughton (1951, 1957) and Baker and Garwood (1959) found similar seasonal patterns in some agricultural grass species. Most roots occur at Llyn Llydaw in the surface soil layers: 35% in the top 0–1 cm, 45% in the 1–8.5 cm layer and only 20% from 8.5 cm to 30 cm. Evdokimova and Grishina (1968) found that the maximum amount of roots in the 0–25 cm layer in natural grassland was 15–40 times that in deeper layers of the same thickness. Ratios of aerial (excluding mosses): root biomass (Table 1) indicate that there was 2.6 to 5.2 times as much below-ground as above-ground material in the grassland. The mean aerial biomass (including mosses) was 220 g m^{-2} and together with roots constituted a total for the site of 798 g m^{-2} (range 641 to 959).

The biomass of a grazed grassland, like its species composition, reflects the effects of controlling factors, especially grazing. It is from the residual biomass (mainly grasses at Llyn Llydaw) that new growth takes place, providing herbage for animals and tissue for continuation of the plant community. The seasonal pattern of living and attached dead grasses in the sward sampled from turf cores

Table 1. Biomass (g m^{-2} ± se) of living aerial components (standing crop green) and roots in 0–30 cm soil in grazed grassland at Llyn Llydaw during 1969

Sample date	22 Jan.	9 Apr.	12 May	9 June	4 July	5 Aug.	2 Sept.	29 Sept.	5 Nov.	13 Jan.	Mean[b]
Agrostis tenuis	48±10	42±5	67±6	62±6	88±10	125±10	95±8	102±8	87±8	42±4	76±9
Festuca ovina	13±5	14±2	16±3	24±4	30±6	34±6	61±10	34±8	26±4	25±5	28±4
Anthoxanthum odoratum	5±2	3±1	7±1	12±3	11±2	18±4	9±2	10±2	10±3	3±1	9±1
Other grasses	4±2	7±3	3±1	5±2	5±2	7±3	13±6	10±5	7±2	3±2	6±1
Sedges and rushes	(23)[a]	(11)	17±4	7±1	15±2	10±3	8±2	12±4	17±3	21±6	14±2
Herbs	(29)	(14)	17±3	20±3	29±6	32±5	31±4	33±5	28±4	31±5	26±2
Mosses	(64)	(76)	66±10	56±7	58±9	51±11	71±13	56±11	47±8	64±8	61±3
Total aerial biomass	(186)	(167)	194±12	186±8	235±15	276±14	289±15	256±15	221±9	189±12	220±13
Roots (0–30 cm)	588±39	474±46	563±48	609±52	647±51	683±47	574±47	545±43	588±38	507±37	578±20
Total biomass	(774)	(641)	757	795	882	959	863	801	809	696	798±29
Ratio aerial : roots	1:3.2	1:2.8	1:2.9	1:3.3	1:2.8	1:2.5	1:2.0	1:2.1	1:2.7	1:2.7	1:2.6
Aerial (−mosses): roots	1:4.8	1:5.2	1:4.4	1:4.7	1:3.7	1:3.0	1:2.6	1:2.7	1:3.4	1:4.1	1:3.6

[a] Values in brackets derived from incompletely separated components.
[b] se of mean indicates variation between monthly values.

Table 2. Mean number ($n \times 10^{-3}$) of tillers m^{-2}, and weight (mg tiller^{-1} dw) of grasses at Llyn Llydaw during 1969 and January 1970

Sample date		22 Jan.	9 Apr.	12 May	9 June	4 July	5 Aug.	2 Sept.	29 Sept.	5 Nov.	13 Jan.
Agrostis tenuis	*n*	23.7±3.9	21.2±2.3	31.8±2.9	36.4±3.7	41.7±4.2	44.2±3.9	41.0±3.4	45.6±4.2	37.5±3.3	24.5±2.9
	dw	2.23±0.14	1.98±0.11	2.11±0.12	1.70±0.10	2.11±0.18	2.83±0.23	2.32±0.22	2.24±0.14	2.32±0.13	1.71±0.19
Festuca ovina	*n*	7.9 ±2.9	10.8±1.9	10.2±1.6	12.7±2.7	15.0±2.8	13.9±2.3	15.9±2.9	14.1±2.8	12.2±2.2	9.9 ±2.9
	dw	1.77±0.16	1.30±0.10	1.57±0.11	1.88±0.08	2.00±0.15	2.45±0.14	3.85±0.20	2.40±0.18	2.14±0.10	1.81±0.21
Anthoxanthum odoratum	*n*	2.0 ±0.9	2.0 ±0.5	2.3 ±0.4	3.7 ±0.9	2.8 ±0.8	4.0 ±0.7	3.7 ±0.9	2.9 ±0.8	3.4 ±0.9	1.7 ±0.7
	dw	2.43±0.35	1.50±0.37	3.39±0.32	3.29±0.31	3.78±0.19	4.4 ±0.45	2.69±0.47	3.58±0.42	3.52±0.27	2.30±0.31
Other grasses	*n*	1.9 ±0.7	1.9 ±0.7	2.3 ±0.9	2.2 ±0.9	2.2 ±0.8	1.6 ±0.6	2.5 ±0.9	2.4 ±1.1	2.3 ±0.7	2.0 ±0.7
	dw	2.5 ±0.3	2.1 ±0.6	2.1 ±0.4	1.8 ±0.3	2.7 ±0.5	3.9 ±1.1	3.6 ±0.7	3.7 ±0.6	4.5 ±0.9	2.5 ±0.3

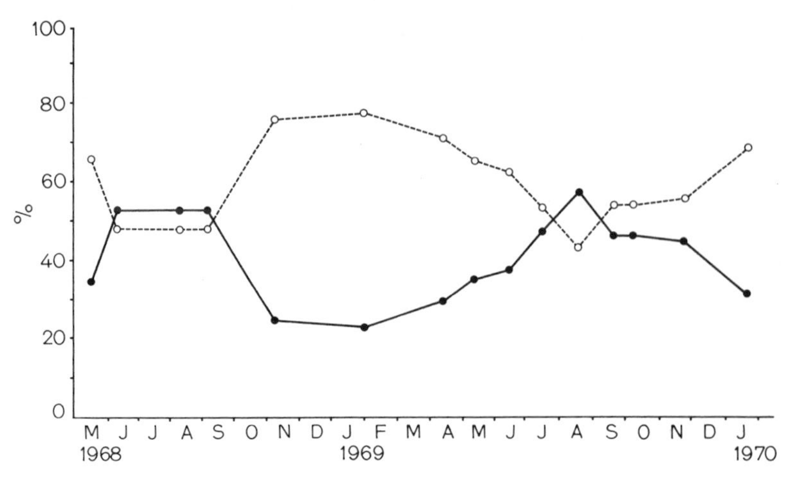

Fig. 1. Seasonal fluctuation in biomass of living (●——●) and attached dead (○------○) grasses as a % of the total biomass of grasses

(Fig. 1) indicates that large quantities of dead material occur throughout the year and that living material increases through the summer months, then decreases from autumn to winter. Overwintering of perennial grass plants depends upon the presence of young tillers in the sward towards autumn. Langer (1956, 1958) showed that grass tillers live about 12 months and Spedding (1971) that relatively few leaves of agricultural species live longer than about eight weeks. Leaf production and death are reported to occur at a rate of one leaf every 5 to 11 days in summer (Jewiss, 1972) so that leaf and tiller replacement is continuous in a grass sward. The general pattern of growth of grasses at Llyn Llydaw appears to parallel that found by Langer and described by Evans et al. (1964) for perennial festucoid grasses. Although determined largely by climatic, edaphic and biotic factors, it is also the result of a distinctive dynamic system of tiller growth. Overwintering tillers commence growth in the spring, extend their laminas and begin to produce daughter tillers; this may not be accompanied, at least initially, by an increase in the living biomass. In May and June a large production of dead material, resulting in part from the death of overwintering tillers, is masked, in terms of numbers m^{-2}, by the rapid production of new tillers. These are subject to defoliation first by slugs (Chap. 17) and later by large herbivores. Some of the earliest produced tillers may die during the ensuing winter. Under the intense grazing at the site tillers initiated to produce inflorescences do not usually complete their reproductive phase. In early summer their internodes elongate rapidly, the growing points are grazed and the remaining part dies. Growth is continued by non-reproductive and newly produced tillers.

A. odoratum and *F. ovina* exhibit the greatest increase in weight during the growing season from a spring minimum to a summer maximum (Table 2). *A. tenuis* tillers, the largest constituent in the sward, have the smallest increase in dry matter per tiller. *A. odoratum* biomass increased $\times 6.0\ m^{-2}$ and $\times 1.4\ tiller^{-1}$. The number and weight of tillers and thus sward production (Sect. 16.2.3) are affected

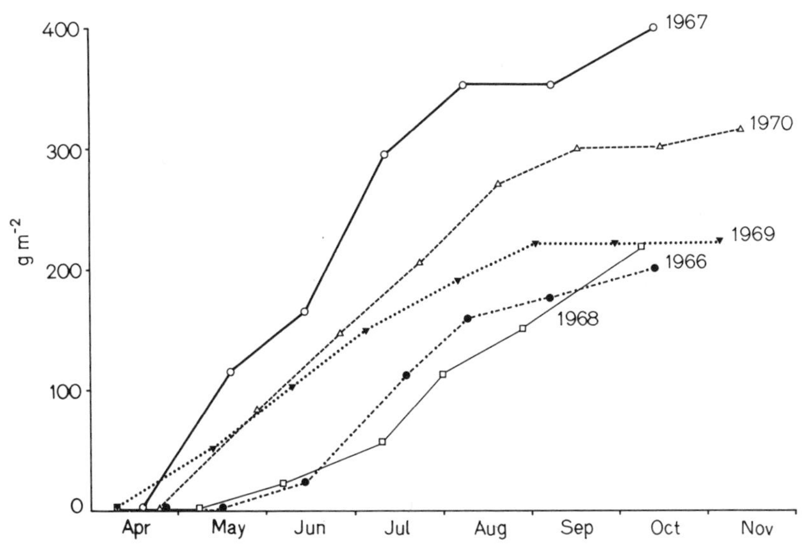

Fig. 2. Variation in cumulative net primary harvestable production, unadjusted for losses to standing dead, between 1966 and 1970 at the *Agrostis-Festuca* site at Llyn Llydaw

by factors such as degree of defoliation, rate of leaf appearance and death, competition (light and nutrients) and growth rate of different species and different tillers of the same species.

16.2.3 Net Primary Production of the Grassland

Net primary production of the pasture was estimated by comparable harvest methods from 1966 to 1970. The mean cumulative production of living and dead material (Fig. 2), above 1–2 cm from the ground in quadrats protected from grazing for *c* 28 d periods throughout the growing season (method 1), was 271 g dw m^{-2} yr^{-1} (range 199–400). Data from 1966 to 1968 were obtained by harvests of clipped quadrats and included attached dead material. The 1969 and 1970 data were derived from turf core samples (total dw excluding litter), adjusted to remove the contribution of the stubble normally left when clipped. A comparison of methods in 1968 indicated that production assessed by clipping (y) was related by regression to that derived by coring (x) using total living and attached dead (but excluding litter) by $y = 0.784x$ ($P < 0.1$). Production in a plot protected from grazing from 10 March to 12 October 1964 was estimated at 233 g dw m^{-2} at the end of the season when clipped 1–2 cm above ground but 975 g dw m^{-2} if all the above-ground living and dead material was included.

Clipping, whilst providing an estimate of herbage yield and consumption by sheep, underestimated production through:

1. material left in the stubble; and
2. losses of dead parts of the plant and loss in weight and disappearance of dead.

Table 3. Cumulative production (g dw m^{-2} yr^{-1}) of *Agrostis-Festuca* grassland components (excluding mosses) during 1969 adjusted for losses in formation of dead material[a]

Component	Living yield	Living + L	Living + L_w	Living + L_w + L'_w
Agrostis tenuis	156	298	381	562
Festuca ovina	42	85	125	192
Anthoxanthum odoratum	33	57	69	112
Other grasses	16	31	41	75
Sedges and rushes	15	22	22	23
Herbs	5	5	5	5
Total	267	498	643	979

[a] L = production of dead material; $L_w = L$ corrected for loss in weight upon senescence; L'_w = correction for underestimate of L due to disappearance of biomass between harvests adjusted for weight loss upon senescence. Only significant ($P < 0.1$) concurrent positive values accepted.

An attempt was made to correct the measurements by:

1. sampling the entire above-ground biomass by the turf core method (by dominant species and groups of minor species); and

2. measuring (L) the appearance of dead material (dead production) more precisely, estimating the amount disappearing (L') during the sample period and adjusting for weight loss (L_w) during senescence.

Mosses were excluded from the calculations because throughout the period April to October negative values were recorded in Equation (2) (Sect. 16.2.1.2) and spatial heterogeneity led to enhanced variation in the data. The cumulative aerial production of vascular plants for 1969 (Table 3) indicates the effect of adjustments made to the harvestable living material. Living material $+L$ increased the total production from 267 g m^{-2} to 498 g m^{-2} ($\times$ 1.9), and correction of dead material to live weight (L_w) increased the estimate to 643 g m^{-2} ($\times$ 2.4). Breakdown, leaching and translocation of soluble metabolites and mobile minerals occur in the moribund leaf. Thus inclusion of dead production would underestimate total primary production unless the original living dw were reconstituted. Whilst mobile elements are lost during senescence the bulk of plant ash consists of relatively non mobile minerals such as silicon and calcium, and a comparison of the ash contents of the current living and dead components gives a factor for adjustment to the former living weight. The underestimate due to removal of ash during senescence is small compared with changes in the organic matter content of leaf or stem material. Weighted mean conversion factors were $\times$ 1.63 for L_w and $\times$ 1.42 for L'_w ($\times$ 1.52 overall) for vascular plants and $\times$ 1.18 for mosses, a factor of $\times$ 1.48 for the entire vegetation.

A final adjustment adding the production which was not recorded due to the rapid disappearance rates at the site ($L_w + L'_w$) corrected for weight loss upon senescence, gave a production of 979 g m^{-2} ($\times$ 3.7). Similarly Coupland (1973) estimated production at the Canadian grassland site at Matador: production of living shoots, 133 g m^{-2}, increased to 349 g m^{-2} ($\times$ 2.6) when dead shoots were added and to 469 g m^{-2} ($\times$ 3.5) when corrected for weight loss. Under much drier conditions on the prairie higher initial recovery of dead material took place probably because dead disappearance rates during growth periods were much lower than at Llyn Llydaw. It appears therefore that production estimated by harvestable yield of living or living plus dead material (method 1) grossly underestimates net primary production. The values suggest that the dynamics of leaf growth and death in grassland operate at a much higher level and provide a greater input to the decomposer system than previously indicated.

Production was greatest in *A. tenuis* ($562\ g\ m^{-2}$) and *F. ovina* ($192\ g\ m^{-2}$) during a 210 day period, 9 April to 5 November, which corresponds well with the generally accepted threshold temperature for growth of $>5.6°$ C. The rate of aerial production for the site averaged $4.66\ g\ m^{-2}\ d^{-1}$ (Table 4), *A. tenuis* contributing $2.68\ g\ m^{-2}\ d^{-1}$. High rates of production, $13.07\ g\ m^{-2}\ d^{-1}$ overall, *A. tenuis* $9.97\ g\ m^{-2}\ d^{-1}$, occurred from 12 May to 9 June. Much of this production went into sward tiller replacement and only a small part was available for grazing. The high production rates can be compared with those of S53 Perennial ryegrass at Pant-y-dŵr, Aberystwyth, which with added fertilisers was *c* $9.0\ g\ m^{-2}$ on an upland mineral soil and *c* $6.5\ g\ m^{-2}\ d^{-1}$ on a peaty soil (Munro et al., 1972). These values, unadjusted for losses, indicate the high potential production of agricultural species.

Growth of mosses in the sward was low when the biomass of vascular plants was high and rainfall low during the summer months. Poor moss growth within cages was possibly due to reduced light penetration in the ungrazed sward. During one period in summer and from November 1969 to April 1970, moss biomass increased significantly in the grazed sward and production was estimated as $58\ g\ m^{-2}$ (adjusted for losses to standing dead). An alternative estimate of $56\ g\ m^{-2}\ yr^{-1}$ ($66\ g\ m^{-2}\ yr^{-1}$ when corrected to live weight) assumed steady state conditions and no grazing. It was based on the production required to replace measured production of dead moss material from January 1969 to January 1970. Measurement of herbage consumed by sheep (Chap. 19) indicated that some moss could be grazed at the beginning and end of the grazing season when vascular plant biomass was low. Mosses were abundant in faecal samples from Soay sheep (Milner and Gwynne, 1974) in late winter when available herbage was low under the continuous natural grazing system of St. Kilda. Under restricted April to October grazing at Llyn Llydaw the small quantities recorded by harvest methods were not significant.

An estimate of the mean adjusted production for 1966–1970 was based on the assumption that the correction factors used for the 1969 data would apply. Thus the mean harvestable yield of $271\ g\ dw\ m^{-2}$ is equivalent to $569\ g\ m^{-2}$ of dead production, or 865 g living material (using the ash correction factor of $\times$ 1.52). Insertion of this value in Equation (6) ($P = L + G$) where G consists of a mean of 199 g of living material consumed by sheep (Chap. 19) and 13 g by slugs (Chap. 20), gives a total production of $1077\ g\ m^{-2}$ (1143 g if the estimated moss production of $66\ g\ m^{-2}$ is added).

In the third method of estimating production, L is calculated from Equation (5) using the mean biomass of standing dead (213 g) and plant litter (260 g) for the period 1968 to 1970 and a mean disappearance rate, determined in situ within exclosures during 1969 and 1970, of $3.38\ mg\ g^{-1}\ d^{-1}$. This gives a mean litter

Table 4. Rate of net primary production (g dw m^{-2} d^{-1}) of *Agrostis-Festuca* grassland components (excluding mosses) during 1969

Sample date Days of growth		9 April —	12 May 33	9 June 28	4 July 25	5 Aug. 32	2 Sept. 28	29 Sept. 27	5 Nov. 37	Mean 210
Agrostis tenuis	(1)	0	1.61	2.48	2.29	0.93	1.11	0.77	0	1.25
	(2)	0	2.72	9.75	2.29	0.96	1.55	2.54	0	2.68
Festuca ovina	(1)	0	0.53	0.06	0.69	0.37	0.31	0	0.30	0.32
	(2)	0	1.32	1.79	0.69	0.37	0.78	0	1.27	0.91
Anthoxanthum odoratum	(1)	0	0.29	0.38	0.20	0.22	0.38	0.04	0.20	0.24
	(2)	0	0.48	1.14	0.20	0.22	0.68	0.53	0.50	0.53
Other grasses	(1)	0	0.02	0.13	0.34	0.27	0.03	0	0.13	0.13
	(2)	0	0.02	0.39	0.78	0.67	0.08	0	0.54	0.36
Sedges and rushes	(1)	0	0	0	0.59	0.21	0	0	0	0.10
	(2)	0	0	0	0.94	0.29	0	0	0	0.16
Herbs	(1)	0	0	0	0.06	0.10	0	0	0	0.02
	(2)	0	0	0	0.06	0.10	0	0	0	0.02
Total	(1)	0	2.45	3.05	4.17	2.10	1.83	0.81	0.63	2.06
	(2)	0	4.54	13.07	4.96	2.61	3.09	3.07	2.31	4.66

(1) Adjusted for losses (L) sustained from living to dead.
(2) Corrected losses ($L_w + L'_w$) taking account of disappearance during the sampling period and weight loss upon senescence.

disappearance rate of 583 g m^{-2} yr^{-1}. Under steady state conditions production would equal input; the addition of the mean 49 g of standing dead grazed by sheep gives an estimate of dead production of 632 g m^{-2} yr^{-1}. Since this would include mosses the × 1.48 ash correction factor is applied, giving a living weight of 1145 g m^{-2} yr^{-1} which compares well with the above estimate. Root production was estimated from biomass differences in the 0–8.5 cm soil horizon. Turnover rates calculated in 1969 and 1970 gave values of 0.38 (0–1 cm 0.36; 1–8.5 cm 0.40) and 0.46 respectively. Total root production in 1969 was estimated at 270 g m^{-2}: 86 g m^{-2} 0–1 cm, 126 g m^{-2} 1–8.5 cm, 58 g m^{-2} 8.5–30 cm. In 1970 the top 1 cm was not separated but total root production was estimated at 386 g m^{-2} from the complete 0–8.5 cm layer. For comparison the 1969 data were reconstituted as if the sample were from the complete 0–8.5 cm layer giving an estimate of 288 g m^{-2}.

16.2.4 Factors Affecting Biomass and Production

A complex combination of factors, especially temperature, sunlight and nutrient supply (Perkins, 1967), influences plant growth, becoming less favourable as altitude increases. The annual variation in production at Llyn Llydaw (Fig. 2) mainly reflects changes in climate, particularly temperature. The marked seasonal pattern of growth of grasses within years, however, is not entirely related to changes in temperature and solar radiation (Anslow and Green, 1967). Higher rates of production were found at the same relatively low temperatures in spring than in autumn. Alcock (1969) concluded that deviations from a seasonal trend can only be attributed to complex interactions between various climatic elements and different growth processes in the upland grass plant.

The relationship between production of grasses and climatological parameters is illustrated for the 1969 results (Fig. 3). Mean air temperature (Stevenson screen at 122 cm) until 9 April was $<2.5°$ C and soil temperature $<2.9°$ C at 5 cm and 2.0° C at 15 cm. When the first measurable production occurred, between 10 April and 12 May, air temperature had reached a mean of 5.7° C and soil temperature 6.5° C at 5 cm and 5.3° C at 15 cm. Temperatures at the end of the season before 5 November were $>5.6°$ C, subsequently falling to 2.8° C when no further growth was measured. Two periods of growth related to inherent characteristics of tiller phenology (Sect. 16.2.2) lead to peaks of production whose magnitude loosely reflects climate. Temperature is more important in spring when moisture is usually adequate, but moisture is sometimes limiting in late summer. The first peak occurs in April and May, when tillers initiated to produce inflorescences grow rapidly, followed by a decline as they become defoliated by sheep grazing. The second, in August, is due partly to continued growth of non-reproductive tillers but mainly to the assumption of assimilatory independence by new tillers.

Only general conclusions can be drawn from the limited data. The rate of dry matter production at Llyn Llydaw (g m^{-2} d^{-1}) was best positively correlated ($r = 0.54$, $P<0.01$) with mean maximum air temperature 122 cm above ground in a Stevenson screen and mean soil temperature at 5 cm ($r = 0.53$, $P<0.01$). No production was recorded below *c* 5.6° C and little at higher temperatures during periods of low rainfall and moisture stress. In 1966 and 1968 the onset of produc-

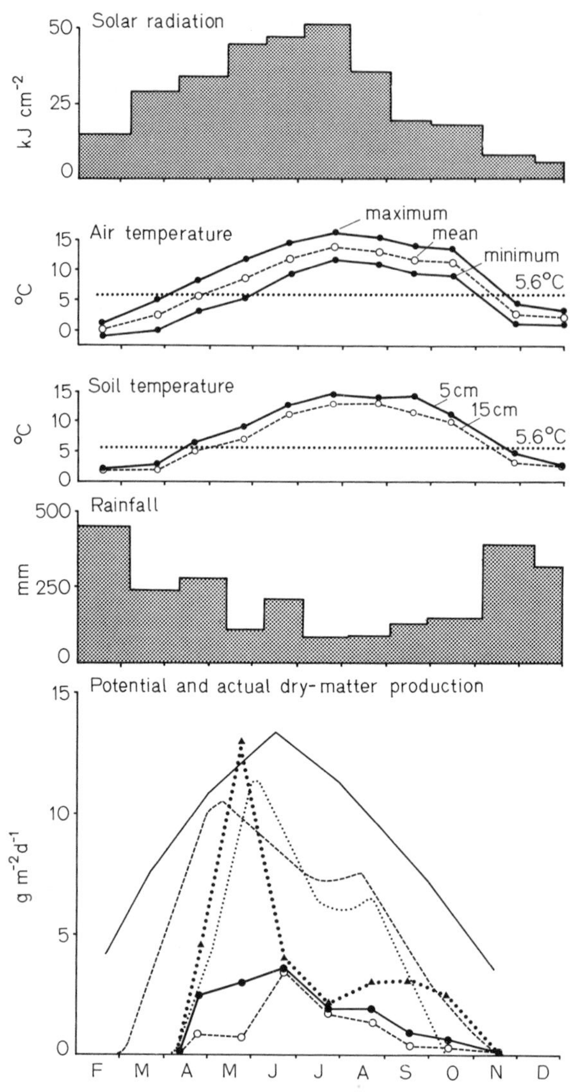

Fig. 3. Rate of production of grasses at Llyn Llydaw during 1969. Living (○------○) and living + dead (●——●) production yields (method 1), and living + $L_w + L_w'$ (▲.....▲) adjusted production (method 2) together with the mean air temperature (°C) (Stevenson screen 122 cm above ground), soil temperatures (at 5 and 15 cm depth) plotted at the mid point of the growth period, and total solar radiation and rainfall. Redrawn from Alcock (1975) is the potential production = 0.6 maximum estimated growth of Perennial ryegrass with a leaf area index of 6 (———); and the actual production at Pant-y-dŵr, upland site at 305 m (·········) and Aber, lowland site at 6 m (-------)

tion was very late (Fig. 2), appreciable amounts not being recorded until July; dry matter not accumulated in the spring was not made up later in the year. Temperature and moisture being adequate in most years, mid season production is reasonably constant and predictable, *c* 2.2 g m^{-2} d^{-1}. In an analysis of several climatological factors and three agricultural grasses at different growth stages in a lowland and an upland location in north west Snowdonia, Alcock and Lovett (1968) showed that when nutrients were not limiting growth climatic factors accounted for 86% of the variation in the lowland and 52–88% in the upland dry matter production in the early part of the year. The temperature of the soil at 10 cm was found to be the most significant parameter positively correlated with growth. Solar radiation varies little between upland sites except in the spring when cloud cover may reduce the cumulative value (and also temperature in degree hours) in the central mountainous area in Snowdonia. It is then that differences in production between sites within years are more pronounced. Alcock et al. (1974) and Alcock (1975), predicting the growth of S23 Perennial ryegrass, concluded that with no other constraints solar radiation was unlikely to be a limiting factor in hill climate. Data redrawn from Alcock (1975) are included in Figure 3 for comparison with the Llyn Llydaw results. Munro and Davies (1973), studying the growth of heavily fertilised swards of Perennial ryegrass in a lowland and an upland situation in mid Wales, concluded that the lower productivity in the uplands, though partially attributable to a short growing season and poor summer, was also the result of damage by low temperatures during the previous winter. Extremely poor growth of Perennial ryegrass in 1969 compared with 1968 was due to severe winter conditions. Although moisture may have limited production at Llyn Llydaw in the latter part of 1969, growth in the spring was comparable to that in 1967 and 1970 and better than that in 1968; the semi-natural *Agrostis-Festuca* grassland was apparently more successful in withstanding the winter conditions of 1969 which killed sown grasses.

Production of above-ground dry matter at 32 other sites in Snowdonia was investigated in 1966 and related to environmental gradients. Above-ground production was assessed on vegetation types grouped by reciprocal averaging ordination (Hill, 1973) and ranged along axis 1 from a mean of 366 ± 46 g m^{-2} on *Agrostis-Festuca* types, including Llyn Llydaw; 203 ± 12 g m^{-2} on *Festuca-Agrostis*; 93 ± 28 g m^{-2} on *Nardus-Festuca-Agrostis*; to 217 ± 66 g m^{-2} on *Nardus-Molinia-Juncus*. The variation in botanical composition of sites was related on axis 1 to soil pH (range 3.95–5.25), soil moisture (range 35–91%) and organic matter content (range of loss on ignition 11.9–89.7%). *Agrostis-Festuca* vegetation types were associated with higher pH, lower soil organic matter content, lower moisture and soil parent material of higher base status, generally those brown earth soils developed from pumice-tuff materials in contrast to the harder and less weathered igneous rhyolitic materials and glacial drift.

The sites were grouped on axis 2 into those at higher altitude (mean 484 m, range 349–909 m) and those at lower altitude (mean 349 m, range 242–424 m), where there was less rainfall and higher temperature, especially in spring and autumn, when these parameters are more critical. In both groups a range of production was associated with the vegetation type (itself a reflection of abiotic factors) but the mean production of all types at low altitude (320 ± 43 g m^{-2}) was nearly twice that at higher altitudes (182 ± 20 g m^{-2}), indicating a close relationship between production, rainfall and temperature. Rawes and Welch (1969) obtained similar results for grazed *Agrostis-Festuca* swards at Moor House (Chap. 2).

16.3 Appearance and Disappearance of Dead Plant Material

16.3.1 The Appearance of Dead Material

Dead plant material and litter in the biomass at Llyn Llydaw exceeds the above-ground living material at any time of the year. The mean monthly value of the standing dead during 1969 was 213 ± 9 g m^{-2} and litter 378 ± 21 g m^{-2} and the ratio of living to total dead and litter = 1:2.69. The appearance of standing dead (dead production) was estimated from the difference between the weight inside cages of the current harvest and outside cages of the previous harvest. Data from within exclosures gave better results as the crop was protected from treading by grazing animals, normally an important factor in incorporating standing dead into litter. Litter consisted of a range of decomposing and decomposed dead material of all species on the surface of the soil core after removal of living and standing dead material. The standing dead, being mainly recently formed, comprised a smaller range of decomposing states and was separable into its components. The rate of incorporation of standing dead into litter is not known with certainty; it is a continuous process affected by animal treading and climatic conditions.

The standing crop of dead and litter increased from 8 April to 9 June, after which standing dead declined to August, rose in September, and decreased through to January (Table 5). The litter biomass was lower in July and rose again in August before decreasing through to January. The greatest rate and quantity of production of standing dead, from 9 April to 9 July, results from death of over-wintering tillers and potential flowering tillers of grasses through defoliation (Sect. 16.2.1.2). By contrast, little dead production occurred during midsummer or from January until the spring when tiller replacement commenced. There is no large dieback of flowering tillers later in the season; because of early defoliation few, if any, tillers attain flowering stage in the sward.

16.3.2 The Disappearance (Decomposition) of Litter

16.3.2.1 Estimation by Weight Loss

The rate of disappearance of plant litter is an estimation of the rate of decomposition taking place in the system. It provides an estimate of the flow of energy and transfer of mineral nutrients through the decomposers integrating the activity of the microflora and fauna of the site. Whilst the total amounts of standing dead and litter can be determined with reasonable precision, they consist of material in various stages of decomposition and age which may have differing rates of disappearance. Decomposition processes have been studied by weight loss of samples inside exclosures. Two methods gave essentially similar results: one by enclosing weighed quantities of recently dead material in nylon mesh bags (mesh 4 mm × 6 mm), and the other an adaptation of the turf core sample. A turf core (6.0 cm diameter × 8.5 cm) was removed from the sward and in the laboratory the vegetation and litter were replaced by a weighed sample of dead or litter material. The sample was held in close contact with the soil by a mesh ring on top of the core which was replaced in the sward.

Table 5. Biomass and production of dead material in the grassland at Llyn Llydaw during 1969 and January 1970[a]

Sample date	22 Jan.	8 Apr.	12 May	9 June	4 July	5 Aug.	2 Sept.	29 Sept.	5 Nov.	13 Jan.
Biomass ($g\,m^{-2}$)										
Standing dead	232 ± 27	180 ± 21	211 ± 17	237 ± 12	206 ± 13	197 ± 11	269 ± 17	223 ± 15	193 ± 11	182 ± 16
Litter	367 ± 29	378 ± 38	417 ± 33	498 ± 54	388 ± 41	438 ± 45	404 ± 47	316 ± 26	299 ± 29	274 ± 31
Cumulative dead production ($g\,m^{-2}$)	—	19	72	195	202	205	220	236	250	283
Cumulative dead production ($g\,m^{-2}$)[b]	—	19	88	338	356	370	399	444	487	520
Rate of adjusted dead production since last harvest ($g\,m^{-2}\,d^{-1}$)	—	0.3	2.09	8.90	0.72	0.44	1.04	1.67	1.16	0.50

[a] During the growing season April to November 468 $g\,m^{-2}$ of adjusted dead material was produced. During winter months 52 $g\,m^{-2}$ of dead accrued from the biomass.
[b] Adjusted for understimate due to disappearance between harvests.
Data are derived from turf core sampling ($n = 32$).

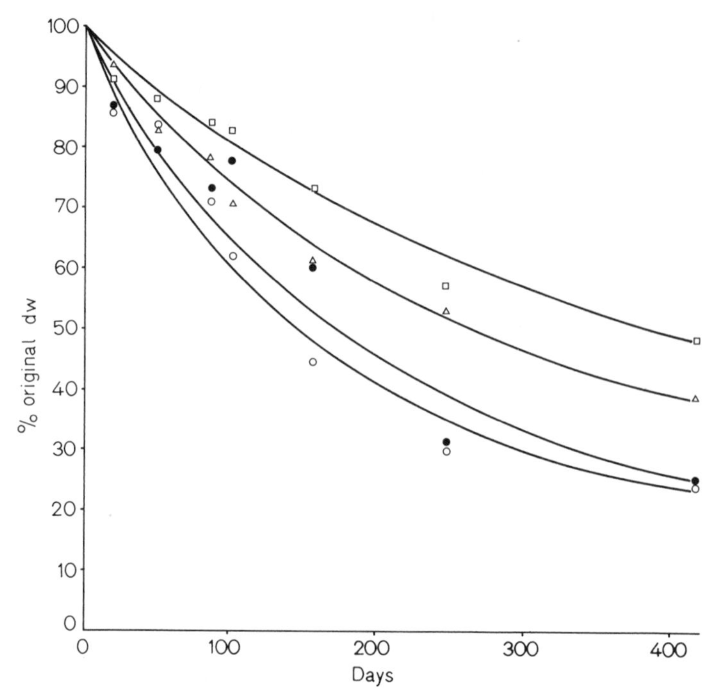

Fig. 4. Rates of disappearance of plant litters in the *Agrostis-Festuca* grassland. Mixed vegetation ●; grasses (*Agrostis tenuis* and *Festuca ovina*) ○; *Nardus stricta* △; and mosses □

16.3.2.2 Rates of Litter and Nutrient Disappearance

The nylon mesh bag experiment, 10 April 1968 to 15 January 1970, resulted in 67% weight losses from the mixed dead vegetation samples from April 1968 to April 1969, and 70% from January 1969 to January 1970. The turf core method was used to study the disappearance of different types of litter: (1) mixed vegetation typical of the sites; (2) grasses—*A. tenuis*, *F. ovina* and *Nardus stricta*—a species present in limited amounts but representing more oligotrophic conditions; (3) a mixture of the mosses *Hypnum cupressiforme*, *Rhytidiadelphus squarrosus* and *Polytrichum commune*. The rates of disappearance (Fig. 4) from May 1970 to July 1971 were higher overall for mixed litters (75%) and grasses (76%), than for *Nardus* leaves (61%) and mosses (52%).

The material remaining was chemically analysed to determine the rates of nutrient and energy loss. In samples of mixed vegetation after the 418 day period high losses were observed in K (88%), Ca (85%), N (75%) and kJ (80%). Grasses generally exhibited similar losses but mosses, whilst having similar rates of K loss (81%), showed slower Ca (55%), N (45%) and kJ (58%) losses. *Nardus* material, however, lost 65% kJ and exhibited a very slow loss of K (33%) but gained Ca (36%) and N (5%). Similar results have been recorded for phosphorus in *Nardus* material from other Snowdonia sites and are believed to result from the activity of microorganisms which retain and acquire P and N during decomposition, possibly from rainwater and plant leachates.

16.3.3 Factors Affecting Decomposition

A negative exponential curve often provides a reasonable fit to weight loss data in the initial stages of decomposition (Heal and Perkins, 1976), although there is good theoretical evidence that decay rate declines as organic matter ages (Mindermann, 1968). Some of the mineral loss data tend to fit a more concave type of curve, increasing in concentration and content with time, perhaps associated with an apparent decline in rate of loss of weight. Whilst weight loss data for both mesh bags and modified cores fit a negative exponential, differences in rate of loss throughout the period can be identified. Thus for mixed vegetation decomposing on modified cores, the rate (Fig. 4) varies from 6.5 mg g^{-1} d^{-1} initially to 2.1 mg g^{-1} d^{-1} in July and to 5.3 mg g^{-1} d^{-1} in the autumn.

In another experiment, from May 1969 to April 1970, samples were put in position for periods of one and two months in sequence throughout the year. Decomposition was positively correlated with temperature and moisture. Rates of loss of between 2 and 10 mg g^{-1} d^{-1} were positively correlated with moisture (rainfall as well as soil moisture) when air temperatures were above *c* 9° C and soil temperature (recorded at 5 cm) above 7° C. In summer, while the temperatures are higher (mean 11° C), rainfall is often low (5 mm d^{-1}) and the average rate of loss is 4.9 mg g^{-1} d^{-1}. When rainfall is higher (11.2 mm d^{-1}) in winter, temperatures are low (mean 3.2° C) and the average rate of loss is 1.28 mg g^{-1} d^{-1}, some 25% of the summer value. Thus, even in one of the highest rainfall areas in Britain, decomposition can be limited by available moisture when summer temperatures are optimal for microorganism activity; when moisture is adequate in winter, temperatures are limiting. Decomposition of recently dead material before incorporation into the litter is likely to be more affected by dry conditions.

16.4. Nutrient Content of Plant Biomass and Production

The dynamics of some major mineral nutrients were investigated to give not only insight into growth processes leading to organic matter production but also parameters for nutrient transfer models of the ecosystem (Chap. 20). The concentration of minerals (mg g^{-1} dw) in biomass and production samples was determined by standard chemical methods, atomic absorption and X-ray fluorescence spectrometry. Analyses included total ash content (material ashed at 520° C in an electric muffle furnace), K, P, Ca, Mg, and N.

16.4.1 Concentration of Ash and Mineral Nutrients in the Biomass

Nutrient concentrations in vegetation depend on the nutrient supplying power of soil, rate of defoliation and manuring by herbivores and fluctuations in temperature and moisture. Variation occurs between species and with the stage of growth within a species; supply and demand within a plant results in variation in distribution between different tissues.

Table 6. Concentration (mg g^{-1} dw) of mineral nutrients and total ash in the biomass at Llyn Llydaw

Constituent	Ash	K	P	Ca	Mg	N[a]
Living biomass						
Agrostis tenuis	77.59 ± 3.78	11.30 ± 0.44	1.75 ± 0.10	1.70 ± 0.07	3.40 ± 0.22	22.1
Festuca ovina	61.13 ± 3.26	12.90 ± 0.46	1.86 ± 0.10	1.40 ± 0.07	2.75 ± 0.22	13.2
Anthoxanthum odoratum	78.58 ± 4.87	14.65 ± 0.75	2.63 ± 0.15	1.88 ± 0.17	4.46 ± 0.31	26.9
Other grasses	89.36 ± 6.79	14.47 ± 0.95	2.09 ± 0.16	2.13 ± 0.27	4.30 ± 0.35	} 20.4
Sedges, rushes and herbs	103.36 ± 6.12	11.69 ± 0.62	1.63 ± 0.10	4.38 ± 0.40	5.31 ± 0.30	}
Mosses	136.59 ± 14.92	6.88 ± 0.24	1.30 ± 0.07	2.88 ± 0.15	7.86 ± 0.53	17.6
Roots	101.71 ± 4.48	0.51 ± 0.02	0.61 ± 0.06	1.61 ± 0.07	5.79 ± 0.42	9.0
Dead biomass						
Grasses	127.83 ± 10.87	4.50 ± 0.48	1.10 ± 0.10	1.66 ± 0.05	6.64 ± 0.53	11.7
Sedges, rushes and herbs	107.30 ± 7.52	6.87 ± 0.36	1.04 ± 0.10	3.38 ± 0.29	6.41 ± 0.43	17.9
Mosses	159.88 ± 11.73	4.91 ± 0.18	1.06 ± 0.09	2.74 ± 0.15	9.38 ± 0.84	12.9
Litter	134.90 ± 17.93	1.55 ± 0.15	0.85 ± 0.07	2.18 ± 0.10	7.17 ± 0.70	11.4

[a] N determined on bulked samples.
Values are a mean of the 1969 turf core samples ($n = 32$).

Mean nutrient concentrations in the biomass (Table 6) indicate considerable variation between components: $\times 30$ in K, $\times 4$ in P and $\times 3$ in Ca, Mg, and ash. Grasses, particularly *Anthoxanthum odoratum*, had high concentrations of K (11.3–14.7 mg g^{-1}) and P (1.8–2.6 mg g^{-1}) but sedges, rushes and herbs had greater concentrations of Ca (4.4 mg g^{-1}) and Mg (5.3 mg g^{-1}). Roots had low concentrations of K (0.5 mg g^{-1}) and P (0.6 mg g^{-1}) but high Mg (5.8 mg g^{-1}). Differences were shown between dead and living material; dead grass, sedge and rush, and herb components had lower concentrations of K (mean 6.0 mg g^{-1}) and P (1.1 mg g^{-1}) through active translocation to younger leaf material, and leaching of these highly mobile elements from moribund tissue. In contrast concentrations of Mg and Ca, which are relatively immobile and complexed to the leaf structure, remained similar or higher in dead material and litter. Ash consisted of mainly immobile elements (at least 60% silica) and concentrations were higher in dead than living materials, reflecting the change in weight from living to dead material following the initial breakdown of organic matter.

16.4.2 Nutrient and Ash Content of the Plant Biomass and Production

The nutrient content of plant material (g m^{-2}) depends on the concentration of the constituent (mg g^{-1}) and the quantity of organic matter of each component per unit area (g m^{-2}). The values must be correctly weighted for the concentration and amount of dry matter of each component of the ecosystem. Whilst the nutrient content of the biomass in general reflects the quantity of organic matter (Table 7), differences in concentration result in some disproportion. Within the living biomass there were large quantities of K (58%) and P (29%) in the aerial parts but more Mg (37%) and Ca (34%) in roots. *Agrostis tenuis* contained high amounts of K (22%) and P (11%) as did the combined sedges, rushes and herbs (K, 11% and P, 5%). Living mosses contained 5–10% and dead mosses 2–3% of nutrients. In view of their slow disappearance rate, mosses constitute an important part of the litter component, retaining nutrients in the above-ground components of the ecosystem.

The dead dry matter biomass, 43% of the total, contained 35–49% of the nutrients and ash. Amounts of the mobile elements K (35%) and P (44%) retained in the dead biomass are important in the nutrient economy of the ecosystem (Chap. 20). Although the rate of decomposition is high at the site the amount of P retained in standing dead and litter (44%) may to some extent be limiting production.

The quantities of nutrients contained in production depend on factors similar to those which influence biomass. Nutrients tend to be more concentrated in actively growing tissues and therefore new production would be expected to have a greater nutrient content. Although nutrient concentrations tended to be higher in material harvested from clipped quadrats inside cages the greatest difference, exhibited by K (17.6 ± 1.9 inside and 15.5 ± 1.5 outside), was not significant. The concentration in the biomass at the time of sampling was, therefore, used to calculate the quantity of nutrients in production (Chap. 20) but this could underestimate the more mobile elements whose concentration rapidly fluctuates.

Table 7. Distribution (g m^{-2} and % of total) of dry matter, ash and mineral nutrients in the biomass at Llyn Llydaw

Constituent	Dry matter		Total ash		K		P		Ca		Mg		N	
	g m^{-2}	%	g m^{-2}	%	g m^{-2}	%	g m^{-2}	%	g m^{-2}	%	g m^{-2}	%	g m^{-2}	%
Living biomass														
Agrostis tenuis	76	5.5	5.9	3.7	0.965	22.1	0.148	11.0	0.143	5.2	0.270	3.3	1.68	10.2
Festuca ovina	28	2.0	1.7	1.1	0.399	9.2	0.058	4.3	0.044	1.6	0.075	0.9	0.37	2.2
Anthoxanthum odoratum	9	0.7	0.7	0.5	0.160	3.7	0.027	2.0	0.023	0.8	0.046	0.6	0.40	2.4
Other grasses	6	0.4	0.6	0.4	0.107	2.5	0.014	1.0	0.017	0.6	0.035	0.4		
Sedges, herbs and rushes	40	2.9	4.1	2.6	0.468	10.8	0.068	5.1	0.170	6.2	0.225	2.7	0.82	5.0
Mosses	61	4.4	8.3	5.3	0.418	9.6	0.080	6.0	0.179	6.5	0.480	5.8	1.07	6.5
Total aerial biomass	220	15.9	21.3	13.6	2.517	57.9	0.395	29.4	0.575	20.9	1.131	13.7	4.34	26.3
Roots (0–30 cm)	578	41.5	58.8	37.0	0.290	6.7	0.360	26.7	0.920	33.7	3.080	37.2	5.20	31.5
Total living biomass	798	57.4	80.1	50.6	2.807	64.6	0.755	56.1	1.495	54.6	4.211	50.9	9.54	57.8
Dead biomass														
Grasses	166	12.0	21.2	13.4	0.750	17.2	0.190	14.1	0.280	10.2	1.120	13.5	1.94	11.8
Sedges, herbs and rushes	23	1.7	2.5	1.6	0.130	3.0	0.080	6.0	0.070	2.6	0.150	1.8	0.41	2.5
Mosses	24	1.7	3.8	2.4	0.120	2.8	0.030	2.2	0.080	2.9	0.230	2.8	0.31	1.9
Total standing dead	213	15.4	27.5	17.4	1.000	23.0	0.300	22.3	0.430	15.7	1.500	18.1	2.66	16.1
Litter	378	27.2	51.0	32.0	0.540	12.4	0.290	21.6	0.810	29.7	2.560	31.0	4.31	26.0
Total dead biomass	591	42.6	78.5	49.4	1.540	35.4	0.590	43.9	1.240	45.4	4.060	49.1	6.97	42.2
Total	1,389	100	158.6	100	4.347	100	1.345	100	2.735	100	8.271	100	16.51	100

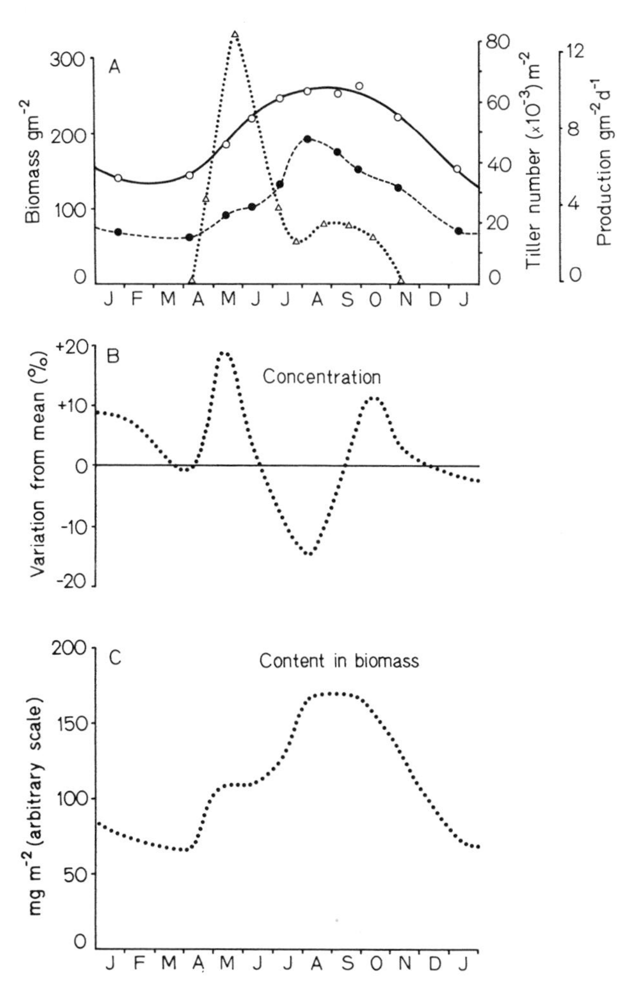

Fig. 5. Generalised diagrammatic model of the variation in concentration (graph *B*) and content (graph *C*) of nutrients (as the average % difference from mean values), in relation to tiller numbers (○——○), biomass (●------●) and dry matter production (△·········△) at Llyn Llydaw (graph *A*)

16.4.3 Factors Affecting Nutrient Concentration and Content

Nutrient concentrations fluctuated throughout the year and between years, probably largely reflecting variations in temperature and soil moisture which are known to affect the mineral composition of herbage (Fleming, 1973). There was, however, a pattern within years which was related to the annual growth of the dominant grass component (Fig. 5). Concentrations (mg g^{-1}, graph B) were derived from the average of the differences from the unweighted mean of N, P, K, Ca, and Mg in the four grass components sampled by turf cores in 1969. Content (mg m^{-2}, graph C) was calculated from the average concentration (assigning zero difference = 1 mg) × the biomass of the grass components (graph A). The curve thus reflects the individual component and constituent content data. The model elucidates the main features:

1. Concentrations (except K), above or near average during the winter, decreased at the onset of growth in the spring as growth demand exceeded uptake. K decreased in concentration markedly during the winter and when growth commenced was 20% below average whereas P and N were 20% above average.

2. Concentrations increased rapidly when the rate of production was high during the first growth period, due mainly to growth of reproductive tillers.

3. Defoliation of reproductive tillers caused a rapid decline to minimum values in July and August. The second period of production and an increase in the biomass resulted from continued growth of non-reproductive tillers with lower concentrations of nutrients and a smaller but increasing number of new tillers. Although largely dependent upon parent tillers, new tillers contained higher concentrations of nutrients and assimilates, translocated from the larger biomass of older leaf material. As they developed and became more independent, constituting a greater proportion of the biomass, concentrations were restored by further uptake which continued until a decline in some elements occurred in the autumn.

4. The content of minerals (mg m^{-2}, graph C), despite low concentrations in July and August, reflected the dry matter biomass graph A, increasing through the grazing season.

Concentrations often vary between years. In 1969 concentrations of Ca and Mg were higher at the beginning than the end of the year, largely accounting for the lower average on graph B. Low concentrations in clipped quadrat samples of July and August in 1969 and 1970 were not detected in 1968, perhaps because of the late onset of growth in spring and consequent fusion of the first and second growth periods, rapid changes in concentrations not being detectable by the infrequent sampling.

Concentration and content of nutrients in roots fluctuated less than those in above-ground material, but low summer values reflected slow uptake, especially of P, and translocation to the shoot system at a time of peak demand. Concentrations in recently dead material, although always lower for mobile elements, paralleled the concentrations in corresponding living material.

16.5 Variation in Biomass and Production in Relation to Species Composition, Soil and Plant Chemical Composition

An objective of the IBP study at Llyn Llydaw was to relate the spatial variation in productivity within the site to species composition, soil chemistry (Chap. 15) and intensity of sheep grazing (Chap. 19).

Production/biomass ratios (Table 8) in grasslands show a wide range within and between vegetation types. The ratios for different species within the *Agrostis-Festuca* site indicate their relative contributions to site production; thus bryophytes, while comprising 28% of the living biomass, contribute little to production ($P/B = 0.35$). Sedges and rushes (0.88) and herbs (0.15) also have low ratios,

Table 8. Production/biomass ratios for components of grassland vegetation at Llyn Llydaw and other sites in Snowdonia (values are totals of living and standing dead material for the period April to November)

Vegetation	Production ($g\ m^{-2}\ yr^{-1}$)	Biomass ($g\ m^{-2}$)	P/B
Llyn Llydaw[a]			
Agrostis tenuis	261	195	1.34
Festuca ovina	68	69	0.99
Anthoxanthum odoratum	51	19	2.68
Other grasses	27	15	1.80
Sedges and rushes	22	25	0.88
Herbs	5	33	0.15
Mosses	30	86	0.35
Total aerial	464	442	1.05
Total roots (0–30 cm)	270	578	0.50
Total aerial + roots	734	1,020	0.72
Snowdonia[b]			
Agrostis-Festuca	366	205	1.79
Festuca-Agrostis	203	240	0.85
Nardus-Festuca-Agrostis	93	348	0.27
Nardus-Molinia-Juncus	217	352	0.62

[a] Unadjusted (for L) yield determined by turf core sampling in 1969.
[b] Unadjusted (for L) yield determined by clipping quadrats in 1966 (values are a mean of several sites). Llyn Llydaw P/B in 1966 assessed by clipping quadrats was $199/179 = 1.11$.

but constitute only 6% of the biomass. In contrast, the high ratios of grasses, e.g. *A. odoratum* (2.68), indicate their major contribution to site production. Similar differences occur between sites: on the wide range of grazed grasslands in Snowdonia, P/B ratios are positively correlated with the numbers of sheep grazing the swards. Heavy grazing results in high ratios by reducing biomass but not production, while light grazing or its absence produces low ratios; similar variation within the *Agrostis-Festuca* site at Llyn Llydaw is much smaller and more difficult to detect. The total P/B ratio (including roots to 30.0 cm) at Llyn Llydaw was 0.72 in 1969, and total aerial P/B 1.05 (averaging 1.23 over the period 1966 to 1970).

A major factor related to the distribution of biomass on the site is species composition. The vegetation grouped by ordination, using % cover as a quantitative measure, is related to variations in production (see later). Within the plots, however, a scale of pattern exists which could be detected by turf core samples at 10 cm intervals (Perkins and Hill, unpubl.). Samples for biomass determination were taken across and up and down the slope along four 11 m transects adjacent to the main site. Ordination, by reciprocal averaging (Hill, 1973), revealed that the main trend was related to the balance between the mosses *Polytrichum* and *Pleurozium*, and *A. tenuis* and a number of eutrophic species, with *F. ovina* in an intermediate position. Much of the heterogeneity of biomass was related to variations in bryophyte cover, high moss samples varying by at least a factor of 2.5

whereas the vascular plant biomass varied by 1.5. Analysis of the differences between successive cores indicated that *F. ovina* was balanced by the remainder of the species, reflecting the micro tussocking nature of the vegetation below 10 cm scale. This was at least partly due to preferential intake of *A. tenuis* by grazing sheep (Chap. 19).

The characteristics of the grassland vegetation at Llyn Llydaw result from interactions between environmental factors and the plant species. The conditions in which the plants are growing, particularly the chemical properties of the soil, may vary considerably within quite small distances (Ball and Williams, 1968; Chap. 15). The response of *F. ovina* to variations in soil chemistry has been demonstrated by Snaydon and Bradshaw (1961) and considerable variation exists in the genotypes of an *Agrostis-Festuca* grassland (Smith, 1972). Thus at the species level of organisation within the 32 plots there is much interaction. For this synthesis, however, the pattern of distribution of vegetation types distinguished at the plot level (Chap. 14) will be related to production and variations in local controlling factors.

The main feature of the 32 plot ordination (groups A to E) was the higher *A. tenuis* cover towards group A and an increase in species such as *N. stricta* towards group E plots (Chap. 14). The bryophyte variation described above tended to be within plots although there was a greater cover of *Rhytidiadelphus squarrosus* towards A and *Hypnum cupressiforme* towards E. An analysis of the soil properties of the 32 plots, based on Ball's data for Table 5 in Chap. 15, provides mean values for each vegetation group (Table 9). Correlation coefficients for the 32 plot data revealed that loss on ignition (and similarly %C) was positively correlated with K ($r = 0.83$), Na ($r = 0.51$), Mg ($r = 0.59$), N ($r = 0.88$) all at $P < 0.001$. Phosphorus was negatively correlated with loss on ignition, %C, N, and K ($P < 0.01$). The grouped data indicate an increase in loss on ignition and %C from vegetation groups A to E, associated with an increase in the more unpalatable species for sheep grazing, e.g. *N. stricta*, and a decrease in % cover of the preferentially selected and more productive *A. tenuis*. Smaller increases from A to E are seen in K and N, while soil P_2O_5 decreases. The increase in extractable minerals and N results from the increased amounts held on the exchange complex of the surface organic material derived from the slowly decomposing plant material (see Sect. 16.3.3.2). There is little difference in the soil parent material or soil at deeper horizons at the plot level (Chap. 15), the main variation, particularly in K and Ca, being well within the plots at the cm and m scale. Thus variation in soil chemistry between plots largely results from variation in litter quality and decomposition. Rates of disappearance of litter are related to the species composition of the litter (Sect. 16.3.3.2), variation in which is sufficient to cause small but significant variation throughout the plots. This was investigated by assigning the results of litter weight losses in core samples described previously to the plots grouped by ordination. Higher weight losses are associated with the combined groups A+B (16·5%) compared with groups C+D (13·5%) or E (13·2%); A+B is significantly different from E ($P < 0.1$). Groups C, D and E show increasing cover and biomass of the more oligotrophic species, e.g. *N. stricta* and *Juncus squarrosus*, and of mosses e.g. *H. cuppressiforme*. The resulting litters have a higher proportion of species resistant to decomposition. Thus in the series A–E the rate of litter disap-

Table 9. Variation in soil chemical properties within the plot vegetation groups

Vegetation groups (plots)	pH	Loss on ignition (wt%)	C (wt%)	Exchangeable cations (me 100 g^{-1})					Extractable P_2O_5 (mg 100 g^{-1})	Total N (wt%)
				K	Na	Ca	Mg	Mn		
A	4.69±0.01	9.47±0.34	3.94±0.15	0.28±0.03	0.47±0.01	1.92±0.10	0.74±0.05	0.26±0.05	4.87±0.99	0.40±0.03
B	4.78±0.05	12.18±0.28	5.17±0.13	0.43±0.04	0.50±0.02	2.12±0.25	0.85±0.04	0.22±0.03	3.48±0.79	0.55±0.03
C	4.69±0.06	11.75±0.45	4.98±0.21	0.40±0.02	0.49±0.01	2.03±0.10	0.84±0.03	0.30±0.06	2.55±0.22	0.49±0.03
D	4.80±0.04	13.30±0.22	5.68±0.10	0.43±0.02	0.49±0.01	1.82±0.15	0.79±0.05	0.12±0.01	1.76±0.18	0.54±0.01
E	4.75±0.10	14.73±0.62	6.35±0.04	0.54±0.04	0.52±0.01	1.97±0.16	0.96±0.03	0.24±0.06	2.03±0.22	0.60±0.05

The standard errors indicate variation at the plot level; many samples comprise each plot value (Chap. 15).

Table 10. Variation between plot vegetation groups of biomass and its nutrient concentration; net primary production; numbers of grazing sheep and amount of herbage consumed[a]

Vegetation group (plots)	Living aerial biomass[b] ($g\,m^{-2}$)	Root biomass[c] ($g\,m^{-2}$)	Aerial production[d] ($g\,m^{-2}\,yr^{-1}$)	Aerial P/B	Nutrient concentration in aerial biomass[e] ($mg\,g^{-1}$ dw) K	P	N	Ca	Mg	Grazing sheep[f] ($eu\,ha^{-1}$)	Herbage removed[g] ($g\,m^{-2}\,yr^{-1}$)
A ($n=6$)	211±29	517±22	339±72 (1,433)	1.61	18.4±0.5	2.5±0.1	25.5±1.0	4.2±0.4	4.2±0.4	11.2±0.6	311±73
B ($n=6$)	207±25	565±30	314±87 (1,327)	1.52	18.9±0.9	2.2±0.2	25.5±1.4	4.1±0.3	5.3±0.6	11.1±0.9	270±62
C ($n=10$)	228±12	592±23	256±51 (1,082)	1.12	17.5±0.5	2.0±0.1	22.0±0.7	4.5±0.3	3.7±0.3	8.6±0.5	230±32
D ($n=5$)	237±29	637±43	246±80 (1,039)	1.04	16.1±0.8	1.8±0.1	20.8±1.4	3.7±0.2	3.3±0.7	7.5±0.8	178±37
E ($n=5$)	220±47	704±20	199±47 (841)	0.90	15.9±0.7	1.7±0.1	20.9±0.8	4.5±0.4	3.5±0.4	7.0±0.8	253±33
Site mean	220±12	603±20	271±60 (1,145)	1.23	17.4±0.4	2.1±0.1	23.0±0.6	4.3±0.2	4.0±0.2	9.1±0.4	248±21
Range (yr)	190—250	578—627	199—400 —	—	15.5—19.0	1.9—2.2	21.3—25.0	3.9—5.5	3.3—4.4	9.1—9.2	185—332

[a] The values for plot vegetation groups obtained in specific years have been adjusted in proportion with the overall site mean. Standard errors indicate variation between plots within years, the range indicates variation of overall mean between years.
[b] Based on 1969 turf core sampling.
[c] Based on 1969–1970 mean (0–30 cm soil depth).
[d] Cumulative aerial production (clipped yield) based on 1968 data adjusted to 1966–1970 average. Values in brackets are production adjusted for losses to standing dead.
[e] Based on 1968 data adjusted to 1968–1971 average.
[f] Based on 1970–1971 average (Chap. 19).
[g] Based on 1968 data adjusted to 1966–1970 average.

pearance is negatively correlated with loss on ignition, carbon, and also the moisture content of surface soil. The otherwise favourable conditions for decomposition (higher soil moisture) of plot groups C, D, and E are offset by the presence of a more decay-resistant litter. The direct effects of variation in soil composition are not known because the litter is probably affected more by older litter deposited on the surface and influencing the surface soil horizon. The rate of decomposition may also be increased by heavier grazing pressure on groups A and B (Chap. 19), through treading and enhanced nutrient availability from dung (Chap. 20).

Variations in biomass and production in the plots (Table 10) were best demonstrated in the 1968 data, derived from clipped 0.25 m^2 quadrats. These were considered more representative of the plot, the turf core data being affected by within-plot heterogeneity. There was little variation in the biomass but a greater variation and clear trend in production, greatest in plot group A (339 ± 72 g m^{-2}) and lowest in group E (199 ± 47 g m^{-2}). The more productive plots were 26% more, and the least productive 29% less, than average. Production/biomass ratios (1.6 on A, 0.9 on E) showed a similar trend, positively correlated with sheep grazing pressure, and similar to that observed between vegetation types in the more extensive Snowdonia grasslands data.

Root concentration (Table 10) tended to increase from A (517 ± 22 g m^{-2}) to E (704 ± 20 g m^{-2}), 14% less and 17% more than the average (603 ± 20 g m^{-2}) respectively. The trend is negatively correlated with production and sheep numbers, and weakly positively correlated with aerial biomass. Most studies indicate an increase in root biomass with increased grazing pressure, but Bartos and Sims (1974) found no significant differences under different experimental grazing treatments in their US IBP grassland study. At Llyn Llydaw, where grazing has been in operation for a long time, no significant difference could be found in root biomass values in areas where sheep had been excluded for *c* 10 and 15 years, though aerial biomass markedly increased compared with grazed areas. There was, however, a significant change in the distribution of roots, the proportion in the 0–1 cm layer decreasing, and in the 1–8.5 cm layer increasing in ungrazed plots. It appears that root biomass changes take place at the site very slowly.

Correlation coefficients were calculated for the mineral nutrients and nitrogen concentrations (Table 10) of herbage sampled from the groups throughout the 1968 season. N, P, and K all showed positive correlations and N–K ($r = 0.81$), N–P ($r = 0.71$) and P–K ($r = 0.59$) were significant ($P < 0.001$). Ca and Mg overall did not show any correlation although in June and August samples they were positively correlated ($r = 0.60$ and 0.63). N, P, and K all exhibit strong trends from plot group A (high) to E (low), correlated positively with net primary production and negatively with soil loss on ignition, % C, K, N and root concentration. Herbage P concentration is, however, positively correlated with soil P; the other major source of P is sheep dung (Chaps. 19, 20). The cycling of P (and to a less extent K and N) by sheep grazing is seen as an important controlling factor, largely responsible for maintaining the variation in biomass and production at the plot level of organisation in the *Agrostis-Festuca* ecosystem at Llyn Llydaw.

Acknowledgement. We are grateful to Mr. S. E. Allen and colleagues at Merlewood Research Station for undertaking the nitrogen analyses.

17. The Role of Slugs in an Agrostis-Festuca Grassland

J. LUTMAN

Slugs were the main invertebrate herbivores studied on the montane grassland site. From a population analysis their productivity and their influence on primary production was determined.

17.1 Introduction

The study of the ecology of slugs poses problems as they are relatively sparsely distributed, partly subterranean and have aggregated populations. Although an account of the physiology, behaviour, ecology and economic importance of slugs is available (Runham and Hunter, 1970) most ecological investigations have been carried out on agricultural land where slugs are recognised as pests. The distribution and abundance of slugs has been studied on a 7–10 year old ley and in a market garden (South, 1965; Hunter, 1966, 1968b) and the life cycles of most species have been described (Bett, 1960). Slugs feed mainly on plant material and information about their food preferences has been collected by Frömming (1962) and Pallant (1969, 1972). Stern (1970) gives an energy budget for *Arion rufus* in the laboratory, but similar work on field populations has not been reported.

The aims of this study were to determine the biomass of slugs at the Llyn Llydaw site and estimate their productivity and food consumption. Seasonal variation in size and structure of the population was recorded over 21 months, and growth and defecation rates experimentally determined. When combined with published respiration rates (Newell, 1967) these data enabled a tentative energy budget to be established.

17.2 Sampling and Extraction Methods

As slug activity above ground is dependent on humidity, neither collecting nor trapping techniques give reliable estimates of population size and structure. The only satisfactory method is extraction from soil samples (Hunter, 1968a). Although slugs may occur down to 30 cm in arable soils (Hunter, 1966), the generally compact and frequently shallow soils in the study area, and the absence of markedly subterranean species (e.g. *Arion hortensis* and *Milax* spp) allowed sampling to be limited to 10 cm depth.

Commencing in April 1970 the study area of 0.28 ha was sampled at monthly intervals. On the sampling date one turf (0.3 m × 0.3 m × *c* 10 cm deep) was taken from the same randomly selected position in each of the 12 plots (15.2 m × 15.2 m).

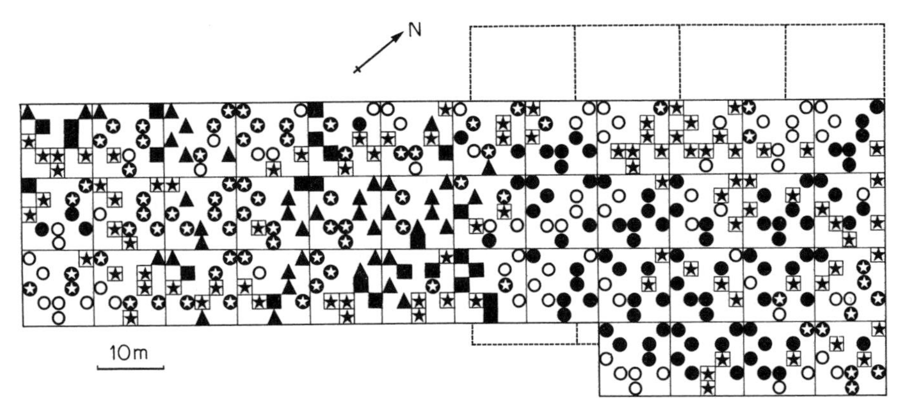

Fig. 1. Diagrammatic map of the slug study area, showing the outline of the 1970 sampling area *(dotted line)* and the vegetation of the 1971 sampling area, between April and December. Each symbol represents a sample (0.3 m × 0.3 m) taken at the centre of a square (2.1 m × 2.1 m). Vegetation types (see Section 17.3 for description): ○: A; ●: B: ✪: C; ⊠: D; ▲: E; ■: F

The minimum distance between any sample and a previous one was at least 3 m. In 1971 the study area was extended south westwards to 0.45 ha (Fig. 1) and 40 samples were taken every four weeks. The size of the 40 plots was reduced to 10.7 m × 10.7 m and consequently the minimum distance between succeeding samples to 2.1 m. The sampling pattern was analogous to the centric systematic system (Milne, 1959) and for statistical purposes the samples were treated as random.

The turf samples were collected into tins with perforated bases. After the vegetation had been recorded and clipped to litter level and any visible slugs and eggs removed, the tins were progressively flooded from below over a period of six days. This method of extraction, which relies on the behaviour of slugs in escaping from a rising water table, was chosen because it is cheaper and easier than soil washing and flotation (Salt and Hollick, 1944) and the turves can be replaced afterwards to prevent erosion of the site. Hunter (1968a) has shown that this technique recovers 90% of the numbers extracted by washing and flotation.

All slugs were identified and weighed, and a proportion was killed by immersion in liquid nitrogen and dried at 80° C for dry matter determinations and chemical analysis.

17.3 The Vegetation of the Study Area

The slug study area adjoined the south western end of the main 32 plot area which could not be sampled as the removal of large turves would have interfered with other projects. A description of the vegetation was needed for comparison with the main area and in relation to the analysis of slug distribution.

Vegetation ordination by reciprocal averaging (Hill, 1973), based on species presence or absence data from the sample turves collected April–December 1971, revealed no well-defined groups of stands. The ordination diagram was therefore divided arbitrarily into six segments (at scores 30 and 60 on axis 1 and score 50 on

Table 1. Percentage occurrence of plant species in each vegetation type

Score		Species with occurrence greater than 36% in at least one type	Percentage occurrence in type					
Axis 1	Axis 2		A	B	C	D	E	F
18	71	*Agrostis canina*	39	4	20	1	4	0
23	29	*Achillea millefolium*	28	58	31	34	9	19
23	44	*Polytrichum* spp	92	75	85	45	29	11
23	59	*Dicranum scoparium*	75	25	50	9	24	3
24	37	*Galium saxatile*	100	98	72	76	21	30
24	42	*Pleurozium schreberi*	96	69	50	40	36	30
25	31	*Luzula campestris*	72	94	52	72	26	46
28	61	*Rhacomitrium lanuginosum*	30	9	39	9	14	0
29	35	*Rhytidiadelphus squarrosus*	89	100	89	92	73	65
31	34	*Anthoxanthum odoratum*	67	90	79	80	70	73
31	36	*Agrostis tenuis*	100	100	95	98	90	100
31	37	*Hypnum cupressiforme*	88	90	93	83	92	84
31	38	*Festuca ovina*	100	98	98	92	95	88
32	36	*Pseudoscleropodium purum*	60	30	35	38	0	19
35	33	*Cirsium* spp	42	68	91	89	78	80
35	47	*Carex pilulifera*	36	24	71	28	56	34
36	61	*Thymus drucei*	44	8	83	11	53	23
37	38	*Lophocolea bidentata*	14	13	56	42	21	34
37	49	*Viola riviniana*	22	9	70	28	41	23
38	7	*Poa* spp	1	24	9	54	19	42
38	52	*Hylocomium splendens*	15	9	39	7	24	26
39	20	*Ranunculus repens*	3	36	21	58	46	57
39	31	*Trifolium repens*	17	51	70	86	92	84
39	73	*Sieglingia decumbens*	30	1	51	5	63	0
42	20	*Cerastium holosteoides*	7	9	25	59	24	57
44	20	*Mnium undulatum*	0	8	18	38	24	42
44	35	*Atrichum undulatum*	0	2	37	40	21	53
47	49	*Thuidium tamariscinum*	7	4	44	22	58	42
48	32	*Holcus lanatus*	7	4	24	44	53	61
49	34	*Cynosurus cristatus*	6	5	22	22	51	61
50	45	*Plantago lanceolata*	2	2	55	26	68	69
52	11	*Cardamine pratensis*	0	0	5	30	14	57
54	43	*Prunella vulgaris*	3	0	25	20	63	61
54	44	*Festuca rubra*	0	1	20	10	41	19
55	12	*Sagina procumbens*	0	1	1	18	17	46
62	44	*Ranunculus acris*	0	1	12	4	58	42
62	57	*Nardus stricta*	0	0	21	3	73	23
64	36	*Scapania undulata*	0	0	6	5	24	46
64	43	*Bellis perennis*	0	0	9	8	80	65
64	52	*Lysimachia nemorum*	0	0	5	1	36	11
72	55	*Fissidens bryoides*	0	0	2	2	70	19

axis 2), and the resulting groups of samples (A–F) described in terms of the percentage occurrence of plant species in them (Table 1). Species typical of *Agrostis-Festuca* grassland occurred at an almost uniformly high level in all groups. The change in species composition of vegetation types along axis 1 suggested increasing soil disturbance due to small scale erosion (A, B < C, D < E, F). Along axis 2 the groups were differentiated mainly in terms of changes in the proportion

Table 2. Percentage of samples containing given species of slugs, in each vegetation type, in the north eastern and south western halves of the study area (April–December 1971)

Vegetation type	A		B		C		D		E		F	
Half of study area	NE	SW	NE	SW	NE	SW	NE	SW	NE	SW	NE	SW
Number of samples	47	29	83	8	13	62	47	45	0	40	0	26
Agriolimax reticulatus	13	17	22	25	61	39	51	44	—	50	—	19
Agriolimax laevis	0	0	0	0	0	13	2	11	—	45	—	23
Arion intermedius	19	21	46	50	54	55	68	71	—	35	—	38
Arion fasciatus	0	3	0	0	0	3	11	7	—	5	—	15
Arion subfuscus	2	3	0	0	0	8	4	11	—	2	—	0
Arion ater	0	3	0	0	8	8	0	2	—	2	—	8
Percentage of each type containing slugs	29		57		73		79		82		65	

of species, indicating a gradient of decreasing moisture, nutrients and base status (B, D, F > A, C, E).

The distribution of the six vegetation types (Fig. 1) shows that the north eastern half was the more uniform, with half the samples consisting of grassy moss swards (type B) and most of the rest of more acidic mossy swards (type A) or damp *Agrostis-Festuca* grassland (type D). The south western half was more irregular, with steeper slopes and incipient erosion gullies as well as a few areas of impeded drainage. The most disturbed vegetation types (E and F) were restricted to this half where they accounted for a third of the samples, while a further third consisted of more typical herb rich *Agrostis-Festuca* grassland (type C). Overall, the slug study area appeared to contain most of the vegetation categories characteristic of the herb rich *Agrostis-Festuca* grassland of the main 32 plot area (Chap. 14), except for the *Juncus effusus* wet flushes. The other difference in the vegetation of the slug study area was the disturbed element in the south west.

17.4 The Distribution of Slugs in Relation to the Vegetation of the Study Area

Slugs were not distributed uniformly throughout the study area; the vegetation groups showed clear differences both in the proportion of turves containing slugs and in the average number of slugs per sample. The dry mossy vegetation of type A had the fewest slugs (mean 0.6 per sample) with more than 70% of the samples containing no slugs at all. In the moister, herb rich and somewhat disturbed areas with *Nardus stricta*, types D and E, almost 80% of the samples contained slugs, with respective means of 3.5 and 3.3 compared with an overall mean of 2.1 slugs per sample.

Although the south western half of the study area contained 57% of the slug population and the north eastern half 43%, the percentage occurrence of the slug

Table 3. Population density of slugs (mean number $m^{-2} \pm$ se)

	Agriolimax reticulatus	*Agriolimax laevis*	*Arion intermedius*	*Arion fasciatus*	*Arion subfuscus*	*Arion ater*
1970						
14 Apr.	9.9± 5.4	0	31.4±14.2	0	0.9±0.9	0
13 May	0.9± 0.9	0	8.1± 3.3	0	0.9±0.9	0
10 June	2.7± 1.9	0	13.4± 7.6	0	0	0
15 July	8.1± 4.3	0	13.4± 4.4	0	0.9±0.9	0
12 Aug.	2.7± 1.9	0	2.7± 2.7	0.9±0.9	1.8±1.2	0
16 Sept.	16.1±12.3	0	9.0± 3.5	1.8±1.2	0	0
14 Oct.	21.5± 9.5	0	19.7±11.4	0.9±0.9	0	0
10 Nov.	25.1±12.7	0	4.5± 3.6	0.9±0.9	0	0
8 Dec.	17.9± 8.7	0.9±0.9	9.9± 3.6	4.5±2.5	1.8±1.2	0
1971						
12 Jan.	26.9± 6.1	3.5±1.2	21.2±16.8	1.6±0.9	0.5±0.4	0.5±0.4
9 Feb.	19.9± 5.4	4.0±1.9	14.7± 3.5	1.6±0.9	0.8±0.4	1.3±0.6
9 Mar.	31.5± 8.8	1.9±0.9	23.7± 5.7	1.6±0.9	0.5±0.4	0.3±0.3
6 Apr.	21.5± 6.0	1.3±0.6	22.6± 5.2	0.5±0.4	0.8±0.4	0.3±0.3
4 May	17.8± 5.5	1.9±1.0	14.7± 4.3	0	0.3±0.3	0
1 June	5.4± 2.1	1.3±0.9	13.1± 3.0	0	0.3±0.3	1.1±0.8
29 June	5.4± 1.9	0.3±0.3	13.7± 2.5	0.5±0.4	0.5±0.4	0.5±0.4
27 July	2.4± 0.7	0.3±0.3	9.1± 2.0	0.3±0.3	1.3±0.8	0
24 Aug.	6.5± 2.1	3.0±1.0	7.3± 1.9	1.1±0.5	0.8±0.4	0.3±0.3
21 Sept.	4.0± 1.5	1.9±1.2	5.9± 2.0	0.3±0.3	0.3±0.3	0.3±0.3
21 Oct.	3.5± 1.4	0.3±0.3	14.3± 3.8	1.1±0.8	0.5±0.4	0.3±0.3
17 Nov.	5.6± 2.3	3.8±1.5	11.8± 2.7	1.3±0.6	0.5±0.4	0.3±0.3
14 Dec.	4.6± 1.7	2.7±1.3	13.7± 3.6	0.5±0.4	1.1±0.5	0.3±0.3

species in the various vegetation types was different even when the two halves of the study area were considered separately (Table 2). *Arion intermedius* Normand[1] occurred more frequently in types B, C and D with a peak in type D, the fairly moist, herb rich swards, the more disturbed ground being less favoured by this species. *Agriolimax reticulatus* (Müller) was more frequent in the herb rich and somewhat disturbed types, C, D and E, with a peak in E. *Agriolimax laevis* (Müller) had a very pronounced peak in type E indicating a very restricted distribution in the study area. This species is reported to favour damp conditions (Quick, 1960) and it has been mentioned (Chap. 14) that the *N. stricta* vegetation types tended to retain more moisture during dry periods. The occurrence of *Arion ater* (Linnaeus), *Arion fasciatus* (Nilsson) and *Arion subfuscus* (Draparnaud) was sporadic and there was insufficient information to determine their main habitats.

17.5 Abundance and Biomass of Slugs

Population density and biomass data (Tables 3 and 4) indicate that in 1970 sampling intensity was too low, with large standard errors and considerable

[1] Nomenclature follows Quick (1960).

Table 4. Biomass of slugs (mg dw $m^{-2} \pm se$)

	Agriolimax reticulatus	*Agriolimax laevis*	*Arion intermedius*	*Arion fasciatus*	*Arion subfuscus*	*Arion ater*
1970						
14 Apr.	248±200	0	61±31	0	34± 34	0
13 May	19± 19	0	24±11	0	29± 29	0
10 June	46± 43	0	34±19	0	0	0
15 July	293±173	0	100±37	0	103±103	0
12 Aug.	13± 11	0	151±21	1± 1	14± 12	0
16 Sept.	48± 44	0	144±60	20± 2	0	0
14 Oct.	168± 89	0	65±34	5± 5	0	0
10 Nov.	294±162	0	8± 6	28±28	0	0
8 Dec.	191± 92	4± 4	6± 2	56±32	49± 33	0
1971						
12 Jan.	477±114	34±11	26± 7	11± 6	6± 4	15± 12
9 Feb.	340± 98	31±15	22± 5	18± 9	21± 14	217±163
9 Mar.	629±161	21±10	31± 8	23±15	27± 23	6± 6
6 Apr.	583±178	16± 7	47±10	5± 3	54± 48	8± 8
4 May	554±166	16± 9	44±14	0	3± 3	0
1 June	153± 62	10± 6	66±17	0	18± 18	236±208
29 June	62± 27	1± 1	79±15	8± 6	38± 36	401±294
27 July	31± 20	1± 1	80±20	1± 1	60± 44	0
24 Aug.	28± 10	7± 3	111±30	8± 5	3± 2	2± 2
21 Sept.	52± 25	9± 6	89±28	1± 1	25± 25	143±143
21 Oct.	41± 37	1± 1	87±20	7± 5	11± 8	4± 4
17 Nov.	108± 48	22± 9	44±16	18± 8	67± 47	2± 2
14 Dec.	123± 51	24±13	22± 6	16±15	59± 48	9± 9

fluctuation between months making it difficult to describe population structure adequately. Most of these deficiencies were remedied by expanding the sampling programme in 1971, but the standard errors of mean population size were still large. This might be partly the result of aggregation which seemed very marked at Llyn Llydaw, probably reflecting the heterogeneity of the site.

Agriolimax reticulatus and *Arion intermedius* were the most abundant species (12 m^{-2} and 14 m^{-2} respectively in 1971). *A. reticulatus* had the highest biomass early in the year and accounted for 56% of the mean annual biomass. *A. intermedius* contributed only 13% to the mean annual, but up to 50% of the monthly biomass between July and October.

Arion ater was the rarest species in the study area (0.4 m^{-2}) and also the largest, the live weights of adults being in excess of 4 g (cf *A. reticulatus* 0.4 g and *A. intermedius* 0.1 g). Its contribution to the mean annual biomass was 19% but when occasional mature specimens were present in the samples, as in June and September 1971, its share of the monthly biomass rose to over 40%. *Arion fasciatus* and *A. subfuscus* were also comparatively rare (0.8 m^{-2} and 0.6 m^{-2} respectively) and with *Agriolimax laevis* accounted for 12% of the mean annual biomass. The low numbers of the latter (2 m^{-2}) probably reflect its restricted distribution in the study area rather than its rarity.

17.6 Methods of Estimating Recruitment and Production

Laboratory data on egg laying, hatching and growth of *A. intermedius* and *A. reticulatus* cultures at *c* 8° C were used to estimate monthly egg production in the field and hence recruitment to the next generation, by determining:

1. the time between egg laying and hatching;
2. the hatching success of eggs laid in the laboratory compared with those collected in the field;
3. the daily laying rate (from the number of eggs found in the field samples, taking account of the hatching time);
4. recruitment, i.e. the potential number of slugs hatching (from egg production × hatching success).

Production due to growth by a generation was estimated from the area under a curve of numbers plotted against weight at given ages, from hatching to maturity (Southwood, 1966). The curve was constructed as follows:

1. weight limits of the age classes were determined from the growth curves of slugs cultured in the laboratory;
2. slugs from the field samples were allocated to monthly age classes on a weight basis (from hatching to three months was one class owing to small weight differences involved);
3. lower numbers of young slugs in the samples than the calculated recruitment were assumed to be due to a high mortality rate up to the age of three months;
4. as recruitment was spread over several months the total number of slugs reaching each age was calculated. A smoothed survivorship curve was then drawn for the generation;
5. the values read from the smoothed survivorship curve were plotted against the mean weights of each age class.

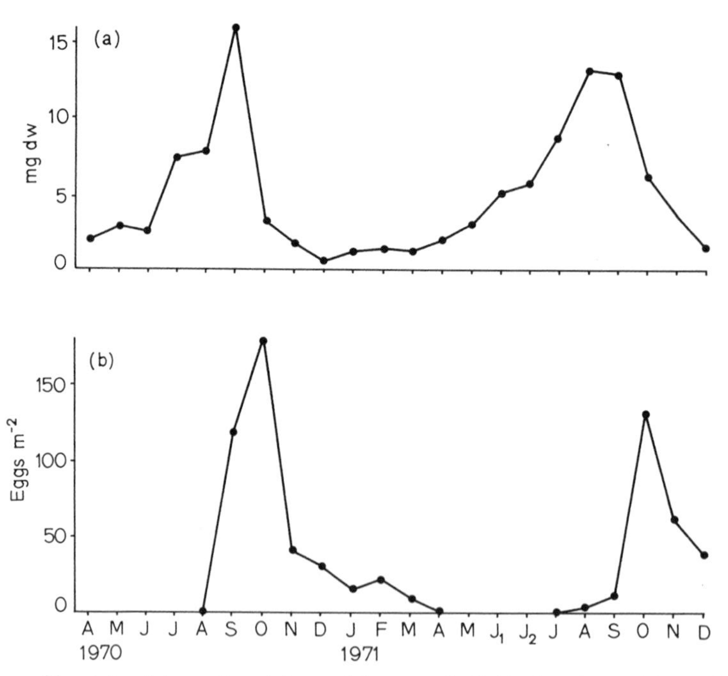

Fig. 2a and b. *Arion intermedius:* (a) monthly mean individual weights; (b) numbers of eggs present in the monthly samples

17.7 Life Cycles and Production

17.7.1 Arion intermedius

This species had a clearly defined annual life cycle and its abundance (Table 3) and phenology were very similar in 1970 and 1971. Although eggs were present in the samples between August and March (Fig. 2b), the main laying period was between August and October, after which slugs of the old generation disappeared. Hatching success in the laboratory was 60%. In the field considerable mortality of slugs occurred during the first few months, but their numbers remained around 20 m^{-2} over the winter. There was a slow decline in the spring and early summer as the slugs increased in weight (Fig. 2a) prior to maturation. The life history of *A. intermedius* is summarised in Figure 3 which shows the changes in the frequency of age classes in the population throughout the life of the 1970/71 generation.

Production by the 1970/71 generation of *A. intermedius* (Fig. 4), based on a survivorship curve derived from the data given in Figure 3, was estimated at 0.41 g m^{-2}. Egg production in the autumn of 1971 was 0.04 g m^{-2}, and the production/biomass ratio 7.1.

17.7.2 Agriolimax reticulatus

There is considerable overlap in generations of *A. reticulatus* and its life history is complex. Various workers (e.g. Bett, 1960; Hunter, 1966, 1968b) have reported that although some eggs could be found at most seasons, this species has breeding peaks in spring and autumn with two main generations per year. This may result from alternating generations with a nine month generation interval (Hunter and Runham, 1970).

The results of sampling during 1970 and 1971 suggest that at Llyn Llydaw the life cycle approximated to an annual one with the main breeding period at about midsummer and some reproduction occurring throughout the year. While the population structure was similar during comparable periods (Fig. 5), the data indicate the range of between-year variation in population size (Table 3). Between April and July 1970 the population, mainly large individuals, was *c* 5 m^{-2}. There was a pronounced peak of egg laying at the end of June, followed by a sharp drop in mean individual weight from 0.036 g in July to 0.005 g in August (Fig. 6a), as the older slugs disappeared and new ones hatched. Between September 1970 and May 1971 the population remained fairly stable at *c* 20 m^{-2}, with the slugs growing steadily, but by June their numbers had dropped to 5.4 m^{-2} and by the end of July to 2.4 m^{-2}. The estimated maximum laying rate was 1.5 eggs m^{-2} d^{-1} on 12 June 1971, compared with the peak of 5.9 eggs m^{-2} d^{-1} on 25 June 1970. Relatively few slugs were found in autumn 1971 and the population remained *c* 5 m^{-2} until December when sampling ceased. The decrease in population size during May–June 1971 and the lower rate of recruitment which followed must have been due to premature death of much of the potential breeding population. If the slugs had merely retreated below the 10 cm sampling depth, they or their offspring would have been expected to reappear in later samples. Although the spring

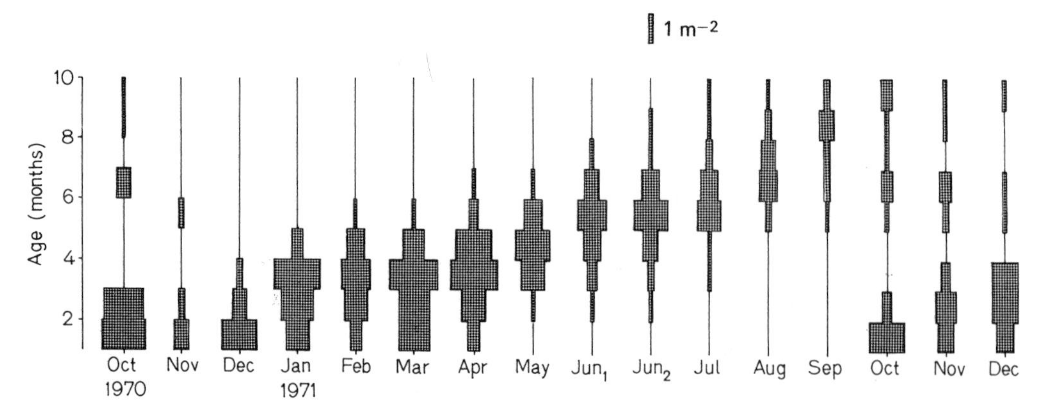

Fig. 3. *Arion intermedius:* the frequency of age classes in the monthly samples

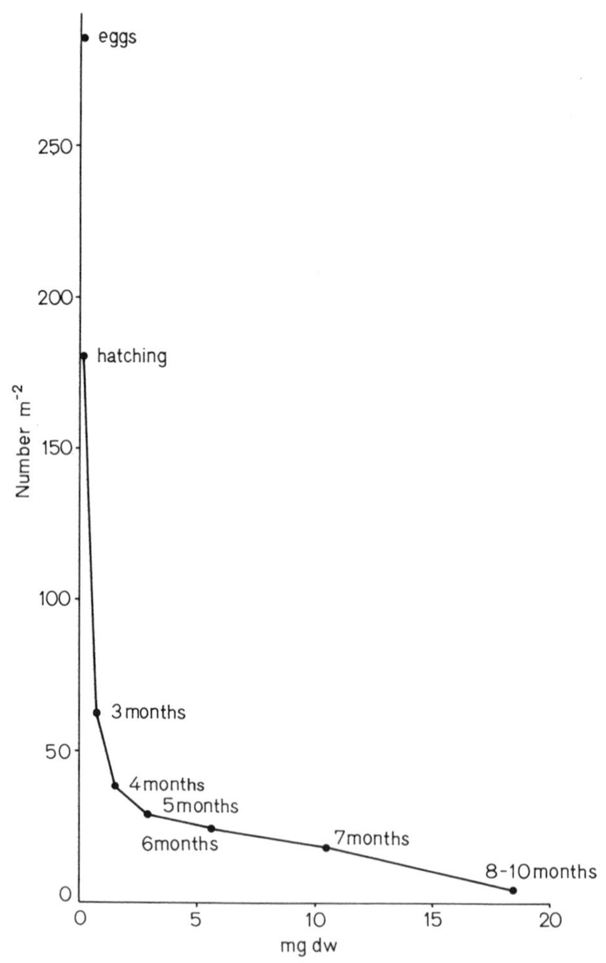

Fig. 4. Production curve for the 1970/71 generation of *Arion intermedius*. The area under the curve is proportional to production

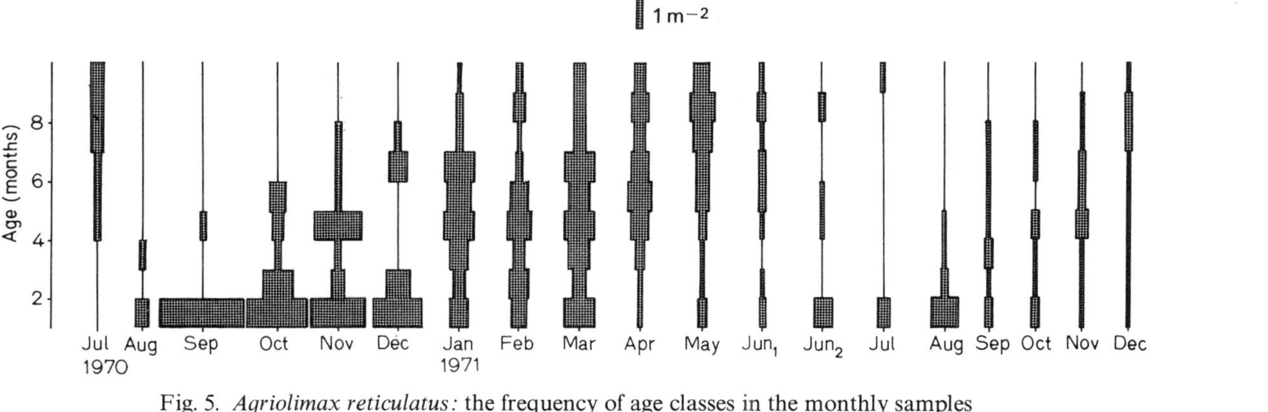

Fig. 5. *Agriolimax reticulatus*: the frequency of age classes in the monthly samples

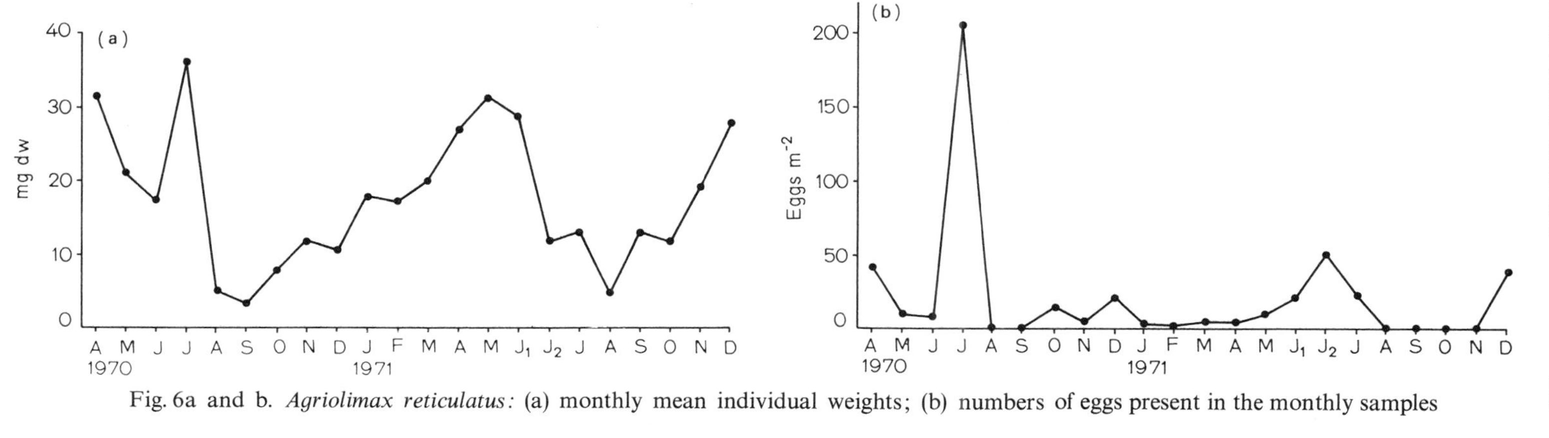

Fig. 6a and b. *Agriolimax reticulatus*: (a) monthly mean individual weights; (b) numbers of eggs present in the monthly samples

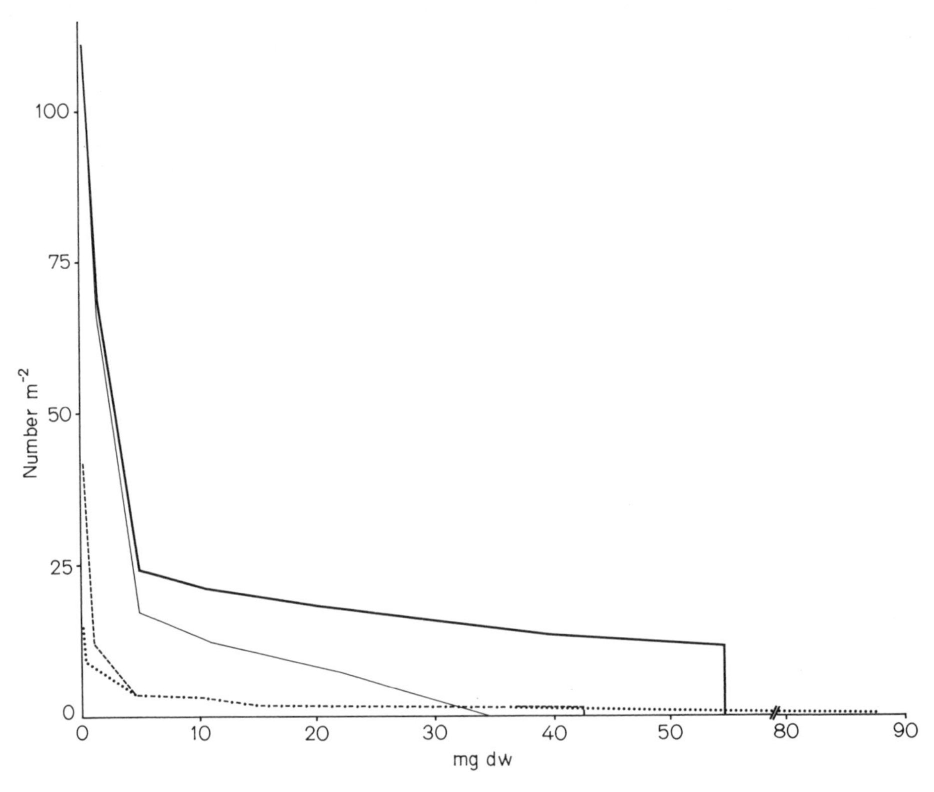

Fig. 7. Production curves for cohorts of *Agriolimax reticulatus*. The total area under each curve is proportional to production by a particular cohort. ———: cohort hatching May–December 1970, production during 1970. ———: cohort hatching May–December 1970, production during 1971. ·········: cohort hatching January–April 1971, production during 1971. ---------: cohort hatching May–December 1971, production during 1971

weather was somewhat drier than in 1970, there is no direct evidence to suggest that it was the cause of mortality.

Production curves for successive cohorts of *A. reticulatus* (Fig. 7) enable differences in performance between the two years to be compared. Estimated production by the cohort hatching May–December 1970 in its year of hatching (0.46 g m^{-2}) was four times that of the 1971 May–December cohort (0.12 g m^{-2}). In the year following their hatching, the estimated production of the 1970 cohort was 0.62 g m^{-2} and, although no comparable data are available for the 1971 cohort, their production clearly could not have attained similar levels. Production during 1971 by slugs hatching January–April 1971 was estimated as 0.12 g m^{-2} but there are no data for a similar cohort in 1970.

Total production by *A. reticulatus* in 1971, 0.85 g m^{-2}, was estimated by summing the production of the three cohorts hatching May–December 1970, January–April 1971 and May–December 1971. Egg production was equivalent to 0.06 g m^{-2}, and the production/biomass ratio 3.5 in 1971.

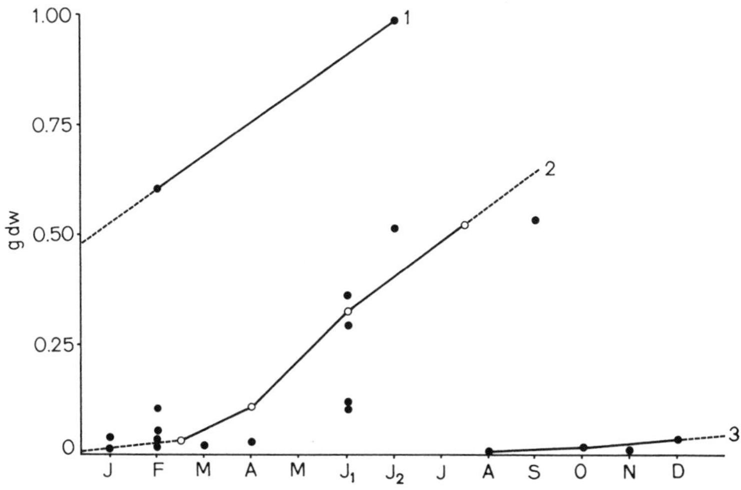

Fig. 8. Growth curves of *Arion ater*, fitted to weights of individual slugs from the 1971 samples. See Section 17.7.3 for a description of cohorts 1, 2, and 3. ●: individual weights; ○: interpolated mean weights

17.7.3 Arion ater

An attempt was made to estimate production by *A. ater* because of its large contribution to biomass in some months; the scanty field data being interpreted with the help of information in the literature. In a population of *A. ater* in Bangor (North Wales) egg laying in the late summer and early autumn was followed by high mortality, few animals overwintering except as eggs or very young stages (Smith, 1966). It has been reported that in laboratory cultures of *A. rufus*, a closely related species, there was a continuous and almost linear increase in weight over 52 weeks, followed by reproduction (Stern, 1970).

The 1971 data were divided by weight into three cohorts of slugs, which had

1. overwintered at a fairly high weight, perhaps having failed to breed the previous autumn (1970);
2. hatched during winter 1970/71;
3. hatched early autumn 1971.

Growth curves were interpolated and extrapolated, by eye, to the earliest and latest dates when the three cohorts could have been present (Fig. 8), and survivorship and production curves drawn for each. Production by *A. ater* during 1971 was tentatively estimated at 0.58 g m^{-2}, giving a production/biomass ratio of 7.1.

17.7.4 Agriolimax laevis, Arion fasciatus and Arion subfuscus

Production by these sporadically occurring species, whose contribution to the mean annual biomass was small, would be expected to be relatively insignificant. An order of magnitude was obtained by multiplying the average biomass of each species by the production/biomass ratio of the most closely related species for which data were available (*Agriolimax reticulatus* for *A. laevis*, and *Arion interme-*

dius for *A. fasciatus* and *A. subfuscus*). The approximate estimates of production in 1971 were: *A. laevis*, 0.05 g m^{-2}; *A. fasciatus*, 0.06 g m^{-2}; *A. subfuscus*, 0.22 g m^{-2}.

17.8 Estimation of Consumption

Total consumption by the slug population (Table 5) was estimated by summing amounts converted to production, lost in respiration and defecated by each species. Production was based on a calorific value for slugs of 20.08 kJ g^{-1} (Petrusewicz and Macfadyen, 1970, give 19.25 kJ g^{-1} for molluscs; Stern, 1970, gives 20.59 kJ g^{-1} for *Arion rufus*).

As respiration was not measured directly in this study, estimates were based on Newell's (1967) data for daily oxygen uptake. At 10° C, allowing for 6 h activity and 18 h quiescence, these were equivalent to: *Agriolimax reticulatus* 141.6 and *Arion hortensis* 63.9 kJ g^{-1} yr^{-1}. The respiratory requirements of the field population were derived by multiplying the species' mean annual biomass by the metabolic value of the more closely related species (*A. reticulatus* for both *Agriolimax* species and *A. hortensis* for all *Arion* species). Newell (1967) suggested that temperature corrections applied to oxygen uptake during activity, which was the significantly temperature dependent component, would add little to the accuracy of the overall value, since crawling takes place mainly at night when seasonal temperature variations are at their lowest. Thus a correction for temperature of the Llyn Llydaw data (7.3° C mean annual) was considered superfluous.

An average rate of defecation was obtained by weighing the faeces produced in 24 h by slugs of various sizes removed from the samples at different times during the day, but a satisfactory estimate was obtained only for *A. reticulatus*: 0.07 g g^{-1} d^{-1}. This value, multiplied by the mean annual biomass of each species, gave an estimate of faeces production by the population. The calorific value of slug faeces and dry plant material was taken as 18.70 kJ g^{-1} (Chap. 19).

17.9 Qualitative Aspects of Feeding by Slugs

There is little information on slug feeding behaviour in natural habitats, but the wide range of agricultural situations in which they have been recorded as pests (Runham and Hunter, 1970) suggests that they may be relatively unspecialised in their food requirements. *Ranunculus repens* and *Urtica dioica* formes a large proportion of the gut contents in *Agriolimax reticulatus* from woodland (Pallant, 1969), while in rough grassland the dominant *Holcus lanatus* was the principal food, most of the leaf fragments showing no signs of senescence. Small arthropods appeared to have been ingested, possibly fortuitously, and occasionally parts of earthworms. It was concluded that *A. reticulatus* was primarily a herbivore but also a versatile opportunist feeder (Pallant, 1972).

In the present study faeces of the various species were preserved but not analysed in detail. The faeces of *A. reticulatus* were bright green more frequently than those of other species, suggesting that *A. reticulatus* tends to ingest live plant material, whilst other species take senescent or dead material. *Arion ater* had a wide range of foods, from living plants to various forms of dead organic matter. Its faeces sometimes contained masses of hairs of *Cirsium* sp and it was also observed feeding on sheep dung.

17.10 Discussion

Although the several components from which annual consumption (C) by slugs was determined (Table 5) were estimated with varying degrees of accuracy, the results for the more important species (except *Arion ater*, data for which were scanty) seem reasonably consistent with published information. Production/respiration ratios (P/R) for *Agriolimax reticulatus*, 0.49, and *Arion intermedius*, 2.27, were close to the range of 0.14 to 1,86 for various woodland snails (Mason, 1971). Laboratory studies of *Arion rufus* gave an average P/R of 0.37 (Stern, 1970), which suggests that the value for *Arion ater* of 2.26 may have been excessive and production overestimated.

The production/biomass ratios (P/B) of *A. reticulatus*, 3.5, and *A. intermedius*, 7.1, estimated directly, are of the same order as Mason's (1971) ratios for woodland snails: within the range 1 to 34.3, with few greater than 10. They also tend to confirm his view that smaller species have higher P/B ratios, or greater population turnover.

The average assimilation efficiency, $\frac{P+R}{C} \times 100$, of slugs at Llyn Llydaw, calculated from the data given in Table 5, was 30%. This is lower than the efficiencies obtained by other workers for woodland snails and considerably lower than those for slugs (Table 6). It is not to be expected that values calculated on a population basis would be in complete agreement with experimentally determined results. Consumption by a woodland snail population, 193 kJ m^{-2} yr^{-1}, calculated as the sum of production and respiration and assuming an assimilation efficiency of 50%, was approximately double the value based on food consumption as a percentage of body weight (Mason, 1971).

Compared with densities found in some other habitats the size of the total slug population at Llyn Llydaw, 30 m^{-2}, was small. Averages of 47 and 65 m^{-2} *A. reticulatus*, 60 and 51 m^{-2} *Arion hortensis* and 32 and 31 m^{-2} *Milax budapestensis* were found on an arable plot in Northumberland during 1963 and 1964 respectively (Hunter, 1966); but densities of only 12 m^{-2} *A. reticulatus* and 4 m^{-2} *A. intermedius* in an adjacent pasture in 1961 (South, 1964). Hunter's (1966) data tend to confirm that the between-year fluctuations in numbers of *A. reticulatus*, which were indicated by the present study, are a feature of the ecology of this species. The consumption of 171 kJ m^{-2} yr^{-1}, more than half the annual total for slugs, by this primarily herbivorous species demonstrates its importance, even in a year when its numbers were known to have been depressed.

The estimated consumption by slugs at Llyn Llydaw of 304 kJ m^{-2} yr^{-1} (16.3 g m^{-2} yr^{-1}) in 1971, was a small proportion of the total intake by herbivores, amounting to 6.6% of the estimated sheep intake from a biomass of *c* 5%. Nevertheless, their impact on the system may be greater than is suggested by this

Table 5. Mean biomass and the components of consumption by slugs during 1971 (kJ m^{-2} yr^{-1})

	Biomass	Production	Respiration	Defecation	Consumption
Agriolimax reticulatus	4.89	17.07	34.68	119.49	171.28
Agriolimax laevis	0.29	1.05	2.18	7.33	10.55
Arion intermedius	1.17	8.28	3.64	27.80	39.73
Arion fasciatus	0.17	1.30	0.59	4.40	6.28
Arion subfuscus	0.59	4.35	1.92	14.61	20.89
Arion ater	1.63	11.63	5.15	39.02	55.81
Total	8.74	43.68	48.16	212.65	304.54

Table 6. The assimilation efficiency of various terrestrial molluscs, from the literature, determined experimentally as $\frac{C-F}{C} \times 100$

Species	Fed on	% Assimilation	References
Agriolimax reticulatus	*Ranunculus repens*	78.4	Pallant (1970)
Arion rufus	*Lactuca sativa*	74.0	Stern (1970)
Oxychilus cellarius	*Lactuca sativa*	70.2	Mason (1970)
Helix aspersa	*Lactuca sativa*	53.5	Mason (1970)
Hygromia striolata	*Urtica dioica*	52.4	Mason (1970)
Discus rotundatus	*Urtica dioica*	47.7	Mason (1970)
Discus rotundatus	*Circaea lutetiana*	46.0	Mason (1970)
Discus rotundatus	*Mercurialis perennis*	40.4	Mason (1970)

Table 7. Estimated monthly consumption by the slug population at Llyn Llydaw during 1971

Month	g dw m^{-2} 28 d^{-1}
Jan.	1.6
Feb.	1.8
Mar.	2.1
Apr.	2.0
May	1.8
June (beginning)	1.4
June (end)	1.7
July	0.5
Aug.	0.4
Sept.	0.9
Oct.	0.7
Nov.	0.7
Dec.	0.7

annual value. Slug feeding intensity was at its highest from mid October–April (Table 7) and was probably of the same order of magnitude as that of the few sheep overwintering on the site. Low temperatures in this period cause primary production to be at a standstill and the standing crop to decrease due to die back of vegetation. Thus slugs could remove photosynthetic "capital" at a critical stage in the sward growth, and delay the onset of measurable growth in the spring. Moreover, in a study of the effect of slug damage on the establishment of ryegrass (*Lolium perenne*) swards, *A. reticulatus* this author found that to cut through young leaf or stem bases, while leaving the leaves uneaten on the soil (Hatto, unpubl.; Harper, 1977).

The correlation between numbers of slugs and vegetation type could have had various causes. Slug populations are characteristically aggregated, partly because their eggs are laid in clusters and they have rather poor dispersive powers. South (1965) demonstrated an association between *A. reticulatus* and cocksfoot tussocks (*Dactylis glomerata*) in grassland, perhaps related to the greater shelter they provided for eggs in the dry spring period. At Llyn Llydaw the structure of the vegetation cover may have had a similar direct effect, or it may have integrated several environmental attributes, each of which influenced slug survival. Palatability of plant species in the herb rich parts of the site may also have influenced slug distribution. It is, however, clear that the impact of slugs on the vegetation must be greater in the herb rich than in the mossy areas.

Acknowledgements. I am indebted to the Wardens of the Nature Conservancy, Bangor, and to other colleagues without whose assistance the field work would not have been possible. Mrs. V. Jones and Mrs. P. E. Neep painstakingly carried out the botanical identifications. I thank Dr. D. F. Perkins and Dr. D. C. Seel for their advice and encouragement and Mr. M. O. Hill for his help with analysing and interpreting the botanical data.

18. Sheep Population Studies in Relation to the Snowdonian Environment

J. Dale and R. E. Hughes

The montane grasslands in Snowdonia are used by man for sheep grazing. Repeated censuses of a range of sites shows the pattern of distribution of sheep and their relationship to environmental factors, particularly soil parent material and vegetation type.

18.1 Introduction

Sheep are the dominant grazing animals in Snowdonia and their effect on the upland ecosystem has increased profoundly since Neolithic times. Since the Mediaeval period sheep have been present in greater numbers than other classes of domestic livestock (Emery, 1967; Hughes et al., 1973). It is usual for flocks of sheep to graze specific sections of mountain land and each is managed by one farmer. Some sections, or sheepwalks, are enclosed, e.g. in the uplands of the Valley of the Conwy, but in the Snowdon area, flocks have been trained to occupy specific un-enclosed areas of mountain land.

During the summer period, April to October inclusive, sheep occupy the sheepwalks. They selectively graze broad categories of vegetation but the precise location of individual groups of sheep varies diurnally and is affected by weather. Management is confined to periodic gatherings of flocks for dipping and shearing. At the end of the season they are withdrawn, the breeding flocks moved to winter quarters and a proportion sold as store lambs. In favourable situations in the uplands of the Conwy Valley the winter withdrawal may be less complete, stray sheep and some wethers remaining, but in general the mountain pastures are only intensively grazed during the summer period.

Grazing intensity plays an important role in determining the vegetation of an area. The relationship is complex, however, since the vegetation itself will influence the selective grazing behaviour of sheep and their population densities (Boulet, 1939; Hunter, 1954; Hughes, 1958a; Hughes et al., 1964). Local differences in complexes of abiotic factors such as rainfall, altitude and soil also have a major influence on sheep population densities and the intensity of grazing. Thus the aim of the work summarised here was to study the variation in sheep population densities relative to the complex environmental background of northern Snowdonia (The Royal Society, 1967) and to define the broad ecological context within which the IBP site is placed at Llyn Llydaw.

18.2 Earlier Studies

In earlier work in northern Snowdonia the units of study were whole sheep-walks, hence only broad generalisations and approximations could be made on the factors influencing sheep populations (Hughes, 1954, 1958a). It was recognised that the sheep-carrying capacity of grazings in the area studied could be influenced by the local zonal distribution of complexes of abiotic and biotic factors related to altitude, rainfall, tendencies in soil development and general features of the vegetation.

The conclusions of this earlier work suggested the need to obtain estimates of sheep population densities characteristic of specific plant communities and habitats.

18.3 Present Studies

As a consequence of the earlier work a weekly census of sheep numbers found on sites with different vegetation and soil types was started in 1956 and continues to the present time.

A gradient in mean annual rainfall exists from 889 mm at the north coast near Conwy to 3810 mm or more over the Snowdon massif to the south. Initially, there were seventy census sites distributed along this gradient over a distance of 24 km, extending over an altitudinal range of *c* 914 m, on a ridge of Crib Goch near Snowdon, to 244 m in the uplands of the Conwy Valley.

The geological and physiographical variability of the area result in few tracts of uniform vegetation and soils; instead they occur as a complex mosaic. Sheep census plots were selected from a range of these, representative of the most frequently occurring types, on the basis of their comparative homogeneity. *Agrostis-Festuca-*, *Nardus-*, *Molinia-*, *Juncus-*, *Calluna-* and *Pteridium-*dominated communities were all represented on a range of soil types ranging from well-developed acidic brown earths to peaty podzols and peat rankers. The plots varied in area from 0.2 to 6.8 ha and were selected so that all the ground within them was visible to the observer. Counting of sheep, recorded as ewe units (Hughes, 1958a) took place on the same day of the week and at approximately the same time of day (between 1100 and 1430 h) at intervals of one week throughout the year. During the spring up to 1 July lambs were counted separately and two lambs were taken as being equivalent to one ewe unit. It was considered that sheep were occupying their main daylight grazing areas by midday and that the density recorded would give a reasonable estimate of the distribution of sheep during the major part of the day. Though there is a diurnal pattern of numbers of sheep, observations made were representative of the period from at least 1000 h and perhaps earlier, up to 1700 h (Hughes et al., 1975).

The abiotic environment and vegetational features of the census plots are summarised in Table 1A and B. Details of the vegetational features of the IBP site are given in Chapter 14, and those for the census plots generally by Hughes et al. (1975). Bracken (*Pteridium aquilinum*)-dominated areas have not been included in the range of census plots, since insufficient is known about the grazing intensi-

Table 1. Abiotic environment and vegetation of sheep census plots in north west Wales

A.

Abiotic environments	Vegetational types[a]					
	I	II	III	IV	V	VI
A		1			1	2
B		3				1
C			6		1	1
D	2[b]	8	1	1		
E		1	3	1		
F				1	4	
G	2	3		1		
H			3	5	1	

B.

Abiotic environments				Vegetational types	
High rainfall	Altitude[c]	Soil parent material[d]	Drainage[e]		
A	H	A	F	I	Herb rich *Agrostis/Festuca* grassland
B	H	B	F	II	*Agrostis/Festuca/Nardus* grassland
C	M	A	F	III	*Nardus stricta* grassland
D	M	B	F	IV	*Molinia/Juncus* grassland
E	M	A	I	V	*Vaccinium/Calluna* heath
Medium rainfall					
F	M	A	F	VI	*Agrostis/Festuca/Cryptogramma crispa* communities on unstable sites and screes
G	M	B	F		
H	M	A	I		

[a] The numerals give the number of plots within each abiotic/vegetational type.
[b] Census plot in which the IBP 32 plot site was situated.
[c] H = high (>427 m OD); M = medium (<427 m OD).
[d] A: acidic; B: basic.
[e] F: free; I: impeded.

ties to which they are subject during the summer months when few animals graze them and since the density and height of vegetation make observation difficult.

The mean sheep population densities per plot for April–October inclusive, 1956–1968, were used to determine thirteen internally homogeneous population groups (Table 2). These were isodemic, and statistically significant differences existed between the groups at the 5% level. The procedure used was that proposed by Gabriel (1964), and the groups established are discussed in detail by Hughes et al. (1975).

The mean sheep population densities of isodemic groups 1 to 7 represent the situation at higher levels of rainfall, i.e. above 2540 mm. Vegetation and soil types

Table 2. Mean sheep population densities (eu ha^{-1}) of the isodemic groups

	Group number	Mean sheep population density (eu ha^{-1})	Abiotic/vegetational types present	
High rainfall (>2,540 mm)	1	0.53	A, B, C, D	VI
	2	1.45	A	III
	3	1.74	B, C	II
	4	3.12	E	II, III, IV
	5	4.98	D	I
	6	4.53	B, D	II
	7	3.65	C, D	III
Medium rainfall (<2,540 mm)	8	2.83	F	IV, V
	9	3.35	H	III, IV, V
	10	6.90	G	II, IV
	11	5.63	G	I
	12	7.14	F, H	III, IV
	13	18.17	G	I

range from sparse open communities associated with screes (type VI) through predominantly *Nardus stricta* grassland (type III) on peaty podzolic soils to communities on soils of impeded drainage, with a high proportion of *Juncus* spp and *Molinia caerulea* present (type IV). On the deeper free-draining acid brown earth soils, which have a restricted distribution, herb rich *Agrostis-Festuca* grasslands occur (type I). This group of plots occurred over the full altitude range investigated.

At medium rainfall (isodemic groups 8–13) a similar range of soils and associated vegetation occurs. On peaty podzolic soils the vegetation includes *Calluna-Vaccinium* (type V), *N. stricta* (type III) and *N. stricta* with *Agrostis-Festuca* (type II) communities. *Juncus* spp and *M. caerulea* (type IV) predominate on soils of impeded drainage. The free draining acidic brown earth soils are more widespread and the associated vegetation is herb rich *Agrostis-Festuca* (type I). All the census sites in the medium rainfall division lie within the medium altitudinal range.

Isodemic group 1, with its very low sheep population density of 0.53 ewe units per hectare (eu ha^{-1}; 1 lamb = 0.5 eu) occurs in relation to the most extreme conditions of site instability, with poor soils, high altitude and rainfall, and sparse vegetation.

Isodemic group 13 (18.17 eu ha^{-1}) has the most favourable habitat conditions, with medium rainfall and altitude, and free-draining brown earth soils bearing herb rich *Agrostis-Festuca* grassland. The constituent plots of this group, however, occupy sites of former cultivation, and high sheep population densities may in part reflect the residual effects of these former cultivations. Group 11 plots (5.63 eu ha^{-1}) have not been cultivated, though they occur under conditions similar to those of group 13, and their sheep population densities are more comparable with those of other groups.

Isodemic group 5, with its mean sheep population density of 4.98 eu ha^{-1}, is of particular interest since one of its two constituent plots supporting 6.66 eu ha^{-1} during 1100 and 1430 h and 4.44 eu ha^{-1} during the major part of daylight hours includes the IBP site at Llyn Llydaw. This site supports local aggregates of sheep estimated at 9.21 eu ha^{-1} during daylight hours. The dynamic nature of the grazing system is evident: population densities are in a constant state of flux, varying temporally, spacially and seasonally; animals move into, within and out of the area concerned; aggregates develop, the whole being influenced by interaction with the species population and environment generally. In view of these considerations estimates of sheep densities within the herb rich *Agrostis-Festuca* grassland give only an indication of the range of variation that can exist during the summer months. The second plot is at a higher altitude, is very steep with rather precipitous rock outcrops and has a sheep population density of 3.31 eu ha^{-1}. These high densities reflect the importance of basic soil parent material, on which deep acidic brown earth soils (mean pH 5.4) develop, bearing herb rich *Agrostis-Festuca* grassland, in a mountain environment with rainfall in excess of 2540 mm. The major influence of basic soil materials in determining high sheep populations is well illustrated by the contrast between the densities characteristic of group 5 at high rainfall (4.98 eu ha^{-1}) and group 11 at medium rainfall (5.63 eu ha^{-1}) on the one hand and the population density of 2.91 eu ha^{-1} on *N. stricta* grassland on a soil derived from acidic parent material only a few metres from the IBP site on the other. The sheepwalk within which the IBP site was located had an overall mean summer sheep population density of 2.98 eu ha^{-1}.

Expressed in terms of biomass, the summer sheep population densities provide an additional basis for comparison of these grassland types. The mean summer biomass of sheep grazing the IBP 32 plot site was estimated to be 302 kg ha^{-1} (Chap. 19) and for the Llyn Llydaw site in general, within which the IBP site is located, 175 kg ha^{-1}. These figures contrast sharply with an estimate of 79 kg ha^{-1} for biomass of the *Nardus* grassland site adjacent to the IBP area.

18.4 Conclusions

The local intensity of rainfall, altitude, soil parent material, drainage, type of vegetation and social behaviour have an influence on the numbers of sheep likely to be present within a particular area. The link between soil parent material and type of vegetation is of fundamental importance; in particular, herb rich *Agrostis-Festuca* grassland in northern Snowdonia, associated with soils of basic origin, carries the highest sheep population densities. The IBP site at Llyn Llydaw near Snowdon is an example of a habitat of this kind and the elucidation of some of the factors relating to its ability to maintain high sheep population densities (9.21 eu ha^{-1} during daylight hours) is of intrinsic scientific interest and of significance in furthering understanding of the abiotic and biotic circumstances which contribute to high economic output of animal material. Immediately adjacent *Nardus* pastures support only 2.91 eu ha^{-1}, comparable with the population density over the sheepwalk generally.

Biomass estimates for the two grassland types, herb rich *Agrostis-Festuca* and *N. stricta*, reflect the wide differences between them in terms of their species composition, herbage production and nutritional value. In the case of the Llyn Llydaw site itself, these figures also demonstrate that such areas are not completely uniform and variation exists, not only in the quality of vegetation but also in the distribution of the sheep within an area. There may often be local concentration of sheep on particularly favourable segments within an area which otherwise forms, in general terms, a distinct vegetational, pedological and climatic unit. These figures re-emphasise the important influence of basic soil parent materials in these upland areas.

19. The Grazing Intensity and Productivity of Sheep in the Grassland Ecosystem

S. BRASHER and D. F. PERKINS

Within the *Agrostis-Festuca* grassland of the main site, the variation in numbers of sheep, their food consumption, egestion and production are described as part of the analysis of the relationship between primary production and herbivores.

19.1 Introduction

Sheep are the dominant herbivore in upland grasslands and play an important dynamic role in the ecosystem. The aim of this investigation was to examine some of the ways in which the sheep contribute to the structure and functioning of the ecosystem, the principal requirement being to estimate their density and productivity. The *Agrostis-Festuca* grassland within which the site is situated is highly favoured by sheep for grazing (Chap. 18). Because of the botanical variation within this one vegetation type (Chap. 14), and the fine degree of selectivity shown by sheep, it was necessary to investigate the sheep within the study area in more detail. The density and distribution of sheep present and grazing on the site and also fluctuations in time and space in relation to environmental factors have been determined. To assess the importance of the grazing regime, attention was focused on the effects of defoliation and faecal deposition (dung) on the site. The relationship between sheep density and herbage intake and dung return have been determined by separate field experiments in order to calculate the energy flow and nutrient transfer caused by the sheep at the primary consumer level.

This chapter is divided into three main sections. The first examines the overall density of sheep on the site in relation to natural environmental changes and management practice, consideration being given to the way in which the sheep utilise the surrounding grassland. In the second section, the distribution of grazing within the site is examined and its significance assessed in terms of botanical composition, herbage consumption and faecal return. The productivity of sheep grazing the site is estimated and discussed in the third section.

19.2 Methods of Recording Sheep

The main 32 plot study area of 0.74 ha (Chap. 14) represents only a small portion of a sheep's home range. It was not enclosed and sheep moved in and out of the area throughout the day. The movements of sheep on the surrounding area were observed throughout the study.

The study extended during the two grazing seasons 1970 to 1971, beginning in May when ewes with lambs had been returned to the uplands, and terminating in mid-October when all sheep were taken to lower ground for winter. Using binoculars, the numbers and location of sheep in the 32 plots were recorded at 30 min intervals from dawn to dusk, one day a week. Sheep were recorded as grazing (selecting or consuming herbage or moving), and resting (idling, standing, or lying down). Ewes and lambs were recorded separately until September, when it became too difficult to distinguish between them. No counts were made during the weeks when shearing and sales took place, there being very few sheep on the hillside and negligible grazing. No observations were made at night, when walking round the study area was unsafe and caused too much disturbance; but during 1970, observations made just after dusk and just before dawn on the next day provided some indication of the extent to which the site was utilised at night.

19.3 Density of Sheep on the Site

19.3.1 Seasonal and Diurnal Fluctuations

The density of sheep is expressed in ewe units (eu) ha^{-1}, a lamb counting as 0.5 eu in May and June, 0.75 in July and one eu from August onwards. These values allow for the increased food demand of the ewe during pregnancy and just after parturition (Rawes and Welch, 1969). Mean monthly densities (Table 1) have been estimated from daily averages and results are very similar for both years. Grazing pressure remained high on the site throughout the season, ranging from 5.4 ha^{-1} grazing and 7.6 ha^{-1} present in June 1971 to 15.5 ha^{-1} grazing and 19.4 ha^{-1} present in October 1971. A slight but significant increase in grazing intensity occurred, 70% of the flock grazing in the first half of the season and 80% in the second half. Hunter and Milner (1963) and Jones (1967) have demonstrated increased activity in sheep flocks with decreasing daylength, but because the grazing season at Llyn Llydaw is relatively short, and the present study represents only a small proportion of the flock, changes are far less pronounced. Inter-monthly variation (Table 1) is considerably less than day-to-day variation and factors most important in effecting any great changes are of short-term duration.

The mean annual pattern of diurnal activity (Fig. 1), shows a buildup of numbers towards midday and a decline towards dusk, largely due to the pattern of sheep movement on the surrounding pastures. The number of sheep on the herb rich *Agrostis-Festuca* sward, including the 32 plot site, gradually built up during the day but declined towards evening. At dusk the sheep settled in resting grounds, favouring dry rocky areas above the plots.

Around dawn a general movement to wetter areas and lakeside for grazing and drinking resulted in lowest density on the site in the early morning. Assuming that the average sheep density (7.2 ha^{-1}) at 0500 h and 2000 h remained constant at night, the estimate for a 24 h period is 10.6 ha^{-1}. Hughes and Reid (1951) have shown that 95–100% of grazing occurs during daylight hours, with more night grazing in spring and autumn, while Tribe (1949) found more night grazing in summer than winter. In view of declining numbers at dusk and the preference for night camping areas, minimal grazing would be expected at night.

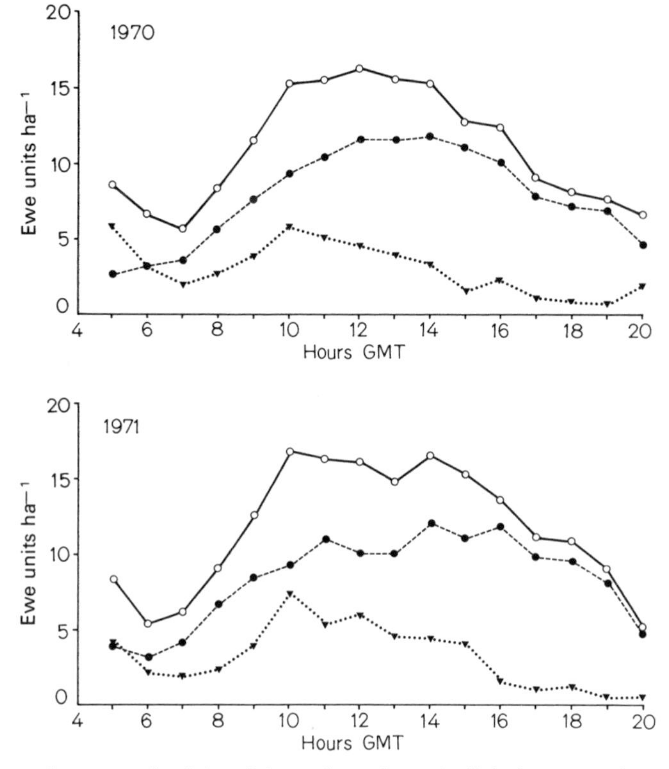

Fig. 1. Annual pattern of activity of sheep throughout daylight hours at Llyn Llydaw in 1970 and 1971. Grazing sheep ●-----●; resting sheep ▼······▼; total sheep ○———○

Table 1. Mean monthly density of sheep (eu ha^{-1} ± se) present and grazing on the site at Llyn Llydaw

	1970		1971	
	Present	Grazing	Present	Grazing
May	13.6 ± 1.4	10.3 ± 1.2	11.6 ± 0.8	8.6 ± 0.6
June	13.5 ± 1.5	8.8 ± 1.0	10.7 ± 1.8	7.4 ± 1.2
July	14.0 ± 1.5	9.4 ± 1.2	16.7 ± 3.8	10.7 ± 2.0
Aug.	11.6 ± 2.0	8.8 ± 1.6	11.3 ± 1.0	8.5 ± 1.1
Sept.	13.2 ± 1.5	10.3 ± 1.6	12.4 ± 2.4	9.6 ± 1.0
Oct.[a]	8.6 ± 2.7	6.9 ± 1.4	14.0 ± 5.2	11.3 ± 4.2
Seasonal average	12.7 ± 0.7	9.2 ± 0.6	12.6 ± 1.0	9.1 ± 0.7

[a] During October 1970 some sheep were removed for experimental purposes.

19.3.2 The Effects of Weather

Weather conditions in the uplands are very changeable, even during the course of a day's recording, and site microenvironment is likely to be of importance (Blaxter, 1964). Shelter in the form of rocks, grass tussocks and ground undulations modifies wind velocity and temperature. Daily climatological data were limited to records of the mesoenvironment, and no attempt was made to relate sheep density to microenvironment.

Throughout the study period the most noticeable change in activity on the site occurred during the post-shearing period in July, when sheep are most susceptible and responsive to changes in weather. The highest densities occurred in this month on the study area, probably a result of general migration from more exposed areas to the sheltered slopes around the Llydaw site.

Over the rest of the season the sheep appeared to be less affected by the weather. There was a slight increase in numbers on the site with the onset of north easterly gales, and when winds were from the south west, especially when accompanied by heavy rain storms, there was a general movement up the slopes or to sheltered resting places on the site. On hot days sheep grazed morning and afternoon, resting at midday; during showers they stood around and resumed grazing when the rain ceased. It may be concluded that the influence of change in the weather on sheep density at the site was mainly short-term.

19.3.3 The Effects of Management

Sheep numbers on the site fall markedly, and grazing pressure is negligible, when sheep are rounded off the area for shearing, sales and drafting. Following sales there is a change in the age structure of the population. A significant drop occurred in the number of lambs present before and after August wether sales, from 5.6 ± 0.4 to 2.9 ± 0.4 in 1970 and from 5.4 ± 0.2 to 3.0 ± 0.5 in 1971. A similar reduction in numbers of ewes after drafting in September cannot be demonstrated because ewes and lambs were not distinguished after August. Neither lamb nor draft ewe sales have any lasting effect on the density remaining on the site in September and October (Table 1), numbers gradually being restored by sheep from surrounding pastures. It has been shown (Hunter and Milner, 1963) that a gradual change in sheep social behaviour occurs as the season progresses. In summer when ewes are feeding lambs, there is considerable competition for the better swards, the more aggressive sheep forcing others to graze on poorer vegetation. Towards autumn there is less dispersion, and the reduced aggressiveness in the flock together with a general decline in numbers after the sales, allows less aggressive sheep to move on to the favoured *Agrostis-Festuca* swards. Thus a consistent grazing pressure is maintained, independent of management.

19.4 Distribution of Sheep Within the Site

19.4.1 Selection of Plots for Grazing and Resting

Analyses on the 32 plot sites were performed by M. Mountford (pers. comm.) to examine whether or not the plots were selectively grazed or rested on by sheep. To eliminate the variation arising from different numbers of observations and animals recorded at each sampling, the data were ranked. Friedman's multiple-sample test (Friedman, 1937) was then applied to determine any plot preference.

The values of chi-square obtained for between-monthly samples were:

	Grazing	Resting
1970	107.8	88.8
1971	96.0	91.5

The four values are significant ($P<0.001$) indicating that the distribution of sheep is not random but certain plots are consistently selected. Spearman's rank correlation coefficient was calculated as 0.88 for grazing and 0.84 for resting sheep ($P<0.1$) indicating similar preference each year. It was also evident that sheep generally preferred the same plots for both grazing and resting [Spearman coefficient 1970 (0.79); 1971 (0.54); and 1970+1971 (0.72), all significant at $P<0.1$] although the preference is less strict than within-grazing or within-resting comparisons.

19.4.2 Distribution of Grazing in Relation to Sward Characteristics

The relationship between sheep distribution on the site and botanical variation was examined by superimposing the combined 1970 and 1971 data on the ordination scatter diagram (Chap. 14). Plots ranked 1 to 10 were defined as high, 11 to 20 as medium, and 21 to 32 as low preference areas. High preference plots had a mean grazing sheep density of 11.9 eu ha^{-1}, medium 9.3 ha^{-1}, and low 6.6 ha^{-1}. The major trend (Fig. 2) is decreasing density from plot vegetation group A to E. Groups A and B contain a greater percent cover of species such as *Agrostis tenuis*, *Holcus lanatus*, *Trifolium repens* and *Poa annua* which have been shown to be highly preferred by sheep (Hunter, 1962). These plots also contain tussocks of *Juncus effusus* which, though not grazed in summer, are thought to provide a shelter belt with a favourable environment for sheep. Herbage between

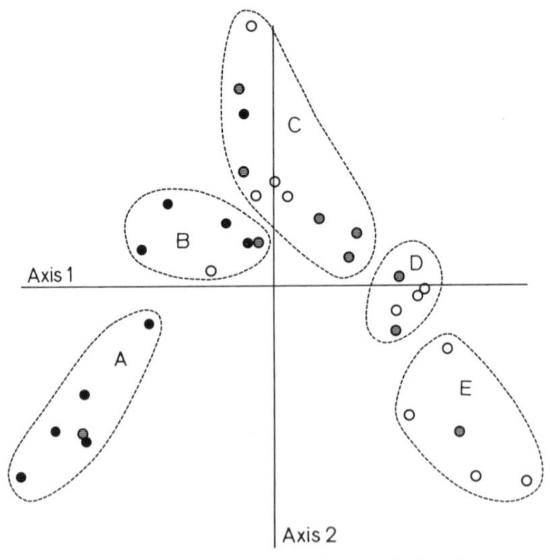

Fig. 2. Relationship between grazing sheep density and the 32 plot ordination (vegetation). High density 11.9 eu ha^{-1} ●; medium density 9.3 eu ha^{-1} ⊜; low density 6.6 eu ha^{-1} ○

the tussocks is comparatively rich in nitrogen, phosphorus and potassium (Chap. 16), high concentrations of which are positively correlated ($P<0.01$) with sheep density ($r=0.75$, 0.66 and 0.62 respectively). The more intensively grazed plots (A and B) tend to have lower standing crops, and the percentage living material present is significantly correlated with sheep grazing densities ($r=0.392$, $P<0.05$). In general, sheep appear to select food which is rich in phosphorus and nitrogen and young or green, in preference to old or senescent (Arnold, 1964).

In the remaining ordination groups, grazing density is average or low. Whilst this is generally attributable to a decline in the above factors and the more frequent occurrence of *Nardus stricta*, *Sieglingia decumbens* and *Juncus squarrosus*, a number of other factors may contribute. Vegetation group C is comparatively species-rich but some plots have only a low sheep density. Below plots 16 to 18 there is a sheer drop to *Eriophorum*-dominated gullies and close to the boundaries is a footpath. Sheep are frequently disturbed by its use as an alternative route by hikers when the usual lake causeway is flooded in August and September due to maintenance work on the pipeline from the lake to the Cwm Dyli hydro-electric station. On plots 23 to 25 steepness of slope and litter accumulation in the hollows are more likely to determine preference than botanical composition. In groups D and E plots 19, 23, 27, 28 and 32 have low sheep density and frequent, though small, patches of *N. stricta* which is of limited nutritive value and one of the least preferred species in the summer. This is, however, unlikely to be the single determining factor and the increase in dead material in this part of the sward (Chap. 16), a factor known to influence grazing preference (Arnold, 1964), may effectively reduce the numbers of sheep grazing.

19.5 Herbage Consumption and Egestion

19.5.1 Quantities of Dry Matter Consumed

Estimates of dry matter consumption on each of the 32 plots were obtained by monthly harvesting within cages. Sampling in 1966 to 1968 was by clipping quadrats, and in 1969 and 1970 by the turf core method (Chap. 16). Although grazing in both years occurred from April to October, the number of grazing days depends on management and seasonal estimates of consumption assume a total grazing period of *c* 150 days.

The mean cumulative consumption on the site for the five year period 1966 to 1970 was 248 ± 30 g m^{-2} yr^{-1} (dw including mineral ash). Consumption (y) was closely correlated with the clipped yield of net primary production (x) between years (Fig. 3), and similarly within years between the 32 plots ($y=0.891x$, $P<0.001$). Sheep removed most, *c* 92%, of the available yield as measured by harvest methods. Estimates were based on the total amount of herbage dry matter including attached (standing) dead but excluding litter. Botanical separation of core samples in 1969 allowed an estimate to be based on sward components (Table 2). Total consumption of living and dead components was 206 g m^{-2} yr^{-1} (165 g living and 41 g dead), mainly *Agrostis tenuis* (66%). Smaller amounts of *Festuca ovina* and *Anthoxanthum odoratum*, and negligible amounts of sedges,

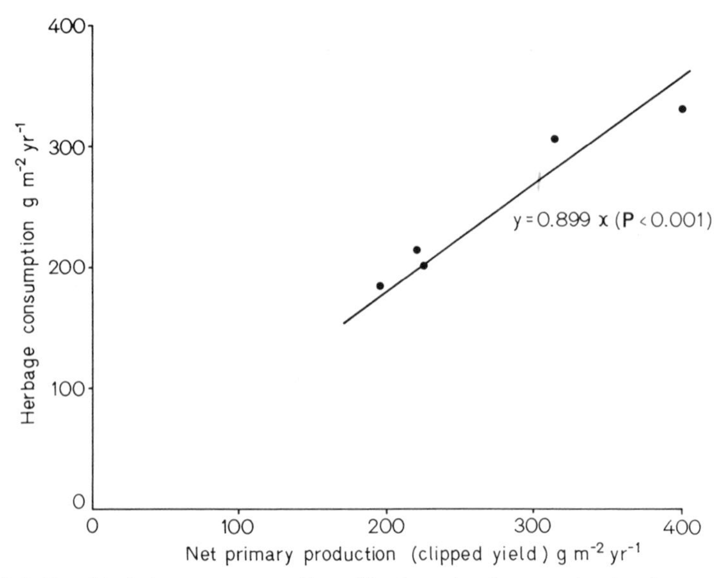

Fig. 3. Relationship between consumption of herbage by sheep and net primary production (clipped yield) of the *Agrostis-Festuca* site between 1966 and 1970

Table 2. Consumption, preference and utilisation by sheep of living and dead components of *Agrostis-Festuca* grassland

Component	Consumption[a]		Production (yield)[a]		Index of preference[b]	% utilisation
	$g\,m^{-2}\,yr^{-1}$	(%)	$g\,m^{-2}\,yr^{-1}$	(%)	(IOP)	
Agrostis tenuis	136	(66)	261	(56)	1.17	52
Festuca ovina	24	(12)	68	(15)	0.79	35
Anthoxanthum odoratum	31	(15)	51	(11)	1.37	61
Other grasses	11	(5)	27	(6)	0.92	41
Sedges and rushes	4	(2)	22	(5)	0.41	19
Herbs	Trace		5	(1)	0	0
Mosses	Trace		30	(6)	0	0
Total	206	(100)	464	(100)	—	44

[a] Determined from turf core samples, includes living and dead material. Production is the yield of living and dead material unadjusted for loss in weight or disappearance of biomass during sample period.

[b] Calculated from $IOP = \frac{c}{p} \times \frac{P}{C}$ where c = consumption of component, p = production of component, C = total consumption, P = total production. IOP > 1.0 = preference, < 1.0 = rejection.

rushes and herbs were consumed. Rawes and Welch (1969) and Grant and Milne (1973) similarly found grass to be a substantial component of the diet of sheep in comparable swards.

The mean ratio of living to dead (attached) material in the sward biomass was *c* 1:1, whereas sheep consumed living material at *c* 3:1. The proportion of dead material in clipped herbage varied from *c* 78–80% in spring (when most dead was consumed) to only 20–30% in mid summer and 55–65% in autumn. Some mosses were also removed from short swards but the small amounts were not statistically significant. An index of preference indicates selection of *A. odoratum* and *A. tenuis* and rejection of *F. ovina*, sedges, rushes and herbs. The total utilisation of living and dead production was 44%; the most utilised grass species was *A. odoratum* (61%) and *F. ovina* the least (35%). Eadie (1970) estimated that less than 30% of pasture production is utilised over the whole year, but that the efficiency would be greater where upland grazing is limited to the summer months. Rawes (1971) showed that utilisation of *Agrostis-Festuca* swards varies from 46 to 73% depending on food available, which in turn is associated with altitude and sheep density.

The measurement of herbage removal by harvest methods gives an estimate of the amount of organic matter (and its contained nutrients) consumed by a known number of sheep from the site (Milner and Hughes, 1968; Chap. 16). The average rate of herbage consumption based on a total of 12.5 sheep ha^{-1} on the 32 plot site was 1.3 kg $sheep^{-1}$ d^{-1}, within the range (0.6–1.6 kg) given by Rawes and Welch (1969) for *Agrostis-Festuca* swards.

19.5.2 Nutrient and Energy Content of Consumed Herbage

The concentrations of nutrients and energy values of vegetation available to sheep were determined throughout the grazing season in samples harvested from clipped 0.25 m^2 quadrats. These samples excluded plant bases but contained living leaves and attached dead material in the proportion that was probably consumed. Concentrations varied throughout the season (Fig. 4), reflecting the response to inherent growth characteristics of the dominant grass component in the samples (Chap. 16). At the onset of growth in April, concentrations were low, but N, P and K increased rapidly as most of the new production was channelled into the reproductive tillers which predominated in the sward. Dying overwintering tillers and, later, defoliated reproductive tillers were also being replaced, so that although the nutrient content was high, the quantity of living material tall enough to graze was smaller than would be expected from the estimated production (adjusted for *L*). Concentrations decreased in midsummer but increased again during the second phase of growth. Concentrations of Ca and Mg tended to be higher in cut herbage samples which contained more mature leaf material than in turf cores (Chap. 16) which had a greater proportion of younger material. Variation within years was paralleled by variation in some elements between years (Table 3), believed to be mainly in response to climatic fluctuations which also affected net primary production. The ratio of Ca:P, recognised as being important in animal nutrition (Fleming, 1973) varied in mean value between years from 1.68 to 2.89. The mean energy value of clipped herbage from 1968 to 1971, 18.79 kJ g^{-1} dw, varied little within or between years, variation being greatest in dw samples because of variation in mineral ash content. Generally values were lowest in the spring when the proportion of dead material was highest, and greatest in summer when there was least dead material. Values for recently dead vegetation, 17.92 kJ to 18.28 kJ, decrease as decomposition proceeds.

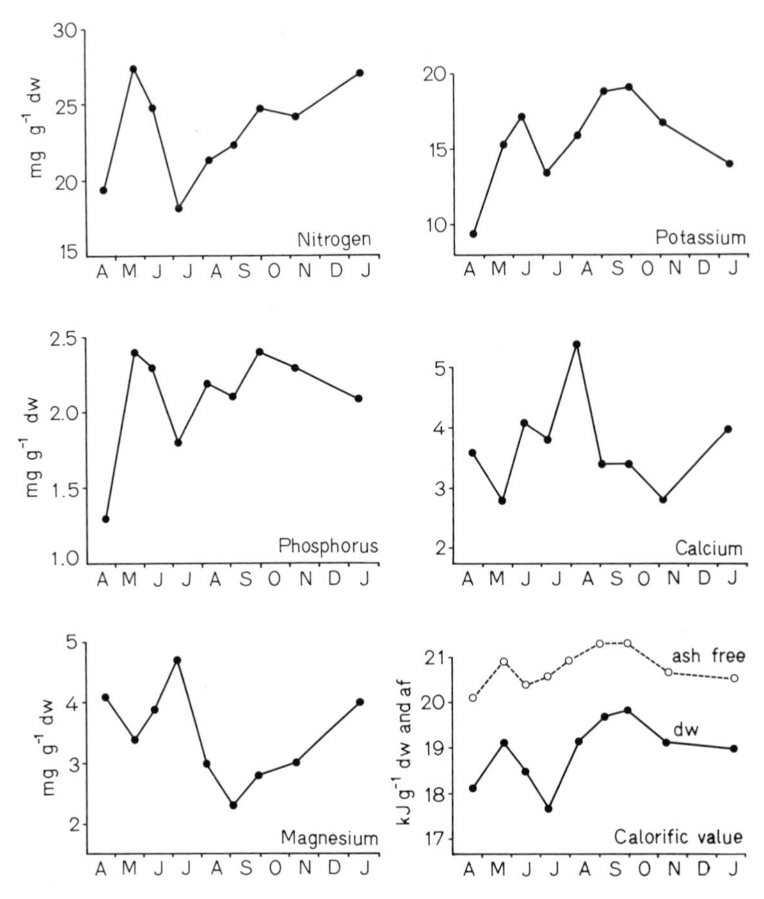

Fig. 4. Variation in nutrient concentration and energy value in clipped *Agrostis-Festuca* herbage samples during 1969

Table 3. Variation in concentration of nutrients (mg g^{-1} dw) and energy (kJ g^{-1})$\pm$se between seasons in grassland vegetation (clipped aerial standing crops) at Llyn Llydaw 1968–1971

Year	1968	1969	1970	1971
Na	0.8$\pm$0.1	0.7$\pm$0.1	1.3$\pm$0.2	0.9$\pm$0.1
K	15.5$\pm$1.5	16.6$\pm$0.7	18.4$\pm$1.4	19.0$\pm$0.4
Ca	3.9$\pm$0.2	3.7$\pm$0.3	4.2$\pm$0.3	5.5$\pm$0.4
Mg	4.2$\pm$0.4	3.3$\pm$0.3	4.4$\pm$0.4	4.1$\pm$0.3
P	2.1$\pm$ $<$0.1	2.2$\pm$0.1	2.0$\pm$0.1	1.9$\pm$0.1
N	21.3$\pm$0.5	23.3$\pm$1.1	25.0$\pm$1.1	22.2$\pm$1.0
Energy				
kJ g^{-1} dw	18.4$\pm$1.7	18.8$\pm$2.5	18.8$\pm$0.4	18.8$\pm$ $<$0.4
kJ g^{-1} ash free	20.1$\pm$1.3	20.5$\pm$1.3	20.5$\pm$0.4	20.1$\pm$0.8
% ash content	9.4$\pm$0.5	9.0$\pm$0.8	7.1$\pm$0.2	6.9$\pm$0.4

Table 4. Standing crop, SC ($g\,m^{-2}$) and daily increment, I ($g\,m^{-2}\,d^{-1}$)$\pm$se of dung in the grassland at Llyn Llydaw between 1968 and 1970

Year		May	June	July	Aug.	Sept.	Oct.	Mean for season
1968	SC	—	4.4 ±0.5	4.7 +0.8	7.5 ±0.8	8.1 ±1.2	2.6 ±0.4	5.7 ±0.4
	I[a]	—	0.13±0.01	0.17±0.02	0.49±0.04	0.28±0.03	0.07±0.01	0.20±0.01
1969	SC	2.4 ±0.3	6.1 ±1.0	10.1 ±1.2	10.4 ±1.1	8.4 ±1.0	6.5 ±0.7	7.2 ±0.4
	I[a]	0.19±0.03	0.15±0.02	0.40±0.04	0.22±0.02	0.22±0.02	0.23±0.03	0.24±0.02
1970	SC	7.0 ±0.7	10.4 ±0.8	7.1 ±0.7	4.3 ±0.4	4.1 ±0.4	5.3 ±0.6	6.2 ±0.3
	I[b]	0.58±0.16	0.59±0.13	0.18±0.06	0.26±0.07	0.37±0.08	0.76±0.29	0.46±0.09

[a] Increments derived from 32 sub plots which had accumulated dung for approximately four weeks.
[b] Increments derived from 64 sub plots which had accumulated dung for single 24 h periods.

19.5.3 Egestion of Faeces

Investigation of egestion was limited to a study of rate of dung production, distribution and chemical composition, no measurements of urine being made. It is assumed, however, that the pattern of urine distribution is similar to that of faeces (Hilder, 1966). Each of the 32 plots was divided into 25 sub plots, each 9.29 m^2, and sampling on these was carried out in phase with the monthly sampling routine on the site, from 1968 to 1970. Samples were weighed, dried separately to determine their dry matter content, and chemically analysed. Any particularly large samples were weighed and subsampled, and the remainder returned to the site. Standing crop was determined on a different sub plot (two in 1970) in each of the 32 plots. The standing crop accumulated in one month on sub plots previously cleared of dung was used in 1968 and 1969 to determine increment and assess the rate of deposition. Owing to rapid disappearance rates, however, this underestimated the deposition rate and in 1970 the dung accumulated over single 24 h periods was used to assess the rate.

The average rate of deposition of dung on the site is 0.46 g dw $m^{-2}\,d^{-1}$ with an intermonthly error of approximately 20% (Table 4). Assuming the full quota of sheep is present on the site from the beginning of May to mid October and allowing for management periods when sheep are absent, the cumulative amount returned to the system (using the 1970 data) over the grazing season (*c* 150d) is estimated at 69 $g\,m^{-2}$. Hilder (1966) showed that about one third of the total dung output could be deposited on night camping areas; dung return to the plots thus underestimated dung output per sheep. Dung output (F) was therefore estimated from:

$$F = \frac{C(100 - D)}{100}$$

where C = herbage consumption and D = digestibility.

A rearranged form of this equation (Herriott and Wells, 1963) was suggested by Milner and Hughes (1968) as an alternative method of estimating consumption when faecal output is known. Consumption of herbage has been determined as

248 g m^{-2} yr^{-1} on the plots and digestibility (mean 63%, seasonal range 58.6 to 65.2%) was estimated by in vitro analysis undertaken by the Welsh Plant Breeding Station, Aberystwyth. Total dung output for the grazing period was therefore estimated as 92 g m^{-2} (0.58 kg sheep^{-1} d^{-1}) of which 25% was deposited on night camping sites remote from the plots.

An alternative method of determining dung output of captive sheep is by the use of inert tracers (Reid, 1962; Milner, 1967). Over a nine day trial period, C. Milner administered known doses of chromic acid to sheep enclosed on the *Agrostis-Festuca* site adjacent to the 32 plots. The concentration of chromium in dung, determined by D. F. Perkins using X-ray fluoresence spectrography, gave a mean estimate for dung output of 0.45 kg sheep^{-1} d^{-1}. In contrast, the value for sheep enclosed on a nearby *Nardus-Festuca* sward was 0.60 kg sheep^{-1} d^{-1}, though the chemical composition of herbage in the two enclosures was similar apart from a higher Ca concentration in the *Agrostis-Festuca* sward (Hughes et al., 1964).

The mean standing crop of dung for the three years was 6.49 g m^{-2}; allowing for a mean increment rate of 0.46 g m^{-2} d^{-1}, dung remains on the surface for *c* 14 days, varying at different times of the season according to sheep density and rate of return and disappearance. The latter was investigated between May and November 1970. Samples of fresh dung were collected from adjacent areas on the *Agrostis-Festuca* sward and placed in 625 cm^2 quadrats of 16 gauge plastic coated wire mesh, 10 cm high, protected by an outer sheep-proof fence. Two experiments were conducted, starting in May and August 1970, the first using 50 g aliquots of dung and the second variable aliquots of 30, 60, 90, and 120 g fresh weight. The aliquots were weighed at 14 day intervals, five samples being retained for dry weight and chemical determinations (Fig 5). As with rates of disappearance of plant litter a negative exponential curve is a reasonable fit to the data. The 90 and

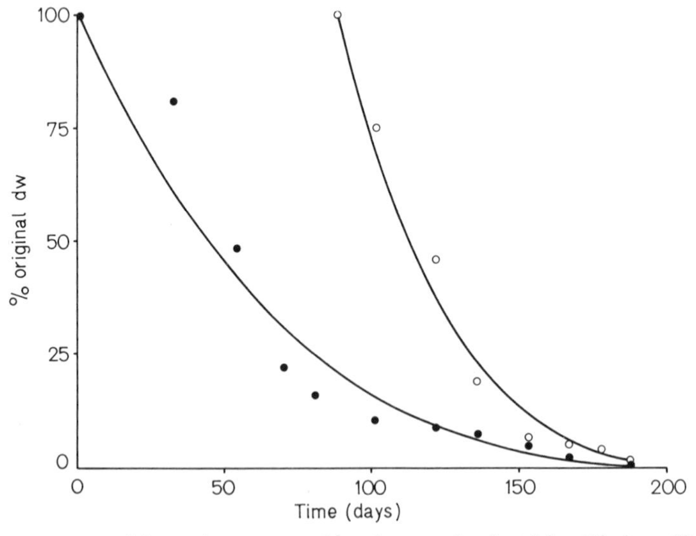

Fig. 5. Disappearance of sheep dung exposed on the grassland at Llyn Llydaw. 50 g aliquot ●——●; 60 g aliquot ○——○.

120 g aliquots yielded similar rates but the smaller 30 g aliquot (not illustrated) a slower disappearance rate in the later stages than the illustrated 60 g aliquot. Differences in the rate of loss of dung throughout the period and between the two starting dates appear to be related to temperature and moisture conditions, the smaller samples tending to dry more quickly. A similar relationship between rate of loss of plant litters and rainfall has been established (Chap. 16). The rates of disappearance of dung varied between 8.9–11.1 mg g^{-1} d^{-1} in drier weather (3.2–4.1 mm rainfall d^{-1}) to 9.9–14.4 mg g^{-1} d^{-1} during wetter weather (5.7–7.0 mm d^{-1}). The study period was too short for the investigation of temperature effects. The rapid rate of loss resulted in an 89.5% and 98.4% disappearance of the May and August aliquots respectively in three months. The dung was protected from sheep and thus not subject to the normal trampling and mechanical translocation effects which enhance the breakdown of dung by a factor of 3 (Chap. 20). Hence most of the dung deposited in a grazing season, even late in the year, will have disappeared from the surface within 12 months, i.e. before the start of the next season.

19.5.4 Chemical Composition and Energy Content of Faeces

The chemical composition and energy value of faeces were determined, in samples collected for assessment of dung standing crop and increment, between 1968 and 1970. Mean values varied little between years (Table 5) particularly in increment samples whether exposed for about one month or 24 h. Standing crop samples exposed for longer periods exhibited greater variation. There was, however, seasonal fluctuation similar to that in herbage, but with reduced amplitude (Fig. 6). Higher concentrations were present earlier and later in the season when faecal output was high and lower in mid summer when faecal output was low.

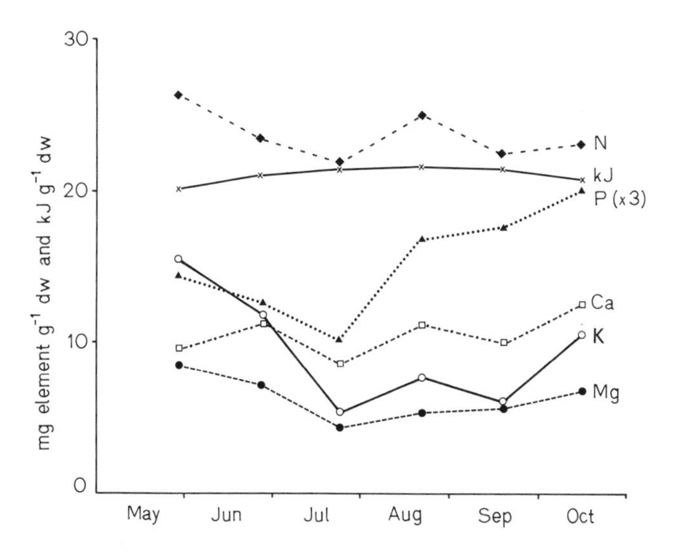

Fig. 6. Variation in chemical composition and energy value of sheep faeces collected over 24 h periods from the *Agrostis-Festuca* site

Table 5. Mean annual concentration of nutrients, ash (mg g^{-1} dw) and energy (kJ g^{-1} dw) ± se in the standing crop (SC) and increment (I) samples of sheep dung during the grazing season at Llyn Llydaw between 1968 and 1970

Constituent		Na	K	Ca	Mg	P	N	Ash	kJ
1968	SC	1.4±0.3	10.1±1.7	10.9±0.9	6.3±0.4	5.1±0.2	21.6±0.9	114.6±7.2	19.83±2.9
	I[a]	1.3±0.3	8.6±2.1	10.2±0.6	6.0±0.3	5.0±0.3	22.2±1.0	104.6±5.0	20.04±2.5
1969	SC	1.0±0.2	7.9±1.1	7.9±1.2	5.6±0.1	4.5±0.4	18.0±3.3	119.2±3.5	20.21±2.5
	I[a]	1.1±0.1	8.4±0.9	10.1±0.6	5.4±0.2	5.0±0.2	23.4±1.0	115.4±5.0	20.33±2.1
1970	SC	1.6±0.4	6.4±1.8	10.8±0.4	7.0±0.7	5.3±0.3	23.6±0.8	102.1±7.1	20.79±2.5
	I[b]	1.7±0.2	9.4±1.5	10.5±0.6	6.3±0.6	5.2±0.4	23.7±0.7	85.2±7.1	21.09±2.1

[a] Increments derived from 32 sub plots which had accumulated dung for approximately four weeks.
[b] Increments derived from 64 sub plots which had accumulated dung for single 24 h periods.

Table 6. Variation in sheep density, herbage consumption and deposition of dung at the *Agrostis-Festuca* grassland at Llyn Llydaw (±se)

Plot vegetation group	Grazing sheep[a] (eu ha^{-1})	Resting sheep[a] (eu ha^{-1})	Total sheep[a] (eu ha^{-1})	Herbage consumption[b] (g m^{-2} yr^{-1})	Standing crop[c] of dung (g m^{-2})	Deposition[d] of dung (g m^{-2} yr^{-1})	Ratio dung deposition: standing crop
A (n=6)	11.2±0.6	2.8±0.5	14.0±1.0	311±73	5.6±0.7	77± 5	14
B (n=6)	11.1±0.9	5.3±0.9	16.4±1.9	270±62	8.4±0.9	90±11	11
C (n=10)	8.6±0.5	3.2±0.6	11.8±1.0	230±32	6.4±0.7	65± 5	10
D (n=5)	7.5±0.8	3.2±0.8	10.7±1.8	178±37	5.6±0.6	59±10	10
E (n=5)	7.0±0.8	2.6±0.5	9.6±1.3	253±33	6.4±0.4	53± 7	8
Site mean	9.2±0.3	3.4±0.4	12.5±0.7	248±21	6.4±0.4	69± 4	11

[a] Sheep data based on observations 1970 and 1971.
[b] Herbage consumption based on average intake 1966 and 1970 but in proportion to the 1968 distribution and error limits.
[c] Standing crop of dung based on average values 1968 and 1970.
[d] Deposition of dung based on 24 h values of 1970.

Bromfield (1961) indicated a similar trend in P concentration which was also the reverse of the seasonal trend in faecal output estimated using harnessed animals.

The energy value of dung varied little between or within years. Midsummer values were slightly higher than spring or autumn values, corresponding to the general fluctuation in consumed herbage.

One set of dung samples collected over 24 h was analysed for inter-plot variation. Analysis of variance indicated significant differences in P and Ca concentration. Plot groups A and B samples contained 20%, and C and D 10% higher concentration of P than E ($P<0.001$). Highest Ca concentration was in plot group C, group A being 23% and E 19% lower. The K concentration of group E was significantly 20% lower than in other groups whilst Mg in group B was 16% higher. These differences are correlated with concentrations in vegetation of the plot groups (Chap. 16). Thus sheep seem to graze and deposit dung on the same range of plots, indicating some within-site group behavioural characteristics, similar to those described by Hunter (1964).

The disappearance from the site surface of plant nutrients and energy in deposited dung was examined in samples from the experiment described in Section 19.5.3. Concentrations ($mg\ g^{-1}$ dw) of Na und K decreased to only 10% of the original value within six to eight weeks after exposure, whereas Ca, Mg, P, and N concentrations remained similar or increased over the period. The nutrient content ($mg\ aliquot^{-1}$) and energy closely followed the disappearance rate of dry matter (Fig. 5), >93% of the most concentrated elements and energy disappearing from the May samples after five months' and from the August samples after three months' exposure.

19.6 Variation in Sheep Density, Herbage Consumption and Dung Deposition in Relation to Plot Vegetation Groups

A non-random distribution of sheep has been shown to exist (Sect. 19.4.2) over the 32 plot site, preference being related to species composition of the sward. Similar trends in herbage consumption, measured at plot level, occurred from 1968 to 1970.

The 1968 herbage data from clipped quadrats were less affected by botanical microheterogeneity than those from the turf core samples of 1969 and 1970. As consistent trends were evident, these data were related to the mean 1970–1971 sheep distributions, assumed to apply equally to 1968–1970 as stocking levels on the farm had remained unchanged for the past 10 years and distribution between and within years was constant. The relationships with plots, grouped according to vegetation data by ordination, is shown in Table 6. More herbage was consumed from the most heavily grazed plots (groups A and B) than from the less grazed plots (C, D, and E). Using the 32 plot data, herbage removal (y) was positively correlated with sheep grazing density (x) by regression as $y=47.22+18.32x$ ($P<0.05$). Similarly net primary production (Chap. 16), was positively correlated with the quantity and nutrient quality of herbage removed by sheep.

The distribution of dung standing crop and rate of deposition were similarly positively correlated with sheep distribution. Using the three year mean the regression of dung standing crop for all 32 plots was significant ($r = 0.44$, $P < 0.01$) on grazing sheep and highly significant on resting sheep, $y = 18.40 + 0.30x$ ($r = 0.71$, $P < 0.001$). There was a similar positive correlation with the rate of dung deposition, plot groups A and B receiving most dung, D and E least. Thus the non-random deposition of dung largely reflected a preference for certain plots for defecation while resting. Since the same plots were preferred for grazing, most dung was deposited in the most heavily grazed plots.

The ratio of dung deposited: standing crop, an index of the disappearance rate, is relatively high in plot group A and low in group E. Thus while more dung was deposited on plot group A, the higher rate of disappearance resulted in a lower average standing crop. Unpalatability of herbage through contamination by dung and urine is thought unlikely to influence the pattern of grazing on the site. Frame (1970/1971) indicated that rejection of contaminated herbage is lessened under competition at high grazing pressures. The high and continuous activity of sheep on the site during daylight hours, together with the high rate of disappearance and high rainfall, reduces the level of unacceptable herbage. The differential deposition of dung is, however, important in the transfer of nutrients. It is concluded that in high density plots, larger amounts of dung, containing a higher concentration of P, will be deposited and rate of turnover will be greater than in the lower sheep density areas.

19.7 Productivity of Sheep on the Site

The management of sheep is such that natural population changes found in wild animals are largely precluded. In 1970, the full quota of sheep were on the site by 1 May and removed by 15 October. There is little variation in this pattern or in sheep density from year to year, in contrast to the fluctuations observed in unmanaged Soay sheep on St. Kilda (Milner and Gwynne, 1974). Here the sheep are solely dependent on the availability throughout the year of the island's plant resources, and fluctuate in response to natural factors. At Llyn Llydaw the yearlings (1 yr old) are returned during April depending upon weather conditions. The main flock with their lambs, born earlier at the lowland farm, soon follow, with approximately equal proportions (20%) of 4, 3, 2 and 1 yr old ewes and lambs (*c* equal numbers of males and females). The feature of management which most affects sheep density (Sect. 19.3.3) is the sale of wether (male) lambs in August and the drafting to lowland areas of 4 yr old full mouthed and six-toothed ewes, and a few 3 yr old ewes in September. These are regarded as no longer able to graze the upland pastures efficiently.

The phase of growth of sheep on the site between May and October can, as pointed out by Petrusewicz and MacFadyen (1970), be considered as stable, production being closely reflected in biomass growth. Mortality is low, and was omitted from the calculations of productivity and biomass losses. Since the period of grazing is not complicated by reproduction, positive biomass changes represent net animal production. Although the sheep range freely, their grazing is restricted

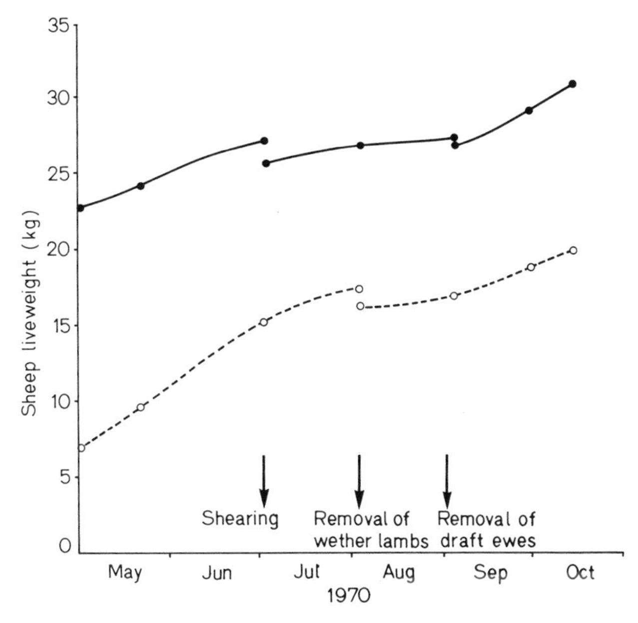

Fig. 7. Variation of mean live weight of sheep present on the *Agrostis-Festuca* grassland during 1970. Ewes ●——●; lambs ○------○ (based on data provided by D.I.Rees).

to vegetation of similar botanical and chemical composition within their home range, and the results are assumed accurate for the site in terms of animal density, increase of biomass and amount of herbage removed. The absolute quantities of herbage consumed per sheep cannot be measured with certainty as individual animals were not identified. Calculations are, therefore, based on the numbers of animals on the site, these being related to the total population within the sheep-walk.

19.7.1 Biomass of Sheep

As part of the 32 plot investigation sheep were gathered and weighed at intervals throughout 1970 by D.I.Rees (pers. comm.). Weights for 1 May (start) and 14 October (end) were estimated from predicted rates of growth before and after the nearest sample date (Fig. 7). The annual cyclical pattern of sheep weight (Russel and Eadie, 1968) indicates that when the ewes are put on to the mountain they are at or near their lowest weight following lambing. Over the grazing period at the site their live weight increases until the autumn, and slowly declines throughout the winter until prior to lambing it may increase. The mean live weight of ewes at Llyn Llydaw in the spring (23 kg ewe^{-1}) was two thirds of that when they were removed in October (32 kg ewe^{-1}). Thus, about one third of their weight is lost during overwintering and lambing.

Removal of wether lambs for sale on 4 August resulted in a decrease of the mean live weight of lambs from 17.3 kg to 16.2 kg for the ewe lambs remaining on the site. The mean weight of wether lambs when removed was 18.9 kg having

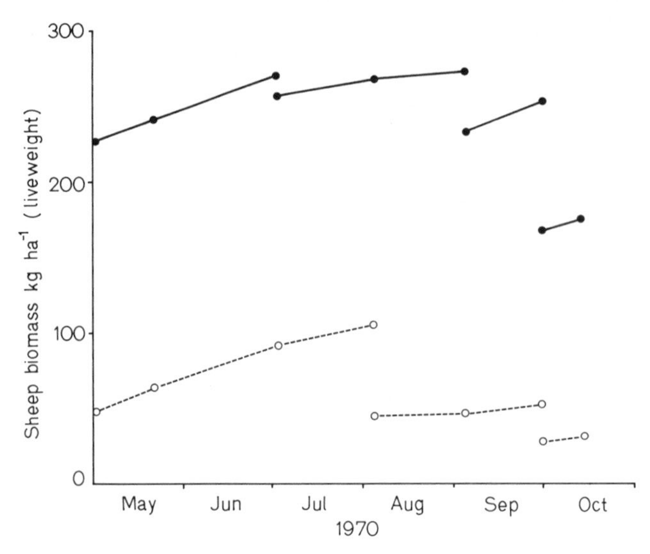

Fig. 8. Variation in biomass of sheep present on the 32 plot site at Llyn Llydaw during 1970. Ewes ●——●; lambs ○-----○

increased from 6.6 kg since being put on the site in the spring. Ewe lambs in the spring were estimated to weigh 7.3 kg, increasing to 16.2 kg by 4 August and 19.8 kg by 14 October. The mean live weights of ewes were reduced by 1.36 kg ewe^{-1} by the shearing on 4 July, and from 27.4 kg to 26.9 kg ewe^{-1} as a result of the sale of the heavier draft ewes on 4 September. The live weight of ewes, and particularly of lambs, tended to increase throughout the summer grazing period but least during midsummer.

The changes in weight and numbers of ewes and lambs as a result of management were taken into account when calculating the biomass of sheep (kg ha^{-1}) at the site throughout the grazing season (Fig. 8). Mean summer biomass of sheep on the 32 plot site was 302 kg ha^{-1} (30·2 g m^{-2}). As a result of their greater number and weight, ewes account for nearly 80% of the biomass with a mean of 239 kg ha^{-1} (23.9 g m^{-2}) without fleece, compared with 63 kg ha^{-1} (6.3 g m^{-2}) for lambs. By August, at the time of the wether sales, the large weight increment of the lambs raised their share of the biomass to about a third. Removal of the fleece from ewes in July resulted in a biomass loss of 13.57 kg ha^{-1}, and the draft ewe sales in September caused a further loss of 71 kg ha^{-1}. This can be accounted for by the removal of 26% of the ewes at an average weight of 27.4 kg or 23% of ewes at the mean draft ewe weight of 29.4 kg. Removal of sheep biomass was calculated when the flock was removed from the site for overwintering in the lowlands on 15 October and a balance sheet for the gains and losses of biomass constructed (Table 7). This indicated that the site gained and subsequently lost 346 kg ewes ha^{-1} and 115 kg lambs ha^{-1} during the grazing season. The initial biomass at the site on 1 May together with the influx following removal of the draft ewes totalled 306 kg ha^{-1}, closely balancing the 317 kg ha^{-1} removed at the end of the season.

Table 7. Annual balance sheet of the biomass and production (lw) of sheep on the Llyn Llydaw site in 1970

Component	Ewes kg ha^{-1}	Lambs kg ha^{-1}
Biomass (gain)		
Initial quota of sheep on 1 May	228	46
Added as production on site	86	69
Added as influx following drafting on 1 Sept.	32[a]	—
Total biomass added to the site	346	115
Biomass (loss)		
Removal of fleece 2 July	14	—
Removal of wether lambs 4 Aug.	—	59
Removal of draft ewes 4 Sept.	71	—
Removal for wintering by 15 Oct.	261	56
Total biomass removed from the site	346	115
Production (gain)		
1 May to 21 May	14	18
21 May to 2 July	28	28
2 July to 4 Aug.	12	14
4 Aug. to 4 Sept.	5	2
4 Sept. to 30 Sept.	20	5
30 Sept. to 14 Oct.	7	2
Total net production on the site	86	69
Production (loss)		
Removal of wether lambs 4 Aug.	—	34
Removal of draft ewes 4 Sept.	12	—
Removal for wintering by 15 Oct.	74	35
Total production loss from the site	86	69

[a] Ewes and lambs could not be distinguished by observation at this time of year.

19.7.2 Estimation of Net Sheep Production

Net production (P) of sheep was estimated from the increase in biomass of ewes and lambs assessed separately during the different phases of management, and calculated in kg ha^{-1} from:

$$P = \Delta B + E$$

where ΔB = change in biomass (biomass increment) and E = biomass loss (biomass eliminated, Petrusewicz and MacFadyen, 1970). The biomass losses included fleece, wether lambs and draft ewes. The cumulative net production (live weight) from 1 May to 14 October (Table 7) was 155 kg ha^{-1} (15.5 g m^{-2}), 86 kg ha^{-1} (8.6 g m^{-2}) from ewes and 69 kg ha^{-1} (6.9 g m^{-2}) from lambs.

The average rate of production (Table 8) of ewes was 0.53 kg ha^{-1} d^{-1} and of lambs 0.40 kg ha^{-1} d^{-1} but lambs had a greater individual rate of production (0.079 kg head^{-1} d^{-1}) compared with an average for all ewes of 0.056 kg head^{-1} d^{-1}. Eadie (1970) gave the initial growth rate of hill lambs as 0.22 kg head^{-1} d^{-1} between 0 and 6 weeks. When lambs were first at Llyn Llydaw, at 8–12 weeks, when they are more dependent upon the pasture for food, a growth rate

Table 8. Rate of net production of sheep present on the 32 plot *Agrostis-Festuca* grassland at Llyn Llydaw during 1970

Sample Date	Lambs		Ewes	
	kg head^{-1} d^{-1}	kg ha^{-1} d^{-1}	kg head^{-1} d^{-1}	kg ha^{-1} d^{-1}
1 May–21 May	0.134	0.89	0.070	0.70
21 May–2 July	0.100	0.66	0.068	0.68
2 July–4 Aug.	0.070	0.43	0.030	0.36
4 Aug.–4 Sept.	0.025	0.07	0.015	0.15
4 Sept.–30 Sept.	0.073	0.21	0.087	0.76
30 Sept.–14 Oct.	0.071	0.11	0.060	0.52
Mean for season	0.079	0.40	0.056	0.53

of 0.134 kg head^{-1} d^{-1} was estimated. The mean live weight gain of lambs of Welsh mountain sheep grazing sown swards of Ryegrass and Red fescue at Pant-y-dŵr in 1970 (Munro et al., 1972) was 0.12 kg head^{-1} d^{-1}, 50% greater than the gain by lambs at Llyn Llydaw (mean 0.08 kg head^{-1} d^{-1}). Lambs exhibited the greatest rates of production per head in the spring, production of both lambs and ewes declining to its lowest level in August and increasing again in September and October. The production: biomass ratio for the season was 0.4 for ewes and 1.0 for lambs. Production losses were calculated to complete the balance sheet and indicate the live weight of sheep leaving the system during the grazing season. About half the production (34 kg ha^{-1}) of lambs (22% of the total site production) was removed as wether lambs while removal of draft ewes included only 8% of the total annual production. Most of the production (70%) was removed as 1–3 year old ewes and ewe lambs at the end of the season.

Rawes and Welch (1969) gave production data for sheep at Moor House which showed that for the entire site, which included a variety of vegetation types, total production was 22.8 kg ha^{-1} (live weight). Robertson and Davies (1965) quoted unpublished data of Hunter indicating a range of production from 7 to 63 kg ha^{-1} for upland areas. The production (for sale) from the Llyn Llydaw site (53 kg ha^{-1}) in 1970 fell within this range (34 kg ha^{-1} from wether lambs, 12 kg ha^{-1} from draft ewes and an estimated 7 kg ha^{-1} from fleece). The fleece weight was calculated as that proportion (*c* 50%) produced on the site during the previous summer, estimated from data of Doney and Smith (1961). Some production was incorporated into the biomass of sheep transferred to the lowlands for wintering; if this is included, total production is 64 kg ha^{-1}. Comparable data for sheep production at Pwllpeiran at 300–400 m in upland mid Wales (Jones, 1967) indicate live weight gains of 70 kg ha^{-1} on *Festuca-Molinia-Nardus* swards, 122 kg ha^{-1} on *Festuca-Agrostis* swards and up to 219 kg ha^{-1} on experimentally re-seeded land. Lambs contributed 80% of the production compared with 64% at Llyn Llydaw. Peart (1970) reported sheep production of 42 kg ha^{-1} over an 11 yr period for free-range Border Cheviot sheep grazing between 212 and 333 m on mainly *Agrostis-Festuca* and *N. stricta* grassland.

The results obtained from the Llyn Llydaw site appear, therefore, to be within the general range of values for upland grassland areas.

19.8 Discussion

It is well established that *Agrostis-Festuca* swards are highly preferred by sheep during the summer grazing period (Hunter, 1962; Hughes, 1958a, b, 1964; Rawes and Welch, 1969) but the degree and period of grazing on a particular sward type may be affected by its location in the pasture (Hunter, 1960). The Llyn Llydaw site is not only herb rich, but is situated on sunny slopes, leeward to cold north easterly winds, close to a constant water supply and shelter in the form of rocks and grass tussocks. It is not surprising that it attracts a comparatively high proportion of the sheep flock. A well-defined spatial and diurnal pattern of grazing is established throughout the season, subject to only short-term environmental change. The importance of the relationship between grazing and botanical variation in plant communities is apparent; a high degree of selectivity results in a two-fold variation in density in a relatively small area of pasture. This is largely responsible for determining the botanical and heterogeneous nature of the sward, and it is concluded, in the light of similar work described by Boulet (1939), Hughes (1958a), Hunter (1962), and Arnold (1964), that sheep are important in determining sward diversity on the site.

High grazing intensities may stimulate growth of grasses and help prevent the accumulation of dead material (Rawes and Welch, 1969; Rawes, 1971). Gradients in primary production, proportion of living and dead, and botanical composition may be due, at least in part, to the current grazing regime. Thus, where *Agrostis tenuis* is dominant, intensive grazing may stimulate the growth of favoured species, attracting further grazing and maintaining a young, palatable sward (Black, 1967). Conversely the transition towards a community type dominated by *Festuca ovina* and characterised by sporadic occurrence of *Nardus*, *Sieglingia*, and *Juncus squarrosus* is accompanied by a decline in grazing which may result in lower overall growth and an increase in the dead fraction of sward. This affects the availability and quality of food, influencing grazing preference and consumption.

Because the present study has been only short-term it is impossible to determine the degree or direction of any change which may occur as a result of such a grazing regime. Utilisation of the natural grassland's primary production is high, sufficient quantity and quality of material being available to meet the sheep's requirements throughout the summer grazing period. The biomass of primary producers generally increased throughout the grazing season at the site, suggesting an efficient producer-consumer system. The site is, however, part of the pastures of the sheepwalk and must also be judged in its contribution towards the ecology of the area as a whole. Eadie (1970) has stressed that efficient pasture utilisation is not only desirable but also, because it determines herbage quality for future grazing, beneficial to the whole grazing complex.

The biotic influence of grazing is also important in its effect on the sward nutrient balance, both through transfer of nutrients within the sward and to a small extent in their removal by the sale of carcasses and fleece (Chap. 20). The importance of faecal deposits to nutrient transfer depends on nutrients contained and their availability to the system (Spedding, 1971). Although most of the constituents in dung and urine are of potential use to plants, net losses are incurred through volatilisation and leaching, and one of the most important factors in determining initial availability is the pattern of excretal distribution.

It is recognised that sheep not only distribute excreta unevenly but also, by virtue of their night camping habits, can deprive certain areas of a valuable source of nutrients (Hilder, 1966). Heavy grazing is associated with a heavier return of excreta, and treading may stimulate biological activity and improve nutrient availability (Rawes, 1971). The overall nutrient budget (Chap. 20) suggests that phosphorus is finely balanced and may limit primary production, and it is concluded that uneven dung deposition by sheep exerts a strong influence upon the spatial variation in productivity of the ecosystem.

Acknowledgements. We are indebted to Mr. D. I. Rees and Mr. I. Ellis Williams for help and useful discussions during the practical stages of investigation. Thanks are also due to Mr. M. Mountford and Mr. M. Hill for help and advice on numerical matters. Chemical analyses were undertaken by Mr. S. E. Allen and colleagues at Merlewood Research Station.

20. The Distribution and Transfer of Energy and Nutrients in the Agrostis-Festuca Grassland Ecosystem

D. F. PERKINS

The results from the component studies are combined to describe the distribution and circulation of dry matter, energy and nutrients in the montane grassland ecosystem of Snowdonia.

20.1 Introduction

The montane grassland ecosystem at Llyn Llydaw has been described in terms of its main components: vegetation (Chaps. 14, 16); soil (Chap. 15); herbivores (sheep: Chaps. 18, 19; slugs: Chap. 17). The IBP study aimed to quantify these components by estimating their biomass and productivity, and the dynamics of the transfer within and between them of materials (organic matter and nutrients) and energy. This chapter synthesises the structure and function of the main units.

Plant and animal populations in an ecosystem are interrelated and interact in a common environment. The grassland ecosystem, dominated by *Agrostis tenuis* and *Festuca ovina*, is a pasture whose development is largely controlled by sheep grazing. Analysis of the distribution and transfer of photosynthetically produced material and its contained energy and nutrients indicates the functioning of an ecosystem across the matrix of its component parts. The two main trophic pathways are examined: the grazing pathway, and the litter (or detritus) pathway.

20.2 Methodology

Distinction is made between the biomass (and reserves of nutrients in the soil) and production. Biomass (B) is the measured quantity of dry matter and its energy and nutrient content present in the ecosystem at any given time. Production (P) estimates the net balance of material accumulating through growth processes during a specified period of time, i.e. the net balance between assimilation (A), that part of food (energy or nutrient) consumed (C) and utilized by the biomass, and respiration (R), that part of the energy consumption liberated during metabolism.

Production, according to Petrusewicz and MacFadyen (1970) may be written:

$$P = A - R \quad (1)$$

but in the present studies primary production was measured (Chap. 16) as the increase in biomass during the growing season and the sum of that quantity of material lost through death of plant parts (*L*) and that grazed by herbivores (*G*), thus:

$$P = \Delta B + L + G\,. \tag{2}$$

In the plagioclimactic grassland there is little change in biomass from year to year and *G* was estimated as the quantity of herbage consumed (*C*). The amounts of herbage grazed but not consumed were not measured, thus:

$$P = L + C\,. \tag{3}$$

Sheep production was estimated from the change in biomass plus that biomass eliminated (*E*) as fleece or animals removed for livestock sales, thus:

$$P = \Delta B + E\,. \tag{4}$$

The energy and nutrient balance of sheep was estimated from:

$$C = P + R + U + F + M = D + F \tag{5}$$

where additionally F = egesta (faeces); U = excreta (urine); M = rumen fermentation (methane); and D = digested matter (energy or nutrient). The quantity of matter (or energy) involved in the metabolism of the component (or trophic unit) is:

$$A = P + R\,. \tag{6}$$

Respiration (*R*) was not measured directly but for sheep was assumed by difference as:

$$R = C - (P + U + F + M) \tag{7}$$

and for other organisms estimated from literature sources.

Nutrient consumption by sheep was estimated as:

$$C = P + U + F\,. \tag{8}$$

Data for consumption, production and faecal output were available for sheep from direct observation whilst partial data for concentrations of nutrients in urine allowed the volume excreted to be estimated, an otherwise difficult procedure in free-range animals.

20.3 Distribution of Dry Matter, Energy and Nutrients

The data have been assembled as diagrammatic models of dry matter, energy and nutrients (See Appendix, Sect. 20.6). Models represent one year, the compartments the biomass of the component (trophic unit) or the nutrient reserves in the soil in g dw (or energy kJ) m^{-2}. Transfers between biomass compartments are in g dw (energy kJ) $m^{-2}\ yr^{-1}$. Variation in some components within years has been

Table 1. Energy values (kJ g^{-1} dw) of materials used in calculating the energy budget of the grassland ecosystem at Llyn Llydaw

Material	Energy content	
Entire aerial biomass, annual weighted means		
Grasses (living)	18.58	
Other vascular plants (living)	18.06	
All vascular plants (living)	18.45	
Mosses (living)	16.84	
All vegetation (living)	18.00	
All grasses (dead)	17.36	
Other vascular plants (dead)	17.30	
Mosses (dead)	16.83	
All standing dead material	17.30	
Litter	18.11	
Other samples		
Roots	17.96	
Clipped herbage samples	18.79	
Earthworms	19.50	
Dung invertebrates	21.46	
Slugs[a]	20.08	
Sheep (biomass)[b]	28.41	
Sheep (production)[c]		
mean weight gain of lambs	35.26	(12.34 g^{-1} live weight gain)
mean weight gain of ewes	39.00	(13.65 g^{-1} live weight gain)
weighted mean	37.59	(13.06 g^{-1} live weight gain)
Sheep faeces		
biomass	20.28	
fresh	21.09	
Sheep urine	15.21	
Sheep wool	21.04	
Soil organic matter[d]	21.40	(ash-free)

[a] Lutman (Chap. 17).
[b] Calculated from body composition data of Panaretto (1963) and Thomson (1965).
[c] Calculated from Agricultural Research Council (1965) data.
[d] S. E. Allen (pers. comm.).

indicated in previous chapters; whilst this variation is important and gives further insight into the dynamics of the ecosystem, it cannot be considered in detail in this summary synthesis.

20.3.1 Dry Matter and Energy

Distribution of energy is closely related to dry matter, being based on dry matter data but modified by the energy value of the material (Table 1). Animal and insect materials tend to be of higher energy content than plant materials, sheep values being particularly high for production gains which contain high amounts of fat. Dry matter data include mineral ash; the energy values are based on dw and whilst varying with ash content are, nevertheless, not identical when expressed on an ash-free basis. Live plant materials have greater energy content than dead; weighted means for the annual biomass of vascular plants were 18.45 kJ (living), 17.35 kJ (standing dead) and 18.11 kJ (litter) g^{-1} dw. Recently dead mixed vegetation declined from 18.28 kJ (19.70 ash-free) to 16.37 kJ (18.24 ash-free) g^{-1} after 418 days decomposition. The higher value of litter (18.11 kJ g^{-1} dw) is the result of the addition of *c* 20% of sheep dung at a higher calorific value (20.28 kJ g^{-1} dw) than incoming dead plant material (17.35 kJ g^{-1} dw).

Soil at Llyn Llydaw contains 12% organic matter in the 0–15 cm and 7% in the 15–30 cm horizon and averaged 8.4×10^3 g dw m^{-2} for each 1 cm depth. Thus organic matter was calculated as 15.1×10^3 (0–15 cm) and 8.8×10^3 (15–30 cm), a total of 23.9×10^3 g dw (498.6×10^3 kJ) m^{-2} for the 30 cm depth. Results were calculated on 30 cm of soil because very few roots were found below that depth. Density of roots at 30 cm was 0.14 mg dw cm^{-3} compared with 28.5 mg dw cm^{-3} at 1 cm. Roots had a mean biomass of 603 g (10,830 kJ) m^{-2} which was only 2.6% of the soil organic matter, the latter constituting 95% of the total organic matter in the ecosystem.

Aerial parts of living plants had an average biomass of 220 g (3960 kJ) m^{-2} composed of 159 g (2933 kJ) m^{-2} vascular plants (75% of which were grasses) and 61 g (1027 kJ) m^{-2} mosses. The biomass, determined using turf cores (Chap. 16), included the stubble normally left when quadrats are clipped. Standing dead (213 g dw, 3684 kJ m^{-2}) comprised 188 g dw (3280 kJ) m^{-2} vascular plants and 25 g dw (404 kJ) m^{-2} mosses. Live biomass varied seasonally with a peak towards midsummer, but there was little variation in this and standing dead between years. There was, however, some variation in litter (fallen dead material recovered from the soil surface) over the period 1968–1970 (mean 321 g dw, 5813 kJ m^{-2}). Litter was composed of 22% finely divided material from disintegrated sheep dung and 78% coarse decomposing plant leaf material.

The live weight biomass of Welsh mountain sheep, the dominant animal in summer, was 30.2 g m^{-2} (Chap. 19). A dry matter content for sheep of 35% was assumed, derived from data of Lawes and Gilbert (1895) and Panaretto (1963) allowing an additional 4% of water as the animals had been fasted before slaughter, giving a biomass of 10.6 g dw (301 kJ) m^{-2}. Energy content for sheep was calculated as 11.09 kJ g^{-1} lw (28.41 kJ g^{-1} dw) assuming sheep composition of 61% water, 20% fat (39.12 kJ g^{-1}), 14% protein (23.30 kJ g^{-1}) and 5% ash (Panaretto, 1963; Thomson, 1965). The derived value compares well with individual component values determined by Thomson e.g. 29.83 for carcass, 30.08 for gut, 23.64 for skin and 23.38 kJ g^{-1} dw for blood. Slugs (Chap. 17) have a biomass of 0.44 g dw (8.7 kJ) m^{-2} and together with sheep constitute a total herbivore biomass of 11.04 g dw (310 kJ) m^{-2} during the sheep grazing period April to October.

During the grazing season sheep dung amounted to a mean biomass of 6.5 g dw (132 kJ) m^{-2}. The dung was rapidly colonized with invertebrates, and after 24 h 20% of samples contained *Aphodius* beetles, 11% Sphaeridinae and 10% Staphylinidae. Later in the season dipterous larvae appeared and increased to 22% infestation in October. A similar sequence has been described by Kajak (1974) and Olechowicz (1974). Earthworms were absent from 24 h old dung but a few appeared in older samples (Fig. 1). The mean biomass of invertebrates was 0.004 g m^{-2} between May and September, dominated by *Aphodius* adults and dipterous larvae. Whilst the biomass is small these insects are important in the system as decomposers of sheep dung (Olechowicz, 1974; Breymeyer, 1974).

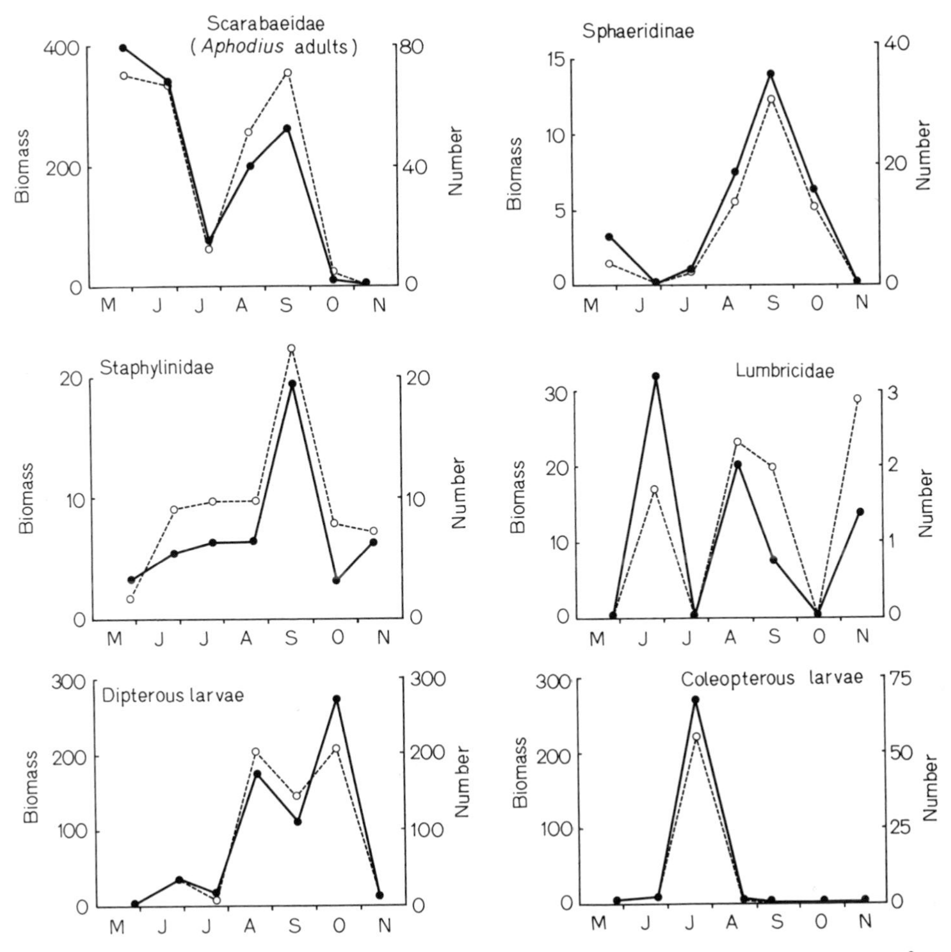

Fig. 1. Variation in number (○- - - - -○, 100 m^{-2}) and biomass (●———●, mg dw 100 m^{-2}) of invertebrates in the standing crop of sheep dung at Llyn Llydaw during 1970 (data provided by S. Brasher)

Earthworms mix consumed surface litter with mineral soil and raise soil casts, enhancing decomposition and mineralization of nutrients (Czerwiński et al., 1974). The mean density of earthworms, estimated by handsorting, was 169 m^{-2} with a biomass of 7.5 g dw (145 kJ) m^{-2} (Table 2) composed mainly of the larger *Lumbricus* spp. Nelson and Satchell (1962) recovered 93% of marked introduced earthworms in handsorting but Raw (1960) recovered only 52% of the number but 84% of the weight, the smaller worms being underestimated. Adjustment of the weight recovered at Llyn Llydaw indicates a biomass of *c* 8.9 g dw m^{-2} which exceeds the biomass of slugs and closely approaches the summer biomass of sheep as in good pastures (Sears and Evans, 1953; Waters, 1955). The annual biomass of sheep and slugs is *c* 6.6 g dw (sheep grazing is only between April and October) and is exceeded by the earthworms on an unadjusted weight.

Table 2. Density (numbers m^{-2}) and biomass (g fw and dw m^{-2}) of earthworms sampled from 12 pits 30 cm^{-3} at Llyn Llydaw during 1970

Sample date	28 April	22 May	13 July	24 Aug.	21 Sept.	7 Dec.	Mean
Density	260 ±51	177 ±48	133 ±49	187 ±57	146 ±44	112 ±33	169 ±52
Biomass fw	53.7± 9.1	37.2±10.4	34.6±13.4	55.2±16.9	30.9±10.6	26.6± 8.6	39.7±11.9
Biomass dw	10.1± 1.7	8.0± 2.2	6.4± 2.4	10.6± 3.2	5.7± 1.9	4.2± 1.2	7.49± 2.5

(Data provided by D. C. Seel, mean se indicates variation between sample dates.)

Table 3. Concentration of mineral nutrients and ash in slugs, dung invertebrates and earthworms at Llyn Llydaw (analyses by V. Jones)

Concentration (mg g^{-1} dw)	P	K	Ca	Mg	Ash %
Agriolimax reticulatus	10.1	12.3	7.5	2.8	8.4
Arion intermedius	11.3	6.6	13.4	2.6	8.5
Arion subfuscus	10.7	7.3	13.3	3.0	8.6
Arion ater	14.7	7.3	14.2	3.3	9.4
Slug faeces	2.2	3.3	12.9	6.8	21.6
Dung invertebrates	7.6	9.0	2.3	1.7	5.2
Earthworms	27.0	28.0	11.6	49.1	19.4

N value for dung invertebrates was 108 mg g^{-1} dw and slugs 101 mg g^{-1} dw.
N value for earthworms of 92.6 mg g^{-1} dw taken from Satchell (1963).

Small mammals trapped in 1969–1970 by K.C. Walton and D.C. Seel on a grid of 160 Longworth live-traps were *Apodemus sylvaticus*, *Microtus agrestis* and *Clethrionomys glareolus*. The number of captures was very small and only for *Apodemus* were there sufficient data to estimate the numbers (1.94 ha^{-1}) and biomass (32.5 g lw ha^{-1}). The effect of these animals as herbivores is considered to be small.

20.3.2 Nutrients

Analysis of soil (Chap. 15) indicated total reserves (non-ignited soil–dw basis) of Ca 0.11%, Mg 4.5%, K 0.54%, and P 0.12%. Nitrogen, which varied down the profile, was taken as 0.51% (0–15 cm) and 0.32% (15–30 cm) over the 32 plot area. No data were available for the composition of soil organic matter, but an estimate derived from the lowest concentration in recorded plant litter decomposition experiments indicated that amounts are small compared with the total reserves: *c* 1% K and Mg, 5% P, 15% Ca, and 22% N. The entire biomass of organisms and above-ground dead materials, however, contained only 6–8% of the nutrients retained in soil organic matter except P (12%) and K (39%). Roots contained most N (5.4 g m^{-2}), Ca (0.97 g m^{-2}), and Mg (3.2 g m^{-2}) whilst living aerial plants contained most P (0.40 g m^{-2}) and K (2.52 g m^{-2}) which are required in high concentrations in leaves for metabolic processes. N, P and K are concentrated in live plant materials much more than in standing dead or litter, but the large biomass of dead material retains much of these elements whose decomposition and mineralization is important in the nutrient cycle. Concentrations in sheep live weight (0.91% Ca, 0.03% Mg, 0.14% K, 0.52% P, and 2.09% N) were derived from Lawes and Gilbert (1895) and Agricultural Research Council (1965), and analyses of slugs, earthworms and dung invertebrates (Table 3) show higher concentrations of N, P and K than plants. Biomass of nutrients in earthworms, even at the unadjusted dw of 7.5 g m^{-2}, exceeds the content in sheep during the grazing season except for Ca.

20.4 Transfer of Energy and Nutrients

20.4.1 Producers

Net primary production above ground of the *Agrostis-Festuca* grassland was 1145 g dw (21.16×10^3 kJ) m^{-2} yr^{-1}. Similar values were derived from either the mean of production measurements from 1966 to 1970 (Chap. 16) adjusted for losses sustained through formation of dead material (L) or estimation from data of mean dead biomass and rate of disappearance. Production determined by harvesting the yield of live and dead material was, however, only 271 g dw (5100 kJ) m^{-2} but increased to 464 g dw (8732 kJ) m^{-2} when stubble was included, ignoring weight loss and disappearance of dead material during the sample period. Production of dead material was estimated as 632 g dw (10.96×10^3 kJ) m^{-2} yr^{-1} which, when corrected for weight loss (*c* 32%), was equivalent to 933 g dw (17.15×10^3 kJ) m^{-2} yr^{-1} of living.

Root production was tentatively estimated as 337 g dw (6.05×10^3 kJ) which together with adjusted aerial production totalled 1482 g dw (27.21×10^3 kJ) m^{-2} yr^{-1}. Efficiency of utilization of photosynthetically active solar radiation (assuming 45% of total) received during the growing season (*c* 210 days) was 2.3%, or 2.1% over the full year.

Concentrations of nutrients in vegetation were determined throughout the year in component species (Chap. 16) and in cut herbage samples to simulate material grazed by sheep (Chap. 19). A seasonal pattern of higher concentrations in spring and autumn was related to the phenology of the dominant grasses. Concentrations differed between tillers, younger tillers and in particular potential flowering tillers containing much higher concentrations of N, P and K but lower Ca and Mg than older tillers. Concentrations of Ca and Mg increased with increasing leaf maturity. Nutrient concentration also declines downwards through the standing crop canopy, leaf material having generally higher concentrations than basal parts of the plants. Weighted mean nutrient concentrations were used to calculate nutrients in production but could underestimate the higher concentrations found in production material, a feature also found in woodland species (Woodwell et al., 1975).

20.4.2 Herbivores

Sheep consumed 248 g dw (4659 kJ) m^{-2} yr^{-1} of which 199 g dw (3763 kJ) was living and 49 g dw (896 kJ) was taken from the standing dead biomass (Chap. 19). This consumption was 92% of the production (harvestable clipped yield) of living and dead material. Similar high consumption by sheep (86%) of the harvestable production has been reported from the Polish IBP study in the Pieniny mountains (Kajak, 1974).

Slugs consume 16 g dw (305 kJ) m^{-2} yr^{-1} (Chap. 17), mostly plant material (1.1% of the adjusted annual production) but some sheep dung is probably consumed (*c* 3 g dw m^{-2} yr^{-1}) as only this material has a phosphorus content which would allow sufficient consumption to balance production and faecal elimination. Total consumption of plant material by sheep and slugs was, therefore, *c* 261 g dw (4903 kJ) m^{-2} yr^{-1}, 23% of the annual production.

The proportion of herbage which is digestible influences the energy balance of the sheep. In general a larger quantity of more digestible herbage is consumed (Blaxter, 1962); thus, quality in addition to availability and selectivity of herbage greatly influences production. Consumption is regulated by the rate at which material is removed from the rumen during digestion and there is a positive relationship between consumption and herbage digestibility. The value of digestible energy is dependent upon the amount of energy lost in rumen fermentation, excretion in urine and the efficiency with which the remaining metabolizable energy can be used (Spedding, 1971). An energy balance for sheep (Table 4) was calculated from Equation (7) (Sect. 20.2). Animals, although free-ranging, were largely restricted within their home range to *Agrostis-Festuca* and *Agrostis-Festuca-Nardus* vegetation which had a small range of variation in chemical composition. Estimates of consumption, therefore, closely approach the consumption per animal grazing the home range. The energy content of production (203 kJ

Table 4. The energy balance of sheep grazing the *Agrostis-Festuca* grassland

Energy parameter		$kJ\ m^{-2}\ yr^{-1}$	% C
Herbage consumed	(C)	4,659[a]	100.0
Production	(P)	203[a]	4.4
Respiration	(R)	1,794[e]	38.5
Dung output—total	(F)	1,940[a]	41.6
Dung output—on plots		1,455[a]	(31.2)
Urine output—total	(U)	396[b]	8.5
Urine output—on plots		300[c]	(6.4)
Rumen fermentation—methane	(M)	326[d]	7.0

[a] Data from Brasher and Perkins (Chap. 19); energy value of sheep live weight gain calculated from Agricultural Research Council (1965).
[b] Urine output calculated from K balance thus:

Consumption K in Herbage	(C)	$4.31\ g\ m^{-2}\ yr^{-1}$	
Dung—total output	(F)	$0.87\ g\ m^{-2}\ yr^{-1}$	(0.65 to plots)
Production	(P)	$0.022\ g\ m^{-2}\ yr^{-1}$	
Urine $U = C—(F+P)$ total	(U)	$3.42\ g\ m^{-2}\ yr^{-1}$	(2.57 to plots)[c]
Concentration K in urine (weighted mean)		$8.3\ mg\ ml^{-1}$	
Output—total		$410\ ml\ m^{-2}\ yr^{-1}$	(310 to plots)[c]
% K excreted in urine		80%	

[c] Assuming similar distribution as dung (Hilder, 1966).
[d] Calculated from data of Blaxter (1960).
[e] Estimated from Equation (7) (Sect. 20.2) $R = C - (P + F + U + M)$.

$m^{-2}\ yr^{-1}$) was calculated from live weight increments and data of the Agricultural Research Council (1965) which take into account the relatively high energy content of fat and protein associated with gains in live weight (Table 1) which are positively related to age and body weight.

Dung output from the plots was estimated as 92 g (1940 kJ), 25% being deposited on night camping sites (Chap. 19). Urine output, calculated from the estimated excretion of K (3.42 g $m^{-2}\ yr^{-1}$) averaged 410 ml (26 g dw) $m^{-2}\ yr^{-1}$, corresponding to 2170 ml $sheep^{-1} d^{-1}$ and within the range 1000–5000 ml d^{-1} given by Spedding (1971). The energy content of the urine was 15.21 kJ g^{-1} dw giving an energy output of 396 kJ. Assuming that urine is distributed in a similar manner to dung (Hilder, 1966) the proportion expected on the 32 plots is 310 ml (20 g dw, 300 kJ) $m^{-2}\ yr^{-1}$. Energy expended in the rumen as methane varies between 6 and 8% of the energy consumed over a range of digestibility of 50 to 80% (Blaxter, 1960). In vitro determinations indicated a 63% digestibility corresponding to an estimated energy loss (M) of 7% or 326 kJ. Respiration was estimated from Equation (7) as 1794 kJ, indicating the high energy cost of maintenance of mountain sheep. The energy assimilation efficiency (A/C) is 43%, similar to the 42% for Moor House sheep on *Agrostis-Festuca* grassland (Heal and Perkins, 1976) and the 40% generally found for homeothermic animals (Heal and Mac Lean, 1975). The ratio P/C is 4.4%, higher than the 2–3% indicated by Turner (1970) for a range of homeothermic animals; the higher value for sheep is the result of the higher energy content used in estimating production live weight gains.

The nutrient balance of sheep indicates that only a small part (1–15%) of nutrients consumed goes into production and most is returned to the site in dung and urine, an important aspect of the nutrient budget of the grassland.

In the energy balance for slugs (Chap. 17), consumption was 305 kJ, rejecta ($F+U$) 213 kJ, production 44 kJ and respiration 48 kJ m^{-2} yr^{-1}. The assimilation efficiency (A/C) was 30%, less than that for sheep, but the P/C ratio of 14% was 3.3 times greater. Nutrient budgets were based on chemical analyses of slugs and faeces (Table 3). The 16 g dw consumed by slugs when multiplied by the weighted mean plant nutrient concentrations did not allow sufficient P intake to balance that retained in production and faecal output. Selection of material at *c* 3.4 mg P g^{-1} is required whilst the weighted mean concentration in vascular plants was 2.0 mg g^{-1}. The highest concentration in grasses was in *Anthoxanthum odoratum* (2.6 mg g^{-1} annual mean, peaking to 3.5 mg g^{-1}) and mean concentration in herbs was 3.9 mg g^{-1}. *Agriolimax reticulatus* (Chap. 17) was probably the most frequent feeder on live plants, other species tending to take senescent or dead material which is low in P, N and K. *Arion ater*, however, was wide ranging, feeding on sheep dung (rich in P) and *Cirsium* spp (rich in Ca) indicating that allowance should be made for selection of food source. Selection by slugs of a diet of 40% grasses and sedges (2.1 mg P g^{-1}), 40% herbs (3.9 mg P g^{-1}) and 20% sheep dung (5.0 mg P g^{-1}) would result in a consumption of *c* 0.05 mg P m^{-2}, sufficient to balance the estimated output. This suggests that, as the dw proportions are 20:4:1, a high degree of selectivity of grazing is performed by slugs. Slugs may involuntarily ingest small invertebrates on plant material which contain high levels of phosphorus. Other nutrients were calculated on the same basis and results suggest that whilst sheep may consume 15 times more dry matter than slugs, their consumption of P is only 8 times and Ca only 6.5 times greater.

20.4.3 Decomposers

A detailed investigation of the organisms involved in decomposition was not undertaken but an integrated measure of the activity of microflora and fauna is available from studies of disappearance of plant litters (Chap. 16) and sheep dung (Chap. 19).

Two main sources of decomposable material are dead plant material (632 g m^{-2} yr^{-1}) and herbivore faeces (80 g m^{-2} yr^{-1}; 69 g from sheep and 11 g from slugs). Other materials include dead invertebrates, production of which is relatively small. The disappearance of sheep dung from the surface is rapid, a rate of 22.4 mg g^{-1} d^{-1} from May to November in sheep exclosures, 4.8 times faster than the disappearance of litter (4.7 mg g^{-1} d^{-1}) over the same period. Invertebrates quickly inhabit dung and although they consume only about 0.5 mg g^{-1} d^{-1} the dung beetles accelerate the disappearance by mechanical disintegration. In addition Breymeyer (1974) reported that 10% of the dung deposited at the Polish site was transported by beetles directly into the top 10 cm of soil. Decomposition of the biomass of litter and dung occurs at approximately the same rate after the initial rapid rates caused by the disappearance of more readily decomposable substances (Sauerbeck, 1968). The enhanced dung disappearance rate is, therefore, probably the result of both mechanical transportation by dung beetles to the soil and also disintegration leading to translocation of fine particles to the litter layer. The rate of disappearance is even faster in the presence of sheep, a direct result of treading leading to further disintegration. With a mean dung biomass of 6.5 g during the grazing season and an input of 0.46 g m^{-2} d^{-1}, the disappearance rate is 66 mg g^{-1} d^{-1}, three times the rate determined by weight loss. Whilst the surface loss of dung is very rapid, decomposition (assumed equal to plant litter disappearance rate) is relatively small (8%) and 92% is mechanically translocated to the soil and litter layer. Assuming that 10% of deposited dung is transported to the soil by dung beetles as reported by Breymeyer (1974), *c* 56 g would be expected to appear in the litter. This, together with 11 g from slugs, totals 57 g of faecal input into the litter compartment, and largely maintains the 71 g of finely divided particles in the litter biomass.

Input of plant material to the standing dead and litter compartments was 632 g m^{-2} yr^{-1} of which 49 g was consumed by sheep. Together with the 67 g input from faeces, total decomposable input was 650 g m^{-2} yr^{-1}. Disappearance rates were rapid, measured by weight loss in litter bags (2.96 mg g^{-1} d^{-1}) or in modified turf cores (3.38–3.52 mg g^{-1} d^{-1}) and, apart from faster initial rates, a negative exponential curve was a reasonable fit to the data (Chap. 16). Decomposition of dead plant material begins when the material is still attached to the plant, with a rapid initial weight loss due to removal of easily leached and decomposable substances. Heal and French (1974) estimate that leaching may account for most of the initial 20% weight loss in some litters. The standing dead material has largely undergone the initial and more rapid rate of loss and is disappearing at a similar rate to litter. Considering the combined standing dead and litter compartments on an annual basis under steady state conditions, the instantaneous rate of disappearance $k = 3.34$ mg g^{-1} d^{-1} (where $k = I/B$, input $I = 650$ and biomass $B = 534$) compared with the measured rates of 3.38–3.52 mg g^{-1} d^{-1} in weight loss experiments. Jenny et al. (1949) pointed out that under steady state conditions the fractional loss (k') of litter is balanced by the input. In grassland maintained as a plagioclimax by sheep grazing the input of litter balances the output and the amount disappearing in a year (k') approximates to $k' = I/I + B = 55\%$. As material ages in the litter the rate of loss will decline (Mindermann, 1968; Heal and French, 1974). Different materials decay at different rates and the proportion of material with a slower rate, e.g. mosses, increases with time resulting in a slowing of the rate. As the time for 95% decomposition is only 2.3 years, however, the overall effect of a slowing down will be small. It is perhaps because of the rapid loss rates at the site that the input of dead material so closely matches the output, errors of longer-term weight loss experiments being largely avoided.

Earthworms are important decomposers at Llyn Llydaw; their activity was estimated from Satchell (1967) who indicated that activity could be expected when grass minimum temperatures were above 2° C, when soil temperature did not exceed 10.5° C and there had been rain within the previous four days. During 1969–1970 about 100 d yr^{-1} met those requirements. Consumption was estimated as 27 mg g^{-1} fw of worm (van Rhee, 1963; Satchell, 1967) which with a mean weight of 39.7 g fw m^{-2} gave an estimated consumption of *c* 107 g m^{-2} yr^{-1}. Using data of Lakhani and Satchell (1970) which indicated a relative production of 0.428 g g^{-1} fw yr^{-1}, production was estimated at 17.0 g fw [3.2 g dw (62 kJ)

$m^{-2}\ yr^{-1}$]. Respiration (R) was estimated at 180 kJ from McNeill and Lawton (1970) by substitution of the production estimate in energy terms in their regression equation for poikilotherms. Material passing through the earthworm's gut (F) was thus 1663 kJ [$F = C - (P + R)$]. Earthworms could therefore process at least a sixth of the dead material and litter at the site, considerably enhancing decomposition and maintaining soil fertility by incorporation of litter into soil and raising to the surface soil from lower horizons. Soil processed by earthworms (Czerwiński et al., 1974) becomes enriched with exchangeable cations as a result of enhanced decomposition, the P content increasing 4–10 times which is indicated as being of particular importance in mountain soils deficient in this element.

The biomass of dung invertebrates was small but their feeding and activity may be important in decomposition. As well as transporting dung the Scarabaeidae stimulate the development of ammonifying bacteria in dung (Breymeyer, 1974). *Aphodius* spp form 55% of the mean biomass of adult insects in dung from May to November and 98% in the early part of the season. Coleopterous larvae, mainly *Aphodius*, were very numerous in July forming 72% of the entire faunal biomass. Consumption by litter layer saprovores (excluding earthworms) is within the range 1–5% of dry body weight (Heal and MacLean, 1975) but that of coprophagous saprovores may be much higher. *Aphodius* larvae in the laboratory have an assimilation efficiency (A/C) of 7.0–10.4%, production efficiency (P/A) of 23–45% and a consumption of 3.8–5.3 mg $g^{-1}\ d^{-1}$, resulting in total growth (from 1 to 30 mg per individual) in six weeks (Holter, 1974). Assuming that annual P equals B (0.086 kJ m^{-2}), respiration was estimated using McNeill and Lawton's (1970) data for short-lived poikilotherms as $R = 0.0628$ kJ m^{-2}. Then, assuming an A/C of 10%, consumption was estimated as 1.49 kJ m^{-2} and egestion (F) as 1.34 kJ m^{-2}. Growth efficiency (P/A) was then 58%, somewhat higher than for other saprovores.

20.4.4 Grazing and Litter Pathways

In the pasture ecosystem part of the annual production of plant material and its contained energy and nutrients is consumed by herbivores. The quantities entering the grazing and litter pathways (Table 5) constitute 74% of the adjusted primary production, 26% being leached, broken down or recirculated from senescent material. Grazing by sheep and slugs accounted for 23% of the production whilst 51% went into the litter pathway. A similarly small proportion of the energy flow through sheep occurred in native and sown pasture ecosystems in Australia (Moule, 1968).

The dynamics of the live plant biomass is such that there is an almost continual production of both new living and dead. During the summer, production of living and dead exceeds disappearance of dead material, resulting in a biomass increase which is grazed by sheep. The fine balance between biomass accumulation and that consumed is indicated by the relatively high (92%) consumption by sheep of the production yield determined by harvesting, the amount grazed being related positively to the production yield (Chap. 19). The extent of the formation of dead material was revealed only by estimation of dead production allowing for the rapid disappearance rate in the sward. It appears that the quantities of dead

Table 5. Dry matter (g dw m^{-2} yr^{-1}), energy (kJ m^{-2} yr^{-1}) and nutrients (g m^{-2} yr^{-1}) in the grazing and litter pathways

	dw	kJ	N	P	K	Ca	Mg
Dead material entering litter pathway from primary source (dead production)	632	10,963	7.88	0.87	2.96	1.28	4.45
Living and dead material entering the grazing pathway (herbivore consumption)	261	4,903	5.96	0.59	4.47	1.32	1.06
Rejecta and eliminated material from herbivores	100	1,836	4.30	0.39	3.38	0.92	0.81
Dead production less dead consumed	583	10,067	7.27	0.80	2.73	1.13	4.15
Total material available to decomposers from above-ground sources	683	11,903	11.57	1.19	6.11	2.05	4.96

which enter the litter pathway are not available for grazing by sheep because of the dynamics of leaf production and death. Grazing takes place from the biomass, the amount being replenished by the excess production of living and dead material over the amount of dead disappearing. The sheep, therefore, very closely regulate the biomass on a more or less instantaneous basis throughout the growing season; at the same time quantities of dead are formed and transferred to the litter and rapidly disappear. As sheep are removed during the winter months at Llyn Llydaw, grazing does not have to take place from the capital of biomass remaining at the end of the season, or from vegetation types not selected during the summer as would a continuously grazed pasture.

The partition of energy between grazing and litter pathways is similar to that of dry matter, but for nutrients 45% N, 43% P, 62% K, 54% Ca, and 20% Mg passes through herbivores compared with 31% of dry matter as a result of the different concentrations in live and dead material. Most of those nutrients (99% K and Mg, 94% N, 88% Ca, 86% P) are output as dung and urine, compared with dry matter (45%) and energy (48%) because less is utilized in maintenance and growth. Only 75% of the dw output is, however, returned to the 32 plot site, 25% being deposited on night camping areas.

Thus, the main associated effects of the grazing were to:

1. closely regulate the biomass of vegetation and control its development;
2. accelerate the turnover of nutrients in that part of the production which was grazed;
3. remove about 25% of nutrients from the 32 plot area to remote night camping grounds within the site; and
4. contribute to site heterogeneity by uneven distribution of nutrients over the plots.

20.5 Nutrient Budget

Whilst energy is dissipated as heat during metabolism by organisms involved in the transfer, nutrients may be recirculated within the system. The quantitites of nutrients entering the biotic components should equal the quantities transferred, which are potentially available for recirculation. In some ecosystems, e.g. the bog and woodland, nutrients are accumulated in peat and timber and do not recirculate for a considerable time. In grasslands the turnover of nutrients is more rapid and there is no large accumulation of nutrients in organic matter. Nutrients may, however, be lost or gained and the balance is important in the long-term ecology of the mountain ecosystem where there is a finite supply of elements in some soils. Weathering of parent material, which is rapid in the case of the pumice-tuff at the site (Chap. 15), would result in a continuous input of nutrients to the soil. Partial data are available which allow a tentative nutrient budget to be constructed (Table 6). Complete budgets are difficult to achieve unless an isolated watertight catchment is examined and the input and output of water and material are closely monitored.

Nutrients in the organisms at Llyn Llydaw contained only small quantities of the total nutrients in the soil (0.04–1.1%) and little more (0.07–1.7%) in the total living and dead organic matter (excluding soil organic matter). Nutrients in circulation through the primary production were greater than the quantities in the living biomass; nevertheless only 0.7% P, 1.2% K, and 2% N were in circulation. The soil provides the greatest reserve of nutrients, Mg > K > N > P > Ca, most being inorganic rather than organic in origin. Quantities of extractable nutrients (Chap. 15) indicate, however, that P is of relatively low availability to plants, a small part of the total reserve and organic fraction. Larger amounts of K and Ca were extractable suggesting relative ease of availability from the inorganic exchange complex.

The nutrient-supplying power of the soil is important to grassland plants as, with low reserves, roots must take up large quantities to replace in new growth that lost by defoliation and death. To investigate soil nutrient supply a continuous percolation system was used where water was passed through soil in series with a column of ion exchange resin for collection of released nutrients which is periodically exchanged and analysed (Lees, 1948; Matthews and Smith, 1957). Percolation of soil at a rate of *c* 300 ml h^{-1} for 300–400 h (Fig. 2) showed that the release varies with the element and depth in the profile. The initial rate of release indicates the readily exchangeable content which ceases for some elements at an early stage when the curve flattens, an indication of a lower supplying power. K and P in particular exhibit low levels of both initial and final supplying power quickly attaining a flat type of curve. Over the time of percolation employed, the Mg and Ca curves showed no tendency to flatten, an indication of relatively high supplying power. Inital supply of K in 0–15 cm soil was 12.6 g m^{-2} and in 15–30 cm soil 10.7 g m^{-2}, a total of 23.3 g m^{-2}. Supply of P in 0–15 cm and 15–30 cm soil was 0.63 and 0.44 g m^{-2} respectively, a total of 1.07 g m^{-2}. In continuous cropping experiments Matthews and Smith (1957) demonstrated a close positive correlation between uptake of nutrient and release by percolation. The closeness of the amounts of P and K percolated and the amounts in circulation suggests a

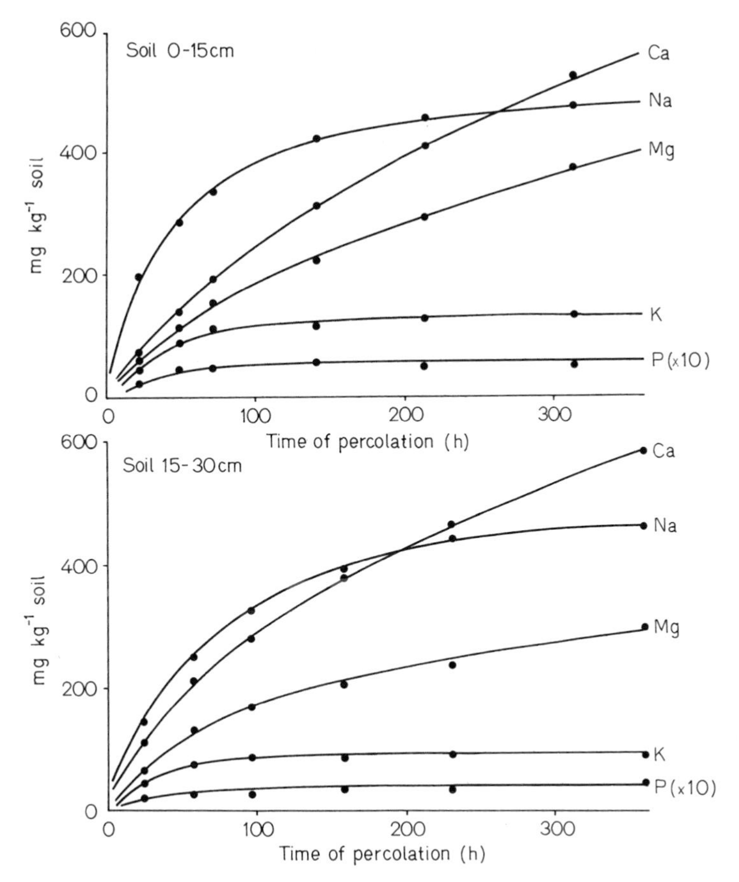

Fig. 2. Release of mineral nutrients during continuous percolation of soil from Llyn Llydaw (experimental work carried out by D. E. Edwards)

similar relationship. As P appears to be retained by living material the 1.39 g transfer to grazing and litter pathways, i.e. the minimum uptake, more closely balances the 1.07 g indicated supply. These results suggest that production may be limited by the P supply from the soil.

The nutrient economy of plants is related to the balance between nutrient uptake and loss in dead material, grazing, and leaching from living leaves. Meristematic regions and expanding leaves require high concentrations, especially N, P and K, and receive supplies via the roots and from old leaf tissue. Infection of leaves by microorganisms begins at an early stage (Webster, 1957) and whilst this may not lead to a reduction in quantity of nutrients the rate of leaching may be enhanced. It is well established that nutrients are leached from leaves (Tukey et al., 1958); the quantitites leached from young leaves are small but increase as leaves approach senescence. Phosphorus was only slightly leached ($<1\%$) from young leaves compared with Ca, Mg and K (1–10%). The quantitiy of the more mobile elements, e.g. K, in circulation may be considerably underestimated if leaching occurs from live leaves; in view of the high rainfall this must be consid-

Table 6. Nutrient budget (g m^{-2} yr^{-1}) for the *Agrostis-Festuca* grassland ecosystem (32 plot site)

Nutrient	N	P	K	Ca	Mg
Input					
rainfall (average year)	1.84	0.17	0.30	2.60	1.13
In system (to 30 cm depth, g m^{-2})					
organisms (living)	11.14	0.73	3.08	1.92	4.71
living and dead organic matter (ex soil)	17.60	1.73	4.63	3.12	8.42
soil (total)	1,046	302	1,361	277	11,340
In circulation					
in primary production	27.59	2.37	16.30	4.08	7.00
return of dead leaves and roots	10.30	1.01	2.90	1.67	5.94
return of dung and urine	4.03	0.35	3.22	0.81	0.75
return by slugs (vegetation grazed)	0.27	0.04	0.16	0.12	0.06
recirculation and leaching from canopy	11.33	0.77	8.93	1.08	0
Output					
in sheep production	0.32	0.08	0.02	0.14	0.01
exported dung and urine	1.34	0.12	1.07	0.26	0.25
Balance: input—output	+0.18	−0.03	−0.79	+2.20	+0.87
Potential leaching (0–15 cm)	nd	0.15	0.82	4.05	2.34

nd: not determined.

ered an important factor. As leaves mature the chemical composition changes; the concentrations of K, P and N decrease whereas Ca and Mg may increase. This change is due to the shift in the balance of nutrient supply in the leaf, demand within the plant, and leaching. Thus in grasses new tillers and potential flowering tillers have higher concentrations and contain more nutrients than older tillers. As live material becomes senescent there is translocation of the mobile and soluble substances, but leaching may become increasingly important. In intact younger leaves, where leaching may occur to a greater or lesser extent depending on the species, supply of nutrients is usually sufficient to replace losses; in older leaves the supply may be less and not able to satisfy the leaching demand and the concentration of mobile elements decreases.

The difference between the quantity of nutrients in the production and that transferred through the grazing and litter pathways may be due partly to leaching and partly to redistribution within the plant. Nutrient leaching has not been measured in grassland but the scale can be assessed by reference to woodland work where the canopy through-fall can be determined by the placement of rain gauges within and outside the canopy (Carlisle et al., 1966). Data abstracted from July to November (excluding aberrant effects due to addition of pollen before July and in the absence of leaves after November) indicated that inorganic N was absorbed from rainfall but organic N was leached resulting in no net change. Leaching of K was high (90 times content in rainfall) increasing towards autumn, whilst P (2.0 times), Ca (2.6 times) and Mg (1.9 times) showed moderate leaching.

Thus in grassland much leaching could be expected of K but much less P and negligible N. At least part of the 0.77 g m^{-2} P unaccounted for between production and amounts entering the litter pathway in standing dead could be recirculated in the plants, adding to the 1.07 g m^{-2} P supplied by the soil and approaching the 2.37 g m^{-2} in circulation. Quantities of P returned to the site in dung (0.35 g m^{-2} to the 32 plots), and supplied in rainfall (0.17 g m^{-2}), are important to the nutrient economy of the grassland for, added to the soil and recirculated quantities, they total 2.36 g m^{-2} yr^{-1}, equal to the amount in circulation through the primary production. Part of the P in dung is organic and is only slowely available to plants (Bromfield, (1961). Inorganic phosphate is readily available if incorporated in the soil (Gunary, 1968) and availability has been shown by Barrow (1975) to be related to the high solubility, in soil at relatively low pH, of dicalcium phosphate in the dung and its low solubility if remaining on the surface. As incorporation of dung into soil by beetles and decomposition of the remainder in the litter takes place rapidly, it is assumed that most P deposited in sheep dung will be available to plants. The very fine balance of P within the grassland ecosystem suggests that supply may limit primary production and input into the grazing and litter pathways. There is indication that P is low during periods of high growth rate (Chap. 16); variation in production between years suggests that there is sufficient supply to enable climate to exert influence and that P may be regarded as conditionally limiting production. The rate of mineralization from litter decomposition is positively correlated with temperature and moisture conditions. Thus conditions under which high plant production occurs may also be those under which high mineralization occurs leading to enhanced P supply. It would be expected that ecosystem production would adjust to the supply of the most limiting factor within restraints imposed by climatic conditions; thus P may limit production at the site when climatic conditions are optimal for growth whereas climate, particularly temperature, may limit the start of the growing season and annual primary production.

The high rainfall experienced in the mountains is well known as an agency for leaching nutrients from soils (Pearsall, 1950). The scale of leaching at Llyn Llydaw was assessed using twelve polyethylene microlysimeters designed to accommodate turf cores (104 mm diameter × 150 mm length) in situ in the sward. The turf core was replaced at the same level but the lysimeter rim projected *c* 5 cm above the surface. Drainage water was collected from the base and periodically analysed. Water collected constituted 50–85% of the incident rainfall (3800 mm in an average year); losses from the lysimeters exceeded input in rainfall in all elements except P (N was not determined). The amounts must be regarded as potentially leachable rather than actual quantitites lost from the site. It is probable that for the 32 plot site as much or even more soluble nutrients pass in from higher ground as pass out. It is likely that the general trend in the area as a whole is to lose nutrients as soil particles in run-off, as can be seen by the exposure of bare rock on the mountain and the accumulation of alluvium washed down into the bed of the lake below. The losses of Ca and P pose potential long-term problems in that reserves of these elements are low in the soil. Ca, however, is relatively plentiful (7.9%) in the pumice-tuff rocks at Llyn Llydaw (Chap. 15) whereas P is scarce (estimated 0.13%, D. F. Ball, pers comm.). Whilst the quantity of P in the system is low, it is the least leached from the turf core lysimeters, reflecting the retention of the element in live plant material and microorganisms. As P is relatively little leached (together with N) from live plant material, the efficiency of its use in the ecosystem may be greater than the more mobile potassium.

Nutrients leached from plants may be taken up again in situ, absorbed or exchanged on organic matter of litter and soil. Because of their solubility, leachates must be regarded as a potential loss in run-off. Mosses are capable of uptake of nutrients from very dilute solutions and in view of the quantities of nutrients retained in the moss biomass (12% of K and 8% of P) may take up much of the leached nutrients in mountain grassland as does ground flora in woodland.

The 32 plot site has a loss of nutrients in sheep production which is removed during and at the end of the grazing season, and also in dung and urine, 25% of which is deposited on night camping sites. The input of nutrients in rainfall, however, may compensate for the loss of nutrients in sheep production. Input of Ca and Mg more than replaces the total loss in production, dung and urine, but K and to a lesser extent N and P are not completely replaced. Loss of K is the greatest and is largely the result of the high transference to night camping sites. Nutrients in dung and urine are not lost to the grassland site as a whole, the transfer being part of the wider effect of sheep grazing and reflecting the uneven distribution observed within the 32 plots. Redistribution within the plots largely maintains the gradients in production and soil organic matter which, together with selective grazing, impose a spatial pattern. This, as well as additional nutrients washed downwards from weathering rocks and night camping sites, appears to be the main cause of the heterogeneity observed in many of the parameters recorded in previous chapters. It cannot be accounted for by inherent soil differences as the soil is formed in situ by weathering of the single rock type. It is expected that the observed gradients will continue. Whether or not the trend will intensify, leading to a decrease in fertility and production of those plots already of lower productivity, and lower decomposition rate of litter resulting in an increase of soil organic matter, could not be detected during the time scale of IBP. It appears in other grassland studies (Kajak, 1974) that decomposition in soil does not equal input through the primary producers. A trend towards accumulation of organic matter in some soils in Snowdonia is indicated as *Agrostis-Festuca* grassland tends towards domination by *Nardus stricta* with accompanying decrease in productivity. The *Agrostis-Festuca* site at Llyn Llydaw is subject to minor soil disturbance; landslips have occurred during the time of IBP observations, and the continual process of exposure of the quickly weathered pumice-tuff rock particles may be at least partly responsible for the high fertility of the site. In addition, the high earthworm population (and the dung beetles) mixing surface organic matter and soil also maintains soil fertility.

Acknowledgements. I acknowledge the help of colleagues participating in the IBP study in providing data in addition to those incorporated in previous chapters. Mr. D.E. Edwards assisted in chemical analyses of rain waters and the lysimeter and soil percolation experiments. Other analyses were undertaken by Mrs. V. Jones and Mrs. P. Neep at Bangor and Mr. S.E. Allen and colleagues at Merlewood Research Station. Dung invertebrates were recorded by Miss. S. Brasher and data of earthworms and small mammals provided by Dr. D.C. Seel and Mr. K.C. Walton. Diagrams throughout the Snowdonia chapters were carefully prepared by Mrs. P. Neep.

20.6 Appendix

Dry matter, energy and nutrient models of the montane *Agrostis-Festuca* grassland at Llyn Llydaw, Snowdonia, Wales. Boxes represent the mean biomass of grassland components and reserves of soil (g m^{-2} or kJ m^{-2}) throughout the year except sheep and sheep dung and invertebrates which are summer values. Transfers, on the arrows connecting boxes, represent the flows (g m^{-2} yr^{-1} or kJ m^{-2} yr^{-1}) between components. *P*: production; *E*: material eliminated; *U*: urine; *F*: faeces; *M*: methane (rumen fermentation); *R*: respiration; *D*: sheep dung decomposed on the soil surface; and *T*: sheep dung estimated to have been transferred to soil by dung beetles (Fig. 3 a–g).

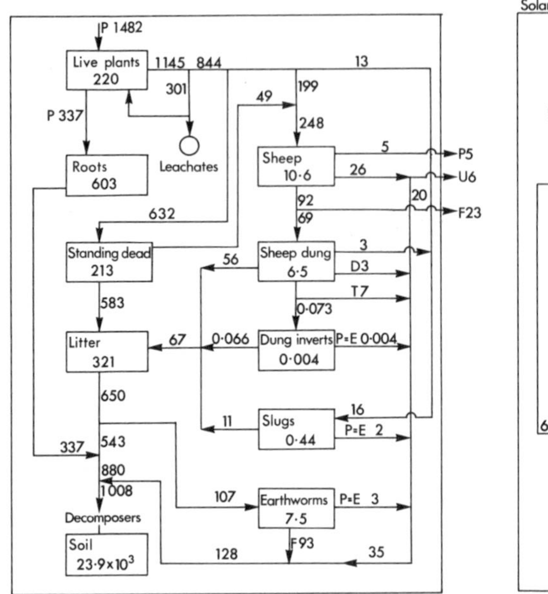

Fig. 3a. Dry matter

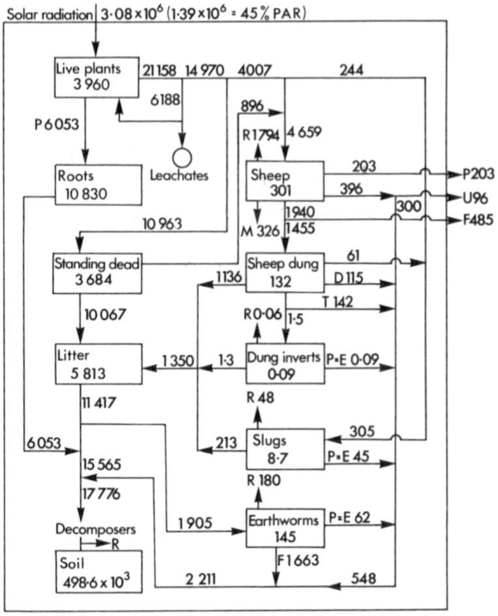

Fig. 3b. Energy

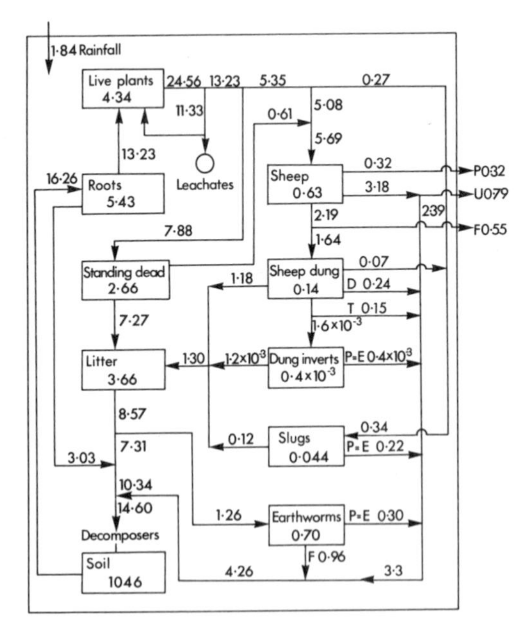

Fig. 3c. Nitrogen (N)

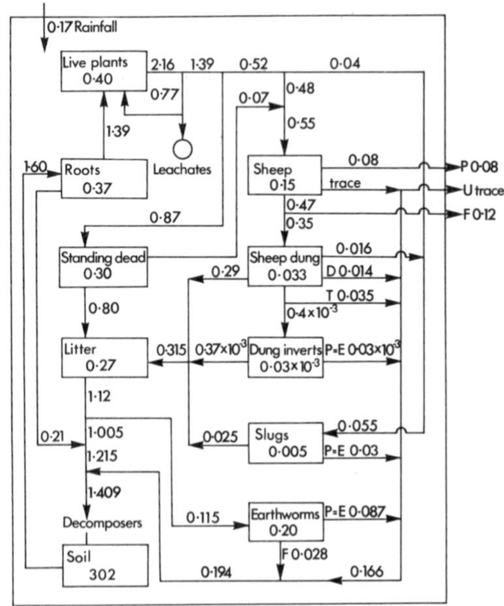

Fig. 3d. Phosphorus (P)

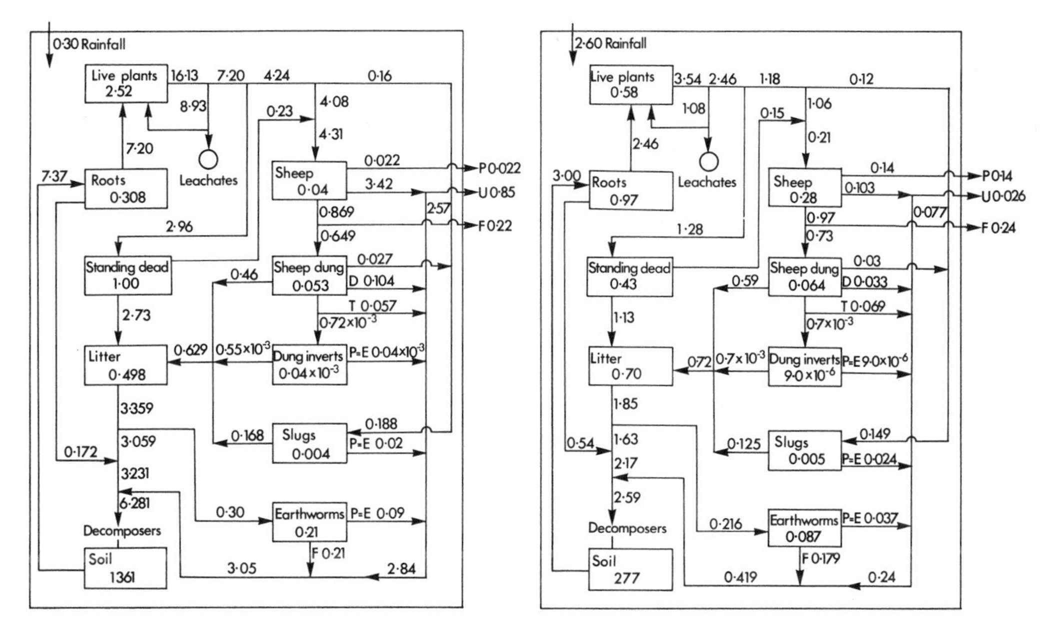

Fig. 3e. Potassium (K)

Fig. 3f. Calcium (Ca)

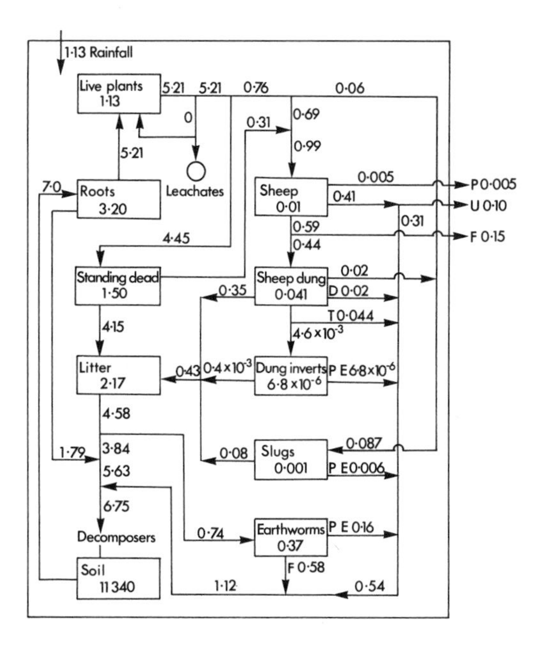

Fig. 3g. Magnesium (Mg)

References

Adams, W. A., Evans, L. J., Abdulla, H. H.: Quantitative pedological studies on soils derived from Silurian mudstones III. Laboratory and in situ weathering of chlorite. J. Soil Sci. **22**, 158–165 (1971)

Agricultural Research Council: The nutrient requirements of farm livestock, No. 2: Ruminants. London: Agr. Res. Coun., 1965

Alcock, M. B.: The effect of climate on primary production of temperate grassland with particular reference to the upland climate and the possible role of shelter as a modifying influence. Joint Shelter Res. Coun. Symp. No. 3, pp. 49–75. Cambridge: Min. Agr. Fish. Food, 1969

Alcock, M. B.: The contribution of science to the improvement of hill pasture production. Proc. Brit. Soc. Anim. Prod. **4**, 25–43 (1975)

Alcock, M. B., Harvey, G., Thomas, J.: Measurement, evaluation and management of climatic resources for grassland production in hill and lowland areas. In: Climatic Resources and Economic Activity. Taylor, J. A. (ed.). Newton Abbot: David and Charles, 1974

Alcock, M. B., Lovett, J. V.: Analysis of environmental influence on productivity. In: Proc. Europ. Grassld. Fed. Symp. Hill-land Productivity. Hunt, I. V. (ed.). Aberdeen: Brit. Grassld. Soc., 1968, pp. 20–29

Alexander, M.: Introduction to Soil Microbiology. New York: Wiley, 1961

Alexander, V.: A synthesis of the IBP Tundra Biome circumpolar study of nitrogen fixation. In: Soil Organisms and Decomposition in Tundra. Holding, A. J., Heal, O. W., MacLean, S. F., Flanagan, P. W. (eds.). Stockholm: Tundra Biome Steering Committee, 1974, pp. 109–121

Allen, S. E.: Chemical aspects of heather burning. J. Appl. Ecol. **1**, 347–367 (1964)

Allen, S. E., Carlisle, A., White, E. J., Evans, C. C.: The plant nutrient content of rainwater. J. Ecol. **56**, 497–504 (1968)

Allen, S. E., Evans, C. C., Grimshaw, H. M.: The distribution of mineral nutrients in soil after heather burning. Oikos **20**, 16–25 (1969)

Anon: Proc. 3rd Intern. Peat Cong. Quebec 18–23 August 1968. Depart. Energy, Mines and Resources and Natl. Res. Council Canada (undated)

Anslow, R. C., Green, J. O.: The seasonal growth of pasture grasses. J. Agr. Sci. Camb. **68**, 109–122 (1967)

Arnold, G. W.: Some principles in the investigation of selective grazing. Proc. Australian Soc. Anim. Prod. **5**, 258–310 (1964)

Baalsrud, K., Baalsrud, K. S.: Studies on *Thiobacillus denitrificans*. Arch. Microbiol. Bd. **20**, 34–62 (1954)

Babiuk, L. A., Paul, E. A.: The use of fluorescin isothyocyanate in the determination of the bacterial biomass of grassland soil. Can. J. Microbiol. **16**, 57–62 (1970)

Baird-Parker, A. C.: The classification of staphylococci and micrococci from world wide sources. J. Gen. Microbiol. **38**, 363 (1965a)

Baird-Parker, A. C.: Staphylococci and their classification. Ann. N.Y. Acad. Sci. **128**, 4 (1965b)

Baker, H. K., Garwood, E. A.: Studies on the root development of herbage plants IV. Seasonal changes in the root and stubble weights of various leys. J. Brit. Grassld Soc. **14**, 94–104 (1959)

Baker, J. H.: The rate of production and decomposition of *Chorisodontium aciphyllum* (Hook. f. and Wils.) Broth. Bull. Brit. Antarc. Surv. **27**, 123–129 (1972)

Ball, D. F.: The Soils and Land Use of the District around Bangor and Beaumaris. Memoirs of the Soil Survey. HMSO 1963

Ball, D. F.: Loss-on-ignition as an estimate of organic matter and organic carbon in soils. J. Soil Sci. **15**, 84–92 (1964)

Ball, D. F.: Chlorite clay minerals in Ordovician pumice-tuff and derived soils in Snowdonia, North Wales. Clay Miner. **6**, 195–209 (1966)

Ball, D. F., Mew, G., MacPhee, W. S. G.: Soils of Snowdon. Fld. Stud. **3**, 69–107 (1969)

Ball, D. F., Williams, W. M.: Variability of soil chemical properties in two uncultivated Brown Earths. J. Soil Sci. **19**, 379—391 (1968)

Ball, D. F., Williams, W. M.: Further studies on variability of soil chemical properties: efficiency of sampling programmes on an uncultivated Brown Earth. J. Soil Sci. **22**, 60–68 (1971)

Banage, W. B.: Studies on the nematode fauna of moorland soils. Ph. D. thesis, Univ. Durham, 1960

Banage, W. B.: Some nematodes from the Moor House National Nature Reserve, Westmorland. Nematologica **7**, 32–36 (1962)

Banage, W. B.: The ecological importance of free-living soil nematodes with special reference to those of moorland soil. J. Animal Ecol. **32**, 133–140 (1963)

Bannister, P.: The use of subjective estimates of cover-abundance as the basis for ordination. J. Ecol. **54**, 665–674 (1966)

Barclay-Estrup, P.: The description and interpretation of cyclical processes in a heath community II. Changes in biomass and shoot production during the *Calluna* cycle. J. Ecol. **58**, 243–249 (1970)

Barclay-Estrup, P., Gimingham, C. H.: The description and interpretation of cyclical processes in a heath community I. Vegetational change in relation to the *Calluna* cycle. J. Ecol. **57**, 737–758 (1969)

Barden, L., Berry, P. L.: Model of the consolidation process in peat soils. In: Proc. 3rd Intern. Peat Congr. Quebec 18–23 August 1968, pp. 119–127. Depart. Energy, Mines and Resources and Natl. Res. Council Canada (undated)

Barrow, N. J.: Chemical form of inorganic phosphate in sheep faeces. Australian J. Soil Res. **13**, 63–67 (1975)

Bartlett, B. O.: An improved relationship between the deposition of Sr-90 and the contamination of milk in the United Kingdom. Ann. Rep. Agr. Res. Coun. Letcombe Lab., 26–31 (1971)

Bartos, D. L., Sims, P. L.: Root dynamics of a shortgrass ecosystem. J. Range Manage. **27**, 33–36 (1974)

Benoit, R. E., Starkey, R. L.: Enzyme inactivation as a factor in the inhibition of the decomposition of organic matter by tannins. Soil Sci. **105**, 203–208 (1968a)

Benoit, R. E., Starkey, R. L.: Inhibition of decomposition of cellulose and some other carbohydrates by tannins. Soil Sci. **105**, 291–296 (1968b)

Benoit, R. E., Starkey, R. L., Basaraba, J.: Effect of purified plant tannins on the decomposition of some organic compounds and plant materials. Soil Sci. **105**, 153–158 (1968)

Bergey: Bergey's Manual of Determinative Bacteriology, 6th ed. Breed, R.S., Murray, E. G. D., Hitchens, A. P. (eds.). London: Balliere, Tindall and Cox, 1948

Berry, P. L., Poskitt, T. J.: The consolidation of peat. Géotechnique **22**, 27–52 (1972)

Berthet, P.: L'activité des oribatides d'une chenaie. Mém. Inst. Sci. nat. Belg. **151**, 1–152 (1964)

Bett, J. A.: The breeding seasons of slugs in gardens. Proc. Zool. Soc. London **135**, 559–568 (1960)

Biggins, D. R., Postgate, J. R.: Nitrogen fixation by cultures and cell-free extracts of *Mycobacterium flavum* 301. J. Gen. Microbiol. **56**, 181–193 (1969)

Black, J. S.: The digestibility of indigenous hill pasture species. Rep. Hill Fmg. Res. Org. **4**, 33–37 (1967)

Blackman, V. H.: The compound interest law and plant growth. Ann. Bot. **23**, 353–360 (1919)

Blaxter, K. L.: The utilization of the energy of grassland products. Proc. 8th Intern. Grassld Cong., Reading: Brit. Grassld. Soc., 1960

Blaxter, K. L.: The Energy Metabolism of Ruminants. London: Hutchingson, 1962, 329 pp.

Blaxter,K.L.: The effect of outdoor climate in Scotland on sheep and cattle. Vet. Rec. **76**, 1445–1454 (1964)
Bliss,L.C.: Adaptations of arctic and alpine plants to environmental conditions. Arctic **15**, 117–144 (1962)
Bliss,L.C.: Plant productivity in alpine microenvironments on Mount Washington, N. H. Ecol. Monogr. **36**, 125–155 (1966)
Bliss,L.C.: Devon Island, Canada. In: Structure and Function of Tundra Ecosystems. Ecol. Bull. (Stockholm) **20**, 17–60 (1975)
Bliss,L.C., Wielgolaski,F.E. (ed.): Primary Production and Production Processes, Tundra Biome. Edmonton: IBP Tundra Biome Steering Committee, 1973, 256 pp.
Block.W.C.: Distribution of soil mites (Acarina) on the Moor House National Nature Reserve, Westmorland, with notes on their numerical abundance. Pedobiologia **5**, 244–251 (1965)
Block,W.C.: Seasonal fluctuations and distribution of mite populations in moorland soils, with a note on biomass. J. Animal Ecol. **35**, 487–503 (1966)
Boelter,D.H.: Hydraulic conductivity of peats. Soil Sci. **100**, 227–231 (1965)
Boggie,R., Hunter,R.F., Knight,A.H.: Studies of the root development of plants in the field using radioactive tracers. J. Ecol. **46**, 621–639 (1958)
Boulet,L.J.: The ecology of a Welsh mountain sheep walk. Unpubl. thesis, Univ. Wales 1939
Bovard,P., Grauby,A.: The fixation of radionuclides from atmospheric fallout in peat-bog *Sphagnum* spp, *Polytrichum* and *Myriophyllum*. In: Radioecological Concentration Processes. Åuberg,B., Hungate,F.P.H. (eds.). Oxford: Pergamon, 1967, pp. 533–537
Bower,M.M.: A summary of available evidence and a further investigation of the causes, methods and results of erosion in blanket peat. M. Sc. thesis, Univ. London, 1959
Bower,M.M.: The cause of erosion in blanket peat bogs. A review of evidence in the light of recent work in the Pennines. Scott. Geogr. Mag. **78**, 33–43 (1962)
Boysen Jensen,P.: The production of matter in agricultural plants and its limitation. Biol. Meddr. **21**, 1–28 (1949)
Bray,J.R., Gorham,E.: Litter production in forests of the world. Advan. Ecol. Res. **2**, 101–157 (1964)
Bremeyer,A.: Analysis of a sheep pasture ecosystem in the Pieniny mountains (the Carpathians) XI. The role of coprophagous beetles (Coleoptera, Scarabaeidae) in the utilization of sheep dung. Ekol. Pol. **22**, 617–634 (1974)
Bremeyer,A., van Dyne,G.M. (ed.): Grasslands: Comparative Analysis. London: Cambridge University Press, in prep.
Brian,M.V.: Ant distribution in a southern English heath. J. Anim. Ecol. **33**, 451–461 (1964)
Brinck,P.: Studies on Swedish stoneflies. Opusc. Entom. Suppl. **11**, 250 pp. (1949)
Broadbent,F.E.: Nutrient release and carbon loss from soil organic matter during decomposition of added plant residue. Proc. Soil Sci. Soc. Am. **12**, 246–249 (1947)
Bromfield,S.M.: Sheep faeces in relation to the phosphorus cycle under pastures. Australian J. Agr. Res. **12**, 111–123 (1961)
Brown,A.H.F., Carlisle,A., White,E.J.: Nutrient deficiencies of Scots pine (*Pinus sylvestris* L.) on peat at 1800 ft in the Northern Pennines. Commonw. For. Rev. **43**, 292–302 (1964)
Brown,A.H.F., White,E.J.: Establishment and growth of trees at high elevation. Merlewood Research Station, Report for 1967–1969. Grange-over-Sands, Cumbria: The Nature Conservancy, 1970, pp. 17–19
Bunnell,F.L., MacLean,S.F., Brown,J.: Barrow, Alaska, USA. In: Structure and Function of Tundra Ecosystems. Ecol. Bull. (Stockholm) **20**, 73–124 (1975)
Bunnell,F.L., Scoullar,K.A.: Between site comparison of carbon flux using simulation models. In: Tundra: Comparative Analysis of Ecosystems. Moore,J.J. (ed.). London: Cambridge University Press, in press
Bunnell,F.L., Tait,D.E.N.: Mathematical simulation models of decomposition processes. In: Soil Organisms and Decomposition in Tundra. Holding,A.J., Heal,O.W., MacLean,S.F., Flanagan,P.W. (eds.). Stockholm: Tundra Biome Steering Committee, 1974, pp. 363–373
Burgeff,H.: Mikrobiologie des Hochmoores. Stuttgart: Fischer, 1961
Burges,A.: The ecology of the Cairngorms III. The *Empetrum: Vaccinium* zone. J. Ecol. **39**, 271–284 (1951)

Burnett,T.: Effects of temperature and host density on the rate of increase of an insect parasite. Am. Naturalist **85**, 337–352 (1951)
Burnett,T.: Influences of natural temperatures and controlled host densities on oviposition of an insect parasite. Physiol. Zool. **27**, 239–248 (1954)
Burrows,W.H.: Productivity of an arid zone shrub *(Eremophila gilesii)* community in southwestern Queensland. Australian J. Botany **20**, 317–329 (1972)
Busta,S.S., Ordal,Z.J.: Use of calcium-dipicolinate for enumeration of total viable endospore population without heat application. Appl. Microbiol. **12**, 106–110 (1964)
Butterfield,J.E.L.: Biological studies on a number of moorland Tipulidae. Ph. D. thesis, Univ. Durham, 1973
Butterfield,J.E.L.: The response of development rate to temperature in the univoltine cranefly, *Tipula subnodicornis* Zetterstedt. Oecologia **25**, 89–100 (1976)
Cambray,R.S., Fisher,E.M.R., Brooks,W.L., Peirson,D.H.: Radioactive fallout in air and rain. Results to the middle of 1971. UKAEA Research Group Report. Health Physics and Medical Division, AERE, Harwell. AERE R 6923. HMSO 46 (1971)
Campbell,L.L.,Jr, Williams,O.B.: A study of chitin-decomposing microorganisms of marine origin. J. Gen. Microbiol. **5**, 894–905 (1951)
Cappenberg,T.E.: Ecological observations on heterotrophic, methane oxidizing and sulphate reducing bacteria in a pond. Hydrobiologia **40**, 471–485 (1972)
Carlisle,A., Brown,A.H.F., White,E.J.: The organic matter and nutrient elements in the precipitation beneath a sessile oak *(Quercus petraea)* canopy. J. Ecol. **54**, 65–85 (1966)
Chamberlain,A.C.: Interception and retention of radioactive aerosols by vegetation. Atmos. Environ. **4**, 47–78 (1970)
Chang,P.C., Knowles,R.: Nonsymbiotic nitrogen fixation in some Quebec soils. Can. J. Microbiol. **11**, 29–38 (1965)
Chapman,S.B.: The ecology of Coom Rigg Moss, Northumberland III. Some water relations of the bog system. J. Ecol. **53**, 371–384 (1965)
Chapman,S.B.: Nutrient budgets for a dry heath ecosystem in the south of England. J. Ecol. **55**, 677–689 (1967)
Chapman,S.B.: The nutrient content of the soil and root system of a dry heath ecosystem. J. Ecol. **58**, 445–452 (1970)
Chapman,S.B.: A simple conductimetric soil respirometer for field use. Oikos **22**, 348–353 (1971)
Chapman,S.B., Hibble,J., Rafarel,C.R.: Net aerial production by *Calluna vulgaris* on lowland heath in Britain. J. Ecol. **63**, 233–258 (1975a)
Chapman,S.B., Hibble,J., Rafarel,C.R.: Litter accumulation under *Calluna vulgaris* on a lowland heathland in Britain. J. Ecol. **63**, 259–271 (1975b)
Chase,R.E., Gray,P.H.H.: Application of the Warburg respirometer in studying respiratory activity in soil. Can. J. Microbiol. **3**, 335–349 (1957)
Cherrett,J.M.: Ecological research on spiders associated with moorlands. Ph. D. thesis, Univ. Durham, 1961
Cherrett,J.M.: The distribution of spiders on the Moor House National Nature Reserve, Westmorland. J. Anim. Ecol. **33**, 27–48 (1964)
Christiansen,K.: Bionomics of Collembola. Ann. Rev. Entom. **9**, 147–178 (1964)
Chu,J.P.-H., Knowles,R.: Mineralisation and immobilisation of nitrogen in bacterial cells and in certain soil organic fractions. Proc. Soil Sci. Soc. Am. **27**, 312–316 (1966)
Clark,F.E.: The growth of bacteria in soil. In: Ecology of Soil Bacteria. Gray,R.R., Parkinson,D. (eds.). Liverpool: Liverpool Univ., 1967, pp. 441–457
Cleve,K.van: Organic matter quality in relation to decomposition. In: Soil Organisms and Decomposition in Tundra. Holding, A.J., Heal,O.W., MacLean,S.F., Flanagan,P.W. (eds.). Stockholm: Tundra Biome Steering Committee, 1974, pp. 311–324
Clymo,R.S.: Ion exchange in *Sphagnum* and its relation to bog ecology. Ann. Bot. NS **27**, 309–324 (1963)
Clymo,R.S.: Experiments on breakdown of *Sphagnum* in two bogs. J. Ecol. **53**, 747–758 (1965)
Clymo,R.S.: The growth of *Sphagnum:* methods of measurement. J. Ecol. **58**, 13–49 (1970)
Clymo,R.S.: The growth of *Sphagnum:* some effects of environment. J. Ecol. **61**, 849–869 (1973)

Clymo, R. S., Reddaway, E. J. F.: Productivity of *Sphagnum* (Bog moss) and peat accumulation. Hidrobiologia **12**, 181–192 (1971) Reproduced, without arbitrary cuts, as: A tentative dry matter balance sheet for the wet blanket bog on Burnt Hill, Moor House NNR. Moor House Occasional Paper No 3 (1972)

Clymo, R. S., Reddaway, E. J. F.: Growth rate of *Sphagnum rubellum* Wils. on Pennine blanket bog. J. Ecol. **62**, 191–196 (1974)

Collins, V. G., Willoughby, L. G.: The distribution of bacteria and fungal spores in Blelham Tarn with particular reference to an experimental overturn. Arch. Mikrobiol. **43**, 294–307 (1962)

Coulson, J. C.: Biological studies on the Meadow pipit *(Anthus pratensis)* and moorland Tipulidae: members of a food chain. Ph. D. thesis, Univ. Durham, 1956a

Coulson, J. C.: Mortality and egg production of the Meadow pipit with special reference to altitude. Bird Study **3**, 119–132 (1956b)

Coulson, J. C.: Observations on the Tipulidae (Diptera) of the Moor House Nature Reserve, Westmorland. Trans. Roy. Entom. Soc. London **3**, 157–174 (1959)

Coulson, J. C.: The biology of *Tipula subnodicornis* Zetterstedt. with comparative observations on *Tipula paludosa* Meigen. J. Animal Ecol. **31**, 1–21 (1962)

Coulson, J. C., Horobin, J. C., Butterfield, J., Smith, G. R. J.: The maintenance of annual life cycles in two species of Tipulidae (Diptera): a field study relating development, temperature and altitude. J. Animal Ecol. **45**, 215–233 (1976)

Coupland, R. T.: Producers III. Rates of dry matter production and of nutrient and energy flow through shoots. Matador Project, Technical Report No 33 (1973)

Coupland, R. T. (ed.): Grasslands: their Structure and Function. London: Cambridge University Press, in prep.

Cragg, J. B.: Some aspects of the ecology of moorland animals. J. Ecol. **49**, 477–506 (1961)

Crapo, N. L., Coleman, D. C.: Root distribution and respiration in a Carolina old field. Oikos **23**, 137–139 (1972)

Crisp, D. T.: A preliminary survey of Brown trout (*Salmo trutta* L.) and Bullheads (*Cottus gobio* L.) in high altitude becks. Salm. Trout Mag. **167**, 45–59 (1963)

Crisp, D. T.: Input and output of minerals for an area of Pennine moorland; the importance of precipitation, drainage, erosion and animals. J. Appl. Ecol. **3**, 327–348 (1966)

Crisp, D. T.: Input and output of minerals for a small watercress bed fed by chalk water. J. Appl. Ecol. **7**, 117–140 (1970)

Crisp, D. T., Gledhill, T.: A quantitative description of the recovery of bottom fauna in a muddy reach of a mill stream in southern England after draining and dredging. Arch. Hydrobiol. **67**, 502–541 (1970)

Crisp, D. T., Le Cren, E. D.: The temperature of three different small streams in north west England. Hydrobiologia **35**, 305–323 (1970)

Crisp, D. T., Mann, R. H. K., McCormack, J. C.: The population of fish at Cow Green, Upper Teesdale, before impoundment. J. Appl. Ecol. **11**, 969–996 (1974)

Crisp, D. T., Rawes, M., Welch, D.: A Pennine peat slide. Geog. J. **130**, 519–524 (1964)

Czerwiński, A., Jakubczyk, H., Nowak, E.: Analysis of a sheep pasture ecosystem in the Pieniny mountains (the Carpathians) XII. The effect of earthworms on the pasture soil. Ekol. Pol. **22**, 635–650 (1974)

Dahlman, R. C.: Root production and turnover of carbon in the root-soil matrix of a grassland ecosystem. In: Methods of Productivity Studies in Root Systems and Rhizosphere Organisms. Leningrad: Nauka, 1968, pp. 11–21

Dahlman, R. C., Kucera, C. L.: Root productivity and turnover in native prairie. Ecology **46**, 84–89 (1965)

Dalton, J.: Experimental enquiry into the proportion of the several gases or elastic fluids constituting the atmosphere. Manchester Philosophical Society, 1802

Davidon, W. C.: Variance algorithm for minimization. Comput. J. **10**, 406–410 (1968)

Davis, W. M.: Glacial erosion in North Wales. Quart. J. Geol. Soc. **65**, 281–350 (1909)

Delany, M. J.: Studies on the microclimate of *Calluna* heathland. J. Animal Ecol. **22**, 227–239 (1953)

Diver, C.: The physiography of South Haven Peninsula, Studland Heath, Dorset. Geog. J. **81**, 404–427 (1933)

Doney, J. M., Smith, W. F.: The fleece of the Scottish Blackface sheep I. Seasonal changes in wool production and fleece structure. J. Agr. Sci. **56**, 365–374 (1961)

Dowding, P.: Nutrient losses from litter on IBP tundra sites. In: Soil Organisms and Decomposition in Tundra. Holding, A. J., Heal, O. W., MacLean, S. F., Flanagan, P. W. (eds.). Stockholm: Tundra Biome Steering Committee, 1974, pp. 363–373

Eadie, J.: The nutrition of grazing hill sheep. Rep. Hill Farming Res. Organ. **4**, 38–45 (1967)

Eadie, J.: Sheep production and pastoral resources. In: Animal Populations in Relation to their Food Resources. Watson, A. (ed.). Oxford: Blackwell, 1970, pp. 7–24

Eastman, D. S., Jenkins, D.: Comparative food habits of Red grouse in north east Scotland, using fecal analysis. J. Wildl. Mgmt **34**, 612–620 (1970)

Eddy, A., Welch, D., Rawes, M.: The vegetation of the Moor House National Nature Reserve in the northern Pennines, England. Vegetatio, **16**, 239–284 (1969)

Egunjobi, J. K.: Ecosystem processes in a stand of *Ulex europaeus* L. I. Dry matter production, litter fall and efficiency of solar energy utilisation. J. Ecol. **59**, 31–38 (1971)

Eisenhart, C., Wilson, P. W.: Statistical methods and control in bacteriology. Bact. Rev. **7**, 27–41 (1943)

Ellenberg, H. (ed.): Integrated Experimental Ecology. Ecological Studies. Berlin: Springer, 1971, Vol. II, 214 pp.

Elton, C.: The Pattern of Animal Communities. London: Methuen, 1966

Emery, F. V.: The farming regions of Wales. In: The Agrarian History of England and Wales. Thirsk, J. (ed.). London: Cambridge University Press, 1967, Vol. IV, pp. 1500–1640

Evans, G. C.: The quantitative analysis of plant growth. In: Studies in Ecology. Oxford: Blackwell, 1972, Vol. I

Evans, C. C., Allen, S. E.: Nutrient losses in smoke produced during heather burning. Oikos **22**, 149–154 (1971)

Evans, L. T., Wardlow, I. F., Williams, C. N.: Environmental control of growth. In: Grasses and Grasslands. Barnard, C. (ed.). London: MacMillan, 1964, pp. 102–125

Evdokimova, T. I., Grishina, L. A.: Productivity of root systems of herbaceous vegetation on flood plain meadows and methods for its study. In: Methods of Productivity Studies in Root Systems and Rhizosphere Organisms. Leningrad: Nauka, 1968, pp. 24–35

Farnham, R. S., Finney, H. R.: Classification and properties of organic soils. Advan. Agron. **17**, 115–162 (1965)

Flanagan, P. W., Scarborough, A. M.: Physiological groups of decomposer fungi on tundra plant remains. In: Soil Organisms and Decomposition in Tundra. Holding, A. J., Heal, O. W., MacLean, S. F., Flanagan, P. W. (eds.). Stockholm: Tundra Biome Steering Committee, 1974, pp. 159–181

Flanagan, P. W., Veum, A. K.: Relationships between respiration, weight loss, temperature, and moisture in organic residues in tundra. In: Soil Organisms and Decomposition in Tundra. Holding, A. J., Heal, O. W., MacLean, S. F., Flanagan, P. W. (eds.). Stockholm: Tundra Biome Steering Committee, 1974, pp. 249–277

Fleming, G. A.: Mineral composition of herbage. In: Chemistry and Biochemistry of Herbage. Butler, G. W., Bailey, R. W. (eds.). New York-London: Academic, 1973, pp. 529–566

Flint, P. S., Gersper, P. L.: Nitrogen nutrient levels in arctic tundra soils. In: Soil Organisms and Decomposition in Tundra. Holding, A. J., Heal, O. W., MacLean, S. F., Flanagan, P. W. (eds.). Stockholm: Tundra Biome Steering Committee, 1974, pp. 249–277

Fogg, G. E.: Actual and potential yields in photosynthesis. Advmt Sci., London **14**, 395–400 (1958)

Forrest, G. I.: Structure and production of north Pennine blanket bog vegetation. J. Ecol. **59**, 453–579 (1971)

Forrest, G. I., Smith, R. A. H.: The productivity of a range of blanket bog types in the northern Pennines. J. Ecol. **63**, 173–202 (1975)

Forrester, J. W.: Industrial Dynamics. London: MIT Press, 1961

Frame, J.: Fundamentals of grassland management X. The grazing animal. Scott. Agr. **50**, 28–44 (1970/1971)

Frankland, J. C.: Estimation of live fungal biomass. Soil Biol. Biochem. **7**, 339–340 (1975)

Fridrikson, S.: Grass and grass utilisation in Iceland. Ecology **53**, 785–796 (1972)

Friedman, M.: The use of ranks to avoid the assumption of normality implicit in the analysis of variance. J. Am. Stat. Assoc. **32**, 675–701 (1937)

Frömming, E.: Das Verhalten unserer Schnecken zu den Pflanzen ihrer Umgebung. Berlin: Dunker u. Humblot, 1962

Gaastra, P.: Photosynthesis of crop plants as influenced by light, carbon dioxide, temperature and stomatal diffusion resistance. Meded. LandbHoogesh. Wageningen **59**, 1–68 (1959)

Gabriel, K.P.: A procedure for testing the homogeneity of all sets of means in an analysis of variance. Biometrics **20**, 459–477 (1964)

Gimingham, C.H.: Biological flora of the British Isles. *Calluna vulgaris* (L.) Hull. J. Ecol. **48**, 455–483 (1960)

Gimingham, C.H.: Ecology of Heathlands. London: Chapman and Hall, 1972, 266 pp

Gimingham, C.H., Miller, G.R.: Measurement of the primary production of dwarf shrub heaths. In: Methods for the Measurement of Primary Production of Grassland. Milner, C., Hughes, R.E. (eds.). Oxford-Edinburgh: Blackwell, 1968, pp. 43–51

Gloyne, R.W.: On the growing season. Agric. Mem. XVIII, Edinburgh: Met. Office, 1958

Goode, D.W.: Ecological studies on the Silver Flowe Nature Reserve. Ph. D. thesis, Univ. Hull, 1970

Goodman, G.T. Perkins, D.F.: The role of mineral nutrients in *Eriophorum* communities III. Growth response to added inorganic elements in two *E. vaginatum* communities. J. Ecol. **56**, 667–683 (1968a)

Goodman, G.T., Perkins, D.F.: The role of mineral nutrients in *Eriophorum* communities IV. Potassium supply as a limiting factor in an *E. vaginatum* community. J. Ecol. **56**, 685–696 (1968b)

Goodman, G.T., Roberts, T.M.: Plants and soils as indicators of metals in the air. Nature (London) **231**, 287–292 (1971)

Gore, A.J.P.: Factors limiting plant growth on high-level blanket peat I. Calcium and phosphate. J. Ecol. **49**, 399–402 (1961a)

Gore, A.J.P.: Factors limiting plant growth on high-level blanket peat II. Nitrogen and phosphate in the first year of growth. J. Ecol. **49**, 605–616 (1961b)

Gore, A.J.P.: Factors limiting plant growth on high-level blanket peat III. An analysis of growth of *Molinia caerulea* (L) Moench. in the second year. J. Ecol. **51**, 481–491 (1963)

Gore, A.J.P.: The supply of six elements by rain to an upland peat area. J. Ecol. **56**, 483–495 (1968)

Gore, A.J.P.: A field experiment, a small computer and model simulation. In: Mathematical Models in Ecology. Jeffers, J.N.R. (ed.). Oxford: Blackwell, 1972, pp. 309–325

Gore, A.J.P., Allen, S.E.: Measurements of exchangeable and total cation contents for H^+, Na^+, K^+, Mg^{++}, Ca^{++} and iron in high level blanket peat. Oikos **7**, 48–55 (1956)

Gore, A.J.P., Olson, J.S.: Preliminary models for accumulation of organic matter in an *Eriophorum/Calluna* ecosystem. Aquilo, Ser. Botanica **6**, 297–313 (1967)

Gore, A.J.P., Urquhart, C.: The effects of waterlogging on the growth of *Molinia caerulea* and *Eriophorum vaginatum*. J. Ecol. **54**, 617–633 (1966)

Gorham, E.: On the chemical composition of some waters from the Moor House National Nature Reserve. J. Ecol. **44**, 375–382 (1956)

Gorham, E.: Accumulation of radioactive fallout by plants in the English Lake District. Nature (London) **181**, 1523–1524 (1958a)

Gorham, E.: The influence and importance of daily weather conditions in the supply of chloride, sulphate and other ions to freshwaters from atmospheric precipitation. Phil. Trans. Roy. Soc. **241B**, 147–178 (1958b)

Grace, J.: The growth-physiology of moorland plants in relation to their aerial environment. Ph. D. thesis, Univ. Sheffield, 1970

Grace, J.: An apparatus for the study of leaf canopy optics. J. Appl. Ecol. **10**, 57–61 (1973)

Grace, J., Woolhouse, H.W.: A physiological and mathematical study of the growth and productivity of a *Calluna-Sphagnum* community I. Net photosynthesis of *Calluna vulgaris* L. Hull. J. Appl. Ecol. **7**, 363–381 (1970)

Grace, J., Woolhouse, H.: A physiological and mathematical study of growth and productivity of a *Calluna-Sphagnum* community II. Light interception and photosynthesis in *Calluna*. J. Appl. Ecol. **10**, 63–76 (1973a)

Grace,J., Woolhouse,H.W.: A physiological and mathematical study of growth and productivity of a *Calluna-Sphagnum* community III. Distribution of photosynthate in *Calluna vulgaris* L. Hull. J. Appl. Ecol. **10**, 77–91 (1973b)
Grace,J., Woolhouse,H.W.: A physiological and mathematical study of growth and productivity of a *Calluna-Sphagnum* community IV. A model of growing *Calluna*. J. Appl. Ecol. **11**, 281–295 (1974)
Granhall,H., Sellander,H.: Nitrogen fixation in a subarctic mire. Oikos **24**, 8–15 (1973)
Granlund,E.: De Svenska högmossarnas geologi. Sver. Geol. Unders. Afh. **26**, 1–193 (1932)
Grant,S.A.: Interactions of grazing and burning on heather moors 2. Effects on primary production and level of utilisation. J. Brit. Grassld Soc. **26**, 173–181 (1971)
Grant,S.A., Hunter,R.F.: Ecotypic differentiation of *Calluna vulgaris* in relation to altitude. New Phytologist **61**, 44–55 (1962)
Grant,S.A., Hunter,R.F.: Interactions of grazing and burning on heather moors and their implications in heather management. J. Brit. Grassld Soc. **23**, 285–293 (1968)
Grant,S.A., Milne,J.A.: Factors affecting the role of heather (*Calluna vulgaris*) in grazing systems. Potassium Inst. Ltd., Colloqu. Proc. **3**, 43–46 (1973)
Gresham,C.: The Aberconway Charter. Archaeologia Cambrensis **94**, 123–162 (1939)
Groves,R.H., Specht,R.L.: Growth of heath vegetation I. Annual growth curves of two heath ecosystems in Australia. Australian J. Botany **13**, 261–280 (1965)
Gunary,D.: The availability of phosphate in sheep dung. J. Agr. Sci., Camb. **70**, 33–38 (1968)
Hadley,E.B., Bliss,L.C.: Energy relationships of alpine plants on Mount Washington, N. H. Ecol. Monogr. **34**, 331–370 (1964)
Hadley,M.: The adult biology of the cranefly *Molophilus ater* Meigen. J. Animal Ecol. **38**, 765–790 (1969)
Hadley,M.: Aspects of the larval ecology and population dynamics of *Molophilus ater* Meigen (Diptera: Tipulidae) on Pennine moorland. J. Animal Ecol. **40**, 445–466 (1971)
Hale,W.G.: Observations on the breeding biology of Collembola II. Pedobiologia **5**, 161—177 (1965)
Hale,W.G.: The Collembola of the Moor House National Nature Reserve, Westmorland: a moorland habitat. Rev. Ecol. Biol. Sol. **3**, 97–122 (1966a)
Hale,W.G.: A population study of moorland Collembola. Pedobiologia **6**, 65–99 (1966b)
Handley,W.R.C.: Mull and mor formation in relation to forest soils. Bull. For. Commun. London **23** (1954)
Hanrahan,E.T.: Factors affecting strength and deformation of peat. Intern. Peat Sympos. Dublin 1954. Proc., Sec. B-3 (1954)
Hardy,R.W.F., Holstein,R.D., Jackson,E.K., Burns,R.C.: The acetylene-ethylene assay for N_2 fixation: Laboratory and field evaluation. Plant Physiol. **43**, 1185–1207 (1968)
Harley,N., Fisenne,I., Ong,L.D.Y., Harley,J.: Fission yield and fission produce decay. USAEC HASL-**164**, 251–260 (1965)
Harper,J.L.: Population Biology of Plants. London: Academic, 1977, 892 pp
Hayes,J.B.: Polytypism of chlorite in sedimentary rocks. Clays and Clay Min. **18**, 285–306 (1970)
Heal,O.W., Bailey,A.D., Latter,P.M.: Bacteria, fungi and protozoa in Signy Island soils compared with those from a temperate moorland. Phil. Trans. Roy. Soc., London Ser. B **252**, 191–197 (1967)
Heal,O.W., French,D.D.: Decomposition of organic matter in tundra. In: Soil Organisms and Decomposition in Tundra. Holding,A.J., Heal,O.W., MacLean,S.F., Flanagan,P.W. (eds.). Stockholm: Tundra Biome Steering Committee, 1974, pp. 279–308
Heal,O.W., Howson,G., French,D.D., Jeffers,J.N.R.: Decomposition of cotton strips in tundra. In: Soil Organisms and Decomposition in Tundra. Holding,A.J., Heal,O.W., MacLean,S.F., Flanagan,P.W. (eds.). Stockholm: Tundra Biome Steering Committee, 1974, pp. 341–362
Heal,O.W., Jones,H.E., Whittaker,J.B.: Moor House UK. In: Structure and Function of Tundra Ecosystems. Ecol. Bull. (Stockholm) **20**, 295–320 (1975)
Heal,O.W., MacLean,S.F.: Comparative productivity in ecosystems—secondary productivity. In: Unifying Concepts in Ecology. van Dobben,W.H., Lowe-McConnell,R.H. (eds.). The Hague: Junk, 1975, pp. 89–108

Heal, O. W., Perkins, D. F.: IBP studies on montane grassland and moorlands. Phil. Trans. Roy. Soc., London Ser. B **274**, 295–314 (1976)

Herriott, J. B. D., Wells, D. A.: The grazing animal and sward productivity. J. Agr. Sci. **61**, 89–99 (1963)

Hesketh, J. D., Moss, D. N.: Variation in the response of photosynthesis to light. Crop Sci. **3**, 107–110 (1963)

Hilder, E. J.: Distribution of excreta by sheep at pasture. Proc. 10th Intern. Grassld. Congr. Helsinki: Finnish Grassland Association, 1966, pp. 977–981

Hill, M. O.: Reciprocal averaging: an eigenvector method of ordination. J. Ecol. **61**, 237–249 (1973)

Hodkinson, I. D.: Studies on the ecology of *Strophingia ericae* (Curtis) (Homoptera-Psylloidea). Ph. D. thesis, Univ. Lancaster, 1971

Hodkinson, I. D.: The population dynamics and host plant interactions of *Strophingia ericae* (Curt.) (Homoptera :Psylloidea). J. Animal Ecol. **42**, 565–583 (1973)

Holding, A. J.: The properties and classification of predominant Gram-negative bacteria occurring in soil. J. Appl. Bact. **23**, 515–525 (1960)

Holding, A. J., Franklin, D. A., Watling, R.: The microflora of peat–podzol transitions. J. Soil Sci. **16**, 44–59 (1965)

Holding, A. J., Heal, O. W., MacLean, S. F., Flanagan, P. W. (ed.): Soil Organisms and Decomposition in Tundra. Stockholm: Tundra Biome Steering Committee, 1974, 398 pp.

Holter, P.: Food utilization of dung-eating *Aphodius* larvae (Scarabaeidae). Oikos **25**, 71–79 (1974)

Horobin, J. C.: Studies on the biology of moorland Tipulidae with particular reference to *Molophilus ater* Meigen. Ph. D. thesis, Univ. Durham, 1971

Houston, W. W. K.: Ecological studies on moorland ground beetles. Ph. D. thesis, Univ. Durham, 1970

Houston, W. W. K.: Carabidae (Col.) from two areas of the north Pennines. Entomol. Mon. Mag. **107**, 1–4 (1971)

Houston, W. W. K.: The food of the Common frog, *Rana temporaria*, on high moorland in northern England. J. Zool. Res. London **171**, 153–165 (1973)

Hughes, P. G., Reid, D.: Studies on the behaviour of cattle and sheep in relation to the utilisation of grass. J. Agr. Sci. **41**, 350–366 (1951)

Hughes, R. E.: The vegetation of the northwestern Conway Valley (North Wales) I. J. Ecol. **37**, 306–334 (1949)

Hughes, R. E.: The ecology of some North Wales sheepwalks. European Grassland Conference. OEEC Project 224 (1954)

Hughes, R. E.: Sheep population and environment in Snowdonia (North Wales) J. Ecol. **46**, 169–190 (1958a)

Hughes, R. E.: Report of the Nature Conservancy for year ended 30 September 1958. London: HMSO, 1958b, pp. 63–65

Hughes, R. E.: Studies in sheep population and environment in the mountains of northwest Wales. J. Appl. Ecol. **10**, 107–112 (1973)

Hughes, R. E., Dale, J., Mountford, M. D., Williams, I. E.: Studies in sheep population and environment in the mountains of north west Wales. II. Contemporary distribution of sheep population and environment. J. Appl. Ecol. **12**, 165–178 (1975)

Hughes, R. E., Dale, J., Williams, I. E., Rees, D. I.: Studies in sheep population and environment in the mountains of north west Wales I. The status of the sheep in the mountains of North Wales since mediaeval times. J. Appl. Ecol. **10**, 113–132 (1973)

Hughes, R. E., Milner, C., Dale, J.: Selectivity in grazing. In: Grazing in Terrestrial and Marine Environments. Crisp, D. J. (ed.). London: Blackwell, 1964, pp. 189–202

Hunter, P. J.: Distribution and abundance of slugs on an arable plot in Northumberland. J. Animal Ecol. **35**, 543–547 (1966)

Hunter, P. J.: Studies on slugs of arable ground I. Sampling methods. Malacologia **6**, 369–377 (1968a)

Hunter, P. J.: Studies on slugs of arable ground II. Life cycles. Malacologia **6**, 379–389 (1968b)

Hunter, P. J., Runham, N. W.: Some aspects of recent research on slugs. Proc. Malac. Soc. London **39**, 235–238 (1970)

Hunter, R. F.: The grazing of hill sward types. J. Brit. Grassld Soc. **9**, 195–206 (1954)
Hunter, R. F.: Aims and methods in grazing-behaviour studies on hill pastures. Proc. 8th Intern. Grassld Congr. Reading: Brit. Grassld Soc., 1960, pp. 454–457
Hunter, R. F.: Hill sheep and their pasture: a study of sheep grazing in southeast Scotland. J. Ecol. **50**, 651–680 (1962)
Hunter, R. F.: Home range behaviour in hill sheep. In: Grazing in Terrestrial and Marine Environments. Crisp, D. J. (ed.). Oxford: Blackwell, 1964, pp. 155–171
Hunter, R. F., Milner, C.: The behaviour of individual, related, and groups of south country Cheviot hill sheep. Animal Behav. **11**(4) (1963)
Hutchinson, K. J.: A coring technique for the measurement of pasture of low availability to sheep. J. Brit. Grassld Soc. **22**, 131–134 (1967)
Hynes, H. B. N.: The invertebrate fauna of a Welsh mountain stream. Arch. Hydrobiol. **57**, 344–388 (1961)
Hynes, H. B. N., Coleman, M. J.: A simple method of assessing the annual production of stream benthos. Limnol. Oceanogr. **13**, 569–573 (1968)
Ingram, M.: The ecology of the Cairngorms IV. The *Juncus* zone: *Juncus trifidus* communities. J. Ecol. **46**, 707–737 (1958)
Ingram, H. A. P., Rycroft, D. W., Williams, D. J. A.: Anomalous transmission of water through certain peats J. Hydrol. **22**, 213–218 (1974)
Jannson, S. L.: Tracer studies on nitrogen transformations in soil with special attention to mineralisation-immobilisation relationships. K. LantbrHogsk Annlr **24**, 101–361 (1958)
Jarvis, P. G., Jarvis, M. S.: Growth rates of woody plants. Physiol. Plantarum **17**, 654–666 (1964)
Jaworowski, Z.: Stable and Radioactive Lead in the Environment and the Human Body. Warsaw: Inst. Nuclear Res., 1967
Jenkins, D., Watson, A., Miller, G. R.: Population studies on Red grouse, *Lagopus lagopus scoticus* (Lath.) in northeast Scotland. J. Animal Ecol. **32**, 317–376 (1963)
Jenkins, D., Watson, A., Miller, G. R.: Population fluctuations in the Red grouse *Lagopus lagopus scoticus*. J. Animal Ecol. **36**, 97–122 (1967)
Jenkinson, D. S.: Studies on the decomposition of plant material in soil I. Losses of carbon from C_{14} labelled rye grass inculcated with soil in the field. J. Soil Sci. **16**, 104–115 (1965)
Jenny, H., Gessel, S. P., Bingham, F. T.: Comparative study of decomposition rates of organic matter in temperate and tropical regions. Soil Sci. **68**, 419–432 (1949)
Jewiss, O. R.: The grasses, vegetative growth. In: Grasses and Legumes in British Agriculture. Commonw. Agr. Bur. Bull. **49**, 55–63 (1972)
Johnson, G. A. L., Dunham, K. C.: The Geology of Moor House. Monogr. Nat. Conserv. HMSO, 1963, 182 pp.
Jones, H. E., Gore, A. J. P.: A simulation approach to primary production. In: Analysis of Tundra Ecosystems. IBP Tundra Biome. Moore, J. J. (ed.). London: Cambridge University Press (in press)
Jones, Ll. I.: Studies on hill land in Wales. Welsh Plant Breeding Station Technical Bulletin **2** (1967)
Jones, P. C. T., Mollison, J. E.: A technique for the quantitative estimation of soil microorganisms. J. Gen. Microbiol. **2**, 54–69 (1948)
Jordan, A. M.: *Coleophora alticolella* Zell. (Lepidoptera) and its food plant *Juncus squarrosus* L. in the northern Pennines. J. Animal Ecol. **31**, 293–304 (1962)
Kajak, A.: Analysis of a sheep pasture ecosystem in the Pieniny mountains (the Carpathians). XVII. Analysis of the transfer of carbon. Ekol. Pol. **22**, 711–732 (1974)
Kaye, C. A., Barghoorn, E. S.: Late Quaternary sea-level change and crustal rise at Boston Massachusetts, with notes on the autocompaction of peat. Bull. Geol. Soc. Am. **75**, 63–80 (1964)
Kloet, G. S., Hincks, W. D.: A Check List of British Insects. Stockport: Arbroath, 1945
Knowles, R.: Soil microflora and soil humus. Ph. D. thesis, Univ. London, 1957
Knowles, R.: The significance of nonsymbiotic nitrogen fixation. Proc. Soil Sci. Soc. Am. **29**, 223 (1965)
Knowles, R., Chu, J. P.-H.: Survival and mineralisation and immobilisation of ^{15}N-labelled *Serratia* cells in a boreal forest raw humus. Can. J. Microbiol. **15**, 223–228 (1969)

Krogh, A.: The quantitative relation between temperature and standard metabolism in animals. Intern. Z. Phys.-Chem. Biol. **1**, 491–508 (1914)

Küster, E.: Studies on Irish peat bogs and their microbiology. Microbiol. Esp. **16**, 203–208 (1963)

Kyall, A. J.: Some characteristics of heath fires in north east Scotland. J. Appl. Ecol. **3**, 29–40 (1966)

Ladd, J. M., Brisbane, P. G., Butler, J. H. A.: The susceptibility of the nitrogenous components of humic acids to enzyme attack—inhibition of pronase activity. Trans 9th Intern. Cong. Soil Sci. **3**, 319–327 (1968)

Ladd, J. M., Butler, J. H. A.: Inhibitory effect of soil humic compounds on the proteolytic enzyme pronase. Australian J. Soil Res. **7**, 241–251 (1969a)

Ladd, J. M., Butler, J. H. A.: Inhibition and stimulation of proteolytic enzyme activities by soil humic acids. Australian J. Soil. Res. **7**, 253–261 (1969b)

Ladd, J. M., Butler, J. H. A.: The effect of inorganic cations on the inhibition and stimulation of protease activity by soil humic acids. Soil Biol. Biochem. **2**, 33–40 (1970)

Lakhani, K. H., Satchell, J. E.: Production by *Lumbricus terrestris* (L.) J. Animal Ecol. **39**, 473–492 (1970)

Langer, R. H. M.: Growth and nutrition of Timothy *(Phleum pratense)* 1. The life history of individual tillers. Ann. Appl. Biol. **44**, 166–187 (1956)

Langer, R. H. M.: Changes in the tiller populations of grass swards. Nature (London) **182**, 1817–1818 (1958)

Latter, P. M.: Decomposition of a moorland vegetation, in relation to *Marasmius androsaceus* and soil fauna. Pedobiologia **17** (6) (1977)

Latter, P. M., Cragg, J. B.: The decomposition of *Juncus squarrosus* leaves and microbiological changes in the profile of *Juncus* moor. J. Ecol. **55**, 465–482 (1967)

Latter, P. M., Cragg, J. B., Heal, O. W.: Comparative studies on the microbiology of four moorland soils in the northern Pennines. J. Ecol. **55**, 445–464 (1967)

Latter, P. M., Heal, O. W.: A preliminary study of the growth of fungi and bacteria from temperate and antarctic soils in relation to temperature. Soil Biol. Biochem. **3**, 365–379 (1971)

Latter, P. M., Howson, G.: The use of cotton strips to indicate cellulose decomposition in the field. Pedobiologia **17**, 145–155 (1977)

Latter, P. M., Howson, G.: Studies on the microflora of blanket bog with particular reference to Enchytraeidae II. Growth and survival of *Cognettia sphagnetorum* (Vejd.) on various substrates. J. Animal Ecol. (in press)

Lawes, J. B., Gilbert, J. H.: Experimental enquiry into the composition of animals fed and slaughtered as human food. Trans. Roy. Soc. London **2**, 496–600 (1895)

Le Cren, E. D.: Estimates of fish populations and production in small streams in England. In: Symp. on Salmon and Trout in Streams. Macmillan Lect. Br. Columb. Univ., 1969, pp. 269–280

Lee, J. A., Tallis, J. H.: Regional and historical aspects of lead pollution in Britain. Nature (London) **245**, 216–218 (1973)

Lees, H.: Soil percolation technique. Plant Soil **1**, 221–270 (1948)

Leslie, A. S., Shipley, A. E.: Grouse in Health and Disease. London: Smith Elder, 1912, pp. 79–112

Louw, H. A., Webley, D. M.: Bacteriology of the root region of the oat plant grown under controlled pot culture conditions. J. Appl. Bact. **22**, 216–226 (1959)

Lowe, W. E., Gray, T. R. G.: Ecological studies on coccoid bacteria in a pine forest soil II. Growth of bacteria introduced into soil. Soil Biol. Biochem. **5**, 449–462 (1973)

Ludlow, M. M., Jarvis, P. G.: Photosynthesis in Sitka spruce (*Picea sitchensis* (Bong.) Carr.) I. General characteristics. J. Appl. Ecol. **8**, 925–953 (1971)

Luria, S. E.: Bacterial protoplasm: composition and organisation. In: The Bacteria. Gunsalus, E. C., Stainier, R. Y. (eds.). New York: Academic Press, 1960, Vol. I

Luxton, M.: Studies on the oribatid mites of a Danish beech wood soil. Pedobiologia **12**, 434–463 (1972)

McEvoy, T.: Afforestation of peat soils. International Peat Symposium, 1–8. Dublin: Bord na Mona, 1954

McKerron, D. K. L.: Energy, carbon dioxide and water vapour exchange, and the growth of *Calluna vulgaris* (L) Hull, in relation to environment. Ph. D. thesis, Univ. Aberdeen, 1971
MacLean, S. F.: Ecological adaptation of tundra invertebrates. In: Physiological Adaptation to the Environment. Vernberg, J. (ed.). New York: Intext Press, 1975
McNeill, S., Lawton, J. H.: Annual production and respiration in animal populations. Nature (London) **225**, 472–474 (1970)
McVean, D. N.: Ecology of *Alnus glutinosa* (L.) Gaertn. VII. Establishment of alder by direct seeding of shallow blanket bog. J. Ecol. **47**, 615–618 (1959)
Mani, M. S.: Ecology and Biogeography of High Altitude Insects. The Hague: Junk, 1968
Manley, G.: The effective rate of altitudinal change in temperate climates. Geog. Rev. **35**, 408–417 (1945)
Manley, G.: Climate and the British Scene. London: Collins, 1952, pp. 178–196
Manual of Microbiological Methods. Committee on Bacteriological Technic of American Society for Microbiology (ed.). New York: Mc Graw-Hill, 1957
Marks, T. C.: The effects of moorland management on the growth of *Rubus chamaemorus* L. Ph. D. thesis, Univ. London, 1974
Marks, T. C., Taylor, K.: The mineral nutrient status of *Rubus chamaemorus* L. in relation to burning and sheep grazing. J. Appl. Ecol. **9**, 501–511 (1972)
Martin, N. J.: Microbial activity in peat with reference to the availability and cycling of inorganic ions. Ph. D. thesis, Univ. Edinburgh, 1971
Mason, O. F.: Food, feeding rates and assimilation in woodland snails. Oecologia **4**, 358–373 (1970)
Mason, O. F.: Respiration rates and population metabolism of woodland snails. Oecologia **7**, 80–94 (1971)
Matthews, B. C., Smith, J. A.: A percolation method for measuring potassium supplying power of soils. Can. J. Soil Sci. **37**, 21–28 (1957)
Mattsson, S.: ^{137}Cs in the reindeer lichen *Cladonia alpestris:* deposition, retention and internal distribution, 1961–1970. Lund University Radiophysics Institute Report 1972–08 (1972)
Mattsson, S., Koutler-Andersson, E.: Geochemistry of a raised bog. Annaler. Lantbrukshögskolans **21**, 321–366 (1954)
Merrett, P.: The phenology of spiders on heathland in Dorset. J. Animal Ecol. **36**, 363–374 (1967)
Merrett, P.: The phenology of linyphiid spiders on heathland in Dorset. J. Zool. (London) **157**, 289–307 (1969)
Merrett, P.: Captain Cyril Diver (1892–1969): A Memoir. Wareham: Furzebrook Research Station, 1971
Merson, R. H.: An operation method for the study of integration processes. Proc. Conf. Data Processing and Automatic Computing Machines. Weapons Research Establishment, South Australia (1957)
Metcalfe, G.: The ecology of the Cairngorms II. The mountain *Callunetum*. J. Ecol. **38**, 46–74 (1950)
Meteorological Office: Tables of temperature, relative humidity and precipitation for the world. Part I. Met. Off. 617. London, 1958
Meteorological Office: Estimated soil moisture deficit over Great Britain. Bracknell: Bulletin, Meteorological Office, 1962–1971
Millar, A.: Notes on the climate near the upper forest limit in the northern Pennines. Quart. J. For. **3**, 239–246 (1964)
Millar, A.: The effect of temperature and day length on the height growth of birch *(Betula pubescens)* at 1900 feet in the northern Pennines. J. Appl. Ecol. **2**, 17–29 (1965)
Miller, G. R., Jenkins, D., Watson, A.: Heather performance and Red grouse populations I. Visual estimates of heather performance. J. Appl. Ecol. **3**, 313–326 (1966)
Miller, G. R., Watson, A.: Some effects of fire on vertebrate herbivores in the Scottish Highlands. Proc. Ann. Tall Timbers Fire Ecology Conf. **13**, 39–64 (1974)
Miller, G. R., Watson, A., Jenkins, D.: Responses of Red grouse populations to experimental improvement of their food. In: Animal Populations in Relation to their Food Resources. Watson, W. (ed.). Oxford: Blackwell, 1970, pp. 323–335

Milne,A.: The centric systematic area sample treated as a random sample. Biometrics **15**, 270–297 (1959)
Milner,C.: The estimation of energy flow through populations of large herbivore animals. In: Secondary Productivity of Terrestrial Ecosystems. Petrusewicz,K. (ed.). Warszawa-Krakow: Polish Acad. Sci., 1967, pp. 147–161
Milner,C., Gwynne,D.: The Soay sheep and their food supply. In: Island Survivors: the Ecology of the Soay Sheep of St. Kilda. Jewell,P.A., Milner,C., Morton Boyd,J. (eds.). London: Athlone, 1974, pp. 273–325
Milner,C., Hughes,R.E.: Methods for the Measurement of Primary Production of Grassland. IBP Handbook 6. Oxford: Blackwell, 1968
Minderman,G.: Addition, decomposition and accumulation of organic matter in forests. J. Ecol. **56**, 355–362 (1968)
Mooney,H.A., Billings,W.D.: The annual carbohydrate cycle of alpine plants as related to growth. Am. J. Botany **47**, 594–598 (1960)
Moore,J.J.: Report of the Glenamoy (Ireland) ecosystem study for 1971. In: IBP Tundra Biome Proc. 4th Intern. Meeting on the Biological Productivity of Tundra. Leningrad USSR. Wielgolaski,F.E., Rosswall,T. (eds.). Stockholm: Tundra Biome Steering Committee, 1972, pp. 281–282
Moore,J.J. (ed.): Tundra: Comparative Analysis of Ecosystems. London: Cambridge University Press, in prep
Moore,J.J., Dowding,P., Healy,B.: Glenamoy, Ireland. In: Structure and Function of Tundra Ecosystems. Ecol. Bull. (Stockholm) **20**, 321–343 (1975)
Moore,N.W.: The heaths of Dorset and their conservation. J. Ecol. **50**, 369–391 (1962)
Moore,N.W.: Intra- and interspecific competition among dragonflies (Odonata). J. Animal Ecol. **33**, 49–71 (1964)
Moore,P.D.: The initiation of peat formation and the development of peat deposits in mid Wales. In: Proc. 4th Intern. Peat Cong. Finland: 1972, Vol. I, pp. 89–100
Moore,P.D., Chater,E.H.: The changing vegetation of west central Wales in the light of human history. J. Ecol. **57**, 361–379 (1969)
Moss,R.: A comparison of Red grouse *(Lagopus l. scoticus)* stocks with the production and nutritive value of heather *(Calluna vulgaris)*. J. Animal Ecol. **38**, 103–112 (1969)
Moss,R.: Food selection by Red grouse (*Lagopus lagopus scoticus* (Lath.)) in relation to chemical composition. J. Animal Ecol. **41**, 411–428 (1972)
Moss,R., Miller,G.R., Allen,S.E.: Selection of heather by captive Red grouse in relation to the age of the plant. J. Appl. Ecol. **9**, 771–781 (1972)
Moss,R., Parkinson,J.A.: The digestion of heather *(Calluna vulgaris)* by Red grouse *(Lagopus lagopus scoticus)*. Brit. J. Nutrit. **27**, 285–298 (1972)
Moule,G.R.: Sheep and wool production in semi-arid pastoral Australia. World Rev. Animal Prod. **4**, 46–58 (1968)
Munro,J.M.M., Davies,D.A.: Potential pasture production in the uplands of Wales 2. Climatic limitations on production. J. Brit. Grassld Soc. **28**, 161–169 (1973)
Munro,J.M.M., Davies,D.A., Morgan,T.E.H.: Research on pasture improvement potential at Pant-y-dŵr Hill Centre. Ann. Rept. Welsh Plant Breeding Station, 1972, pp. 209–228
Nelder,J.A., Mead,R.: A simplex method for function minimization. Comput. J. **7**, 308–313 (1965)
Nelson,J.M.: A seasonal study of aerial insects close to a moorland stream. J. Animal Ecol. **34**, 573–579 (1965)
Nelson,J.M.: The invertebrates of an area of Pennine moorland within the Moor House Nature Reserve in northern England. Trans. Soc. Brit. Entom. **19**, 173–235 (1971)
Nelson,J.M., Satchell,J.E.: The extraction of Lumbricidae from soil with special reference to the hand-sort method. In: Progress in Soil Zoology. Murphy,P.W. (ed.). London: Butterworth, 1962, pp. 294–299
Newbould,P.J.: Methods for Estimating the Primary Production of Forests. IBP Handbook 2. Oxford: Blackwell, 1967, pp. 6–9
Newell,P.F.: Molluscs: methods for estimating production and energy flow. In: Proc. UNESCO/IBP Conf. on Methods of Study in Soil Ecology. Phillipson,J. (ed.). Paris: UNESCO, 1967, pp. 285–291

Newton,J. D.: Measurements of carbon dioxide evolved from the roots of various crop plants. Sci. Agr. **4**, 268–274 (1923)
Nicholas, D. J. D.: Inorganic nutrient nutrition of microorganisms. In: Plant Physiology, a Treatise III. Stewart, F. C. (ed.). New York: Academic, 1963
Norris, J. R.: The isolation and identification of azotobacters. Lab. Pract. **8**, 239 (1959)
Odum, E. P.: Fundamentals of Ecology, 3rd ed. Philadelphia and London: Saunders, 1971
Olechowicz, E.: Analysis of a sheep pasture ecosystem in the Pieniny mountains (the Carpathians) X. Sheep dung and the fauna colonizing it. Ekol. Pol. **22**, 589–616 (1974)
Olenin, A. S.: Peat resources of the USSR. In: Trans. 2nd Intern. Peat Cong., Leningrad, 1963. Robertson, R. A. (ed.). Edinburgh: HMSO, 1968, pp. 1–14
Oliver, J.: A study of upland temperatures and humidities in South Wales. Inst. Brit. Geographers **35**, 37–54 (1964)
Olson, J. S.: Energy storage and the balance of producers and decomposers in ecological systems. Ecology **44**, 322–331 (1963)
Østbye, E. (ed.): Hardangervidda, Norway. In: Structure and Function of Tundra Ecosystems. Ecol. Bull. (Stockholm) **20**, 225–264 (1975)
Pallant, D.: The food of the Grey field slug (*Agriolimax reticulatus* (Müller)) in woodland. J. Animal Ecol. **38**, 391–397 (1969)
Pallant, D.: A quantitative study of feeding in woodland by the Grey field slug (*Agriolimax reticulatus* (Müller)). Proc. Malac. Soc. Lond. **39**, 83—87 (1970)
Pallant, D.: The food of the Grey field slug (*Agriolimax reticulatus* (Müller)) on grassland. J. Animal Ecol. **41**, 761—769 (1972)
Panaretto, B. A.: Body composition in vivo III. The composition of living ruminants and its relation to the tritiated water species. Australian J. Agr. Res. **14**, 944–952 (1963)
Park, K. J. F., Rawes, M., Allen, S. E.: Grassland studies on the Moor House National Nature Reserve. J. Ecol. **50**, 53–62 (1962)
Parkinson, J. D., Whittaker, J. B.: A study of two physiological races of the heather psyllid *Strophingia ericae* (Curtis) (Homoptera: Psylloidea). Biol. J. Linn Soc. **7**, 73–81 (1975)
Paul, E. A.: Plant components and soil organic matter. Recent Advan. Phytochem. **3**, 59–104 (1970)
Peachey, J. E.: Studies on the Enchytraeidae (Oligochaeta) of moorland soil. Pedobiologia **2**, 81–95 (1963)
Pearsall, W. H.: Mountains and Moorlands. London: Collins, 1950, 312 pp
Pearson, V., Read, D. J.: The biology of mycorrhiza in *Ericaceae* I. The isolation of the endophyte and synthesis of mycorrhizas in aseptic culture. New Phytologist **72**, 371–379 (1973a)
Pearson, V., Read, D. J.: The biology of mycorrhiza in the *Ericaceae* II. The transport of carbon and phosphorus by the endophyte and the mycorrhiza. New Phytologist **72**, 1325–1331 (1973b)
Peart, J. N.: Sheep production in relation to stocking rates and controlled grazing on hill pastures. Expl. Husb. **19**, 29–39 (1970)
Peirson, D. H.: World-wide deposition of long-lived fission products from nuclear explosions. Nature (London) **234**, 79–80 (1971)
Peirson, D. H., Cawse, P. A., Salmon, L., Cambray, R. S.: Trace elements in the atmospheric environment. Nature (London) **241**, 252–256 (1973)
Perkins, D. F.: The growth of plants in an upland environment. Welsh Soils Disc. Grp **8**, 79–87 (1967)
Perkins, D. F.: Ecology of *Nardus stricta* L. 1. Annual growth in relation to tiller phenology. J. Ecol. **56**, 663–646 (1968)
Petrusewicz, K., MacFadyen, A.: Productivity of Terrestrial Animals: Principles and Methods. IBP Handbook 13. Oxford: Blackwell, 1970
Phillipson, J. (ed.): Methods of Study in Quantitative Soil Ecology: Population, Production and Energy Flow. Oxford: Blackwell, 1971
Picozzi, N.: Grouse bags in relation to the management and geology of heather moors. J. Appl. Ecol. **5**, 483–488 (1968)
Plinston, D. T.: Parameter sensitivity and interdependence in hydrological models. In: Mathematical Models in Ecology. Jeffers, J. N. R. (ed.). Oxford: Blackwell, 1972
Quick, H. E.: British slugs (Pulmonata: Testacellidae, Arionidae, Limacidae). Bull. Brit. Mus. Nat. Hist., **6**(3)D, 103–226 (1960)

Raw,F.: Earthworm population studies: a comparison of sampling methods. Nature (London) **184**, 1661–1662 (1960)

Rawes,M.: The problem of *Nardus* and its productivity in relation to sheep grazing at Moor House, Westmorland. J. Brit. Grassld Soc. **16**, 190–193 (1961)

Rawes,M.: The productivity of a *Festuca-Agrostis* alluvial grassland at 1700 ft in the northern Pennines. J. Brit. Grassld Soc. **18**, 300–309 (1963)

Rawes,M.: Residual effect of a manurial treatment in the northern Pennines. Scott. Agr. **45**, 39–41 (1966)

Rawes,M.: Aspects of the ecology of the northern Pennines 1. The influence of agriculture. Moor House Occasional Papers **1**, 1–15 (1971)

Rawes,M., Welch,D.: Studies on sheep grazing in the northern Pennines. J. Brit. Grassld Soc. **19**, 403–411 (1964)

Rawes,M., Welch,D.: Further studies on sheep grazing in the northern Pennines. J. Brit. Grassld Soc. **21**, 56–61 (1966)

Rawes,M., Welch,D.: Upland productivity of vegetation and sheep at Moor House National Nature Reserve, Westmorland, England. Oikos, Suppl. **11**, 72 pp (1969)

Rawes,M., Williams,R.: Production and utilisation of *Calluna* and *Eriophorum*. In: Hill Pasture Improvement and its Economic Utilisation. Potassium Institute Ltd, Colloq. **3**, 115–119 (1973)

Read,D.J., Stribley,D.P.: Effect of mycorrhizal infection on nitrogen and phosphorus nutrition of *Ericaceae* plants. Nature New Biol. **244**, 81–82 (1973)

Reader,R.J., Stewart,J.M.: The relationship between net primary production and accumulation for a peatland in south eastern Manitoba. Ecology **53**, 1024–1037 (1972)

Reay,R.C.: The numbers of eggs and larvae of *Coleophora alticolella* Zell. (Lep.). J. Animal Ecol. **33**, 117–127 (1964)

Reid,J.T.: Indicator methods in herbage quality studies. In: Pasture and Range Research Techniques Part III. Kennedy,W.K., Reid,J.T. (eds.). Ithaea, New York: Comstock, 1962, pp. 45–56

Rhee,J.A.van: Earthworm activities and the breakdown of organic matter in agricultural soils. In: Soil Organisms. Doeksen,J., van der Drift,J. (eds.). Amsterdam: North-Holland, 1963, pp. 55–59

Richards,O.W.: Studies on the ecology of English heaths III. Animal communities of the felling and burn successions at Oxshott Heath, Surrey. J. Ecol. **14**, 244–281 (1926)

Roberts,R.A.: Trends in semi-natural hill pastures from the 18th Century. 4th Intern. Grassl. Congr. (GB). Aberystwyth: Welsh Plant Breeding Station, 1937, pp. 149–153

Roberts,R.A.: Ecology of human occupation and land use in Snowdonia. J. Ecol. **47**, 317–323 (1959)

Robertson,R.A. (ed.): In: Trans. 2nd Intern. Peat Cong. Leningrad 1963. Edinburgh: HMSO, 1968

Robertson,R.A., Davies,G.E.: Quantities of plant nutrients in heather ecosystems. J. Appl. Ecol. **2**, 211–219 (1965)

Rodin,L.E., Bazilevich,N.I.: Production and Mineral Cycling in Terrestrial Vegetation. Fogg,G.E. (ed.). London: Oliver and Boyd, 1967, 288 pp

Romanov,V.V.: Gidrofizika bolot. English translation Hydrophysics of bogs (1968). Israel program for scientific translations, Jerusalem, 1961

Rosswall,T.: Cellulose decomposition studies on the tundra. In: Soil Organisms and Decomposition in Tundra. Holding,A.J., Heal,O.W., MacLean,S.F., Flanagan,P.W. (eds.). Stockholm: Tundra Biome Steering Committee, 1974, pp. 325–340

Rosswall,T., Berg,B.: Decomposition of cellulose in laboratory pot experiments. In: IBP Swedish Tundra Biome project, Technical Report 14. Sonesson,M. (ed.). Stockholm: Tundra Biome Steering Committee, 1973, pp. 142–153

Rosswall,T., Clarholm,M.: Characteristics of tundra bacterial populations and a comparison with populations from forest and grassland soil. In: Soil Organisms and Decomposition in Tundra. Holding,A.J., Heal,O.W., MacLean,S.F., Flanagan,P.W. (eds.). Stockholm: Tundra Biome Steering Committee, 1974, pp. 93–108

Rosswall, T., Flower-Ellis, J. G. K., Johansson, L. G., Jonsson, S., Ryden, B. E., Sonesson, M.: Stordalen (Abisko) Sweden. In: Structure and Function of Tundra Ecosystems. Ecol. Bull. (Stockholm) **20**, 265–294 (1975a)

Rosswall, T., Heal, O. W. (ed.): Structure and Function of Tundra Ecosystems. Ecol. Bull. (Stockholm) **20**, 450 pp (1975)

Rosswall, T., Veum, A. K., Karenlampi, L.: Plant litter decomposition of Fennoscandian tundra ecosystems. In: Analysis of Fennoscandian Tundra Ecosystems I. Plants and Microorganisms. Wielgolaski, F. E. (ed.). Berlin: Springer, 1975b, pp. 268–278

The Royal Society: International Biological Programme Handbook, 1967

Rudolph, H.: Gaswechselmessungen an *Sphagnum magellanicum*. Ein Beitrag zur Membranchromie der Sphagnen (III). Planta (Berl.) **79**, 35–43 (1968)

Rühling, Å., Tyler, G.: An ecological approach to the lead problem. Botan. Not. **121**, 321–342 (1968)

Rühling, Å, Tyler, G.: Regional differences in the deposition of heavy metals over Scandinavia. J. Appl. Ecol. **8**, 497–507 (1971)

Runham, N. W., Hunter, P. H.: Terrestrial Slugs. London: Hutchinson, 1970

Russel, A. J. F., Eadie, J.: Nutrition of the hill ewe. In: Hill-land Productivity. Proc. Europ. Grassld Fed. Symp. Hunt, I. V. (ed.). Aberdeen: Brit. Grassld. Soc., 1968, pp. 184–190

Salt, G., Hollick, F. S. J.: Studies of wireworm populations I. A census of wireworms in pastures. Ann. Appl. Biol. **31**, 52–64 (1944)

Satchell, J. E.: Nitrogen turnover by a woodland population of *Lumbricus terrestris*. In: Soil Organisms. Doeksen, J., van der Drift, J. (eds.). Amsterdam: North Holland, 1963, pp. 60–66

Satchell, J. E.: Lumbricidae. In: Soil Biology. Burges, A., Raw, F. (eds.). London-New York: Academic, 1967, pp. 259–322

Sauerbeck, D.: Comparison of plant material and animal manure in relation to their decomposition in soil. In: Isotopes and Radiation in Soil Organic Matter Studies. Proc. Symp. Vienna: FAO/IAEA, 1968, pp. 219–225

Savory, C. J.: The feeding ecology of Red grouse in NE Scotland. Ph. D. thesis, Univ. Aberdeen, 1974

Savory, H. N.: Dinas Emrys. Trans. Caernarvonshire Hist. Soc. **17**, 1–8 (1956)

Savory, H. N.: The excavations at Dinas Emrys. Archaeologia Cambrensis **109**, 13–77 (1961)

Scott, D., Billings, W. D.: Effect of environmental factors on the standing crop and productivity of an alpine tundra. Ecol. Monogr. **34**, 243–270 (1964)

Scotter, W.: Growth rates of *Cladonia alpestris*, *C. mitis* and *C. rangiferina* in the Taltson river region NWT. Can. J. Botany **41**, 1199–1202 (1963)

Sears, P. D., Evans, L. T.: The influence of red and white clovers, superphosphate, lime and dung and urine on soil composition, and on earthworm and grass grub populations. N.Z. Jl Sci. Tech. No 1 Suppl. 1 **35**, 42–52 (1953)

Seddon, B.: Late-glacial deposits at Llyn Dwythwch and Nant Ffrancon, Caernarvonshire. Phil. Trans. R. Soc. London **244B**, 451–481 (1962)

Singh, J. S., Lauenroth, W. K., Steinhorst, R. K.: Review and assessment of various techniques for estimating net aerial primary production in grasslands from harvest data. Botan. Rev. **41**, 181–232 (1975)

Siwasin, J.: Anaerobic microorganisms in soil. M. Sc. thesis, Univ. Edinburgh, 1971

Skerman, V. B. D.: A Guide to the Identification of the Genera of Bacteria, 2nd ed. Baltimore: Williams and Wilkins, 1967

Smith, A.: The pattern of distribution of *Agrostis* and *Festuca* plants of various genotypes in a sward. New Phytologist **71**, 937–945 (1972)

Smith, B. J.: Maturation of the reproductive tract of *Arion ater*. Malacologia **4**, 325–349 (1966)

Smith, G. R. J.: Some aspects of the biology of *Molophilus ater* Meigen. M. Sc. thesis, Univ. Durham, 1973

Smith, N. R., Gordon, R. E., Clark, F. E.: Aerobic spore-forming bacteria. Agricultural Monograph 16. Washington DC: US Dept of Agr., 1952

Snaydon, R. W., Bradshaw, A. D.: Differential response to calcium in *Festuca ovina*. New Phytologist **60**, 219–234 (1961)

Snedecor, G. W.: Statistical Methods, 5th ed. Iowa: State Univ. Press, 1961

South, A.: Estimation of slug populations. Ann. Appl. Biol. **53**, 251–258 (1964)
South, A.: Biology and ecology of *Agriolimax reticulatus* (Müller) and other slugs: spatial distribution. J. Animal Ecol. **34**, 403–419 (1965)
Southwood, T.R.E.: Ecological Methods with Particular Reference to the Study of Insect Populations. London: Methuen, 1966
Spedding, C.R.W.: Grassland Ecology. Oxford: Clarendon, 1971
Springett, J.A.: A new species of *Cernosvitoviella* (Enchytraeidae) and records of three species new to the British Isles. Pedobiologia **9**, 459–461 (1969)
Springett, J.A.: The distribution and life histories of some moorland Enchytraeidae (Oligochaeta). J. Animal Ecol. **39**, 725–737 (1970)
Springett, J.A., Brittain, J.E., Springett, B.P.: Vertical movement of Enchytraeidae (Oligochaeta) in moorland soils. Oikos **21**, 16–21 (1970)
Standen, V.: The production and respiration of an enchytraeid population in blanket bog. J. Animal Ecol. **42**, 219–245 (1973)
Standen, V.: The influence of soil fauna on decomposition by microorganisms in blanket bog litter. J. Animal Ecol. (in press)
Stanier, R.Y., Palleroni, N.J., Doudoroff, M.: The aerobic pseudomonads: a taxonomic study. J. Gen. Microbiol. **43**, 159–271 (1966)
Starkey, R.L.: Cultivation of organisms concerned with the oxidation of thiosulphate. J. Bact. **28**, 365 (1934)
Starkey, R.L.: A study of spore formation and other morphological characteristics of *Vibrio desulphuricans*. Arch. Mikrobiol. **9**, 268 (1938)
Stern, G.: Production et bilan energetique chez la limace rouge *(Arion rufus)*. Terre Vie **24**, 403–424 (1970)
Stotzky, G., Mortensen, J.L.: Effect of crop residues and nitrogen additions on decomposition of an Ohio muck soil. Soil Sci. **83**, 165–174 (1957)
Stribley, D.P., Read, D.J.: The biology of mycorrhiza in the Ericaceae IV. The effect of mycorrhizal infections on uptake of ^{15}N from labelled soil by *Vaccinium macrocarpon* AIT. New Phytologist **73**, 1149–1155 (1974)
Summers, C.F.: Aspects of production in montane dwarf shrub heaths. Ph. D. thesis, Univ. Aberdeen, 1972
Svendsen, J.A.: The distribution of Lumbricidae in an area of Pennine moorland (Moor House Nature Reserve). J. Animal Ecol. **26**, 411–421 (1957 a)
Svendsen, J.A.: The behaviour of lumbricids under moorland conditions. J. Animal Ecol. **26**, 423–439 (1957 b)
Svensson, B.: Methane production in a subarctic mire. In: IBP Swedish Tundra Biome Project Technical Report 14. Sonesson, M. (ed.). Stockholm: Tundra Biome Steering Committee, 1973, pp. 154–166
Swaine, D.J.: The trace-element content of soils. Common. Agr. Bur. Techn. Comm. **48**, 1–157 (1955)
Tamm, C.O.: Some observations on the nutrient turnover in a bog community dominated by *Eriophorum vaginatum* (L.). Oikos **5**, 189–194 (1954)
Tansley, A.G.: The British Islands and their Vegetation. London: Cambridge University Press, 1939
Taylor, K.: Biological flora of the British Isles. *Rubus chamaemorus* L. J. Ecol. **59**, 293–306 (1971)
Taylor, K., Marks, T.C.: The influence of burning and grazing on the growth and development of *Rubus chamaemorus* L. in *Calluna-Eriophorum* bog. In: The Scientific Management of Plant and Animal Communities for Conservation. Duffey, E.A., Watt, A.S. (eds.). Oxford: Blackwell, 1971, pp. 153–166
Taylor, M.M.: A study on the species composition, distribution and possible preference for reduced oxygen conditions of fungal isolates from mixed moor (blanket bog). B. Sc. final project, Liverpool Polytechnic, 1970
Taylor, P., Rawes, M.: Aspects of the ecology of the northern Pennines VI. The ecology of the Red grouse. Moor House Occasional Papers **6**, 1–26 (1974)
Thomson, D.J.: Energy retention in lambs by slaughter technique. In: Energy Metabolism. Blaxter, K.L. (ed.). London-New York: Academic, 1965, pp. 319–326

Thomson,J.W.: The Lichen Genus *Cladonia* in N America. Univ. Toronto Press, 1967

Tibbetts,T.E.: Peat resources of the world—a review. In: Proc. 3rd Intern. Peat Cong., Quebec 18–23 August 1968. Quebec: Department of Energy, Mines and Resources and Nat. Res. Council Canada, pp. 8–22 (undated)

Tieszen,L.L.: Carbon dioxide exchange in the Alaskan arctic tundra. Measured course of photosynthesis, 29–35. Univ. Washington: Proc. 1972 Tundra Biome IBP Symp., 1972

Traczyk,T.: Studies on herb layer production estimate and the size of plant fall. Ekol. Pol. Ser. A **15**, 837–867 (1967)

Tranquillini,W.: The physiology of plants at high altitudes. A. Rev. Plant Physiol. **15**, 345–362 (1964)

Tribe,D.E.: Some seasonal observations on the grazing habits of sheep. Emp. J. Expl. Agr. **17**, 105–115 (1949)

Troughton,A.: Studies on the roots and storage organs of herbage plants. J. Brit. Grassld Soc. **6**, 197–206 (1951)

Troughton,A.: The underground organs of herbage grasses. C.A.B. Bull. **44**, 1957

Tukey,H.B.,Jr., Tukey,H.B., Wittwer,S.H.: Loss of nutrients by foliar leaching as determined by radioisotopes. Proc. Am. Soc. Hort. Sci. **71**, 496–506 (1958)

Turner,F.B.: The ecological efficiency of consumer populations. Ecology **51**, 741–742 (1970)

Turner,J.: The anthropogenic factor in vegetational history I. Tregaron and Whixall mosses. New Phytologist **63**, 73–90 (1964)

Turner,J., Chambers,K., Harkness,D.D.: Measurements by Harkness on cores collected by Chambers, and reported by Turner and Chambers in the Moor House 13th Annual Report **17** (1972)

Tyler,G.: Heavy metals pollute Nature, may reduce productivity. Ambio **1**, 52–59 (1972)

Urquhart,C.: An improved method of demonstrating the distribution of sulphide in peat soils. Nature (London) **211**, 550 (1966)

Urquhart,C.: The effect of waterlogging on the growth and nutrition of some moorland plant species with special reference to the soil redox potential. M. Sc. thesis, Univ. Newcastle-upon-Tyne, 1969

Urquhart,C., Gore,A.J.P.: The redox potentials of four peat profiles. Soil Biol. Biochem. **5**, 659–672 (1973)

USDA: Soil classification: a comprehensive system: 7th Approximation. Washington: US Dept. Agr., 1960

USDA: Supplement to soil classification system: 7th Approximation. Washington: US Dept. Agr., 1967

Uvarov,B.: Grasshoppers and Locusts. London: Cambridge University Press, 1966, Vol. I, 482 pp.

Vries,D.M.de: Methods used in scientific plant sociology and in agricultural botanical grassland research. Herbage Reviews **5**, 187–193 (1937)

Waksman,S.A., Purvis,E.R.: The microbial population of peat. Soil Sci. **34**, 95–114 (1932)

Waksman,S.A., Stevens,K.R.: Contribution to the chemical composition of peat V. The role of microorganisms in peat formation and decomposition. Soil Sci. **27**, 271–281 (1929)

Walker,D.: Direction and rate in some British post-glacial hydroseres. In: Studies in the Vegetation History of the British Isles. Walker,D., West,R.G. (eds.). London: Cambridge University Press, 1970, pp. 117–139

Walker,D., Walker,P.M.: Stratigraphic evidence of regeneration in some Irish bogs. J. Ecol. **49**, 169–185 (1961)

Wallwork,J.A.: Ecology of Soil Animals. London: McGraw-Hill, 1970

Warren-Wilson,J.: An analysis of plant growth and its control in arctic environments. Ann. Botany **30**, 383–402 (1966)

Washburn,E.W. (ed.): International Critical Tables. New York: Mc Graw-Hill, 1926

Waters,R.A.S.: Numbers and weights of earthworms under a highly productive pasture. N.Z. J. Sci. Tech. No 1 **36**, 516–526 (1955)

Watson,A.: Social status and population regulation in the Red grouse *(Lagopus lagopus scoticus)*. Proc. Roy. Soc. Populations Study Group **2**, 22–30 (1967)

Watson,A., Jenkins,D.: Notes on the behaviour of Red grouse. Brit. Birds **57**, 137–170 (1964)

Watson,A., Jenkins,D.: Experiments on population control by territorial behaviour in Red grouse. J. Animal Ecol. **37**, 595–614 (1968)

Watson,A., Miller,G.R.: Territory size in a fluctuating Red grouse population. J. Animal Ecol. **40**, 367–383 (1971)

Watson,A., Moss,R.: A current model of population dynamics in Red grouse. In: Proc. 15th Intern. Ornithol. Cong. 1972, pp. 134–149

Watt,A.S.: Bracken versus heather: a study in plant sociology. J. Ecol. **43**, 490–506 (1955)

Watt,A.S., Jones,E.W.: The ecology of the Cairngorms I. The environment and altitudinal zonation of the vegetation. J. Ecol. **36**, 283–304 (1948)

Webb,N.R.: Comparative studies on population metabolism in soil arthropods. Ph. D. thesis, Univ. Wales, 1968

Webb,N.R.: The respiratory metabolism of *Nothrus silvestris* Nic. Oikos **20**, 294–299 (1969)

Webb,N.R.: Population metabolism of *Nothrus silvestris* Nic. Oikos **21**, 155–159 (1970)

Webb,N.R.: Cryptostigmatid mites recorded from heathland in Dorset. Entomol. Mon. Mag. **107**, 228–229 (1972)

Webb,N.R., Elmes,G.W.: Energy budget for adult *Steganacarus magnus* Nic. Oikos **23**, 359–365 (1972)

Webster,J.: Succession of fungi on decaying Cocksfoot culms. J. Ecol. **45**, 1–30 (1957)

Weigert,R.G., Evans,F.C.: Primary production and the disappearance of dead vegetation on an old field in south eastern Michigan. Ecology **45**, 49–63 (1964)

Weigert,R.G., McGinnis,J.T.: Annual production and disappearance of detritus on three south Carolina old fields. Ecology **56**, 129–140 (1975)

Welch,D.: A change in the upper altitudinal limit of *Coleophora alticolella* Zell. (Lep.). J. Animal Ecol. **34**, 725–729 (1965)

Welch,D., Rawes,M.: The early effects of excluding sheep from high level grasslands in the north Pennines. J. Appl. Ecol. **1**, 281–300 (1964)

Welch,D., Rawes,M.: The herbage production of some Pennine grasslands. Oikos **16**, 39–47 (1965)

Welch,D., Rawes,M.: The intensity of sheep grazing on high-level blanket bog in Upper Teesdale. Ir. J. Agr. Res. **5**, 185–196 (1966)

West,R.G.: Pleistocene Geology and Biology. London: Longmans, 1968

White,E.: The distribution and subsequent disappearance of sheep dung on Pennine moorland. J. Animal Ecol. **29**, 243–250 (1960a)

White,E.: The natural history of some species of *Aphodius* (Col. Scarabaeidae) in the northern Pennines. Entomol. Mon. Mag. **96** (1960b)

Whittaker,J.B.: Studies on the Auchenorrhyncha (Hemiptera: Insecta) of Pennine moorland with special reference to the Cercopidae. Ph. D. thesis, Univ. Durham, 1963

Whittaker,J.B.: Auchenorrhyncha (Homoptera) of the Moor House National Nature Reserve, Westmorland, with notes on *Macrosteles alpinus* (Zett) a species new to Britain. Entomol. Mon. Mag. **100**, 168–171 (1965a)

Whittaker,J.B.: The distribution and population dynamics of *Neophilaenus lineatus* (L.) and *N. exclamationis* (Thun.) (Homoptera: Cercopidae) on Pennine moorland. J. Animal Ecol. **34**, 277–297 (1965b)

Whittaker,J.B.: The distribution and survival of two Cercopidae (Homoptera) near to the edge of their range in northern England. Proc. 12th Intern. Cong. Entom. London 1964. (1965c)

Whittaker,J.B.: Population changes in *Neophilaenus lineatus* (L.) (Homoptera: Cercopidae) in different parts of its range J. Animal Ecol. **40**, 425–443 (1971)

Wickman,F.E.: The maximum height of raised bogs. Geol. För. Stockh. Förh. **73**, 413–422 (1951)

Wielgolaski,F.E. (ed.): Fennoscandian Tundra Ecosystems. Berlin: Springer, 1975, Part 1, 366 pp, Part 2, 337 pp

Wielgolaski,F.E., Rosswall,T. (ed.): Proc. 4th Intern. Meeting on Biological Productivity of Tundra, Leningrad, USSR. Stockholm: IBP Tundra Biome Steering Committee, 1972, 320 pp

Williams,H.: The geology of Snowdon. Quart. J. Geol. Soc. **83**, 346–431 (1927)

Williams, S.T., Mayfield, C.I.: Studies on the ecology of actinomycetes in soil III. The behaviour of neutrophilic streptomycetes in acid soils. Soil Biol. Biochem. **3**, 197–208 (1971)

Williams, W.T., Lambert, J.M., Lance, G.N.: Multivariate methods in plant ecology V. Similarity analyses and information analysis. J. Ecol. **54**, 427–445 (1966)

Willoughby, L.G.: Aquatic actinomycetales with particular reference to the Actinoplanaceae. Veröff. Inst. Meeresforsch. Bremerh. **3**, 19–26 (1968)

Wilson, K.: The time factor in the development of dune soils at South Haven Peninsula, Dorset. J. Ecol. **48**, 341–360 (1960)

Winberg, G.C.: Rate of metabolism and food requirements of fishes. Fish. Res. Bd. Canada Translation Series No **194** (1960)

Woodwell, G.M., Whittaker, R.H., Houghton, R.A.: Nutrient concentrations in plants in the Brookhaven oak-pine forest. Ecology **56**, 318–332 (1975)

Worthington, E.B. (ed.): The Evolution of IBP. London: Cambridge University Press, 1975, Vol. I, 268 pp

Zinkler, D.: Vergleichende Untersuchungen zur Atmungsphysiologie von Collembolen (Apterygota) und anderen Bodenkleinarthropoden. Z. Vergl. Physiol. **52**, 99–144 (1966)

Subject Index

Page numbers in bold face refer to whole chapters.

Ecological Studies

Analysis and Synthesis
Editors: W.D. Billings, F. Golley, O.L. Lange, J.S. Olson

Volume 4

Physical Aspects of Soil Water and Salts in Ecosystems

Editors: A. Hadas, D. Swartzendruber, P.E. Rijtema, M. Fuchs, B. Yaron
221 figures, 61 tables. XVI, 460 pages. 1973
ISBN 3-540-06109-6

Contents: Water Status and Flow in Soils: Water Movement in Soils. Energy of Soil Water and Soil-Water Interactions. – Evapotranspiration and Crop-Water Requirements: Evaporation from Soils and Plants. Crop-Water Requirements. – Salinity Control.

Volume 6
K. STERN, L. ROCHE

Genetics of Forest Ecosystems

70 figures. X, 330 pages. 1974
ISBN 3-540-06095-2

Contents: The Ecological Niche. – Adaptations. – Genetic Systems. – Adaptive Strategies. – Forest Ecosystems. – How Man Affects Forest Ecosystems.

Volume 8

Phenology and Seasonality Modeling

Editor: H. Lieth
120 figures. XV, 444 pages. 1974
ISBN 3-540-06524-5

Contents: Introduction to Phenology and the Modeling of Seasonality. – Methods for Phenological Studies. – Seasonality in Trophic Levels. – Representative Biome Studies. – Modeling Phenology and Seasonality. – Applications of Phenology.

Volume 9
B. SLAVÍK

Methods of Studying Plant Water Relations

With contributions by B. Slavík, J. Catský, J. Solárová, H.R. Oppenheimer, J. Hrbáček, J. Slavíková, V. Kozinka, J. Úlehla, P.G. Jarvis, M.S. Jarvis
181 figures. XVIII, 449 pages. 1974
ISBN 3-540-06686-1
Distribution rights for the Socialist countries: Academia Publishing House of the Czechoslovak Academy of Sciences, Prague

Contents: Water in Cells and Tissues. – Water Content. – Water Exchange between Plant Roots and Soil. – Liquid Water Movement in Plants. – Water Exchange between Plant and Atmosphere. – Table Appendix.

Volume 13

Epidemics of Plant Diseases

Mathematical Analysis and Modeling
Editor: J. Kranz
46 figures. X, 170 pages. 1974
ISBN 3-540-06896-1

Contents: The Role and Scope of Mathematical Analysis and Modeling in Epidemiology. – Automatic Data Processing in Analysis of Epidemics. – Multiple Regression Analysis in the Epidemiology of Plant Diseases. – Nonlinear Disease Progress Curves. – Simulation of Epidemics.

Volume 19

Water and Plant Life

Problems and Modern Approaches
Editors: O.L. Lange, L. Kappen, E.-D. Schulze
178 figures, 66 tables. XX, 536 pages. 1976
ISBN 3-540-07838-X

Contents: Fundamentals of Plant Water Relations. – Water Uptake and Soil Water Relations. – Transpiration and its Regulation. – Direct and Indirect Water Stress. – Water Relations and CO_2 Fixation Types. – Water Relations and Productivity. – Water and Vegetation Patterns.

Springer-Verlag Berlin Heidelberg New York

Advanced Series in Agricultural Sciences

Co-ordinating Editor: B. Yaron
Editors: G. W. Thomas, B. R. Sabey, Y. Vaadia, L. D. Van Vleck

Volume 1
A. P. A. Vink

Land Use in Advancing Agriculture

94 figures, 115 tables. X, 394 pages. 1975
ISBN 3-540-07091-5
Distribution rights for India:
Allied Publishers Private Ltd., New Delhi

Contents: Land Use Surveys. – Land Utilization Types. – Land Resources. – Landscape Ecology and Land Conditions. – Land Evaluation. – Development of Land Use in Advancing Agriculture.

Volume 2
H. Wheeler

Plant Pathogenesis

19 figures, 5 tables. X, 106 pages. 1975
ISBN 3-540-07358-2
Distribution rights for India:
Allied Publishers Private Ltd., New Delhi

Contents: Concepts and Definitions. – Mechanisms of Pathogenesis. – Responses of Plants to Pathogens. – Disease-Resistance Mechanisms. – Genetics of Pathogenesis. – Nature of the Physiological Syndrome.

Volume 3
R. A. Robinson

Plant Pathosystems

15 figures, 2 tables. X, 184 pages. 1976
ISBN 3-540-07712-X
Distribution rights for India:
Allied Publishers Private Ltd., New Delhi

Contents: Systems. – Plant Pathosystems. – Vertical Pathosystem Analysis. – Vertical Pathosystem Management. – Horizontal Pathosystem Analysis. – Horizontal Pathosystem Management. – Polyphyletic Pathosystems. – Crop Vulnerability. – Conclusions. – Terminology.

Volume 4
H. C. Coppel, J. W. Mertins

Biological Insect Pest Suppression

46 figures, 1 table, 314 pages. 1977
ISBN 3-540-07931-9
Distribution rights for India:
Allied Publishers Private Ltd., New Delhi

Contents: Glossary. – Historical, Theoretical, and Philosophical Bases of Biological Insect Suppression. – Organisms Used in Classical Biological Insect Pest Suppression. – Manipulation of the Biological Environment for Insect Pest Suppression. – A Fusion of Ideas.

Springer-Verlag
Berlin
Heidelberg
New York